Two-Dimensional Wavelets and their Relatives

Two-dimensional wavelets offer a number of advantages over discrete wavelet transforms when processing rapidly varying functions and signals. In particular, they offer benefits for real-time applications such as medical imaging, fluid dynamics, shape recognition, image enhancement and target tracking. This book introduces the reader to 2-D wavelets via 1-D continuous wavelet transforms, and includes a long list of useful applications. The authors then describe in detail the underlying mathematics before moving on to more advanced topics such as matrix geometry of wavelet analysis, three-dimensional wavelets and wavelets on a sphere. Throughout the book, practical applications and illustrative examples are used extensively, ensuring the book's value to engineers, physicists and mathematicians alike.

Jean-Pierre Antoine is a Professor of Mathematical Physics at the Institut de Physique Théorique, Université Catholique de Louvain.

Romain Murenzi is currently Minister of Education, Science, Technology, and Scientific Research of the Republic of Rwanda, on leave of absence from the Department of Physics, Clark Atlanta University, Atlanta, Georgia.

Pierre Vandergheynst is a Professor at the Signal Processing Institute, Swiss Federal Institute of Technology, Lausanne.

Syed Twareque Ali is a Professor at the Department of Mathematics and Statistics, Concordia University, Montréal.

Two-Dimensional Wavelets and their Relatives

Jean-Pierre Antoine

Institut de Physique Théorique, Université Catholique de Louvain

Romain Murenzi

CTSPS, Clark Atlanta University, Ministry of Education, Science, Technology and Scientific Research, Rwanda

Pierre Vandergheynst

Signal Processing Laboratory, Swiss Federal Institute of Technology

Syed Twareque Ali

Department of Mathematics and Statistics, Concordia University

CAMBRIDGE
UNIVERSITY PRESS

CAMBRIDGE UNIVERSITY PRESS
Cambridge, New York, Melbourne, Madrid, Cape Town, Singapore, São Paulo

Cambridge University Press
The Edinburgh Building, Cambridge CB2 8RU, UK

Published in the United States of America by Cambridge University Press, New York

www.cambridge.org
Information on this title: www.cambridge.org/9780521624060

First published 2004
This digitally printed version 2008

A catalogue record for this publication is available from the British Library

Library of Congress Cataloguing in Publication data

Two-dimensional wavelets and their relatives / Jean-Pierre Antoine . . . [et al.].
     p.   cm.
Includes bibliographical references and index.
ISBN 0 521 62406 1 (hardback)
1. Wavelets (Mathematics)   I. Antoine, Jean-Pierre.
QA403.3.T86   2004
515′2433 – dc22   2003061134

ISBN 978-0-521-62406-0 hardback
ISBN 978-0-521-06519-1 paperback

Contents

Prologue

Wavelets are everywhere nowadays. Be it in signal or image processing, in astronomy, in fluid dynamics (turbulence), in condensed matter physics, wavelets have found applications in almost every corner of physics. In addition, wavelet methods have become standard in applied mathematics, numerical analysis, approximation theory, etc. It is hardly possible to attend a conference on any of these fields without encountering several contributions dealing with them. Correspondingly, hundreds of papers appear every year and new books on the topic get published at a sustained pace, with publishers strongly competing with each other. So, why bother to publish an additional one?

The answer lies in the finer distinction between various types of wavelet transforms. There is, indeed, a crucial difference between two approaches, namely, the *continuous* wavelet transform (CWT) and the *discrete* wavelet transform (DWT). Furthermore, one has to distinguish between problems in one dimension (signal analysis) and problems in two dimensions (image processing), since the status of the literature is very different in the two cases.

Take first the one-dimensional case. Beginning with the classic textbook of Ingrid Daubechies [Dau92], several books, such as those of M. Holschneider [Hol95], B. Torrésani [Tor95] or A. Arnéodo *et al.* [Arn95], cover the continuous wavelet transform, in a more or less mathematically oriented approach. On the other hand, the discrete wavelet transform is treated in many textbooks, more in the signal processing style, such as M. V. Wickerhauser [Wic94], M. Vetterli and J. Kovačević [Vet95], P. Wojtaszczyk [Woj97], or S. G. Mallat [Mal99], whereas others emphasize the algorithmic aspects, sometimes in a rather abstract way, for example, C. K. Chui [Chu92] or Y. Meyer [Mey94] (of course, there are many more on the market). Altogether these books tell a fascinating story, that is ideally depicted in the highly popular volume of B. Burke Hubbard [Bur98], which is based on interviews by the author with all the founding "fathers" of the theory (J. Morlet, A. Grossmann, I. Daubechies, Y. Meyer, etc.).

It is a fact that DWT-inspired methods (multiresolution, lifting scheme, etc., that we shall describe in due time) constitute the overwhelming majority among the wavelet community, under the joint influence of electrical engineering (signal processing with filters and subband coding) and applied mathematics (numerical and algorithmic methods). Yet the CWT and, more generally, redundant representations of signals, offer distinct advantages in certain cases, as we shall see later.

In two dimensions, that is, application to image processing, the situation is clearer. Discrete methods are somewhat trivial, since the basic structure is that of a tensor product, 2-D = 1-D $\otimes$ 1-D, enforcing a Cartesian geometry (x and y coordinates). Thus most textbooks on the DWT will cover, although briefly in general, the 2-D case as a straightforward extension of the 1-D setup. As for the 2-D CWT, it receives at best a cursory treatment in most cases. The raison d'être of the present volume is precisely to fill this gap in the literature and give a thorough treatment of the 2-D CWT and some of its applications in image processing and in various branches of physics. As a byproduct, we will also discuss in detail several extensions, such as 3-D wavelets, wavelets on the sphere or wavelets in space-time.

A historical note

Before entering the subject proper, it may not be uninteresting to give some details on its origin, without pretension to completeness, of course; we are not historians. The first extension of the wavelet transform to imaging is due to Mallat [259,260], who developed systematically a 2-D discrete (but redundant) WT, combining the traditional concept of filter bank and the analogy with human vision. In fact, most of the concepts are indeed already present in the pioneering work of Marr [Mar82] on vision modeling, in particular the idea of multiresolution. Indeed, when we look at an object, our visual system works by registering first a global, low-resolution, image and then focusing systematically to finer and finer details. Thus, contrary to the 1-D case, the 2-D discrete WT preceded the continuous version.

The 2-D continuous WT was born in a quite different way. The story starts in the coffee room of the Institut de Physique Théorique in UCL, Louvain-la-Neuve (LLN), in Spring 1987. Alex Grossmann from Marseille, one of the founding fathers of wavelets, was visiting J.-P. A., indeed they had already started to collaborate on the application of 1-D wavelets in NMR spectroscopy. Thus the two were discussing a possible Ph.D. topic for a young African student, called Romain Murenzi (R.M.). The latter had just concluded a Master's thesis on five-dimensional quantum field theory, a subject hardly practical for a developing country! So the idea came up, why not try to do in two dimensions what had been so successful in 1-D, namely, wavelet analysis? The topic seemed tractable, involving moderate amounts of mathematics and some simple computing technology, and if it worked out, there could be very interesting practical applications. The problem was that nobody knew how to do it! The next summer, R.M. went down to Marseille and started to work with Grossmann and Ingrid Daubechies who happened to be there too. And when he came back 3 months later, the solution was clear. The key is to start from the operations that one wants to apply to an image, namely, translations in the image plane, rotations for choosing a direction of sight, and global magnification (zooming in and out). The problem is to combine these three elements in such a way that the wavelet machine could start rolling (there are mathematical conditions to satisfy here). The result of R.M. was that the so-called similitude group yields a solution (actually, the only one). There remained to put it all together, to turn the mathematical crank and to apply the resulting formalism to a real problem, namely, 2-D fractals (the outcome of a visit of R.M. to Arnéodo in Bordeaux), and the Ph.D. thesis was within reach [Mur90]. Several papers followed [12,13], more M.Sc. or Ph.D. students got involved over the years. We may cite Pierre Carrette, Stéphane Maes, Canisius Cishahayo, Pierre Vandergheynst, Yébéni B. Kouagou, Laurent Jacques, Laurent Demanet. Each of them has brought his contribution to the edifice, small or big, but always useful.

This is probably a good place for asking, why wavelets? After all, there are plenty of methods available for processing images. What is new here? A key fact is probably that wavelets are somehow a byproduct of quantum thinking. More precisely, it is an application of the quantum idea of a *probe* for testing an object, the result being given by the scalar product of the two functions (indeed the framework is a Hilbert space, that of finite energy signals). To get the transform, the probe is translated and scaled (zoom), and turned around in the 2-D case, and the result is plotted as a function of the corresponding parameters. (Actually the same could be said of the so-called Gabor or Windowed Fourier transform.) One gets in this way a highly flexible and efficient tool for signal/image processing, that sheds a different light and offers an alternative approach to many standard problems, in particular those involving the detection of singularities or discontinuities in signals. As somebody once remarked, wavelets do not solve all the problems, but they often help asking the right questions.

Another sign of the quantum influence is the crucial role played by a unitary group representation, a tool largely absent in classical physics – and thus from signal processing as well. And it is no accident, in our opinion, that the crucial steps in developing wavelets were made by Alex Grossmann and Ingrid Daubechies, both educated as theoretical (quantum) physicists. Otherwise, it might have taken much longer for electrical engineers and mathematicians to meet!

About the contents of the book

Now it is time to give some indications on the contents of the book. One can divide it into several stages. In a first part (Chapters 1–3), we develop systematically the continuous wavelet transform, first in one dimension (briefly), then in two dimensions. The emphasis here is on the practical use of the tool, with a minimum of mathematics. Then we devote two long chapters, 4 and 5, to applications. Three short chapters, 6–8, set the general mathematical scene. This allows us, in Chapters 9 and 10, to describe wavelets in more general settings (3-D, sphere, space–time). In Chapter 11, finally, we discuss some recent developments that actually go beyond wavelets. This gradual structure is one of the original aspects of the book, in comparison with those on the market.

Let us go into more details. As a warming up exercise, we begin, in Chapter 1, with a rather concise overview of the 1-D WT. This allows the reader to develop a feeling about the wavelet transform and to understand its success in signal processing. All aspects will be touched upon: the continuous WT, multiresolution and the discrete WT, various generalizations of the latter, some applications. One of the leitmotives is the role of *redundancy*, especially with respect to stability of the representation.

Chapter 2, which forms the hard core of the first part, presents in a systematical way the theory of the 2-D CWT. As said above already, the starting point is to decide which elementary operations one wants to apply to an image. Choosing translations in the image plane, rotations (direction of sight), and global magnification (zooming in and out), together with the probe idea, leads uniquely to the 2-D CWT. We study in detail its basic properties: energy conservation, reconstruction formula, reproducing property, covariance under the chosen operations. Then we describe the interpretation of the WT as a singularity scanner and as a phase space representation of signals. Since the WT of a 2-D image is a function of four variables, visualizing it inevitably becomes problematic. Hence the need to reduce the number of parameters, either by fixing some of them, or integrating over them. This introduces a tool that will prove very useful in the applications, namely, the various partial energy densities, that is, the function obtained by integrating the squared modulus of the CWT over a subset of the parameters. In other words, various types of *wavelet spectra*, the analogs of the familiar power spectrum of a signal.

As is well known in 1-D, the CWT is highly redundant, as one can expect from a transform that doubles the number of variables: one to two in 1-D, two to four in 2-D. This fact may be exploited in two ways. Either one limits oneself to a small subset of the transform, where most of the energy is concentrated, and thus one is led to the notions of local maxima, ridges and skeleton; or one discretizes the CWT and obtains *wavelet frames*. Such a representation is still redundant, but much less than the full CWT, and in many instances is a good substitute for a genuine orthonormal basis. An alternative is the so-called *dyadic WT*, originally due to Mallat, in which only the scale variable is discretized. Together with the latter, we also describe briefly the standard DWT, based on the multiresolution idea, and several generalizations, mostly the so-called *lifting scheme*. We conclude the chapter with a thorough discussion of a different scheme, called directional dyadic wavelet frames. Here, as in 1-D, there are two conflicting requirements: redundancy of the transform, which brings stability, and computing economy, that seeks fast algorithms. The formalism described here offers a good compromise.

When it comes to treating a precise problem, the first question to ask is, which wavelet should one use? Thus there is a need for a sizable collection of them, well documented and calibrated. The aim of Chapter 3 is to provide this. The crucial distinction here is whether directions in the image are relevant or not. If they are not, a pointwise analysis suffices, and one can use rotation invariant (isotropic, radial) wavelets, the best known being the Mexican hat or LOG wavelet (already introduced by Marr [Mar82]). On the contrary, if directions must be detected, one needs a wavelet with a good orientation selectivity. The most efficient result is obtained with the so-called *directional wavelets*. These are filters living in a convex cone, with apex at the origin, in Fourier space. Examples are the 2-D Morlet wavelet and the family of conical wavelets. All these wavelets, and some more, are discussed in detail in Chapter 3, and their performances determined quantitatively.

At this stage, the tool is ready and we turn to applications. Many of them are not easy to find, because they have appeared only in conference proceedings or in (unpublished) Ph.D. theses. For that reason, we have decided to present them in a rather detailed fashion, always giving original references, including personal websites when available. In each case, we emphasize the rationale for using wavelets in the particular problem at hand, rather than go into the technicalities.

It is convenient (although not always unambiguous) to distinguish between two different fields of applications, image processing and physics. To the first type, the subject matter of Chapter 4, belong contour detection and character recognition; automatic target detection and recognition (for instance, in infrared radar imagery); image retrieval from data banks; medical imaging; detection of symmetries in patterns, in particular quasicrystals and other quasiperiodic patterns; and image denoising. The chapter concludes with two nonlinear extensions of the CWT, which both have important applications. The first one is *contrast enhancement* in images through an adaptive normalization. This technique, based on analogy with our visual system, may be of interest in medical imaging. Indeed typical images, such as those obtained by radiography or by NMR imaging, have rather weak contrast, which makes their interpretation sometimes difficult. The other problem we deal with is *watermarking* of images, which consists in adding an invisible "signature" (the watermark) to an image, that only the owner can recognize and is robust to manipulations. Clearly the field of image copyright offers a good market for such techniques. The novel method we present is based on the contrast analysis described previously, exploiting directional wavelets, and it turns out to be particularly efficient.

The second class of applications, described in Chapter 5, concerns various fields of physics. Characteristically, they all belong to classical physics, as opposed to quantum physics, because the former relies much more on images. Indeed, there are very few applications of wavelet analysis in quantum problems.

The first domain on which 2-D wavelets have made a substantial impact is astronomy and astrophysics, for several reasons. The Universe has a marked hierarchical structure.

Nearby stars, galaxies, quasars, galaxy clusters and superclusters have very different sizes and live at very different distances. Thus the scale variable is essential and a multiscale analysis is in order. This, of course, suggests wavelet analysis, and indeed many authors have used it in problems such as determination of the large-scale structure of the Universe, galaxy or void counting, or analysis of the cosmic background radiation. In addition, we describe more in depth two applications of our own, namely, the detection of various magnetic features of the Sun, from satellite images, and the detection of distant gamma-ray sources in the Universe. In the latter case, difficult statistics problems arise, because of the extreme weakness of the signal (such a source emits very few high energy photons).

The next topic is Earth physics: fault detection in geophysics, seismology, climatology (notably, thunderstorm prevision). A number of successful applications pertain to fluid dynamics, from the detection of coherent strucures in fully developed 2-D turbulence (a domain pioneered forcefully by Marie Farge [164]) to the measurement of the velocity field in a turbulent fluid, or the disentangling of a 2-D (or 3-D) wave train. Next comes the world of fractals. These are structures that are solely characterized by their behavior under a scaling transformation: ideal ground for wavelets! However, the self-reproducing properties of physical fractals are in general only approximate, so that methods from statistical mechanics are needed. Thus, a thermodynamical formalism has been designed by Arnéodo and his group in Bordeaux for treating such problems, and we give a brief account of it. Finally we touch upon the problem of shape recognition, where wavelet descriptors have proven useful too.

At this point, the book undergoes a sort of phase transition. Up to here, everything was done by hand, so to speak. The properties of the CWT have been derived by explicit calculations and very few mathematical prerequisites have been asked for. But now it is time to look over the hill and notice that the whole theory is firmly grounded in group theory. Indeed the wavelet transform and all its properties may be entirely derived from an appropriate representation of the affine group, both in one and in two dimensions. A mathematical condition, called square integrability of the representation, ensures the validity of the derivation, in particular the possibility of inverting the wavelet transform, that is, of obtaining reconstruction formulas. We devote two rather short chapters, 6 and 7 to these developments, with a double benefit. First, on the pedagogical level, we want to convince the reader that the group-theoretical approach is not only mathematically correct and pleasant, it is also natural and easy. It allows us indeed to understand in a simple and unified language the deeper mathematical structures involved. It is also quite efficient, in that it yields a general formalism (in fact, a special case of the coherent state formalism, well known in quantum physics, in particular, in quantum optics) that permits us to extend the CWT to more general manifolds, such as $\mathbb{R}^3$, the two-sphere, or space–time, all generalizations that will be discussed in later chapters. Of course, we do not expect our reader to be fully conversant with group theory, and we will define all the needed ingredients along the way. Actually we will essentially restrict our treatment to 2×2 or 3×3 matrices, without resort to abstract notions. Nevertheless, we found it convenient to gather all the group-theoretical information in a separate appendix.

We begin, in Chapter 6, by revisiting the 1-D CWT in the light of the so-called $ax + b$ or restricted affine group of the line, that is, the set of all translations and positive dilations. It turns out that the CWT may also be interpreted as a phase space representation of signals, in the sense of Hamiltonian mechanics, and the group-theoretical language makes this evident. The same treatment is then applied to the Gabor transform, also called Short Time or Windowed Fourier transform, simply replacing the affine group by the Weyl–Heisenberg group, that is, the group of phase space translations (this point of view has also been emphasized by Daubechies [Dau92]). Next, in Chapter 7, we repeat the procedure in two dimensions. Here the relevant group is the similitude group SIM(2), which consists of translations, rotations and dilations of the plane, that is, precisely all the transformations we have chosen to apply to images. Here, as in the 1-D case, the basic tool is a representation of the group by unitary operators acting in the space of finite energy signals, a natural representation that possesses the property of square integrability, meaning roughly that its matrix elements are square integrable functions of the group parameters. Here too, the CWT is a phase space realization of signals, and we spend some time exploring the consequences of this fact.

In a third chapter with a mathematical flavor, Chapter 8, we discuss two less known properties of wavelets. First, some of them have minimal uncertainty, in the sense that they saturate some uncertainty relations linked to the Lie algebra of the wavelet group, exactly as Gaussians saturate those associated to the canonical commutation relations. Then we explore the relationship between wavelet transforms and the Wigner transform, well–known in physics and in radar theory (under the name of the closely related ambiguity function).

The next two chapters are devoted to various extensions of the standard CWT, that can be derived with help of the general formalism just developed. First we treat, in Chapter 9, the higher dimensional cases. We begin with the 3-D CWT, which is a straightforward extension of the 2-D case. Then we examine in depth the CWT over the 2-sphere. Here, of course, there is a strong motivation from several domains, from geophysics to astrophysics. The former is clear. As for the latter, when one considers the whole Universe, as in the problem of gamma source detection mentioned above, it is necessary to take the curvature into account.

However, there is an equally appealing aspect in the mathematics of the subject. Indeed, the group to consider here is the conformal group of the sphere S^2, which is nothing but the proper Lorentz group $SO_o(3, 1)$. The same group is also the conformal group of the plane $\mathbb{R}^2$, for instance, the tangent plane at the North Pole. The sphere and its tangent plane are mapped onto each other by the stereographic projection from the South Pole and its inverse. This operation is in fact the key to the construction of a spherical CWT. Indeed, the operations to be performed on spherical signals are motions on the sphere, given by rotations, and local dilations around a given point. In order to define these, one first defines dilations around the North Pole by lifting the corresponding ones in the tangent plane by inverse stereographic projection. Then, dilations around any other point of the sphere are obtained by combining the previous ones with an appropriate rotation. As a consequence, the parameter space of the spherical CWT is not the Lorentz group itself, but a homogeneous space of it, containing only rotations and the dilations just defined, that is, the quotient of $SO_o(3, 1)$ by a certain subgroup. Therefore, one needs the general formalism described in Chapter 7 in order to get a genuine spherical CWT. As an additional benefit, one recovers the natural link between the sphere and its tangent plane: the spherical CWT tends to the usual plane CWT when the radius of the sphere increases to infinity (the so-called Euclidean limit). It is gratifying that this aspect too is entirely described by the group-theoretical machinery, in terms of an operation called group contraction. Another byproduct of our spherical CWT is the possibility of designing good wavelet approximations of integrable functions on the sphere, another result previously known in the plane case. Here again practical applications are at hand, in the context of the so-called Geomathematics advertised by Freeden and his school [Fre97].

Then we turn, in Chapter 10, to the extension of the CWT to space–time. The problem of interest here is, of course, motion estimation, more precisely, detection, tracking, and identification of objects in (relative) motion. Examples include traffic monitoring, autonomous vehicle navigation, and tracking of ballistic missile warheads. This is a difficult problem, since the data is huge and often very noisy. As a consequence, most algorithms tend to lose track of the targets after a while, particularly if the latter changes its appearance (e.g., a maneuvering aeroplane) or in the case of an occlusion (one moving object hides another one). From the wavelet point of view, one designs a spatio-temporal CWT, whose parameters are space and time translations, rotations, global space–time dilations, that catch the size of the target, and a speed tuning parameter that measures its speed. The usual formalism goes through almost verbatim and allows one to design an efficient algorithm for motion estimation. One key ingredient again is the successive use of several partial energy densities.

In the final Chapter 11, we turn to another kind of generalizations, namely, transforms specially adapted to the detection and modeling of lines and curves, called the *ridgelet* and the *curvelet* transforms. The motivation for these new transforms, and their superiority over standard wavelets, is that they take much better into account the geometry of the object to be analyzed. A curve in the plane is more 1-D than 2-D, and the conventional 2-D CWT simply ignores this fact – hence it is unnecessarily costly. Here, of course, one experiences the much bigger richness of the 2-D world, in particular, concerning singularities of functions. These transforms naturally lead to new approaches to image compression and various nonlinear approximations, that we also describe.

We conclude the chapter and the book with a topic called 'algebraic wavelets'. These are wavelets adapted to self-similar tilings on the line or the plane obtained by replacing the usual natural numbers by a different system of numeration, for instance, the *golden mean* $\tau = \frac{1}{2}(1 + \sqrt{5})$. This is actually a generalization of the discrete WT, but it provides another example of wavelets adapted to a specific geometry, hence it is not out of place in this volume, and we found it interesting to give a short account of it, both in 1-D and in 2-D. In the latter case, typical examples are the famous Penrose tilings of the plane, with pentagonal symmetry, and this brings us back to the study of aperiodic patterns and to quasicrystals!

The conclusion of the whole story is definitely optimistic. Wavelets, and in particular the continuous WT, have proven to be a versatile and extremely efficient tool for image processing, provided one uses the right wavelet on the right problem. Their future is undoubtly bright, in many fields of science and technology.

Before concluding this introduction, several technical remarks are in order. First, most examples that are not reproduced from original papers have been computed using our own wavelet toolbox, called the YAW (Yet Another Wavelet) Toolbox, and freely accessible on the Louvain-la-Neuve website <http://www.fyma.ucl.ac.be/projects/yawtb/>.

Next, we have found it useful to split the references into two sections, devoted to books and Ph.D. theses, and regular journal articles (with a different presentation, viz. [Ald96] and [2], respectively). As we have already said, theses are an extremely rich source of information, although they are often only accessible on the web. In general, we have tried to trace most of the results to the original papers. Of course, there are omissions and misrepresentations, due to our ignorance and prejudices. We take responsibility for this and apologize in advance to those authors whose work we might have mistreated.

Acknowledgements

The present volume results from some fifteen years of continuing research interest in wavelets in Louvain-la-Neuve, starting with a collaboration between Alex Grossmann (Marseille) and J.-P. A. Throughout these years, all four authors have lectured on wavelets in places as diverse as Louvain-la-Neuve, Paris, Atlanta, Zakopane, Havana, Cotonou, Amsterdam, Lyon, Brussels, many papers have been written, and the book reflects all the experience thus acquired. Many students have been involved and they all deserve thanks for their contribution to the edifice. But most of all, we have to express our gratitude to three people, Bernard Piette, Alain Coron, and Laurent Jacques, for the project could never have been completed without their computer skills and tireless help. As expected, the elaboration process involved many reciprocal visits, and we have to thank our respective institutions (UCL, CTSPS at Clark Atlanta U., EPFL, Concordia U.) for their hospitality and support, as well as the various funding agencies that made these travels possible. Finally, we thank many colleagues for stimulating discussions, such as Roberto Cesar, Alain Arnéodo, Matthias Holschneider, Françoise Bastin, Fabio Bagarello, Emmanuel Van Vyve, to name a few. Special thanks are due to Bruno Torrésani, whose friendly comments and criticisms have helped us considerably, and who in addition proofread a large part of the manuscript.

Jean-Pierre Antoine (Louvain-la-Neuve)
Romain Murenzi (Atlanta and Kigali)
Pierre Vandergheynst (Lausanne)
Syed Twareque Ali (Montréal)

1 Warm-up: the 1-D continuous wavelet transform

1.1 What is wavelet analysis?

Wavelet analysis is a particular time- or space-scale representation of signals that has found a wide range of applications in physics, signal processing and applied mathematics in the last few years. In order to get a feeling for it and to understand its success, we consider first the case of one-dimensional signals. Actually the discussion in this introductory chapter is mostly qualitative. All the mathematically relevant properties will be described precisely and proved systematically in the next chapter for the two-dimensional case, which is the proper subject of this book.

It is a fact that most real life signals are nonstationary (that is, their statistical properties change with time) and they usually cover a wide range of frequencies. Many signals contain transient components, whose appearance and disappearance are physically very significant. Also, characteristic frequencies may drift in time (e.g., in geophysical time series – one calls them pseudo-frequencies). In addition, there is often a direct correlation between the characteristic frequency of a given segment of the signal and the time duration of that segment. Low frequency pieces tend to last for a long interval, whereas high frequencies occur in general for a short moment only. Human speech signals are typical in this respect: vowels have a relatively low mean frequency and last quite a long time, whereas consonants contain a wide spectrum, up to very high frequencies, especially in the attack, but they are very short.

Clearly standard Fourier analysis is inadequate for treating such signals. Strictly speaking, it applies only to stationary signals, and it loses all information about the time localization of a given frequency component. In addition, it is very uneconomical. When the signal is almost flat, and thus uninteresting, one still has to sum an infinite alternating series to reproduce it. Worse yet, Fourier analysis is highly unstable with respect to perturbation, because of its *global* character. For instance, if one adds an extra term, with a very small amplitude, to a linear superposition of sine waves, the signal will barely be modified, but the Fourier spectrum will be completely perturbed. This does not happen if the signal is represented in terms of *localized* components. Indeed, as we shall see shortly, the basic idea of the wavelet transform is to decompose a signal *locally* into contributions living at different scales. This is a marked contrast with the Fourier components, which are sinusoidal waves repeating themselves indefinitely. As such, it is difficult to give them any physical reality. If a piece of audio signal is identically zero, it is because no sound is emitted, not because the Fourier components necessary to represent the zero signal interfere destructively. These components are a mathematical construction, rather than a genuine physical phenomenon. To quote J. Ville [364]:

Si nous considérons en effet un morceau de musique ... et qu'une note, *la* par exemple, figure une fois dans le morceau, l'analyse harmonique [de Fourier] nous présentera la fréquence correspondante avec une certaine amplitude et une certaine phase, sans localiser le *la* dans le temps. Or, il est évident qu'au cours du morceau il est des instants où l'on n'entend pas le *la*. La représentation est néanmoins mathématiquement correcte, parce que les phases des notes voisines du *la* sont agencées de manière à détruire cette note par interférence lorsqu'on ne l'entend pas et à la renforcer, également par interférence, lorsqu'on l'entend; mais s'il y a dans cette conception une habileté qui honore l'analyse mathématique, il ne faut pas se dissimuler qu'il y a également une défiguration de la réalité: en effet, quand on n'entend pas le *la*, la raison véritable est que le *la* n'est pas émis.

That is,

If we consider a piece of music ... and if a note, an *A* for instance, appears once in that piece, Fourier analysis will yield the corresponding frequency with a certain amplitude and a certain phase, without localizing the *A* in time. Clearly the *A* will not be heard at certain instants. Yet the representation is mathematically correct, because the phases of the neighboring notes conspire to suppress the *A* by interference when it is not heard and to enhance it, again by interference, when it is heard. However, although this conception shows a skillfulness that honors mathematical analysis, one should not hide the fact that it also distorts reality: indeed, when the *A* is not heard, the true reason is that the *A* is not emitted.

Another eloquent comment along the same line by L. de Broglie may be found, together with the one above, in [Fla93; p.9].

Facing these problems, signal analysts turn to *time–frequency* representations. The idea is that one needs *two* parameters: one, called a, characterizes the frequency, the other one, b, indicates the position in the signal. This concept of a time–frequency representation is in fact quite old and familiar. The most obvious example is simply a musical score (see Figure 1.1). Clearly, it is not sufficient to give the pitch of a given note, that is, the frequency to which it corresponds, it is also important to know when to play it (time information)!

Let $s(x)$ be a finite energy signal, that is, a square integrable function $s \in L^2(\mathbb{R}, dx)$. In most cases, x will be a time variable and the (Fourier) conjugate quantity a frequency,

Fig. 1.1. A traditional time–frequency representation of a signal (from Mozart's Don Giovanni, Act 1).

but in general x simply represents position in the signal. Thus, following [Dau92], we prefer to keep a neutral notation (x, ξ) for the couple of conjugate variables, instead of the more familiar (t, ω). Accordingly, the Fourier transform of the signal s is defined by

$$\widehat{s}(\xi) = \frac{1}{\sqrt{2\pi}} \int_{-\infty}^{\infty} dx \, e^{-i\xi x} s(x). \tag{1.1}$$

If one requires the transform to be *linear*, a general time–frequency transform of the signal s will take the form:

$$s(x) \mapsto S(b, a) = \int_{-\infty}^{\infty} dx \, \overline{\psi_{b,a}(x)} s(x), \tag{1.2}$$

where $\psi_{b,a}$ is the analyzing function. Within this class, two time–frequency transforms stand out as particularly simple and efficient: the windowed (or short time) Fourier transform (WFT) and the wavelet transform (WT). For both of them, the analyzing function $\psi_{b,a}$ is obtained by acting on a basic (or mother) function ψ, in particular, b is simply a time translation. The essential difference between the two is in the way the frequency parameter a is introduced:

(1) *Windowed Fourier transform:*

$$\psi_{b,a}(x) = e^{i(x-b)/a} \, \psi(x - b). \tag{1.3}$$

Here ψ is a window function and the a-dependence is a modulation ($1/a \sim$ frequency); the window has constant width, but the smaller a, the larger the number of oscillations in the window (see Figure 1.2 (left)).

(2) *Wavelet transform:*

$$\psi_{b,a}(x) = \frac{1}{\sqrt{a}} \, \psi\left(\frac{x - b}{a}\right). \tag{1.4}$$

The action of a on the function ψ (which must be oscillating, see below) is a dilation ($a > 1$) or a contraction ($a < 1$): the shape of the function is unchanged, it is simply spread out or squeezed (see Figure 1.2 (right)). In particular, the effective support of $\psi_{b,a}$ varies as a function of a.

The windowed Fourier transform was originally introduced by Gabor (actually in a discretized version), with the window function ψ taken as a Gaussian; for this reason, it is sometimes called the *Gabor transform*. With this choice, the function $\psi_{b,a}$ is simply a canonical (harmonic oscillator) coherent state [Kla85], as one sees immediately by writing $1/a = p$. Since the new variables are the time (position) b and the frequency $1/a$, the Gabor transform yields a genuine time–frequency representation of the signal. As for the wavelet transform, the variables are b and the scale a (or pitch in the case of music), hence we shall speak rather of a *time-scale* representation.

We may remark here that the resemblance between the windowed Fourier transform and the wavelet transform is not accidental. They are both particular instances of a large

$1/a \approx$ frequency

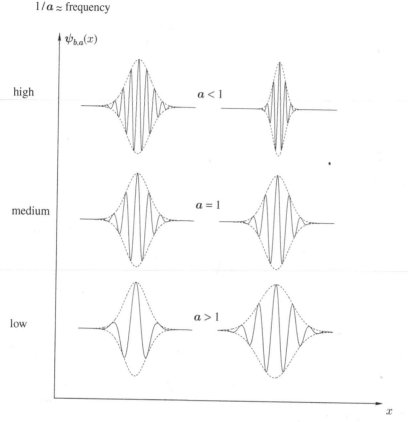

Fig. 1.2. The function $\psi_{b,a}(x)$ for different values of the scale parameter a, in the case of the windowed Fourier transform (left) and the wavelet transform (right). The quantity $1/a$, which corresponds to a frequency, increases from bottom to top.

class of integral transforms constructed by the formalism of coherent states [Ali00]. This general analysis, however, has a more mathematical flavor and is not needed in a first approach, although it clarifies and unifies the picture considerably. Therefore, we postpone it to Chapter 6, since we want to emphasize first the practical aspects of the wavelet transform.

One should note that the assumption of linearity is nontrivial, for there exists a whole class of quadratic or, more properly, sesquilinear time–frequency representations. The prototype is the so-called Wigner–Ville transform, introduced originally by E.P. Wigner [373] in quantum mechanics (in 1932!) and extended by J. Ville [364] to signal analysis:

$$W_s(b, \xi) = \int_{-\infty}^{+\infty} dx \, e^{-i\xi x} \, \overline{s(b - \frac{x}{2})} s(b + \frac{x}{2}), \quad \xi = 1/a. \tag{1.5}$$

Note that the signal $s(x)$ is usually a real function, but, in quantum mechanics, $s(x)$ represents a wave function, and is thus in general complex. This transform is entirely intrinsic to the signal, since it does not contain any extra function (wavelet, window) that inevitably influences the result. On the other hand, it is quadratic, which implies the appearance of interference terms whenever the signal is a superposition of two components. In order to minimize these as much as possible, one usually smoothes the Wigner–Ville transform with some function Φ, thus obtaining a whole class of quadratic transforms, called Cohen's class [109,Fla93], of the general form:

$$C_s(b, \xi) = \iint_{\mathbb{R}^2} db' \, d\xi' \, \Phi(b - b', \xi - \xi') \, W_s(b', \xi'). \tag{1.6}$$

An example is the so-called smoothened pseudo-Wigner–Ville distribution,

$$SPW_s(b, \xi) = \int_{-\infty}^{+\infty} db' \, g(b - b') \int_{-\infty}^{+\infty} dx \, h(x) \, e^{-i\xi x} \, \overline{s(b' - x/2)} \, s(b' + x/2), \tag{1.7}$$

corresponding to a factorizable kernel $\Phi(b, \xi) = (2\pi)^{-1/2} g(b) \widehat{h}(\xi)$, where $\widehat{h}$ denotes the Fourier transform of h. Further information about quadratic transforms may be found in [Fla93], and as a general survey for time–frequency methods, we refer to [Gro01].

1.2   The continuous wavelet transform

Actually one should distinguish two different versions of the wavelet transform, the *continuous* WT (CWT) and the *discrete* (or more properly, discrete time) WT (DWT) [Dau92,Hol95]. The CWT plays the same rôle as the Fourier transform and is mostly used for analysis and feature detection in signals, whereas the DWT is the analog of the Discrete Fourier Transform (see for instance [Bur98] or [326]) and is more appropriate for data compression and signal reconstruction. The situation may be caricatured by saying that the CWT is more natural to the physicist, while the DWT is more congenial to the signal analyst and the numericist. The continuous wavelet transform is the main topic of this book. Nevertheless, for the sake of comparison, we will give short overviews of the discrete WT, both in one and two dimensions.

The two versions of the WT are based on the same transformation formula, which reads, from (1.2) and (1.4):

$$S(b, a) = |a|^{-1/2} \int_{-\infty}^{\infty} dx \, \overline{\psi\left(\frac{x - b}{a}\right)} \, s(x), \tag{1.8}$$

where $a \neq 0$ is a scale parameter and $b \in \mathbb{R}$ a translation parameter (one often imposes only $a > 0$, which is more natural, but makes formulas slightly more complicated; see Chapter 6). Equivalently, in terms of Fourier transforms:

$$S(b, a) = |a|^{1/2} \int_{-\infty}^{\infty} d\xi \, \overline{\widehat{\psi}(a\xi)} \, \widehat{s}(\xi) \, e^{i\xi b}. \tag{1.9}$$

In these relations, s is a square integrable function, representing a finite energy signal, and the function ψ, the analyzing wavelet, is assumed to be well localized *both* in the space (or time) domain and in the frequency domain. In addition ψ must satisfy the following admissibility condition, which guarantees the invertibility of the WT:

$$c_\psi \equiv 2\pi \int_{-\infty}^{\infty} d\xi \, \frac{|\widehat{\psi}(\xi)|^2}{|\xi|} < \infty. \tag{1.10}$$

In most cases, this condition may be reduced to the (only slightly weaker) requirement that ψ has zero mean:

$$\widehat{\psi}(0) = 0 \iff \int_{-\infty}^{\infty} dx \, \psi(x) = 0. \tag{1.11}$$

Intuitively, it expresses the fact that a wavelet must be an oscillating function, real or complex ("little wave"). This is often thought to be the origin of the term "wavelet", but it is *not* the case historically. Indeed the word was widely in use in the geophysics community, with quite a different meaning, when it was introduced by Grossmann and Morlet [205,206] in the present sense, under the name "wavelets of constant shape" – but, of course, this lengthy nomenclature did not survive the very first founding paper!

The wavelet ψ is said to be *progressive* if its Fourier transform $\widehat{\psi}(\xi)$ is real and vanishes identically for $\xi \leqslant 0$. (In the signal processing community, a signal with this property is called *analytic*, following the terminology introduced by J. Ville [364].) In addition, ψ is often required to have a certain number of *vanishing moments*:

$$\int_{-\infty}^{\infty} dx \, x^n \, \psi(x) = 0, \; n = 0, 1, \ldots N. \tag{1.12}$$

This property improves the efficiency of ψ at detecting singularities in the signal, since it is then blind to polynomials up to order N, which constitute the smoothest part of the signal.

Notice that, instead of (1.8), which defines the WT as the scalar product of the signal s with the transformed wavelet $\psi_{b,a}$, $S(b, a)$ may also be seen as the convolution of s with the scaled, flipped and conjugated wavelet $\psi_a^\#(x) = |a|^{-1/2} \, \overline{\psi(-x/a)}$:

$$S(b, a) = (\psi_a^\# * s)(b) = \int_{-\infty}^{\infty} dx \, \psi_a^\#(b - x) \, s(x). \tag{1.13}$$

In other words, the CWT acts as a *filter* with a function of zero mean.

This property is crucial, for the main virtues of the CWT follow from it, combined with the support properties of ψ. Indeed, we must assume that ψ and $\widehat{\psi}$ are as well localized as possible, but respecting, of course, the Fourier uncertainty principle. This means that, up to minute corrections, the product of the lengths of the supports of ψ and $\widehat{\psi}$ is bounded from below by a fixed constant, usually taken as 1/2. Equivalently, the product of the variances of the distributions $|\psi|^2$ and $|\widehat{\psi}|^2$ is bounded from below. More precisely, one defines the centers of gravity (which may in fact be normalized to zero by a suitable redefinition of the coordinates):

$$x_0 = \int_{-\infty}^{\infty} dx \; x \, |\psi(x)|^2, \quad \xi_0 = \int_{-\infty}^{\infty} d\xi \; \xi \, |\widehat{\psi}(\xi)|^2, \tag{1.14}$$

and the corresponding variances

$$(\Delta x)^2 = \|\psi\|^{-2} \int_{-\infty}^{\infty} dx \; (x - x_0)^2 \, |\psi(x)|^2; \tag{1.15}$$

$$(\Delta \xi)^2 = \|\psi\|^{-2} \int_{-\infty}^{\infty} d\xi \; (\xi - \xi_0)^2 \, |\widehat{\psi}(\xi)|^2. \tag{1.16}$$

Then the Fourier uncertainty theorem [Fla93] says that

$$\Delta x \, \Delta \xi \geqslant \frac{1}{2}. \tag{1.17}$$

Under these assumptions, the transformed wavelets $\psi_{b,a}$ and $\widehat{\psi_{b,a}}$ are also well localized. Therefore, the WT $s \mapsto S$ performs a *local filtering*, both in time (b) and in scale (a). The transform $S(b, a)$ is nonnegligible only when the wavelet $\psi_{b,a}$ matches the signal, that is, the WT selects the part of the signal, if any, that lives around the time b and the scale a.

In addition, if $\widehat{\psi}$ has a numerical support (bandwidth) of width $\Delta \xi$, then $\widehat{\psi_{b,a}}$ has a numerical support of width $\Delta \xi / |a|$. Thus, remembering that $1/a$ behaves like a frequency, we conclude that the WT works at constant *relative* bandwidth, that is, $\Delta \xi / \xi = \text{constant}$. This implies that it is very efficient at high frequency, i.e., small scales, in particular for the detection of singularities in the signal. By comparison, in the case of the Gabor transform, the support of $\widehat{\psi_{b,a}}$ keeps the same width $\Delta \xi$ for all a, that is, the WFT works at constant bandwidth, $\Delta \xi = \text{constant}$. This difference in behavior is often the key factor in deciding whether one should choose the WFT or the WT in a given physical problem.

Another crucial fact is that the transformation $s(x) \mapsto S(b, a)$ may be inverted exactly, which yields a reconstruction formula (this is only the simplest one, others are possible, for instance using different wavelets for the decomposition and the reconstruction):

$$s(x) = c_\psi^{-1} \int_{-\infty}^{\infty} db \int_{-\infty}^{\infty} \frac{da}{a^2} \; \psi_{b,a}(x) \, S(b, a), \tag{1.18}$$

where the normalization constant c_ψ is given in (1.10) (incidentally, this relation shows why the admissibility condition $c_\psi < \infty$ is required for the transformation to be invertible). This means that the WT provides a decomposition of the signal as a linear superposition of the wavelets $\psi_{b,a}$ with coefficients $S(b, a)$. Notice that the natural measure on the parameter space (a, b) is $da \, db / a^2$, and it is invariant not only under time translation, but also under dilation. This fact is important, for it suggests that these geometric transformations play an essential rôle in the CWT.

One should emphasize here that the choice of the normalization factor $|a|^{-1/2}$ in (1.4) or (1.8) is not essential. This choice makes the transform unitary: $\|\psi_{b,a}\|_2 = \|\psi\|_2$ and also $\|S\|_2 = \|s\|_2$, where $\|\cdot\|_2$ denotes the L^2 norm in the appropriate variables (the squared norm is interpreted as the total energy of the signal). In practice, one often uses instead a factor a^{-1}, which has the advantage of giving more weight to the small scales, i.e., the high frequency part (which contains the singularities of the signal, if any). Thus, defining

$$\psi_{(b,a)} = \frac{1}{|a|} \psi\left(\frac{x-b}{a}\right),$$

(1.19)

we obtain the so-called L^1-normalized transform:

$$\check{S}(b,a) = \langle \psi_{(b,a)}|s \rangle \equiv |a|^{-1} \int_{-\infty}^{\infty} dx\ \overline{\psi\left(\frac{x-b}{a}\right)} s(x),$$

(1.20)

which preserves the L^1-norm of the signal, as follows immediately from the corresponding convolution formula

$$\check{S}(b,a) = (\psi_a^\# * s)(b),$$

(1.21)

where $\psi_a^\#(x) = |a|^{-1}\overline{\psi(-x/a)}$. Thus indeed $\|\psi_a^\#\|_1 = \|\psi\|_1$ and $\|\check{S}\|_1 = \|s\|_1$, where $\|\cdot\|_1$ denotes the L^1-norm in the corresponding variables.

1.2.1   Examples

In order to fix ideas, we exhibit here two simple examples of wavelets, both in the time domain and in the frequency domain.

(1) *The Mexican hat wavelet*

This wavelet is simply the second derivative of a Gaussian:

$$\psi_H(x) = (1-x^2)\exp(-\tfrac{1}{2}x^2), \quad \widehat{\psi}_H(\xi) = \xi^2 \exp(-\tfrac{1}{2}\xi^2).$$

(1.22)

(2) *The Morlet wavelet*

This wavelet is essentially a plane wave within a Gaussian window:

$$\psi_M(x) = \exp(ik_ox)\exp(-\tfrac{1}{2}x^2) + c(x), \quad \widehat{\psi}_M(\xi) = \exp(-\tfrac{1}{2}(\xi-\xi_o)^2) + \widehat{c}(\xi).$$

(1.23)

Here the correction term c must be added in order to satisfy the admissibility condition (1.11), but in practice one will arrange that this term be numerically negligible ($\leqslant 10^{-4}$) and thus can be omitted (it suffices to choose the basic frequency $|\xi_o|$ large enough, typically $|\xi_o| > 5.5$).

These two wavelets have very different properties and, naturally, they will be used in quite different situations. Typically, the Mexican hat is sensitive to singularities in the signal, and it yields a genuine time-scale analysis. On the other hand, since it is complex, the Morlet wavelet will catch the phase of the signal, hence will be sensitive to frequencies, and will lead to a time-frequency analysis, somewhat closer to a Gabor

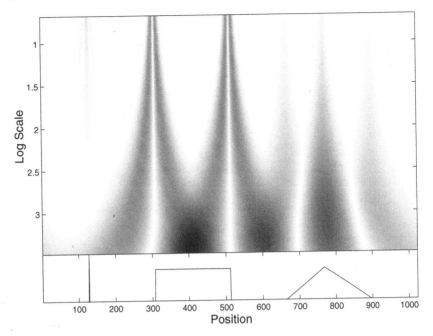

Fig. 1.3. Wavelet analysis with a Mexican hat wavelet of the discontinuous signal *bumps* (shown in the bottom panel).

analysis. In both cases, additional flexibility is obtained by adding a width parameter to the Gaussian (see (3.8) in the equivalent 2-D situation).

   As an illustration of the performance of the CWT as a singularity scanner, we first show in Figure 1.3 the analysis with a Mexican hat wavelet of a discontinuous signal, called *bumps* and consisting of three pieces, a δ function, a boxcar function and a tent function. Clearly the wavelet locates all discontinuities in the signal and in its successive derivatives well. However, if one wants to discriminate between the various types of singularities, one has to invoke the concept of vanishing moment, defined in (1.12). Let us consider the successive derivatives of a Gaussian:

$$\psi_{\text{H}}^{(n)}(x) = -\frac{d^n}{dx^n}\, \exp(-\tfrac{1}{2}x^2). \tag{1.24}$$

For increasing n, these wavelets have more and more vanishing moments, and are thus sensitive to increasingly sharper details. As an example, we consider a continuous signal obtained by glueing together an arc of parabola (the so-called function x_+^2) and a linear piece and we analyze it successively with the first three derivatives of a Gaussian, $\psi_{\text{H}}^{(n)}(x)$, $n = 1, 2, 3$. The result is shown in Figure 1.4. In (a), the first-order wavelet $\psi_{\text{H}}^{(1)}$ has only one vanishing moment, hence it sees the full content of the two pieces of the signal. In (b), the second-order wavelet $\psi_{\text{H}}^{(2)}(x) \equiv \psi_{\text{H}}$ does not see the linear part anymore, only the singularities at the two ends, but still sees the quadratic piece on the left (in technical terms, one would say that this wavelet is blind to a linear trend in the signal). In (c), finally, the third-order wavelet correctly erases both pieces of the

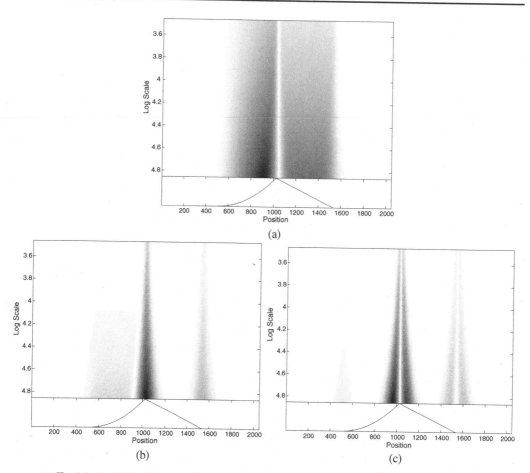

Fig. 1.4. Analysis of a composite signal (bottom panel) with successive derivatives of a Gaussian. (a) First order; (b) second order; (c) third order.

signal, keeping only the three singularities. This example shows the advantage of the *local* filtering effect of the CWT. Notice that a Gabor analysis would be utterly unable to achieve such a discrimination between singularities, let alone to detect them!

As a direct application of this behavior, an interesting technique has been designed by A. Arnéodo *et al.* [49], which consists in analyzing the same signal with several wavelets $\psi_{\rm H}^{(n)}$, for different n. The features common to all the transforms surely belong to the signal, they are not artifacts of the analysis.

1.3   Discretization of the CWT, frames

All this concerns the continuous WT (CWT). But, in practice, for numerical purposes, the transform must be *discretized*, by restricting the parameters a and b in (1.8) to the points of a discrete lattice $\Gamma = \{a_j, b_k, j, k \in \mathbb{Z}\}$ in the (a, b)-(half)-plane. Then we say that Γ yields a good discretization if an arbitrary signal $s(x)$ may be represented as a discrete superposition

$$s(x) = \sum_{j,k \in \mathbb{Z}} \langle \psi_{jk} | s \rangle \, \widetilde{\psi}_{jk}(x), \tag{1.25}$$

instead of the reconstruction formula (1.18). In (1.25), $\psi_{jk} \equiv \psi_{b_k,a_j}$ and $\widetilde{\psi}_{jk}$ should be explicitly constructible from ψ_{jk}. We emphasize that (1.25) must be an *exact* representation, i.e., there is no loss of information as compared to a direct discretization of the continuous reconstruction (1.18). Notice that here also, as in the latter, the reconstruction formula is in general not unique, which offers an additional degree of freedom in a given situation.

One may wonder whether a discrete representation of the type (1.25) is really possible. The answer lies in the reproducing property,

$$S(b', a') = c_\psi^{-1} \iint \frac{da \, db}{a^2} \langle \psi_{b',a'} | \psi_{b,a} \rangle \, S(b, a), \tag{1.26}$$

that every wavelet transform must satisfy. Indeed (1.26) implies that the information content of the wavelet transform $S(b, a)$ is highly redundant. In fact the signal has been unfolded from one to two dimensions, and this explains the practical efficiency of the CWT for disentangling parts of the signal that live at the same time, but on different scales. This redundancy (which is the source of the nonuniqueness of the reconstruction formula) may be eliminated – this is the rationale behind the discrete wavelet transform. It may also be exploited, for instance, by observing that it must be possible to obtain the full information about the signal from a small subset of the values of the transform $S(b, a)$. In particular, the validity of a representation (1.25) means that a discrete subset will do the job, and this is precisely what is needed for the reconstruction of a signal from its wavelet transform.

The problem is to find the minimal sampling grid ensuring no loss of information. In order to formulate it in mathematical terms, one relies on the theory of discrete *frames* or nonorthogonal expansions, that we now sketch. See [121,Dau92] for a complete treatment.

In fact, the discrete representation (1.25) means that the signal $s(x)$ may be replaced by the set $\{\langle \psi_{jk} | s \rangle\}$ of its wavelet coefficients. Since $s \in L^2$, it is natural to require that the sequence of coefficients be also square integrable and that the map $F : s \mapsto \{\langle \psi_{jk} | s \rangle\}$ be continuous from $L^2(\mathbb{R})$ to ℓ^2, i.e.,

$$\sum_{j,k \in \mathbb{Z}} |\langle \psi_{jk} | s \rangle|^2 \leqslant B \|s\|^2, \quad 0 < B < \infty. \tag{1.27}$$

In addition, one wants the reconstruction of $s(x)$ from its coefficients to be numerically stable, that is, a small error in the coefficients implies a small error in the reconstructed signal. In particular, if the left-hand side of (1.27) is small, $\|s\|^2$ should be small also. Therefore, there must exist a constant $A > 0$ such that

$$A \, \|s\|^2 \; \leqslant \; \sum_{j,k \in \mathbb{Z}} |\langle \psi_{jk} | s \rangle|^2 \; \leqslant \; B \, \|s\|^2 \tag{1.28}$$

(the lower bound indeed guarantees the numerical stability [Dau92]). By definition, this relation means precisely that the set $\{\psi_{jk}\}$ constitutes a (discrete) *frame*, with *frame bounds* A and B. This frame is said to be *tight* if $A = B$. Note that (1.28) is in fact a weakened form of the Parseval relation. The latter is recovered in the case of a tight frame, in particular for an orthonormal basis. For a general frame, however, (1.28) is sufficient for inverting the wavelet representation (we will discuss this in detail in the 2-D case, in Section 2.4).

We will present a detailed analysis of these concepts in Chapter 2. Here we simply observe that, for all practical purposes, a good frame is almost as good as an orthonormal basis. By "good frame," we mean that the expansion (1.25) converges sufficiently fast. The detailed analysis of [121,122] shows this to be the case if $|B/A - 1| \ll 1$, thus in particular if the frame is tight. Many functions ψ satisfying the admissibility condition (1.10) will yield a good frame (of course, this must be proved for every given ψ). However, we will *not* get an orthonormal basis, since the functions $\{\psi_{jk}, \ j, k \in \mathbb{Z}\}$ are in general not orthogonal to each other!

Yet orthonormal bases of wavelets can be constructed, but by a totally different approach, based on the concept of *multiresolution analysis*. We emphasize that the discretized version of the CWT just described is totally different in spirit and method from the genuine discrete wavelet transform, that we will sketch in Section 1.5 below. The full story may be found in [Dau92], for instance.

Of course the practical question is: how does one build a good frame? Clearly, the question of the existence of a discrete frame must take into account the geometry of the parameter space. In the present case, this means that the lattice Γ must be invariant under discrete dilations and translations:

- for scale, one chooses naturally $a_j = a_o \lambda^{-j}, \ j \in \mathbb{Z}$, for some $\lambda > 1$;
- for time, one takes $b_k \equiv b_{k,j} = k \, b_o \, a_o \lambda^{-j}, \ j, k \in \mathbb{Z}$.

Thus we get

$$\psi_{jk}(x) = \lambda^{j/2} \, \psi(a_o^{-1} \lambda^j x - k b_o), \quad j, k \in \mathbb{Z}. \tag{1.29}$$

The most common choice is $\lambda = 2$ (octaves!) and $a_o = b_o = 1$, which results in

$$\psi_{jk}(x) = 2^{j/2} \, \psi(2^j x - k), \quad j, k \in \mathbb{Z}. \tag{1.30}$$

It is worth noticing that this so-called *dyadic* lattice $\{(k2^{-j}, 2^{-j}), \ j, k \in \mathbb{Z}\}$ is exactly the same that indexes the DWT (see Section 1.5), which may create some confusion (and sometimes did so!).

For a given choice of ψ, λ, one finds a range of values of b_o such that $\{\psi_{jk}\}$, as given in (1.29), is a frame. Detailed results may be found in [122,Dau92]. Here we will restrict ourselves to the following simplified version.

Theorem 1.3.1. *Let ψ and a_o be such that:*

(i) $\quad \displaystyle\inf_{1 \leqslant \xi \leqslant \lambda} \sum_{j=-\infty}^{\infty} |\widehat{\psi}(\lambda^{-j}\xi)|^2 > 0;$

(ii) $\quad |\widehat{\psi}(\xi)| \leqslant C\,|\xi|^\alpha\,(1+|\xi|)^{-\gamma}, \ \alpha > 0, \ \gamma > \alpha + 1.$

Then there exists b_{oo} such that $\{\psi_{jk}\}$ constitutes a frame for all choices $b_o < b_{oo}$.

Both the Mexican hat and the Morlet wavelet satisfy the conditions of the theorem for the dyadic case, $\lambda = 2$, $a_o = b_o = 1$, thus they both generate discrete frames on the dyadic lattice. Explicit values for the corresponding frame bounds A, B may be found in [Dau92].

However, the numerical implementation of such a dyadic frame is unwieldy, in particular, the reconstruction works well only if the frame is very redundant. Two variants are used in practice, and both of them amount to increase the redundancy. The first one consists in subdividing further the octaves, introducing what are called additional *voices*. This means that one further subdivides each octave, replacing in (2.77) the exponent j by $\frac{v}{N}j$, $v = 0, 1, \ldots, N-1$ (v is called a voice and N the number of voices). The effect is to "densify" the dyadic lattice, which improves the ratio B/A and thus speeds up convergence of the discrete approximation. By taking sufficiently many voices by octaves, one may even get frames which are numerically tight. For further details, we refer the reader to [Dau92]. However, if the speed is the determining criterion, one can do better by using the continuous wavelet packets described in Section 1.6.1.

The other solution consists in replacing the dyadic lattice with a rectangular one, that is, taking the sampling rate independent of the scale. This has the advantage of making the whole analysis invariant under global discrete translations. The resulting version of the discretized CWT has been advocated by Mallat [Mal99,265] (who, curiously, calls it the dyadic WT!) and Torrésani [Tor95], and we will meet it again repeatedly in the sequel (see Section 1.6.1). Note that similar ideas were already used in the classical Littlewood–Paley analysis [Fra91] and the famous Laplacian pyramid from vision analysis [91]. The scheme was also rediscovered by statisticians, who called it "stationary wavelet transform" [112,295].

To illustrate this technique, we show in Figure 1.5 a five-level decomposition of the signal *bumps* with a translation invariant frame of quadratic spline wavelets, as used in [265]. The figure shows, from bottom up, the low resolution approximation and the five levels of details with increasing resolution.

As we shall see in detail in the next chapter, Section 2.5, the whole machinery of frames extends almost verbatim to the two-dimensional case. In particular, both the 2-D Mexican hat and the 2-D Morlet wavelet yield reasonably good 2-D frames. Besides their simplicity and their efficiency, this explains the widespread use of these two wavelets in image processing, as was already the case in one dimension.

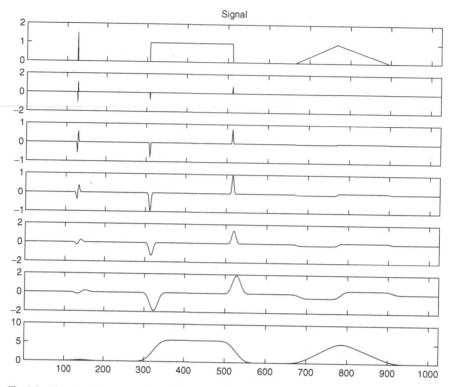

Fig. 1.5. Five-level decomposition of the signal *bumps* with a translation invariant frame of quadratic spline wavelets. The low resolution approximation is shown on the bottom panel and the five levels of details with increasing resolution $a = 2^{-j}$, $j = -5, -4, -3, -2, -1$, in the next five panels.

1.4   Ridges and skeleton

Real life signals are frequently very entangled and noisy, and their WT is difficult to interpret. However, a clever exploitation of the intrinsic redundancy of the CWT is often able to bypass the difficulty and thus to improve the efficiency and the range of applicability of wavelet analysis. The technique consists in using the *skeleton* of the CWT instead of its modulus. Roughly speaking, the skeleton is a collection of lines, called *ridges*, which are approximately lines of local maxima. The concept is easy to visualize in two extreme situations [Mal99,262,358].

Assume first that the signal $s(t)$ consists of a singularity $\gamma_\alpha(x - x_o)$, of order α, at time x_o, superimposed on a smooth background and some stochastic noise [262]. Here the singularity function γ_α is defined as follows:

$$\gamma_\alpha(x - x_o) = \begin{cases} 0, & x \leqslant x_o, \\ (x - x_o)^\alpha, & x > x_o. \end{cases} \tag{1.31}$$

[The index α is in fact an index of homogeneity or a Lipschitz regularity exponent. For instance, a δ function has $\alpha = -1$.] Thus we have

$$\frac{d^{\alpha+1}\gamma_\alpha}{dx^{\alpha+1}}(x - x_o) = \Gamma(\alpha + 1)\delta(x - x_o).$$

Let the wavelet be the nth derivative of a smooth positive function ϕ, that is, $\psi(x) = \frac{d^n}{dx^n}\phi(x)$, with $n \geqslant \alpha + 1$ (typically, a derivative of a Gaussian, like the Mexican hat and its higher order analogs (1.24)). Then the CWT of γ_α with respect to ψ to reads:

$$S_{\gamma_\alpha}(b, a) = \Gamma(\alpha + 1)a^\alpha \frac{d^{n-\alpha-1}\phi}{dx^{n-\alpha-1}}\left(\frac{x_o - b}{a}\right). \tag{1.32}$$

Assume now that the modulus of the $(n - \alpha - 1)$th derivative of ϕ has N maxima $\{\phi_l, l = 1, \ldots, N\}$ at positions $\{x_l, l = 1, \ldots, N\}$. Then, for each a, the modulus $|S_{\gamma_\alpha}(b, a)|$ has N maxima localized at positions $\{b_l = ax_l + x_o, l = 1, \ldots, N\}$, which converge toward x_o as $a \to 0$. Furthermore, the maxima of $|S_{\gamma_\alpha}(b, a)|$ lie on N lines, called (vertical) ridges $\{b = ax_l + x_o, l = 1, \ldots, N\}$, which converge toward the singularity x_o of the signal, and the modulus of $|S_{\gamma_\alpha}(b, a)|$ along the lth ridge behaves as a^α:

$$|S_{\gamma_\alpha}(b = ax_l + x_o, a)| = \Gamma(\alpha + 1)a^\alpha \phi_l. \tag{1.33}$$

Hence the strength α of the singularity may be read off a log–log plot:

$$\ln|S_{\gamma_\alpha}(ax_l + x_o, a)| \sim \alpha \ln a + \ln \phi_l. \tag{1.34}$$

This technique, introduced by Mallat and Hwang [262], has, been developed to a considerable extent for the analysis of fractals by Arnéodo and his collaborators under the name of Wavelet Transform Modulus Maxima (WTMM) (see [Arn95] for a survey). The important point is that the restriction of the WT to its skeleton (the set of its ridges) characterizes the signal completely [264,265]. Thus, in practice, it is enough to compute the skeleton. We will discuss the 2-D extension of the WTMM method in Section 2.3.5 and its application to fractals in Section 5.4.

To give a simple example, we take again our signal *bumps*, and compute the skeleton of its wavelet transform shown in Figure 1.3. The result, presented in Figure 1.6, clearly confirms the analysis above.

As a second example, we consider the analysis of the behavior of a material under impact made in [358]. The physical context is that of a so-called "instrumented falling weight impact" testing. During such a test, a striker falls from a certain height on a clamped disk, so that either the striker rebounds or the disk breaks. In both cases, one records the time and the force on the striker. This type of event occurs on a very short time scale and is thus essentially transient, so that a time–frequency method is required for the analysis. Among the various methods discussed in [358], we focus here on the case of a rebound and a wavelet analysis of the force signal. The Mexican hat detects precisely three discontinuity points, namely, first contact, maximal penetration and last

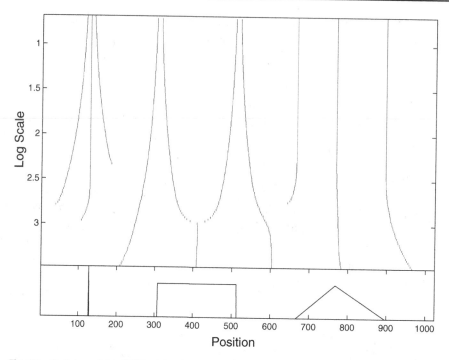

Fig. 1.6. Skeleton of the CWT of the signal *bumps*, as presented in Figure 1.3.

contact. The results are shown in Figure 1.7. The signal is given in panel (a), whereas the CWT and its skeleton are presented in (b) and (c). The latter, in particular, shows the three ridges that point towards the three instants mentioned. A further analysis exploits the behavior of the modulus of the CWT along each ridge. This yields precious insight into the physics of the phenomenon, particularly in the case of the rupture of the sample, not shown here. We refer to [358] for more details.

More generally, vertical ridges allow us to discriminate between genuine signal features and noise. First, noise ridges are usually much shorter, being visible mostly at small scales [21]. Then the modulus of the CWT tends to *increase* for increasing scale a on a noise ridge, whereas it *decreases* along genuine signal ridges. This fact has been exploited, for instance (in a discrete set-up), for the correction of aberrated images of the Hubble Space Telescope [85]: noise and signal have opposite behavior with increasing scale.

As for the second typical situation, the idea is that many signals are well approximated by a superposition of simple spectral lines:

$$s(x) = \sum_{l=1}^{N} s_l(x), \quad s_l(x) = A_l(x) e^{i\xi_l x}, \tag{1.35}$$

where the amplitude $A_l(x)$ varies slowly. By linearity, the WT of this signal is a sum of terms, $S(b, a) = \sum_l S_l(b, a)$, where, from (1.9),

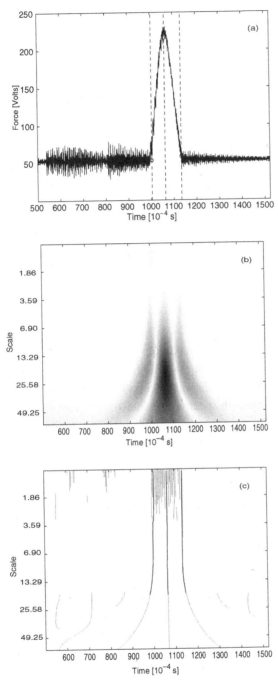

Fig. 1.7. Analysis of the rebound signal, with a Mexican hat wavelet: (a) the signal and the points detected by the respective ridges; (b) the modulus of the CWT; and (c) the corresponding skeleton (from [358]).

$$S_l(b, a) = \sqrt{a} \int_{-\infty}^{+\infty} d\xi \, \widehat{\psi}(a\xi) \, \widehat{A}_l(\xi - \xi_l) \, e^{i\xi b} \tag{1.36}$$

$$= \sqrt{a} \, e^{i\xi_l b} \int_{-\infty}^{+\infty} d\xi \, \widehat{\psi}(a(\xi + \xi_l)) \, \widehat{A}_l(\xi) \, e^{i\xi b}. \tag{1.37}$$

Inserting the Taylor expansion of $\widehat{\psi}$ around $a\xi_l$, one obtains the following expansion for S_l:

$$S_l(b, a) = \widehat{\psi}(a\xi_l) s_l(b) - i a e^{i\xi_l b} \frac{d\widehat{\psi}}{d\xi}(a\xi_l) \frac{dA_l}{db}(b) + \text{ higher order terms} \tag{1.38}$$

Assuming the amplitude to be a smooth function (C^1), all terms beyond the first are easily bound using the intermediate value theorem and one can limit oneself to the lowest order, namely,

$$S_l(b, a) \simeq \widehat{\psi}(a\xi_l) s_l(b), \tag{1.39}$$

and thus

$$S(b, a) \simeq \sum_{l=1}^{N} \widehat{\psi}(a\xi_l) s_l(b). \tag{1.40}$$

Assume that the wavelet $\widehat{\psi}(\xi)$ has a unique maximum in frequency space at $\xi = \xi_o$, like the Morlet wavelet (1.23). Then, if the values of the frequencies ξ_l are sufficiently far away from each other, the factor $\widehat{\psi}(a\xi_l)$ allows to treat each spectral line independently. In this case, the contribution of the lth spectral line to $S(b, a)$ is localized on the scale $a_l = \xi_o/\xi_l$ and, along the line of maxima $a = a_l$, called the lth (horizontal) *ridge*, the CWT is approximately proportional to the lth spectral line:

$$\frac{S(b, a_l)}{\widehat{\psi}(\xi_o)} \simeq s_l(b). \tag{1.41}$$

The set of all the ridges is again called the *skeleton* of the CWT. Thus the restriction of the WT $S(b, a)$ to its skeleton contains the whole information.

The analysis extends to the more general case where the spectral lines have the form

$$s_l(x) = A_l(x) \, e^{i\phi_l(x)}, \tag{1.42}$$

with the amplitude $A_l(x)$ varying slowly with respect to the phase $\phi_l(x)$ (such signals are called *asymptotic*). Typical examples are spectra in NMR spectroscopy [210]. In this case, each term S_l in the CWT (1.8) in the time domain is a rapidly oscillating integral, the essential contribution to which is given by the stationary points of the phase of the integrand. These points are the solutions $x_s(b, a)$ of the equation

$$\frac{d\phi_l}{dx}(x_s) = \frac{\xi_o}{a}. \tag{1.43}$$

Then the corresponding *ridge* of the WT is defined as the set of points (b, a) for which $x_s(b, a) = b$. These constitute a curve in the (b, a)-half-plane, which essentially reduces to a line of local maxima. A detailed analysis [131] shows again that, on this curve, the WT $S(b, a)$ coincides, up to a small correction, with the component $s_l(b)$ of the signal. Taking all ridges together, one obtains the *skeleton* of the CWT, and the analysis shows that the restriction of $S(b, a)$ to it essentially coincides with the analytic signal $Z(b)$ associated to $s(x)$. It follows again that the restriction of the WT $S(b, a)$ to its skeleton contains the whole information. In particular, the so-called frequency modulation law $x^{-1} \arg\{s(x)\}$ of $s(x)$ is easily recovered from it. Thus, it is not necessary to compute the whole CWT, but only its skeleton [99]. This is, of course, much less costly computationally, because there are fast algorithms available. Spectacular applications of this method may be found, for instance, in spectroscopy [131], geomagnetism [4], chirp detection/estimation in gravitational waves [Mor02,277] or shape determination [21]. The last quoted paper, in particular, contains a thorough analysis of the two types of ridges, vertical and horizontal, interpreted in both cases as lines of local maxima (see Section 5.4.2).

1.5   The discrete WT: orthonormal bases of wavelets

One of the successes of the WT was the discovery that it is possible to construct functions ψ for which $\{\psi_{jk}, j, k \in \mathbb{Z}\}$ is indeed an orthonormal basis of $L^2(\mathbb{R})$. In addition, such a basis still has the good properties of wavelets, including space *and* frequency localization. Moreover, it yields fast algorithms, and this is the key to the usefulness of wavelets in many applications.

The construction is based on two facts. First, almost all examples of orthonormal bases of wavelets can be derived from a multiresolution analysis, and then the whole construction may be transcripted into the language of digital filters, familiar in the signal processing literature.

Notice that it is precisely at this point that arises the basic difference between the *discretized* continuous wavelet transform, discussed in the previous section, and the discrete wavelet transform (DWT). In the former case, the wavelet ψ is chosen *a priori* (with very few constraints, as we have seen above), and the question is whether one can find a lattice Γ such that $\{\psi_{jk}\}$ is a frame with decent frame bounds A, B. In the other approach, one imposes from the beginning that the set $\{\psi_{jk}\}$ be an orthonormal basis and tries to construct a function ψ to that effect. The construction is rather indirect and the resulting function is usually very complicated (sometimes it has a fractal behavior).

Definition 1.5.1. *A* multiresolution analysis *of $L^2(\mathbb{R})$ is an increasing sequence of closed subspaces*

$$\ldots \subset V_{-2} \subset V_{-1} \subset V_0 \subset V_1 \subset V_2 \subset \ldots, \qquad (1.44)$$

with $\bigcap_{j \in \mathbb{Z}} V_j = \{0\}$ and $\bigcup_{j \in \mathbb{Z}} V_j$ dense in $L^2(\mathbb{R})$, and such that

(1) $f(x) \in V_j \Leftrightarrow f(2x) \in V_{j+1}$

(2) There exists a function $\phi \in V_0$, called a scaling *function, such that the family* $\{\phi(x-k), k \in \mathbb{Z}\}$ *is an orthonormal basis of V_0.*

Combining conditions (1) and (2), one gets an orthonormal basis of V_j, namely $\{\phi_{jk}(x) \equiv 2^{j/2}\phi(2^j x - k), k \in \mathbb{Z}\}$. Note that the scaling function ϕ is often required only to generate a Riesz basis of V_0, that is, a frame $\{\phi_{0k}, k \in \mathbb{Z}\}$ of linearly independent vectors spanning V_0. However, since the Riesz basis can be orthonormalized, for instance by a Gram–Schmidt procedure, the condition (2) is equivalent, and simpler.

Remark: Some authors (for instance, [Dau92] or [Mal99]) use the opposite convention for the index j, namely $V_j \subset V_{j-1}$ (both have advantages and inconveniences). With the present convention, large j means small scale (of order) 2^{-j} or high frequency 2^j, thus high resolution 2^j.

Each subspace V_j can be interpreted as an *approximation space*. The approximation of $f \in L^2(\mathbb{R})$ at the resolution 2^j is defined by its projection onto V_j, and the larger j, the finer the resolution obtained. Then condition (1) means that no scale is privileged. The additional details needed for increasing the resolution from 2^j to 2^{j+1} are given by the projection of f onto the orthogonal complement W_j of V_j in V_{j+1}:

$$V_j \oplus W_j = V_{j+1}, \tag{1.45}$$

and we have:

$$L^2(\mathbb{R}) = \bigoplus_{j \in \mathbb{Z}} W_j. \tag{1.46}$$

Equivalently, fixing some lowest resolution level j_o, one may write

$$L^2(\mathbb{R}) = V_{j_o} \oplus \left(\bigoplus_{j=j_o}^{\infty} W_j \right). \tag{1.47}$$

The crucial theorem then asserts the existence of a function ψ, sometimes called the *mother wavelet*, explicitly computable from ϕ, such that $\{\psi_{jk}(x) \equiv 2^{j/2}\psi(2^j x - k),$ $j, k \in \mathbb{Z}\}$ constitutes an orthonormal basis of $L^2(\mathbb{R})$: these are the *orthonormal wavelets*.

The construction of ψ proceeds roughly as follows. First, the inclusion $V_0 \subset V_1$ yields the relation (called the *scaling*, or *two-scale*, or *refinement* equation):

$$\phi(x) = \sqrt{2} \sum_{k=-\infty}^{\infty} h_k \phi(2x - k), \quad h_k = \langle \phi_{1,k} | \phi \rangle. \tag{1.48}$$

Taking Fourier transforms, this gives

$$\widehat{\phi}(2\xi) = h(\xi)\widehat{\phi}(\xi), \quad \text{with } h(\xi) = \frac{1}{\sqrt{2}} \sum_{k=-\infty}^{\infty} h_k e^{-ik\xi}. \tag{1.49}$$

Thus h is a 2π-periodic function and it satisfies the relation

$$|h(\xi)|^2 + |h(\xi + \pi)|^2 = 1, \quad \text{a.e.} \tag{1.50}$$

Iterating (1.49), one gets the scaling function as the infinite product

$$\widehat{\phi}(\xi) = (2\pi)^{-1/2} \prod_{j=1}^{\infty} h(2^{-j}\xi), \tag{1.51}$$

which may be proven to be convergent [Dau92]. Then one defines the function $\psi \in W_0 \subset V_1$ by the relation

$$\widehat{\psi}(2\xi) = g(\xi)\,\widehat{\phi}(\xi), \tag{1.52}$$

where g is another 2π-periodic function. By the relation (1.45) and the orthonormality of the functions $\{\phi_{jk}\}$, the functions h, g must satisfy the identity

$$g(\xi)\,\overline{h(\xi)} + g(\xi + \pi)\,\overline{h(\xi + \pi)} = 0, \quad \text{a.e.} \tag{1.53}$$

The simplest solution is to put $g(\xi) = e^{i\xi}\,\overline{h(\xi + \pi)}$, which implies, in particular, $|h(\xi)|^2 + |g(\xi)|^2 = 1$, a.e. Then one obtains

$$\psi(x) = \sqrt{2} \sum_{k=-\infty}^{\infty} (-1)^{k-1} h_{-k-1} \phi(2x - k), \tag{1.54}$$

and one proves that this function indeed generates an orthonormal basis with all the required properties. Another, equivalent, solution is

$$\psi(x) = \sqrt{2} \sum_{k=-\infty}^{\infty} (-1)^k h_{-k+1} \phi(2x - k). \tag{1.55}$$

Various additional conditions may then be imposed on the basic wavelet ψ, such as arbitrary regularity, several vanishing moments (in any case, ψ has always mean zero), symmetry, fast decrease at infinity, and even compact support [Dau92].

Remark: Some authors (for instance, [Dau92] or [Tor95]) denote the functions h, g by m_0, m_1, respectively.

The simplest example of this construction is the Haar basis, which comes from the scaling function $\phi(x) = 1$ for $0 \leqslant x < 1$, and 0 otherwise (boxcar function). The coefficients of the corresponding filter h are $h_0 = h_1 = 1/\sqrt{2}, h_k = 0$, for $k \neq 0, 1$. Applying the recipe (1.55) then yields the Haar wavelet

$$\psi_{\text{Haar}}(x) = \begin{cases} 1, & \text{if } 0 \leqslant x < 1/2, \\ -1, & \text{if } 1/2 \leqslant x < 1, \\ 0, & \text{otherwise.} \end{cases} \tag{1.56}$$

Similarly, various B-spline bases may be obtained along the same line. Other explicit examples may be found in [Chu92] or [Dau92].

In order to set up the discrete WT, the technique consists in translating the multiresolution structure into the language of digital filters, which is precisely what we have just done. Indeed, a filter is simply a multiplication operator in frequency space or a linear convolution in the time variable, and the discussion above amounts to nothing more than expanding (filter) functions in a Fourier series. For instance, $h(\xi)$ is a filter, with Fourier coefficients h_n, $g(\xi)$ is another one, and $\{h, g\}$ are called Quadrature Mirror Filters or QMFs whenever they satisfy the identities (1.50) and (1.53). Then the various restrictions imposed on ψ translate into suitable constraints on the filter coefficients h_n. For instance, ψ has compact support if only finitely many h_n differ from zero (one then speaks of a finite impulse response or FIR filter).

Of course, the goal is to obtain a fast algorithm, and this relies on two aspects, namely, short filters and a pyramidal structure, already familiar in signal processing. Indeed, the rapidity of the algorithms depends crucially on the length of the filters involved, because the pyramidal structure rests on a concatenation of several filters. One major stumbling block is the dilation. This is easy to understand from the very definition of the WT. Indeed, as we have seen above, the WT (1.8) or (1.13) basically a convolution. Once discretized, these formulas become discrete convolutions of digital sequences. Then the point is that, if the sequence $\psi^{\#}(n)$ has length N, then the dilated sequence $\psi_2^{\#}(n)$ has length $2N$, and this leads to an algorithm of exponential increase, clearly not admissible. One trick for avoiding this difficulty is to replace the (natural) dilation by a so-called pseudodilation, which consists in inserting a zero between any two successive entries of $\psi^{\#}(n)$, and then correcting for the distortion so introduced. In this way, one obtains a fast algorithm. Since the sequences resulting from successive dilations have all the same length, but are full of zeros or holes, this algorithm is known as the "algorithme à trous" [157,222]. The interesting fact is that this procedure may be extended to very general situations, involving wavelet transforms on abelian groups, which form a kind of intermediate step ("missing link") between the CWT and the DWT [Kou00,32]. Several other fast algorithms have been designed, mostly along the line proposed by Mallat [Mal99,339]. Altogether, there does exist a *Fast Wavelet Transform*, exactly as a Fast Fourier Transform.

In practical applications, the (sampled) signal is taken in some V_J, and then the decomposition (1.47) is replaced by the finite representation

$$V_J = V_{j_o} \oplus \left(\bigoplus_{j=j_o}^{J-1} W_j \right). \tag{1.57}$$

Figure 1.8 shows an example of a decomposition of order 5, namely our familiar signal *bumps* decomposed over an orthonormal basis of Daubechies d6 wavelets [Dau92]. Thus we take $J = 0$ and $j_o = -5$ in formula (1.57):

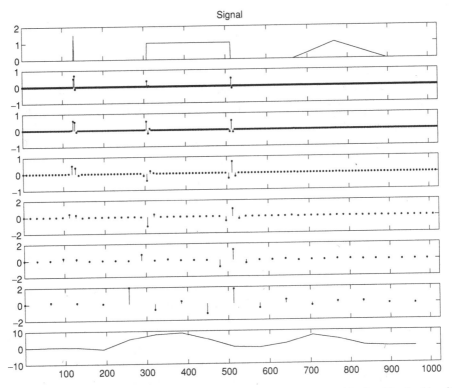

Fig. 1.8. Five-level decomposition of the *bumps* signal on an orthonormal basis of Daubechies d6 wavelets. The low resolution approximation $c_{-5} \in V_{-5}$ is shown on the bottom panel and the five levels of details with increasing resolution, $d_j \in W_j$, $j = -5, \ldots, -1$, in the next five panels.

$$V_0 = V_{-5} \oplus W_{-5} \oplus W_{-4} \oplus W_{-3} \oplus W_{-2} \oplus W_{-1}. \tag{1.58}$$

Correspondingly, the signal $s \in V_0$ is decomposed as

$$s = \sum_{k \in \mathbb{Z}} c_{-5,k} \, \phi_{-5,k} + \sum_{j=-5}^{-1} \sum_{k \in \mathbb{Z}} d_{jk} \, \psi_{jk}$$

$$\equiv c_{-5} + \sum_{j=-5}^{-1} d_j.$$

As in Figure 1.5, the figure shows, from bottom up, the low resolution approximation c_{-5} and the five levels of details with increasing resolution, successively $d_{-5}, d_{-4}, d_{-3}, d_{-2}, d_{-1}$.

At this point, we should add a word of caution concerning the numerical implementation of the reconstruction formula associated to (1.57),

$$s = \sum_{k \in \mathbb{Z}} c_{j_o,k} \, \phi_{j_o,k} + \sum_{j=j_o}^{J-1} \sum_{k \in \mathbb{Z}} d_{jk} \, \psi_{jk}. \tag{1.59}$$

If s is an analog signal, i.e. a function, the approximation coefficients $c_{j_o,k}$ and the wavelet coefficients $d_{j,k}$ are calculated in the standard way,

$$c_{j_o,k} = \langle \phi_{j_o,k} | s \rangle, \quad d_{j,k} = \langle \psi_{jk} | s \rangle.$$

However, if the signal is only accessible through sampled values, then the coefficients must be estimated from the latter. This may be done by making some assumption about the signal (for instance, in the "algorithme à trous" [157,222], the signal is supposed to be a spline function), which amounts to some pre-filtering. Alternatively, one takes for $c_{j_o,k}$ the sampled values themselves. This is a good approximation [Mal99; Section 7.2.3], but without any real theoretical justification. In any case, these procedures generate errors that have to be controlled. See [132] for a comprehensive discussion.

1.6   Generalizations

As we just saw, appropriate filters generate orthonormal wavelet bases. However, this result turns out to be too rigid and various generalizations have been proposed (see also the comments in Section 1.6.2). To name a few: biorthogonal wavelet bases, wavelet packets and the Best Basis Algorithm, the lifting scheme and second generation wavelets. We shall refrain from describing these here. Instead, we will give a rather detailed treatment in Chapter 2, for the two-dimensional case. Further information may also be found in [Mey94].

1.6.1   Continuous wavelet packets

Besides the full discretization described in Section 1.3, and the discrete WT just discussed, there is an intermediate procedure, which consists in discretizing the scale variable alone, on an arbitrary sequence of values (not necessarily powers of a fixed ratio), but leaving translations fully continuous. The resulting transform has the advantage of being completely covariant with respect to translations, a very desirable feature, for instance, in pattern recognition. If we use dyadic scales, the result is called a *dyadic wavelet transform* and was introduced in [264,265], precisely for that reason. A detailed account will be given in the 2-D case, in Section 2.4.4.

An elegant way of deriving this dyadic WT from the CWT was described in [159], under the name of *infinitesimal multiresolution analysis*, or *continuous wavelet packets*. This approach leads to fast algorithms that could put the CWT on the same footing as the discrete WT in terms of speed and efficiency, by extending the advantages of the latter to cases where no exact QMF is available [Tor95,Vdg98,291,360]. While already interesting in 1-D, this method displays its full potential in 2-D, offering a very fast implementation of the so-called directional 2-D wavelets. Accordingly we shall describe it in detail in the next chapter, Section 2.6. Here we shall only sketch briefly the idea, in the version called *linear formalism* in [Tor95,291].

Given a wavelet ψ, one lumps together all low-frequency components in a scaling function (here we take $a > 0$)

$$\Phi(x) = \int_1^\infty \psi\left(\frac{x}{a}\right) \frac{da}{a^2} = \frac{1}{x} \int_0^x \psi(s)\,ds, \qquad \widehat{\Phi}(\xi) = \int_1^\infty \widehat{\psi}(a\xi) \frac{da}{a}, \qquad (1.60)$$

and introduces the integrated wavelet

$$\Psi(x) = \int_{1/2}^1 \psi\left(\frac{x}{a}\right) \frac{da}{a^2} = \frac{1}{x} \int_x^{2x} \psi(s)\,ds, \qquad \widehat{\Psi}(\xi) = \int_{1/2}^1 \widehat{\psi}(a\xi) \frac{da}{a}. \qquad (1.61)$$

These functions satisfy two-scale relations:

$$\Psi(x) = 2\Phi(2x) - \Phi(x), \qquad \widehat{\Psi}(\xi) = \widehat{\Phi}(\xi/2) - \widehat{\Phi}(\xi). \qquad (1.62)$$

Next, one chooses a regular grid, as opposed to the dyadic one used in the discrete case, namely,

$$\Phi_x^j \equiv 2^j \Phi(2^j(\cdot - x)), \qquad \Psi_x^j \equiv 2^j \Psi(2^j(\cdot - x)) \qquad (1.63)$$

[note that one uses here the L^1-normalization (1.19)]. Although the resulting transform will be redundant, it has the great advantage over the conventional DWT of maintaining (integer) translation covariance. Then, exactly as for the DWT, one gets a discrete reconstruction formula:

$$s(x) = \langle \Phi_x^{j_0} | s \rangle + \sum_{j=j_0}^\infty \langle \Psi_x^j | s \rangle. \qquad (1.64)$$

Truncating the summation, as usual, one gets thus a finite sum, to be compared with the decomposition (1.57).

However, there still remains a major problem. Indeed, we may try and mimic the continuous formalism and assume there exist two functions $\check{h}$, $\check{g}$ satisfying the following relations, analogous to (1.49), (1.52):

$$\widehat{\Phi}(2\xi) = \check{h}(\xi)\widehat{\Phi}(\xi), \qquad \widehat{\Psi}(2\xi) = \check{g}(\xi)\widehat{\Phi}(\xi), \qquad \text{a.e.} \qquad (1.65)$$

The difficulty now is that these functions are in general *not* 2π-periodic, which precludes designing any fast (pyramidal) algorithm. There is a way out, however. Since using the regular grid means sampling $\Phi(x)$ at unit rate, we have to assume that the function $\widehat{\Phi}$ is essentially supported in $[-\pi, \pi]$. Therefore, since the functions $\check{h}$, $\check{g}$ always appear in a product with $\widehat{\Phi}$, according to the relations (1.65), it is reasonable to approximate them in a neighborhood of zero by 2π-periodic functions h^a, g^a. In fact [Tor95], there exists a unique pair $\widetilde{h}$, $\widetilde{g}$ that minimizes the distance between $\check{h}$ and h^a, respectively $\check{g}$ and g^a. These approximate filters, called *pseudo-QMFs*, will be described at length in Section 2.6.4. The end result is a very fast implementation of the continuous wavelet transform, truly competitive with the discrete wavelet transform.

1.6.2   Orthogonal or redundant wavelet expansions?

Now we have seen the full spectrum of possible wavelet decompositions, from the min-
imalist, that is, nonredundant, orthonormal bases to increasingly redundant systems,
frames and pseudo-QMFs. Which one to use in practice? Of course, there is no unique
answer, it depends on the problem at hand. If it comes to signal compression, for instance, the most economical representation is certainly preferable, thus orthonormal
bases will be the first choice. In addition, they often yield the least correlation between
wavelet coefficients, and, of course, mathematicians have a long tradition in orthogonal
expansions. This explains the popularity of orthonormal bases in many communities.
However, this is certainly not the last word. In statistical analysis, for instance, one
prefers overcomplete systems to orthonormal bases, because they have higher adaptivity
properties and they allow to control the degree of redundancy. In addition, the lack of
(even discrete) translation invariance is a serious drawback for all applications involving some pattern recognition. In fact, increasing redundancy has many advantages, in
particular, it improves both the quality of reconstruction and the stability with respect
to perturbation, e.g., by noise. We will now comment on these two aspects.

As an illustration, let us compare the three analyses of the *bumps* signal made above,
the full CWT analysis of Figure 1.3, the translation invariant frame decomposition
of Figure 1.5 and the orthonormal basis decomposition of Figure 1.8. The CWT sees
correctly all discontinuities, with intensities depending of their strength, measured by
the singularity index α, defined in (1.31). The three parts of the signal have, respectively,
$\alpha = -1$ for the δ function, $\alpha = 0$ for the boxcar function, and $\alpha = 1$ for the tent or
triangle function. Correspondingly, see (1.32), the wavelet coefficients behave as a^{α},
which fits with Figure 1.3 for a small enough.

Compare now the two finite decompositions. When $a \to 0$, the wavelet coefficients
of the δ function increase, whereas those of the triangle function decrease. As a consequence, both the orthonormal basis and the translation invariant frame barely see the
discontinuities of the triangle. The high resolution (small scale) wavelet coefficients are
so small that one cannot see them on the figure! Indeed, if we redo the same analysis
on the triangular signal alone, using appropriate units, the high resolution coefficients
are perfectly seen (Figure 1.9). The reason is that the δ function singularity is so strong
that it swamps the whole picture. This is a perfect illustration of the fact that wavelet
orthonormal bases are too rigid: the *same* basis cannot analyze correctly the three pieces
of the signal, in a sense it is not local enough! For that reason, substitutes were invented
for improving the local character of the analysis. One of the most popular examples is
that of a *local cosine basis*. The idea is to divide the time axis into arbitrary intervals
(which overlap slightly for continuity), depending on the signal itself, and performing into each of these an independent discrete Fourier-type decomposition (DCT) [see
Figure 1.11 (d)]. We refer to [Mal99] for further information (we quote this technique
here for the sake of comparison only).

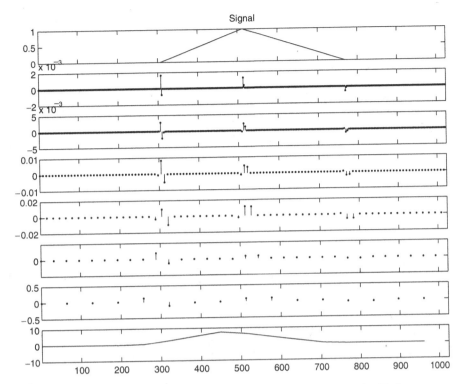

Fig. 1.9. Five-level decomposition of the triangular part of the *bumps* signal, with the same presentation as in Figure 1.8.

On the other hand, the frame of Figure 1.5 performs better. The reason here is spatial (time) resolution. In the orthogonal basis case, the resolution becomes so loose at larger scales that even the wavelet coefficients at $j = -3$ are barely visible on Figure 1.8. But, since the frame is translation invariant, the spatial resolution remains the same at all scales, and the wavelet coefficients are now visible up to $j = -1$.

Another bonus of redundancy is an increased robustness of the representation with respect to small perturbations. Suppose indeed we are working in finite dimension N. Let $s \in \mathbb{R}^N$ be a signal and denote by s_i, $i = 1 \ldots N$ the coefficients of s in a given orthonormal basis. Suppose we perturb each s_i by a random variable n_i (modeling noise, for example). For simplicity, we choose the n_is independent, with zero mean and variance σ^2. It is easy to verify that the mean square error (MSE) of the perturbed signal $\tilde{s}$ is $N\sigma^2$. Now, if we decompose s in a frame composed of $M > N$ elements, the intuition is that this result should be improved because we have *diluted* noise in a much larger space. Actually, if the frame is tight and denoting by $r = M/N$ its redundancy, the MSE reduces to [200]:

$$\text{MSE} = \frac{N\sigma^2}{r}.$$

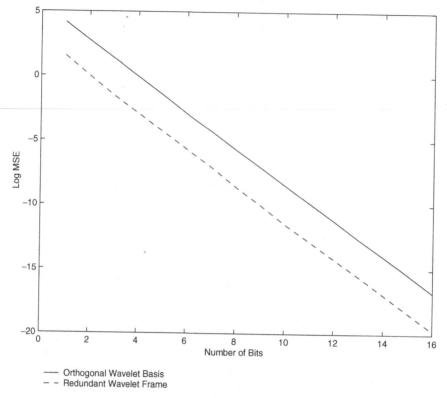

Fig. 1.10. Mean square error of the two reconstructions of the *bumps* signal, using an orthonormal wavelet basis and a redundant translation invariant wavelet frame, respectively.

Redundancy thus implies stability, or robustness to perturbations of the decomposition coefficients. This is illustrated on Figure 1.10. We have decomposed the *bumps* signal using an orthonormal wavelet basis and a redundant translation invariant wavelet frame. We have then quantized the respective coefficients and measured in each case the MSE of the reconstructed signal. The plot shows that the redundant representation always yields a better MSE. Furthermore, the distance between the two curves is constant and equal to the redundancy of the frame used.

Finally, as a visual aid to the various discrete wavelet schemes, it is instructive to characterize each of them by the tiling of the time–frequency plane it induces. We show a schematic presentation of these in Figure 1.11. On the top row, we have, from left to right, the cases of a Gabor transform and of the DWT (dyadic partition). Note, however, that the Gabor tiling is idealized, since this sharp partition can never be realized, because the time–frequency localization is intrinsically limited (the so-called Balian–Low theorem [Dau92]). On the bottom row, we show the cases of wavelet packets and that of a local cosine basis, respectively.

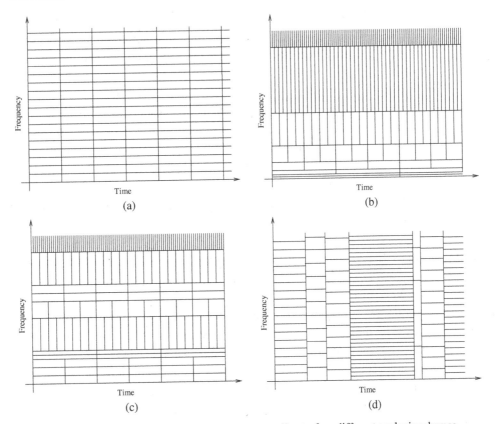

Fig. 1.11. Tilings of the time–frequency plane corresponding to four different analysis schemes. (a) The (idealized) Gabor analysis; (b) the discrete WT; (c) wavelet packets; (d) a local cosine basis.

1.7   Applications of the 1-D CWT

The CWT has found a wide variety of applications in various branches of physics and/or signal processing. Of course, our main concern in this book will be the practical applications of the 2-D wavelet transform, so we will devote two full chapters to them (Chapters 4 and 5). Nevertheless, we will list here a representative selection of one-dimensional applications, in order to convey to the reader a feeling about the scope and richness of the field. Most of the early applications, and the original references, may be found in the proceedings volumes [Com89,Mey91,Mey93]. Another interesting source for applications is the recent volume of Addison [Add02]. In all cases, the CWT is primarily used for analyzing transient phenomena, detecting abrupt changes in a signal or comparing it with a given pattern.

- *Sound and acoustics*
  The first applications of the CWT were in the field of acoustics. A few examples are musical synthesis, speech analysis [123] and modeling of the sonar system

of bats and dolphins. Other examples include various problems in underwater acoustics, such as the disentangling of the different components of an underwater refracted wave (see Section 5.3 for the 2-D case) and the identification of an obstacle.

- *Geophysics*

This is the origin of the method, which was designed in an empirical fashion by J. Morlet for analyzing the recordings of microseisms used in oil prospection. More recently, the CWT has been applied to the analysis of various types of geophysical data, e.g., in gravimetry (fluctuations of the local gravitational field), in seismology (arrival time of the various waves), in geomagnetism (fluctuations of the Earth magnetic field [4]) or in astronomy (fluctuations of the length of the day, variations of solar activity, measured by the sunspots, etc).

- *Fractals, turbulence*

The CWT is an ideal tool for studying fractals, or more generally phenomena with particular properties under scale changes [221]. Thus it is quite natural that the CWT has found many applications in the analysis of (1-D and 2-D) fractals, artificial (diffusion limited aggregates) or natural (arborescent growth phenomena) [Arn95,43,44]. Related to these is the use of the CWT in the analysis of developed turbulence (identification of coherent structures, uncovering of hierarchical structure) [Abr97,163–165]. An interesting example of fractal or self-similar behavior is that of telecommunications network traffic, and here too the WT (although rather the DWT) has given interesting results [1]. We will come back to 2-D fractals and similar objects in Section 5.4.

- *Atomic physics*

When an atom is hit by a short intense laser pulse, it emits radiation that cover a whole spectrum of harmonics of the laser frequency (experimentally, harmonics of order larger than 400 have been observed). This is a fast and complex physical process, which cannot be understood without a time-frequency analysis. This has been done, both with a Gabor analysis and with wavelets (CWT), yielding for instance the time profile of each individual harmonic [30,31] and the effect of the polarization of the laser field on harmonic generation [39,40].

- *Spectroscopy*

This was one of the earliest and most successful applications, in particular for NMR spectroscopy, where the method proves extremely efficient in subtracting unwanted spectral lines or filtering out background noise [60,131,210]. We may note that here, as in the previous application, a Gabor analysis may be fully competitive with the wavelet analysis [33,34].

- *Medical and biological applications*

The CWT has been used for analyzing or monitoring various electrical or mechanical phenomena in the brain (EEG, VEP) or the heart (ECG) [Ald96,Tho98,354]. It also yields good models for the auditory mechanism [123]. Another striking result is the characterization of long-range correlations in DNA sequences (and the solution of a long-standing puzzle) by Arnéodo and his group [49].

- *Analysis of local singularities*

  The strong point of the CWT is to detect singularities in a signal, but it yields also a fine characterization of their strengths (Lipschitz regularity, expressed via the local Hölder exponents), using the homogeneity relations (1.32) and (1.34), in particular in the case of oscillating singularities [48,50,51].

- *Shape characterization*

  A particular case of analysis of local singularities is the determination of the shape of an object, a standard problem in image processing, for instance in robotic vision. A novel approach [Ces97,Cos01,21] consists in treating the contour of the object as a complex curve in the plane and analyzing it with the 1-D CWT. The method benefits from all the good properties of the wavelet transform, for instance its robustness to noise, and looks promising for applications. Since this is in fact a 2-D problem, we will analyze it in detail, in Section 5.4.2.

- *Industrial applications*

  Here again the important aspect is monitoring, for instance in detecting anomalies in the functioning of nuclear, electrical or mechanical installations. A typical application is the analysis of the behavior of materials under impact made in [358] and discussed above for illustrating the concept of ridge and skeleton. In that paper, the same signal is analyzed with a Gabor transform, a CWT with a Morlet wavelet, a CWT with a Mexican hat, a Wigner–Ville transform, and the respective merits of each method are compared. Another, closely related, application is the determination, with a Morlet wavelet, of the vibration normal modes of a high tower excited by wind [259]. The results of this paper fully confirm the previous ones and, in particular, emphasize the role of the width parameter of the Gaussian.

2 The 2-D continuous wavelet transform

Chapter 1 has given us a brief overview of the basic facets of the CWT in the simpler one-dimensional (1-D) case, including its relationship with the various discrete approaches and a glimpse of some applications. Now it is time to enter the proper subject of the book, namely, the two-dimensional (2-D) wavelet analysis.

2.1 Derivation

In 1-D, the CWT (1.8) amounts to projecting the signal onto the wavelet $\psi_{b,a}$, obtained by translation and dilation of the mother wavelet ψ. Thus the transform is fully determined by these elementary operations of the line. Accordingly, in order to derive the CWT in 2-D, a good starting point is to consider first the elementary operations we want to apply to our signals. Actually, as we will see later (Chapter 6), this point of view allows one to extend the CWT to much more general situations, such as wavelets in higher dimensions, wavelets on the sphere, time-dependent wavelets, etc.

2.1.1 Images and elementary operations on them

By an image, we mean a two-dimensional signal of finite energy, represented by a complex-valued function defined on the real plane $\mathbb{R}^2$ and square integrable, i.e., a function $s \in L^2(\mathbb{R}^2, d^2\vec{x})$:

$$\|s\|^2 = \int_{\mathbb{R}^2} d^2\vec{x} \, |s(\vec{x})|^2 < \infty \tag{2.1}$$

(sometimes it is useful to take s integrable as well). In practice, a black and white image will be represented by a bounded non-negative function:

$$0 \leqslant s(\vec{x}) \leqslant M, \ \forall \vec{x} \in \mathbb{R}^2 \ (M > 0), \tag{2.2}$$

the discrete values of $s(\vec{x})$ corresponding to the level of gray of each pixel. However it is useful to keep general functions s as above. In fact, one often considers also as admissible signals generalized functions (distributions), such as a delta function

$\delta(\vec{x} - \vec{x}_o)$, a plane wave $\exp(i\vec{k} \cdot \vec{x})$, a fractal measure, etc. This will be justified below (see the comments after Definition 2.1.3).

The Fourier transform of the signal s is defined, as usual, by

$$\widehat{s}(\vec{k}) \equiv (Fs)(\vec{k}) = \frac{1}{2\pi} \int_{\mathbb{R}^2} d^2\vec{x}\, e^{-i\vec{k}\cdot\vec{x}} s(\vec{x}), \tag{2.3}$$

where $k \in \mathbb{R}^2$ is the spatial frequency and $\vec{k} \cdot \vec{x} = k_1 x_1 + k_2 x_2$ is the Euclidean scalar product. We also write $|\vec{k}|^2 = \vec{k} \cdot \vec{k}$. Of course, the Fourier transform is unitary (Parseval relation):

$$\widehat{s} \in L^2(\mathbb{R}^2, d^2\vec{k}) \quad \text{and} \quad \|\widehat{s}\|^2 = \|s\|^2. \tag{2.4}$$

Given an image s, all the geometric operations we want to apply to it are obtained by combining three elementary transformations of the plane, namely, rigid translations in the plane of the image, dilations or scaling (global zooming in and out) and rotations. (More complicated operations are sometimes applied, such as deformations (shearing), but we will not consider them here. We will come back to this point in Chapter 7.) Explicitly, the transformations act on $\vec{x} \in \mathbb{R}^2$ in the familiar way:

(i) translation by $\vec{b} \in \mathbb{R}^2 : \vec{x} \mapsto \vec{x}' = \vec{x} + \vec{b}$;

(ii) dilation by a factor $a > 0 : \vec{x} \mapsto \vec{x}' = a\vec{x}$;

(iii) rotation by an angle $\theta : \vec{x} \mapsto \vec{x}' = r_\theta(\vec{x})$, where r_θ is the usual 2×2 rotation matrix:

$$r_\theta \equiv \begin{pmatrix} \cos\theta & -\sin\theta \\ \sin\theta & \cos\theta \end{pmatrix}, \quad 0 \leqslant \theta < 2\pi.$$

It will prove convenient to combine a rotation by an angle θ and a dilation by $a > 0$ into a single 2×2 matrix, namely,

$$\mathsf{h} = \mathsf{h}(a, \theta) = \begin{pmatrix} a\cos\theta & -a\sin\theta \\ a\sin\theta & a\cos\theta \end{pmatrix}. \tag{2.5}$$

Using this form, we verify that $\vec{x} \cdot \vec{x}' = a \cos\theta\, |\vec{x}|^2$, demonstrating that the angle between $\vec{x}$ and $\vec{x}'$ is indeed θ and that $\vec{x}'$ is scaled by the amount a relative to $\vec{x}'$. Finally, combining all three operations, we get as general transformation in the plane

$$\vec{x} \mapsto \vec{x}' = \mathsf{h}\vec{x} + \vec{b}. \tag{2.6}$$

In the present context, these transformations are represented by the following unitary operators in the space $L^2(\mathbb{R}^2, d^2\vec{x})$ of finite energy signals:

(i) translation : $(T_{\vec{b}}s)(\vec{x}) = s(\vec{x} - \vec{b})$, $\vec{b} \in \mathbb{R}^2$; $\tag{2.7}$

(ii) dilation : $(D_a s)(\vec{x}) = a^{-1}s(a^{-1}\vec{x})$, $a > 0$; $\tag{2.8}$

(iii) rotation : $(R_\theta s)(\vec{x}) = s(r_{-\theta}(\vec{x}))$, $\theta \in [0, 2\pi)$. $\tag{2.9}$

In addition, we introduce the modulation operator:

$$(E_{\vec{b}}s)(\vec{x}) = e^{i\vec{b}\cdot\vec{x}}\, s(\vec{x}), \ \vec{b} \in \mathbb{R}^2. \tag{2.10}$$

Then a straightforward calculation yields the commutation rules among the operators (2.7)–(2.9), and with the Fourier operator (2.3):

$$
\begin{aligned}
T_{\vec{b}} D_a &= D_a T_{\vec{b}/a}, & F D_a &= D_{1/a} F, \\
T_{\vec{b}} R_\theta &= R_\theta T_{r_{-\theta}(\vec{b})}, & F T_{\vec{b}} &= E_{-\vec{b}} F, \\
R_\theta D_a &= D_a R_\theta, & F R_\theta &= R_\theta F.
\end{aligned}
\tag{2.11}
$$

Combining now the three operators (2.7)–(2.9), we define the unitary operator

$$U(\vec{b}, a, \theta) = T_{\vec{b}} D_a R_\theta, \tag{2.12}$$

which acts on a given function s as

$$\left[U(\vec{b}, a, \theta)s \right](\vec{x}) \equiv s_{\vec{b},a,\theta}(\vec{x}) = a^{-1}s(a^{-1} r_{-\theta}(\vec{x} - \vec{b})), \tag{2.13}$$

or, equivalently, in the space of Fourier transforms,

$$\widehat{s_{\vec{b},a,\theta}}(\vec{k}) = a\, e^{-i\vec{b}\cdot\vec{k}}\, \widehat{s}(ar_{-\theta}(\vec{k})). \tag{2.14}$$

If the function s is rotation invariant, we simply omit the index θ:

$$s_{\vec{b},a}(\vec{x}) = a^{-1}s(a^{-1}(\vec{x} - \vec{b})). \tag{2.15}$$

We may remark that, here, contrary to the 1-D case, it is sufficient to take *positive* dilations, since the effect of a negative dilation $a < 0$ may be obtained by combining a positive one, $a > 0$, with a rotation by $\theta = \pi$.

The geometrical effect of these transformations is easily visualized assuming s and $\widehat{s}$ to be well localized, for instance in an ellipse. An example is shown in Figure 2.1 (in Section 2.3.1).

2.1.2 Wavelets and continuous wavelet transform

As in 1-D, a wavelet is a particular type of finite energy signal, whose properties make it a good analyzing tool. Thus we define, as in (1.10):

Definition 2.1.1. *A two-dimensional wavelet is a complex-valued function $\psi \in L^2(\mathbb{R}^2, d^2\vec{x})$ satisfying the admissibility condition:*

$$c_\psi \equiv (2\pi)^2 \int_{\mathbb{R}^2} d^2\vec{k}\, \frac{|\widehat{\psi}(\vec{k})|^2}{|\vec{k}|^2} < \infty, \tag{2.16}$$

where $\widehat{\psi}$ is the Fourier transform of ψ and $|\vec{k}|^2 = \vec{k} \cdot \vec{k} = (k_1)^2 + (k_2)^2$.

The origin of this condition will be clarified below.

If ψ is regular enough, the admissibility condition (2.16) implies the following easier one, which simply means that the wavelet has zero mean:

$$\widehat{\psi}(\vec{0}) = 0 \quad \Longleftrightarrow \quad \int_{\mathbb{R}^2} d^2\vec{x} \ \psi(\vec{x}) = 0. \tag{2.17}$$

Strictly speaking, the condition (2.17) is only necessary, but in fact it is almost sufficient (see [Dau92] for a precise mathematical statement), and for all practical purposes (2.17) may be taken as admissibility condition. Intuitively, as in one dimension, it expresses the fact that a wavelet must be an oscillating function.

Clearly the three unitary operators $T_{\vec{b}}$, D_a, R_θ preserve the admissibility condition, and so does therefore $U(\vec{b}, a, \theta)$. Hence any function $\psi_{\vec{b}, a, \theta} = U(\vec{b}, a, \theta)\psi$ obtained from a wavelet ψ by translation, rotation or dilation is again a wavelet. Thus the given wavelet ψ generates the whole family $\mathcal{D}_\psi = \{\psi_{\vec{b}, a, \theta}\}$, indexed by the elements $\vec{b} \in \mathbb{R}^2, a > 0, \theta \in [0, 2\pi)$. In the sequel we will denote by G this four-dimensional parameter space.

Proposition 2.1.2. *The linear span of the family* $\mathcal{D}_\psi = \{\psi_{\vec{b}, a, \theta}, \ (\vec{b}, a, \theta) \in G\}$ *is a dense subspace of* $L^2(\mathbb{R}^2)$.

Proof. Let $f \in L^2(\mathbb{R}^2, d^2\vec{x})$ be orthogonal to every vector in the family $\mathcal{D}_\psi$, that is $\langle \psi_{\vec{b}, a, \theta} | f \rangle = 0, \ \forall (\vec{b}, a, \theta) \in G$. This means

$$\langle \widehat{\psi_{\vec{b}, a, \theta}} | \widehat{f} \rangle = a \int_{\mathbb{R}^2} d^2\vec{k} \ e^{i\vec{b}\cdot\vec{k}} \ \overline{\widehat{\psi}(ar_{-\theta}(\vec{k}))} \ \widehat{f}(\vec{k}) = 0, \quad \forall (\vec{b}, a, \theta) \in G,$$

which implies that $\overline{\widehat{\psi}(ar_{-\theta}(\vec{k}))} \ \widehat{f}(k) = 0$ a.e., for all $a > 0, \theta \in [0, 2\pi)$. Now the joint action of rotations and dilations on $\mathbb{R}^2$ is transitive. Thus, if the support of $\widehat{\psi}$ is a "patch" (for instance, a disk or an ellipse) in the $\vec{k}$-plane, the supports of $\widehat{\psi}(ar_{-\theta}(\vec{k}))$ will cover the whole plane when a and θ vary over their range. Therefore, this implies $\widehat{f}(k) = 0$ a.e., that is, $f = 0$. $\square$

Note that rotations are needed to get that result. A naive generalization of the 1-D formalism would consist in combining translations with separate dilations along the x- and the y-axis. But then, if only positive dilations are used, each quadrant in the $\vec{k}$-plane is invariant and additional conditions on the wavelet ψ would be necessary for the argument of the proof above to work – thus spoiling the result. Actually, this is precisely the technique used in the 2-D DWT, where the 2-D multiresolution is obtained by taking the tensor product of two 1-D copies, one in x, one in y (see Section 2.5.1). As we shall see later (see Section 2.5.2), this approach, while commonly used in practice, has severe shortcomings.

As a consequence of Proposition 2.1.2, any vector in $L^2(\mathbb{R}^2, d^2\vec{x})$ is uniquely determined by its projections on the vectors of the family $\mathcal{D}_\psi$. This justifies the basic definition of the CWT.

Definition 2.1.3. *Given an image* $s \in L^2(\mathbb{R}^2, d^2\vec{x})$, *its* continuous wavelet transform *(with respect to the fixed wavelet* ψ), $S \equiv T_\psi s$ *is the scalar product of* s *with the transformed wavelet* $\psi_{\vec{b},a,\theta}$, *considered as a function of* $(\vec{b}, a, \theta)$:

$$S(\vec{b}, a, \theta) = \langle \psi_{\vec{b},a,\theta} | s \rangle \tag{2.18}$$

$$= a^{-1} \int_{\mathbb{R}^2} d^2\vec{x} \; \overline{\psi(a^{-1} r_{-\theta}(\vec{x} - \vec{b}))} \, s(\vec{x}) \tag{2.19}$$

$$= a \int_{\mathbb{R}^2} d^2\vec{k} \; e^{i\vec{b}\cdot\vec{k}} \, \overline{\widehat{\psi}(ar_{-\theta}(\vec{k}))} \, \widehat{s}(\vec{k}). \tag{2.20}$$

The relations (2.18)–(2.20) permit us to extend the formalism beyond the Hilbert space framework. As explained above, the signal *s* may be taken as a singular function (a distribution), provided the wavelet ψ is sufficiently regular (most wavelets used in practice are smooth functions, see below).

Before exploring in detail the mathematical properties of the CWT, it is instructive to exhibit two typical 2-D wavelets (actually the simplest ones).

(1) *The isotropic Mexican hat wavelet*

This wavelet is simply the Laplacian of a Gaussian:

$$\psi_H(\vec{x}) = (2 - |\vec{x}|^2) \exp(-\tfrac{1}{2}|\vec{x}|^2),$$
$$\widehat{\psi}_H(\vec{k}) = |\vec{k}|^2 \exp(-\tfrac{1}{2}|\vec{k}|^2). \tag{2.21}$$

(2) *The Morlet wavelet*

This wavelet is essentially a plane wave within a Gaussian window:

$$\psi_M(\vec{x}) = \exp(i\vec{k}_o \cdot \vec{x}) \exp(-\tfrac{1}{2}|\vec{x}|^2) + \text{corr.};$$
$$\widehat{\psi}_M(\vec{k}) = \exp(-\tfrac{1}{2}|\vec{k} - \vec{k}_o|^2) + \text{corr.} \tag{2.22}$$

As in 1-D, a a correction term must be added in order to satisfy the admissibility condition (2.17), but in practice one will arrange that this term be numerically negligible and thus can be omitted (it suffices to choose the norm $|\vec{k}_o|$ of the wave vector large enough).

The first wavelet is real, the Morlet wavelet is complex. They have very different properties and, naturally, they will be used in quite different situations. Both wavelets, and many more, will be studied in detail in Chapter 3, and many examples of applications in Chapters 4 and 5.

2.2   Basic properties of the 2-D CWT

The main properties of the continuous wavelet transform are conveniently expressed in terms of a linear map W_ψ from the space of finite energy signals $L^2(\mathbb{R}^2, d^2\vec{x})$ into the space of transforms. We summarize them in three propositions [Mur90,13,15,283].

Proposition 2.2.1. *Let the map* $W_\psi : s \mapsto c_\psi^{-1/2} S$ *be defined by*

$$(W_\psi s)(\vec{b}, a, \theta) = c_\psi^{-1/2} \langle \psi_{\vec{b},a,\theta} | s \rangle, \ s \in L^2(\mathbb{R}^2, d^2\vec{x}), \tag{2.23}$$

where c_ψ *is the constant given in (2.16). Then:*

(1) W_ψ *conserves the norm of the signal, thus its total energy:*

$$\iiint_G d^2\vec{b} \, \frac{da}{a^3} \, d\theta \, |S(\vec{b}, a, \theta)|^2 = c_\psi \int_{\mathbb{R}^2} d^2\vec{x} \, |s(x)|^2, \tag{2.24}$$

i.e., it is an isometry from the space of signals into the space of transforms. The latter is a closed subspace $\mathfrak{H}_\psi$ *of* $L^2(G, dg)$, *where* $dg \equiv a^{-3} d^2\vec{b} \, da \, d\theta$ *is the natural measure on* G. *Equivalently, the family of wavelets* $\{\psi_{\vec{b},a,\theta}\}$, *with* $b \in \mathbb{R}^2$, $a > 0$, *and* $0 \leqslant \theta < 2\pi$, *generates a resolution of the identity:*

$$c_\psi^{-1} \iiint_G d^2\vec{b} \, \frac{da}{a^3} \, d\theta \, |\psi_{\vec{b},a,\theta}\rangle \, \langle \psi_{\vec{b},a,\theta}| = I. \tag{2.25}$$

(2) Since it is an isometry, the map W_ψ *is invertible on its range* $\mathfrak{H}_\psi$, *and the inverse transformation is the adjoint of* W_ψ. *This means that the image* $s(\vec{x})$ *may be reconstructed from its wavelet transform* $S(\vec{b}, a, \theta)$ *by the formula:*

$$s(\vec{x}) = c_\psi^{-1} \iiint_G d^2\vec{b} \, \frac{da}{a^3} \, d\theta \, \psi_{\vec{b},a,\theta}(\vec{x}) \, S(\vec{b}, a, \theta). \tag{2.26}$$

Proof. The relation (2.24) follows from a straightforward calculation:

$$\iiint_G d^2\vec{b} \, \frac{da}{a^3} \, d\theta \, |S(\vec{b}, a, \theta)|^2 =$$

$$= \int_{\mathbb{R}^2} d^2\vec{k} \int_{\mathbb{R}^2} d^2\vec{k}' \iiint_G d^2\vec{b} \, \frac{da}{a} \, d\theta$$

$$\times e^{i\vec{b}\cdot(\vec{k}-\vec{k}')} \overline{\widehat{\psi}(ar_{-\theta}(\vec{k}))} \, \widehat{\psi}(ar_{-\theta}(\vec{k}')) \widehat{s}(\vec{k}) \overline{\widehat{s}(\vec{k}')}$$

$$= (2\pi)^2 \int_{\mathbb{R}^2} d^2\vec{k} \int_0^\infty \frac{da}{a} \int_0^{2\pi} d\theta \, |\widehat{\psi}(ar_{-\theta}(\vec{k}))|^2 \, |\widehat{s}(\vec{k})|^2$$

(the exchange of integrals is justified by Fubini's theorem). Introducing polar coordinates: $\widehat{\psi}(\vec{k}) \equiv \widehat{\psi}_p(\rho, \phi)$, with $\rho \equiv |\vec{k}|$, we get

$$\int_0^\infty \frac{da}{a} \int_0^{2\pi} d\theta \, |\widehat{\psi}(ar_{-\theta}(\vec{k}))|^2 = \int_0^\infty \frac{da}{a} \int_0^{2\pi} d\theta \, |\widehat{\psi}_p(a\rho, \phi - \theta)|^2$$

$$= \int_0^\infty \frac{d\rho'}{\rho'} \int_0^{2\pi} d\theta' \, |\widehat{\psi}_p(\rho', \theta')|^2$$

$$= \int_{\mathbb{R}^2} \frac{d^2\vec{k}'}{|\vec{k}'|^2} \, |\widehat{\psi}(\vec{k}')|^2.$$

By comparison with the definition (2.16) of c_ψ, and using Plancherel's theorem, this proves the statement.

Then (2.25) is simply a reformulation of (2.24), in view of the definition (2.23) of $W_\psi : s \mapsto c_\psi^{-1/2} S$. Finally, the reconstruction formula (2.26) follows immediately by applying both sides of (2.25) to a signal $s(\vec{x})$ and taking into account the definition (2.23) of W_ψ.

$\qquad\qquad\qquad\qquad\qquad\qquad\qquad\qquad\qquad\qquad\qquad\qquad\qquad$ $\square$

Three remarks are in order here. First, the relation (2.25) must be taken as a weak integral, that is, both sides are equal when sandwiched between arbitrary vectors. This precisely means that the reconstruction formula (2.26) holds in the weak sense. But, in fact, much more is true, and actually necessary for obtaining good approximation schemes, namely, the relation (2.26) holds in strong L^2 convergence. This will be demonstrated in Section 2.6.1, in two different versions.

Second, the measure $dg \equiv a^{-3} d^2\vec{b}\, da\, d\theta$ on G is precisely the unique measure (up to normalization) that is invariant under all the operations of translation, dilation, and rotation (this is why we have called it natural).

Third, the reconstruction formula (2.26) may also be proven by an explicit calculation of the adjoint map of W_ψ:

$$\langle f|s\rangle_{L^2(\mathbb{R}^2)} = c_\psi^{-1/2}\, \langle f|W_\psi^* S\rangle_{L^2(\mathbb{R}^2)} = c_\psi^{-1/2}\, \langle W_\psi f|S\rangle_{L^2(G)}$$
$$= c_\psi^{-1} \iiint_G d^2\vec{b}\, \frac{da}{a^3}\, d\theta\, \langle f|\psi_{\vec{b},a,\theta}\rangle S(\vec{b}, a, \theta).$$

It is the possibility of having a reconstruction formula (2.26) that justifies the admissibility condition $c_\psi < \infty$ imposed on wavelets. However, (2.26) is not only a reconstruction formula, it also means that the wavelet transform, like its 1-D counterpart, provides a decomposition of the signal in terms of the analyzing wavelets $\psi_{\vec{b},a,\theta}$, with coefficients $S(\vec{b}, a, \theta)$. Under both interpretations, this formula leads in practice to discretization problems (see Section 2.4).

In the same spirit, it is interesting to see the inverse Fourier transform

$$s(\vec{x}) = \frac{1}{2\pi} \int_{\mathbb{R}^2} d^2\vec{k}\, e^{i\vec{k}\cdot\vec{x}}\, \widehat{s}(\vec{k}), \tag{2.27}$$

as the decomposition of the signal into the improper basis $\{e^{i\vec{k}\cdot\vec{x}},\ \vec{k} \in \mathbb{R}^2\}$ of eigenvectors of the translation operators (2.7). In view of the crucial importance of dilations in the wavelet context, it is useful to write down also the polar coordinate version of the Fourier transform, which involves the (improper) eigenvectors $\{e^{in\varphi},\ n \in \mathbb{Z}\}$ of the rotation operator (2.9) and those of the dilation operator (2.8), $\{r_{i\nu},\ \nu \in \mathbb{R}\}$:

$$\widetilde{f}(\nu, n) = \frac{1}{2\pi} \int_0^{2\pi} d\varphi\, e^{-in\varphi} \int_0^\infty \frac{dr}{r}\, r^{-i\nu} rf(r, \varphi), \tag{2.28}$$

$$rf(r, \varphi) = \frac{1}{2\pi} \sum_{n=-\infty}^\infty e^{in\varphi} \int_{-\infty}^\infty d\nu\, r^{i\nu}\, \widetilde{f}(\nu, n). \tag{2.29}$$

Of course, we recover the well-known fact that the polar coordinate version of the Fourier transform is a combination of a Mellin transform in the radial variable r and a Fourier series in the angle φ.

Actually, as in one dimension, the reconstruction formula (2.26) may be generalized in several ways. First, the wavelet used for the analysis, ψ, and the one used for the reconstruction, χ, need not coincide, they have only to satisfy a cross-admissibility condition [Hol95,223], namely, $0 < |c_{\psi\chi}| < \infty$, where

$$c_{\psi\chi} = (2\pi)^2 \int_{\mathbb{R}^2} \frac{d^2\vec{k}}{|\vec{k}|^2} \, \overline{\widehat{\psi}(\vec{k})} \, \widehat{\chi}(\vec{k}) < \infty. \tag{2.30}$$

Then one gets a more general reconstruction formula:

$$s(\vec{x}) = c_{\psi\chi}^{-1} \iiint_G d^2\vec{b} \, \frac{da}{a^3} \, d\theta \, \chi_{\vec{b},a,\theta}(\vec{x}) \, (W_\psi s)(\vec{b}, a, \theta). \tag{2.31}$$

The proof consists in a straightforward verification, including some interchanges of integrals justified by Fubini's theorem. As we shall see later, this is the analog, in the continuous case, of the bilinear scheme commonly used in the discrete approach, namely the construction of biorthogonal wavelet bases (see Section 2.5.2).

In particular, if one takes for the reconstruction wavelet χ a delta function, one obtains the simplified reconstruction formula:

$$s(\vec{x}) = c_{\psi\delta}^{-1} \int_0^\infty \frac{da}{a^2} \int_0^{2\pi} d\theta \, S(\vec{x}, a, \theta), \tag{2.32}$$

where

$$c_{\psi\delta} = \int_{\mathbb{R}^2} \frac{d^2\vec{k}}{|\vec{k}|^2} \, \overline{\widehat{\psi}(\vec{k})}.$$

On the other hand, if ψ is rotation invariant, the wavelet transform S does not depend on θ and we obtain, instead of (2.26), a simpler reconstruction formula:

$$s(\vec{x}) = 2\pi \, c_\psi^{-1} \int_{\mathbb{R}^2} d^2\vec{b} \int_0^\infty \frac{da}{a^3} \, \psi_{\vec{b},a}(\vec{x}) \, S(\vec{b}, a). \tag{2.33}$$

Finally, combining the two preceding points, one obtains the simplified reconstruction formula originally used by Morlet in 1-D [206], in which one reconstructs the original image by summing over scales only:

$$s(\vec{x}) = 2\pi \, c_{\psi\delta}^{-1} \int_0^\infty \frac{da}{a^2} \, S(\vec{x}, a). \tag{2.34}$$

Next, a characteristic feature of the CWT, in fact shared by a large class of transformations, as we shall see later (Chapter 6), is the existence of a so-called *reproducing kernel*, which actually is nothing but the wavelet transform of the wavelet itself, that is, the autocorrelation function of the wavelet. More precisely:

Proposition 2.2.2. *The projection from $L^2(G, dg)$ onto the range $\mathfrak{H}_\psi$ of W_ψ, the space of wavelet transforms, is an integral operator whose kernel $K(\vec{b}', a', \theta'|\vec{b}, a, \theta)$ is the autocorrelation function of ψ, also called reproducing kernel:*

$$K(\vec{b}', a', \theta'|\vec{b}, a, \theta) = c_\psi^{-1} \langle \psi_{\vec{b}', a', \theta'} | \psi_{\vec{b}, a, \theta} \rangle. \tag{2.35}$$

Therefore, a function $f \in L^2(G, dg)$ is the wavelet transform of a certain signal iff it satisfies the reproduction property:

$$f(\vec{b}', a', \theta') = \iiint_G d^2\vec{b} \, \frac{da}{a^3} \, d\theta \, K(\vec{b}', a', \theta'|\vec{b}, a, \theta) \, f(\vec{b}, a, \theta). \tag{2.36}$$

Proof. Since W_ψ is an isometry from $L^2(\mathbb{R}^2, d^2\vec{x})$ into $L^2(G, dg)$, i.e., $W_\psi^* W_\psi = I$, its range $\mathfrak{H}_\psi$ is a closed subspace and the corresponding projection operator is $P_\psi = W_\psi W_\psi^*$. Thus a vector $f \in L^2(G, dg)$ belongs to $\mathfrak{H}_\psi$ iff $f = P_\psi f$. Explicitly, this gives:

$$
\begin{aligned}
f(\vec{b}', a', \theta') &= \left(W_\psi W_\psi^* f \right)(\vec{b}', a', \theta') \\
&= c_\psi^{-1} \iiint_G d^2\vec{b} \, \frac{da}{a^3} \, d\theta \, \langle \psi_{\vec{b}', a', \theta'} | \psi_{\vec{b}, a, \theta} \rangle \, f(\vec{b}, a, \theta),
\end{aligned}
$$

which proves (2.35)–(2.36). $\qquad\qquad\qquad\qquad\qquad\qquad\qquad\qquad\qquad\qquad\square$

Because it may be interpreted as the autocorrelation function of the wavelet, the reproducing kernel leads to the notion of correlation length, that is, it determines the region of influence of a given wavelet in the $\vec{b}, a, \theta$ parameter space. As such, it plays a role in the determination of the capabilities of a given wavelet (calibration), and in particular in the process of discretization. We will discuss these features in Chapter 3, Section 3.4.

Finally, the continuous wavelet transform has the important property of *covariance* (improperly called invariance in the signal processing literature) under all the operations used in its definition.

Proposition 2.2.3. *The map W_ψ is covariant under translations, dilations and rotations, which means that the correspondence $W_\psi : s(\vec{x}) \mapsto S(\vec{b}, a, \theta)$ implies the following ones:*

$$s(\vec{x} - \vec{b}_o) \mapsto S(\vec{b} - \vec{b}_o, a, \theta) \tag{2.37}$$

$$a_o^{-1} s(a_o^{-1}\vec{x}) \mapsto S(a_o^{-1}\vec{b}, a_o^{-1}a, \theta) \tag{2.38}$$

$$s(r_{\theta_o}(\vec{x})) \mapsto S(r_{-\theta_o}(\vec{b}), a, \theta - \theta_o). \tag{2.39}$$

It is worth noting that, conversely, the wavelet transform is uniquely determined by the three conditions of linearity, covariance and energy conservation, plus some continuity [Mur90].

These covariance relations, which are proved by a straightforward calculation, have a crucial importance for the applications. Translation covariance (2.37), often called improperly *shift invariance*, is lost in the standard formulation of the discrete WT, based on multiresolution (see Definition 1.5.1 and Section 2.5.1), and this generates many problems in practice, for instance in pattern recognition. Covariance under dilations, (2.38), is the basis for the application of the wavelet transform to the analysis of fractals (see Section 5.4). Finally, joint covariance under rotations and dilations justifies the use of the CWT for detecting rotation–dilation (inflation) properties of several classes of 2-D patterns, for instance, Penrose tilings of the plane or diffraction patterns of quasicrystals. We will discuss this recent application in Chapter 4, Section 4.5.

As a final remark, we emphasize that, as in the 1-D case, the choice of the normalization factor a^{-1} in (2.8) or (2.12) is not essential and is made mainly for mathematical reasons. It is the only one that makes the dilation operator D_a, and thus the wavelet transform, unitary: $\|\psi_{\vec{b},a,\theta}\|_2 = \|\psi\|_2$ and $\|W_\psi s\|_2 = \|s\|_2$, as stated in Proposition 2.2.1. In practice, one often uses instead a factor a^{-2}, so as to enhance the high-frequency part of the signal, and thus to make more conspicuous its singularities, if any. This amounts to introducing, instead of the unitary D_a, a *nonunitary* dilation operator D^a, which preserves the L^1-norm of the signal: $\|D^a\psi\|_1 = \|\psi\|_1$ and $\|\psi_{(\vec{b},a,\theta)}\|_1 = \|\psi\|_1$, where $\psi_{(\vec{b},a,\theta)} = T_{\vec{b}} D^a R_\theta \psi = a^{-2}\psi(a^{-1} r_{-\theta}(\vec{x} - \vec{b})) = a^{-1}\psi_{\vec{b},a,\theta}$. Correspondingly, one defines the L^1-normalized transform

$$\check{S}(\vec{b}, a, \theta) = \langle \psi_{(\vec{b},a,\theta)} | s \rangle. \tag{2.40}$$

This transform is also useful for making contact with the so-called dyadic wavelet transform (see Section 2.4.4), in particular, for the design of fast algorithms, using the continuous wavelet packets developed in Section 2.6. We will meet it again in Chapter 9, while extending the CWT to the 2-sphere.

2.3   Implementation and interpretation of the 2-D CWT

2.3.1   Interpretation of the CWT as a singularity scanner

In order to get a physical interpretation of the CWT, we notice that in signal analysis, as in classical electromagnetism, the L^2 norm is interpreted as the total energy of the signal. Therefore, the relation (2.24) suggests we interpret $|S(\vec{b}, a, \theta)|^2$ as the energy density in the wavelet parameter space [284].

Assume now, as in 1-D, that the wavelet ψ is fairly well localized both in position space ($\vec{x}$) and in spatial frequency space ($\vec{k}$). Then so is the transformed wavelet $\psi_{\vec{b},a,\theta}$, with effective support suitably translated by $\vec{b}$, rotated by θ and dilated by a. Because (2.19) is essentially a convolution with a function ψ of zero mean, the transform $S(\vec{b}, a, \theta)$ is appreciable only in those regions of parameter space $(\vec{b}, a, \theta)$ where the signal is. Thus we get an appreciable value of S only where the wavelet $\psi_{\vec{b},a,\theta}$ "matches" the features of the signal s. In other words, the CWT acts on a signal as a *local* filter in all four variables $\vec{b}, a, \theta$: $S(\vec{b}, a, \theta)$ sees only that portion of the signal that lives around $\vec{b}, a, \theta$ and filters out the rest. Therefore, if the wavelet is well localized, the energy density of the transform will be concentrated on the significant parts of the signal. This is the key to all the approximation schemes that make wavelets such an efficient tool.

In order to clarify the filtering effect in scale and angle variables, we rewrite the expression (2.20) of the CWT in polar coordinates $\vec{k} = (\rho, \phi)$:

$$S(\vec{b}, a, \theta) = a \int_{\mathbb{R}^2} d^2\vec{k}\, e^{i\vec{b}\cdot\vec{k}}\, \overline{\hat{\psi}(ar_{-\theta}(\vec{k}))}\, \hat{s}(\vec{k}) \tag{2.41}$$

$$= a \int_0^\infty \rho\, d\rho \int_0^{2\pi} d\phi\, e^{ib\rho\cos\phi}\, \overline{\hat{\psi}(a\rho, \phi - \theta)}\, \hat{s}(\rho, \phi). \tag{2.42}$$

On the last relation, we see that the CWT amounts to a convolution in the scale-angle variables ρ, ϕ. In order to better appreciate this, we switch throughout to polar coordinates, so the Fourier transform turns into a Mellin transform, as seen in (2.29). In position space, write (2.19) as

$$S(\vec{b}, a, \theta) = \frac{1}{a} \int_{\mathbb{R}^2} d^2\vec{x}\, \overline{\psi(r_{-\theta}(a^{-1}\vec{x}))}\, s_{\vec{b}}(\vec{x}),$$

where $s_{\vec{b}}(\vec{x}) \equiv s(\vec{x} + \vec{b})$. In polar coordinates $\vec{x} = (r, \varphi)$, we get

$$S(\vec{b}, a, \theta) = \int_0^\infty dr \int_0^{2\pi} d\varphi\, a^{-1} r\, \overline{\psi(a^{-1}r, \varphi - \theta)}\, s_{\vec{b}}(r, \varphi), \tag{2.43}$$

or, in conjugate variables [see (2.29)],

$$S(\vec{b}, a, \theta) = \frac{1}{2\pi} \sum_{n=-\infty}^{\infty} e^{in\theta} \int_{-\infty}^{\infty} dv\, a^{iv}\, \overline{\tilde{\psi}(v, n)}\, \tilde{s}_{\vec{b}}(v, n). \tag{2.44}$$

Performing the change of variables

$$r = e^u, \quad a = e^v, \tag{2.45}$$

we obtain

$$S(\vec{b}, e^v, \theta) = \int_{-\infty}^{\infty} du \int_0^{2\pi} d\varphi\, e^{u-v}\, \overline{\psi(e^{u-v}, \varphi - \theta)}\, e^u\, s_{\vec{b}}(e^u, \varphi). \tag{2.46}$$

Upon introducing the functions

$$G(u, \varphi) = \psi(e^u, \varphi), \quad F(\vec{b}, u, \varphi) = e^u s_{\vec{b}}(e^u, \varphi), \tag{2.47}$$

(2.46) turns into

$$S(\vec{b}, e^{v}, \theta) = \int_{-\infty}^{\infty} du \int_{0}^{2\pi} d\varphi \, \overline{G(u - v, \varphi - \theta)} \, F(\vec{b}, u, \varphi). \qquad (2.48)$$

Thus, the problem of computing $S(\vec{b}, a, \theta)$ reduces into that of computing a convolution of the signal F with a function G of zero mean, hence the filtering effect – and the rapidity of the algorithm [286].

Let us make more precise the support properties of ψ. Assume ψ and $\widehat{\psi}$ to be as well localized as possible (but in a way still compatible with the Fourier uncertainty property), namely, ψ has for numerical support (i.e., the region outside of which the function is numerically negligible) a "disk" of diameter T, centered around $\vec{0}$, while $\widehat{\psi}$ has for numerical support a "disk" of diameter Ω, centered around $\vec{k}_o$. Then, for the transformed wavelets $\psi_{\vec{b},a,\theta}$ and $\widehat{\psi}_{\vec{b},a,\theta}$ we have, respectively:

• supp $\psi_{\vec{b},a,\theta}$ is a "disk" of diameter $\simeq aT$ centered around $\vec{b}$ and rotated by an angle θ;

• supp $\widehat{\psi}_{\vec{b},a,\theta}$ is a "disk" of diameter $\simeq \Omega/a$, centered around $r_\theta(\vec{k}_o)/a$ and rotated by θ.

Notice that the product of the two diameters is constant (we know it has to be bounded below by a fixed constant, by Fourier's theorem). These support properties are illustrated in Figure 2.1 for an elliptic shape.

As a consequence, we may characterize the filter properties of the wavelet.

• If $a \gg 1$, $\psi_{\vec{b},a,\theta}$ is a wide window, whereas $\widehat{\psi}_{\vec{b},a,\theta}$ is very peaked around a small spatial frequency $r_\theta(\vec{k}_o)/a$: this transform will be most sensitive to *low spatial frequencies*.

• If $a \ll 1$, $\psi_{\vec{b},a,\theta}$ is a narrow window and $\widehat{\psi}_{\vec{b},a,\theta}$ is wide and centered around a high spatial frequency $r_\theta(\vec{k}_o)/a$: this wavelet has a good localization capability in the space domain and is mostly sensitive to *high spatial frequencies*.

Thus wavelet analysis operates at constant *relative bandwidth*, $\Delta k/k = $ const, where $k \equiv |\vec{k}|$. Therefore, the analysis is most efficient at high spatial frequencies or small scales, and so it is particularly apt at detecting *discontinuities* in images, either point singularities (contours, corners) or directional features (edges, segments). In other words, the CWT is a *singularity scanner*. Actually, we will see later that it is also a singularity *analyzer*. For instance, the CWT allows one to measure fractal dimensions in images (see Chapter 5, Section 5.4).

In addition to these localization properties, one often imposes on the analyzing wavelet ψ a number of additional properties, for instance, restrictions on the support of ψ and of $\widehat{\psi}$. Or ψ may be required, as in the 1-D case, to have a certain number of *vanishing moments*, up to order $N \geqslant 1$ (by the admissibility condition (2.17), the moment of order 0 must always vanish):

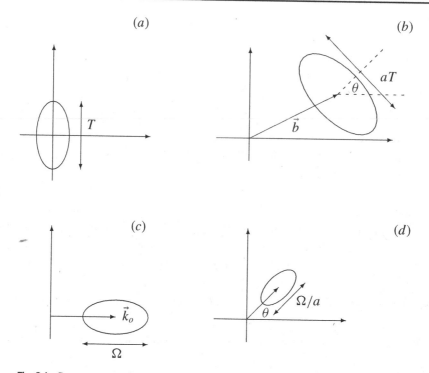

Fig. 2.1. Support properties under the basic operations: (top) in the time domain: (a) the original signal $\psi(\vec{x})$; (b) the modified signal $\psi_{\vec{b},a,\theta}(\vec{x})$, with $\vec{b} = (2.4, 1.2)$, $a = 1.5$, $\theta = 45°$; (bottom) in the frequency domain: (c) the original signal $\widehat{\psi}(\vec{k})$; (d) the modified signal $\psi_{\vec{b},a,\theta}(\vec{k})$.

$$\int d^2\vec{x}\, x^{\alpha}\, y^{\beta}\, \psi(\vec{x}) = 0, \quad \vec{x} = (x, y), \quad 0 \leqslant \alpha + \beta \leqslant N. \tag{2.49}$$

This property improves its efficiency at detecting singularities in the signal. Indeed, the transform (2.19) is then blind to the smoothest part of the signal, that which is polynomial of degree up to N – and less interesting, in general. Only the sharper part remains, including all singularities (jumps in the signal or one of its derivatives, for instance). Equivalently, ψ detects singularities in the $(N + 1)$th derivatives of the signal [264]. For instance, if the first moments ($N = 1$) vanish, the transform will erase any linear *trend* in the signal, such as a linear gradient of luminosity. Conversely, if the signal is rough, *a fortiori* if it is a measure (as in the analysis of fractals [43,221]), it is sufficient to take a wavelet with no nontrivial vanishing moment, i.e., no condition has to be imposed beyond (2.17).

Altogether, as in the 1-D case, the 2-D wavelet transform may be interpreted as a mathematical, direction selective, microscope, with optics ψ, magnification $1/a$ and orientation tuning parameter θ [Arn95,44]. Two features must be emphasized here. First, the magnification (zoom) $1/a$ is *global*, independently of the direction, because we have excluded distortions of the image. Then, there is the additional property of *directivity*, given by the rotation angle θ. This last feature opens the way to a whole new class of applications, in which directions play an essential role. We will detail some of them in Chapters 4 and 5.

2.3.2 The CWT as a phase space representation

In order to get a better insight, it is worth recasting the basic formulas (2.18)–(2.20) into a different form. First, we notice that the CWT is in fact a *phase space* representation (in the usual sense of Hamiltonian mechanics). To see this in a simple way, we observe that the correspondence $\vec{k} \Leftrightarrow (a^{-1}, \theta)$ is a bijection from $\mathbb{R}^2_* \equiv \mathbb{R}^2 \setminus \{\vec{0}\}$ onto itself. Thus, writing $\kappa = a^{-1}$ and $\vec{p} \equiv (a^{-1}, \theta) = (\kappa, \theta) \in \mathbb{R}^2_*$, we get

$$\frac{da}{a^3} d\theta = \kappa \, d\kappa \, d\theta = d^2 \vec{p}, \tag{2.50}$$

so that the measure on G becomes simply the volume element of $\mathbb{R}^2 \times \mathbb{R}^2_*$:

$$d^2\vec{b} \, \frac{da}{a^3} \, d\theta = d^2\vec{b} \, d^2\vec{p}. \tag{2.51}$$

Thus, the full four-dimensional parameter space of the 2-D WT, G, may be interpreted as phase space, with $\vec{q} \equiv \vec{b}$ the position variable and the pair $(a^{-1}, \theta) \equiv (\kappa, \theta)$ playing the rôle of spatial frequency $\vec{p}$, expressed in *polar coordinates*. The same result holds in the 1-D case [122,259]: a^{-1} defines the frequency scale, so that the full parameter space of the 1-D WT, the time-scale half plane, is in fact a time–frequency space, thus a phase space. Of course, this interpretation is borne out by mathematical analysis (see Chapter 7). The variable $\vec{p}$ follows also the common practice in image processing: $a = 0$ is the horizon in spatial frequency plots, corresponding to extremely high frequencies. It is amusing to note that the same interpretation is even supported by some physiological evidence, namely the so-called *orientation hypercolumns* of Hubel and Wiesel [DeV88,Duv91,226]. In certain species, cortical neurons are organized into columns, whose sensitivity to position, orientation, and frequency variables correspond exactly to the geometry of $\mathbb{R}^2 \times \mathbb{R}^2_*$ just described.

In order to manifest the fact that the CWT is really a phase space realization of the signal, we express it explicitly into phase space variables $(\vec{q}, \vec{p})$. For any vector $\vec{x} = (x, y) = (r \cos\varphi, r \sin\varphi)$, with polar coordinates (r, φ), define the matrix

$$\mathfrak{s}(\vec{x}) = \begin{pmatrix} x & -y \\ y & x \end{pmatrix} = r \, r_\varphi \quad (r_\varphi \text{ is the } 2 \times 2 \text{ rotation matrix}). \tag{2.52}$$

One shows immediately that $\mathfrak{s}(\vec{x})\vec{z} = \mathfrak{s}(\vec{z})\vec{x}$ and $r_\theta(\vec{x}) = \mathfrak{s}(\vec{x})\vec{e}_\theta$, where $\vec{e}_\theta$ denotes a unit vector in the direction θ. Then one has

$$ar_{-\theta}(\vec{k}) = \mathfrak{s}(\vec{p})^{-1}\vec{k} = \mathfrak{s}(\vec{k})\frac{\mathsf{R}\vec{p}}{|\vec{p}|^2}, \tag{2.53}$$

where R denotes the reflection with respect to the x-axis. We come back now to the expression (2.20) of the CWT and rewrite it in terms of the phase space variables $(\vec{q}, \vec{p})$:

$$S(\vec{b}, a, \theta) \equiv \tilde{S}(\vec{q}, \vec{p}) = |\vec{p}|^{-1} \int_{\mathbb{R}^2} d^2\vec{k} \, e^{i\vec{q}\cdot\vec{k}} \, \overline{\hat{\psi}(\mathfrak{s}(\vec{p})^{-1}\vec{k})} \, \hat{s}(\vec{k}). \tag{2.54}$$

Alternatively, one may consider the "inverse" phase space variables $\vec{v} = (a, -\theta) \equiv R\vec{p}/|\vec{p}|^2$. Clearly $\mathfrak{s}(\vec{p})\mathfrak{s}(\vec{v}) = I$. Although the variable $\vec{v}$ is less natural, its use sometimes simplifies the computations (see Section 5.3, for instance). Since $ar_{-\theta}(\vec{k}) = \mathfrak{s}(\vec{v})\vec{k} = \mathfrak{s}(\vec{k})\vec{v}$, one gets for the CWT

$$S(\vec{b}, a, \theta) \equiv \check{S}(\vec{b}, \vec{v}) = |\vec{v}| \int_{\mathbb{R}^2} d^2\vec{k} \, e^{i\vec{b}\cdot\vec{k}} \, \overline{\hat{\psi}(\mathfrak{s}(\vec{v})\vec{k})} \, \hat{s}(\vec{k}). \tag{2.55}$$

2.3.3 Visualization of the CWT: the various representations

In practice, once the CWT of a given signal $s(\vec{x})$ has been computed, one immediately faces a problem of visualization. Indeed, $S(\vec{b}, a, \theta)$ is a function of four variables: two position variables $\vec{b} = (b_x, b_y) \in \mathbb{R}^2$, and the pair $(a, \theta) \in \mathbb{R}_*^+ \times [0, 2\pi) \simeq \mathbb{R}_*^2$ (equivalently, (a^{-1}, θ)). Now, to compute and visualize the full CWT in all four variables is hardly possible. Therefore, in order to obtain a manageable tool, some of the variables, a, θ, b_x, b_y must be eliminated. There are two ways of achieving this. The first one consists in fixing the value of some of the variables. In other words, one must restrict oneself to a *section* of the parameter space. Of course, this makes sense only if the variables in question may take arbitrary values in a continuous range.

Alternatively, one may *integrate out* the variables in question. Using the proper part of the natural measure $a^{-3} d^2\vec{b} \, da \, d\theta$, one obtains in this way partial energy densities, since the integral of $|S(\vec{b}, a, \theta)|^2$ over all the variables is interpreted as the total energy of the signal, as results from (2.24). This procedure turns out to be crucial whenever the relevant values of the variables to be eliminated (typically, the scale variable a) take only discrete values. We will see an illuminating example of the difference between the two approaches in the problem of symmetry detection in patterns, discussed in Section 4.5.

Let us treat first the problem of sections, the partial energy densities will be discussed in detail in Section 2.3.4. In general, one considers two- and three-dimensional sections. While there are six possible choices of 2-D sections, the geometrical considerations made above indicate that two of them are more natural. Either (a, θ) or (b_x, b_y) are fixed, and the WT is treated as a function of the two remaining variables. The corresponding representations have the following characteristics [13,19].

(1) The *position representation*: a and θ are fixed and the CWT is considered as a function of position $\vec{b}$ alone (this amounts to take a set of snapshots, one for each value of (a, θ), which may then be collected together into a video sequence). The position representation is the standard one, and it is useful for the general purposes of image processing: detection of position, shape and contours of objects; pattern recognition; image filtering by resynthesis after elimination of unwanted features (noise, for instance). Alternatively, one may use polar coordinates, $\vec{b} = (|\vec{b}|, \alpha)$, in

which case the variables are interpreted as *range* $|\vec{b}|$ and *aspect* or *perception angle* α, another familiar representation of images.

(2) The *scale-angle representation:* for fixed $\vec{b}$, the CWT is considered as a function of scale a and anisotropy angle θ, i.e., of spatial frequency. In other words, one looks at the full CWT as through a keyhole located at $\vec{b}$, and observes all scales and all directions at once. The scale-angle representation will be particularly interesting whenever scaling behavior (as in fractals) or angular selection is important, in particular, when directional wavelets are used.

Clearly, these two representations are complementary, together they provide the full information contained in the signal. Accordingly, both are needed for a full understanding of the properties of the CWT in all four variables, as demonstrated in [13].

In addition to these two familiar representations, there are four other two-dimensional sections, obtained by fixing two of the four variables ($|\vec{b}|, \alpha, a, \theta$), and analyzing the CWT as a function of the remaining two.

(3) The *scale-perception angle representation:* for fixed range $|\vec{b}|$ and anisotropy angle θ, one obtains an analysis at all scales a and all perception angles α.

(4) The *range-anisotropy angle representation:* one fixes the scale a and the perception angle α. This gives an analysis at all ranges $|\vec{b}|$ and all anisotropy angles θ.

(5) The *scale-range representation:* fixing the perception angle α and the anisotropy angle θ gives an analysis at all scales a and all ranges $|\vec{b}|$.

(6) The *angle-angle representation:* on the contrary, if one fixes the range $|\vec{b}|$ and the scale a, one gets an analysis at all perception angles α and all anisotropy angles θ. This case is particularly interesting, because the parameter space is now compact (it is a torus) and the discretization easy (linear) in both variables. This representation will be used in Section 3.4, for illustrating the difference in angular selectivity between two standard wavelets.

For the numerical evaluation, in particular for exploiting the reconstruction formula (2.26), one has to discretize the CWT. In any of these representations, a systematic use of the FFT algorithm will lead to a numerical complexity of $3N_1 N_2 \log_2(N_1 N_2)$, where N_1, N_2 denote the number of sampling points in the two free variables. In the case of the position representation, where (b_x, b_y) are free, the geometry is Cartesian and a square lattice will give an adequate sampling grid. In the scale-angle representation, the CWT is naturally expressed in polar coordinates, like (a, θ) or (a^{-1}, θ), and the discretization must be logarithmic in the scale variable a and linear in the anisotropy angle θ. For each variable, the size of the sampling mesh may be estimated from the support properties of the reproducing kernel K, which plays the rôle of a correlation length. We shall come back to this discussion in Section 3.4 of Chapter 3 (see also [13] and [18]). Similar considerations apply to the remaining four representations.

In addition, one may also consider three-dimensional sections, for which a single variable is fixed. Two of them look promising for applications.

(1) The *position-scale representation:* suppose the anisotropy angle θ is fixed, or that it is irrelevant, because the wavelet is rotation invariant. Then the transform is a function of position $\vec{b}$ and scale a. This representation is optimal for detecting the presence of coherent structures, that is, structures that survive through a whole range of scales. Examples may be found, for instance, in astrophysics (hierarchical structure of galaxy clusters and superclusters) [343] or in the analysis of turbulence in fluid dynamics [164,165]. Further information on these two topics will be found in Chapter 5.

(2) The *position-anisotropy representation:* here the scale a is fixed, and the transform is viewed as a function of position $\vec{b}$ and anisotropy angle θ. If the latter is plotted on the vertical axis of a three-dimensional graph, this means that the plane $\theta = \theta_o$ selects all features (targets) that live in the corresponding line of sight. Similarly, an angular sector of opening $\Delta\theta$ is represented in such a plot by a horizontal slice of thickness $\Delta\theta$. This visualization may offer distinct advantages over the conventional ones.

2.3.4   Partial energy densities of the CWT

As explained in the previous section, the visualization problem of the CWT is solved by eliminating a certain number of variables, either by fixing their values, or by integrating them out. The principal example is the scale variable a. If the signal has significant features for a discrete set of scales only, $\{a_j, \; j \in J\}$, the corresponding properties will be visible only if one chooses one of these values a_j. Otherwise, nothing will be seen, and the transform is useless. A typical example is the problem of dilation symmetry in patterns, discussed in Section 4.5. In such a situation, clearly one should not fix the scale variable, but integrate over all scales (exactly as in the construction of continuous wavelet packets discussed in Section 2.6). Of course, the measure to use is the dilation invariant one, that is, the scale part $a^{-3}da$ of the natural measure of the parameter space G. Proceeding in this way with the squared modulus of the wavelet transform, one obtains a quantity which has the physical meaning of a partial energy density. The same reasoning applies to any combination of "ignorable" variables. Thus, one gets such a partial energy density for each of the representations described in the previous section. The most important ones, of course, are the following.

(1) *Position (or range and aspect) energy density*

In the position representation, $a = a_o$ and θ_o are fixed and the CWT $S(\vec{b}, a_o, \theta_o)$ is considered as a function of position $\vec{b}$ alone, either in Cartesian coordinates b_x, b_y, or in polar coordinates $|\vec{b}|, \alpha$ (range and aspect). Accordingly, if one integrates the phase space energy density $|S(\vec{b}, a, \theta)|^2$ over all scales and orientations, one obtains the position energy density,

$$P[s](\vec{b}) = \int_0^\infty \frac{da}{a^3} \int_0^{2\pi} d\theta \; |S(\vec{b}, a, \theta)|^2, \tag{2.56}$$

either in Cartesian coordinates (position) or in polar coordinates (range and aspect). This density has been used as the basis of a CWT-based algorithm for automatic detection and recognition of targets (ATR) in forward-looking infrared radar (FLIR) imagery [285]. This application will be discussed in Section 4.2.2.

(2) *Scale-angle energy density*

In the scale-angle representation, the CWT is looked at from a fixed position $\vec{b}_o$ as a function of scale a and anisotropy angle θ, i.e., of spatial frequency. In the phase space language of Section 2.3.2, this means considering $|\tilde{S}(\vec{q}_o, \vec{p})|^2$ as a function of $\vec{p}$ alone, for fixed $\vec{q}_o$. The corresponding partial energy density is obtained by integrating over all positions $\vec{b}$ or $\vec{q}$:

$$M[s](a, \theta) = \int_{\mathbb{R}^2} d^2\vec{b} \, |S(\vec{b}, a, \theta)|^2. \tag{2.57}$$

This energy density, called the *scale-angle measure* or, better, the *scale-angle spectrum* of the signal, may be used, for example, for discriminating objects of interest according to their size and orientation [16], in particular target classification in FLIR imagery [287]. It yields also an efficient technique for detecting symmetries, even local ones, in patterns such as quasicrystals or Penrose tilings, the rationale being that such objects have no exact translation invariance, so that any dependence on position variables must be eliminated [24]. Both of these applications will be discussed in Chapter 4. A related concept, introduced in [249], is the *relative scale-angle spectrum* of the signal, obtained by normalizing M[s] over all angles:

$$Z[s](a, \theta) = \frac{M[s](a, \theta)}{\int_0^{2\pi} d\theta \, M[s](a, \theta)}. \tag{2.58}$$

Whereas M[s] gives the distribution of energy at different scales and directions, Z[s] gives the relative distribution of energy at different directions at a particular scale with respect to the total energy at that scale. It turns out that Z[s] reveals more efficiently the scale-space anisotropic behavior of the signal.

Similar partial energy densities may be introduced for the other representations, corresponding to other choices of "ignorable" variables. For instance, one may write the *anisotropy angle and aspect energy density* as

$$A[s](\alpha, \theta) = \int_0^\infty \frac{da}{a^3} \int_0^\infty |\vec{p}| \, d|\vec{p}| \, |S(|\vec{p}|, \alpha, a, \theta)|^2. \tag{2.59}$$

Altogether, there are four one-dimensional partial energy densities, six two-dimensional and four three-dimensional ones. Besides the two main ones $P[s](\vec{b})$ and $M[s](a, \theta)$, only two have found an application so far. One is the *angular spectrum* (called angular measure in [24]):

$$\alpha[s](\theta) = \int_0^\infty \frac{da}{a^3} \, M[s](a, \theta), \tag{2.60}$$

which is used for detecting discrete rotation invariances of patterns (see Section 4.5). The other one is the *wavelet spectrum*, or *scale spectrum*, a function of scale only, obtained by integrating the scale-angle spectrum $\mathsf{M}[s](a, \theta)$ over all angles, or, more often, taking the latter in a rotation-invariant situation:

$$\mathsf{W}[s](a) = \int_{\mathbb{R}^2} d^2\vec{b} \, |S(\vec{b}, a)|^2. \tag{2.61}$$

This is the proper generalization to the wavelet setup of the familiar Fourier power spectrum. It has been used, under various names (wavelet spectrum, wavelet (auto-)power spectrum, wavelet variance, scalogram), by many authors both in 1-D [211,227,273] and in 2-D [175].

2.3.5   Ridges in the 2-D CWT

As we saw in Chapter 1 for the 1-D case, the reproduction property (2.36) means that the information contained in the WT $S(\vec{b}, a, \theta)$ is highly redundant. As a consequence, we might hope that no content will be lost if we restrict the WT to a subset of the parameter space. As in 1-D again, there are two ways of achieving this. The first possibility is to take as determining subset the regions where the energy of the signal is concentrated, that is, essentially the lines of local maxima or *ridges*, the set of all ridges being called again the *skeleton* of the WT. This we will do in the present section. The alternative is to choose a discrete subset of the parameter space (a lattice), and this leads to the theory of frames that we shall discuss in detail in Section 2.4.

When trying to extend the notion of ridge to 2-D signals, one faces again the two extreme situations described for the 1-D case in Section 1.4. We begin with the vertical ridges. Let $s(\vec{x})$ be a 2-D signal (an image), with CWT $S(\vec{b}, a, \theta)$. The square modulus of the latter is to be interpreted, as we have seen already, as the energy density of the signal, that we shall denote

$$\mathsf{E}[s](\vec{b}, a, \theta) = |S(\vec{b}, a, \theta)|^2. \tag{2.62}$$

In the case of a rotation-invariant wavelet, the θ-dependence drops out, and we write simply $\mathsf{E}[s](\vec{b}, a) = |S(\vec{b}, a)|^2$. This is the case we will meet in applications, in Chapter 5, for instance, in the analysis of astrophysical images (Section 5.1).

In this situation, we define the ridges as the lines of local maxima of $\mathsf{E}[s](\vec{b}, a)$ [272]. More precisely, we will define a (vertical) ridge $\mathcal{R}$ as a 3-D curve $(\vec{r}(a), a)$ such that, for each scale $a \in \mathbb{R}^+$, $\mathsf{E}[s](\vec{r}(a), a)$ is locally maximum in space and $\vec{r}$ is a continuous function of scale. Figure 2.2 gives a concrete example. The signal (left panel) is a set of singularities in a smooth background, simulating a set of bright points on the surface of the Sun and modeled by a random distribution of Gaussians of small (but random) width. The corresponding vertical ridges of the CWT of that signal are shown on the right panel, clearly each ridge points towards a singularity.

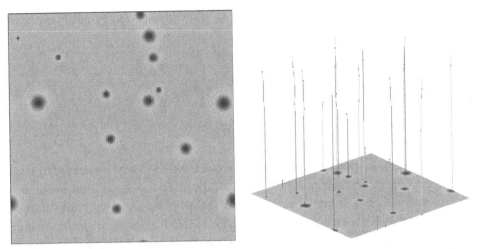

Fig. 2.2. An example of a 2-D ridge: (left) the signal: a field of singularities, simulating a set of bright points on the surface of the Sun; (right) the corresponding vertical ridges of the CWT of that signal.

Given such a vertical ridge, one may distinguish three characteristic features. The first one is the *amplitude* of the ridge, that is, the value of $\mathsf{E}[s]$ on the ridge when a tends to zero,

$$\mathcal{A}_{\mathcal{R}} = \lim_{a \to 0} \mathsf{E}[s](\vec{r}(a), a). \tag{2.63}$$

The second one is the *slope* order, or slope, of $\mathsf{E}[s]$ on the ridge when a is close to 0:

$$\mathcal{S}_{\mathcal{R}} = \lim_{a \to 0} \frac{d \ln \mathsf{E}[s](\vec{r}(a), a)}{d \ln a}. \tag{2.64}$$

The last feature is the ridge energy, that is, the integral of $\mathsf{E}[s]$ along the ridge, assuming the latter to have a finite length, corresponding to the scale interval $[0, a_{\max}]$:

$$\mathcal{E}_{\mathcal{R}} = \int_{0}^{a_{\max}} \frac{da}{a^3} \, \mathsf{E}[s](\vec{r}(a), a). \tag{2.65}$$

Here the measure da/a^3 follows from the L^2 normalization of the wavelet. If one uses the L^1 normalization (see Section 2.6.1), one gets instead da/a. As in 1-D, one can show that the restriction of the CWT to its skeleton characterizes the signal completely [265].

A more sophisticated definition of ridge has been introduced by Mallat and Hwang [262], extending to 2-D the WTMM representation described in Section 1.4. The idea is to consider as (directional) wavelets the partial derivatives of a smoothing function $\zeta(\vec{x})$, a Gaussian, for instance:

$$\psi_1(\vec{x}) = \partial_x \zeta(\vec{x}), \quad \psi_2(\vec{x}) = \partial_y \zeta(\vec{x}), \quad \vec{x} \equiv (x, y).$$

Then, given any function $s \in L^2(\mathbb{R}^2)$, its WT with respect to ψ_1 and ψ_2 can be expressed as a vector

$$\vec{S}(\vec{b}, a) = \vec{\nabla}(S_\zeta(\vec{b}, a)), \tag{2.66}$$

where

$$S_\zeta(\vec{b}, a) = a^{-2} \int_{\mathbb{R}^2} d^2\vec{x} \; \zeta(a^{-1}(\vec{x} - \vec{b})) \, s(\vec{x})$$

(using the L^1-normalization). At a given scale a, the WTMM are defined by the positions $\vec{b}$ where the WT modulus $|\vec{S}(\vec{b}, a)|$ is locally maximum in the direction of the gradient vector $\vec{S}(\vec{b}, a)$. It turns out that the WTMM lie on connected chains in the $\vec{b}$ plane. One then defines the WTMM maxima (WTMMM) as the local maxima of the modulus $|\vec{S}(\vec{b}, a)|$ along the WTMM chains. Globally, these WTMMM live along connected lines across scales, that is, the (vertical) *ridges*. We will describe in Section 5.4.1 the application of this new concept of ridge to the analysis of fractal surfaces. Here again, one can reconstruct from the ridges a good approximation of the original image [262].

An alternate possibility is to introduce horizontal ridges, as in [21]. The algorithm applies to signals which are superpositions of terms of the type

$$s(\vec{x}) = A(\vec{x}) \, e^{i\phi(\vec{x})}, \tag{2.67}$$

where the amplitude $A(\vec{x})$ varies slowly with respect to the phase $\phi(\vec{x})$. A first approximation of the CWT of this signal with the wavelet ψ is obtained by a Taylor expansion, which yields

$$S(\vec{b}, a, \theta) = A(\vec{b}) \, e^{i\phi(\vec{b})} \, \widehat{\psi}(ar_{-\theta}(\vec{\nabla}\phi(\vec{b}))) + R(\vec{b}, a, \theta), \tag{2.68}$$

where the last term is a remainder that can be estimated. Assuming that the wavelet $\widehat{\psi}$ has a unique maximum at a given frequency $\vec{k}_0$, one sees that the CWT is concentrated along a surface in the parameter space, the corresponding ridge (more precisely, the 2-D ridge is a vector field, as we shall see below).

A mathematically more precise approximation may be obtained by a stationary phase argument, following [195,196]. Take again the signal (2.67) and write the wavelet as

$$\psi(\vec{x}) = A_\psi(\vec{x}) \, e^{i\phi_\psi(\vec{x})}.$$

Then the CWT (2.19) of the signal s reads

$$S(\vec{b}, a, \theta) = a^{-1} \int_{\mathbb{R}^2} d^2\vec{x} \; A(\vec{x}) \, A_\psi(a^{-1} r_{-\theta}(\vec{x} - \vec{b})) \, e^{i\Phi_{\vec{b}, a, \theta}(\vec{x})}, \tag{2.69}$$

where

$$\Phi_{\vec{b}, a, \theta}(\vec{x}) = \phi(\vec{x}) - \phi_\psi(a^{-1} r_{-\theta}(\vec{x} - \vec{b})). \tag{2.70}$$

This is an oscillatory integral and, therefore, the main contribution comes from the stationary points $\vec{x}_s$ of the phase $\Phi_{\vec{b},a,\theta}$, that is, the points $\vec{x}_s = \vec{x}_s(\vec{b}, a, \theta)$ such that $\vec{\nabla}\Phi_{\vec{b},a,\theta}(\vec{x}_s) = 0$, or

$$\vec{\nabla}\phi(\vec{x}_s) = a^{-1}r_\theta(\vec{\nabla}\phi_\psi(a^{-1} r_{-\theta}(\vec{x} - \vec{b})))(\vec{x}_s). \tag{2.71}$$

Expanding the phase $\Phi_{\vec{b},a,\theta}$ to second order then yields

$$S(\vec{b}, a, \theta) = s(\vec{x}_s) \overline{\psi(a^{-1} r_{-\theta}(\vec{x}_s - \vec{b}))} C(\vec{b}, a, \theta)^{-1} + R(\vec{b}, a, \theta), \tag{2.72}$$

where $C(\vec{b}, a, \theta)$ is a correction term depending on the phase Φ and $R(\vec{b}, a, \theta)$ is again a remainder that can be estimated. Because of the support properties of the wavelet ψ, we see from (2.72) that the CWT is essentially localized around the points $(\vec{b}, a, \theta)$, if any, such that $\vec{x}_s(\vec{b}, a, \theta) = \vec{b}$. The set of these points is, by definition, the (horizontal) *ridge*. Thus, on the latter, we have

$$\vec{\nabla}\phi(\vec{b}) = a^{-1}r_\theta(\vec{\nabla}\phi_\psi(\vec{0})). \tag{2.73}$$

Writing this relation as

$$\vec{\nabla}\phi(\vec{x}) = a_r(\vec{x})^{-1}r_{\theta_r(\vec{x})}(\vec{\nabla}\phi_\psi(\vec{0})), \tag{2.74}$$

we see that the ridge takes the form of a vector field $\vec{k}_r = \vec{k}_r(\vec{b})$ of polar coordinates $(a_r(\vec{b})^{-1}, \theta_r(\vec{b}))$, to be interpreted as a local wave vector.

We call again *skeleton* of the WT the restriction of the latter to the ridge. Moreover, it may be shown [196] that the restriction of the correction term $C(\vec{b}, a, \theta)$ to the ridge is completely determined by the ridge itself, so that the skeleton of the WT reads as

$$S_s(\vec{b}, a_r(\vec{b}), \theta_r(\vec{b})) = \frac{2\pi \overline{\psi(\vec{0})}}{C(\vec{b}, a_r(\vec{b}), \theta_r(\vec{b}))} s(\vec{b}). \tag{2.75}$$

It follows that the knowledge of the ridge and the skeleton of the WT is sufficient to characterize the signal $s(\vec{x})$.

The discussion extends to a multicomponent signal, i.e., a linear superposition of N terms of the form (2.67). Assuming that the wavelet is sufficiently well localized to prevent any overlap, it will see the ridge corresponding to each term separately, and the skeleton will simply consist of N separate ridges, from which the N components can be extracted and reconstructed individually. Explicit examples may be found in [196]. In general, however, ridges may interact, but this case is much more difficult to handle.

Finally, additional information may be obtained, as in 1-D, from the length of the various ridges (short vertical ridges tend to come from noise) and the behavior of the modulus of the CWT along each ridge, as a function of scale. Techniques based on 2-D ridges have been exploited in the problem of texture determination and, in particular, in the "shape from texture" problem, that we will discuss briefly in Section 5.5.

2.4   Discretization, frames

2.4.1   Generalities on frames

As we saw in Chapter 1 for the 1-D case, the reproduction property (2.36) means that the information contained in the WT $S(\vec{b}, a, \theta)$ is highly redundant. As a consequence, we might hope that no content will be lost if we restrict the WT to a subset of the parameter space, in particular, a discrete subset (for instance, a lattice). Then the integral is replaced by a sum over a discrete (but infinite) family of wavelets $\psi_{\vec{b}_m, a_j, \theta_l}$:

$$s(\vec{x}) = \sum_{mjl} \psi_{\vec{b}_m, a_j, \theta_l}(\vec{x}) \, S(\vec{b}_m, a_j, \theta_l). \tag{2.76}$$

Here too, and by the same reasoning, one is thus led to the introduction of *frames*. Whereas we have barely sketched this topic in Chapter 1, we will now give a fairly detailed treatment. Further information (albeit mostly in 1-D) may be found in [121, 122, Dau92].

Let us start with a precise definition. According to the terminology introduced by Duffin and Schaefer [156] in the context of nonharmonic Fourier series, one has:

Definition 2.4.1. *A countable family of vectors* $\{\psi_n\}$ *in a Hilbert space* $\mathfrak{H}$ *is called a* (discrete) frame *if there are two positive constants* A, B, *with* $0 < A \leqslant B < \infty$, *such that*

$$A \, \|f\|^2 \; \leqslant \; \sum_{n=1}^{\infty} |\langle \psi_n | f \rangle|^2 \; \leqslant \; B \, \|f\|^2, \; \forall \, f \in \mathfrak{H}. \tag{2.77}$$

The two constants A, B *are the* frame bounds. *If* $A = B > 1$, *the frame is said to be* tight. *If* $A = B = 1$, *and* $\|\psi_n\| = 1, \forall \, n$, *the set* $\{\psi_n\}$ *is simply an orthonormal basis.*

The properties of a frame are best discussed in terms of the *frame operator* $F : \mathfrak{H} \to \ell^2$, defined by

$$F : f \mapsto \{\langle \psi_n | f \rangle\}.$$

As discussed in Section 1.3, the upper bound in (2.77) simply means that F is a bounded operator, whereas the left inequality guarantees the numerical stability for the recovery of the signal f from its frame coefficients $\{\langle \psi_n | f \rangle\}$ – in other words, it gives an estimation of the inverse operator F^{-1}. As for the frame bounds A, B, they measure the redundancy of the representation of the signal in terms of its coefficients. For a tight frame, in particular, $A = B > 1$ means that the frame is redundant, and B is its index of redundancy.

In terms of the frame operator F, the inequalities (2.77) may be written as

$$A I \leqslant F^* F \leqslant B I, \tag{2.78}$$

where I is the identity operator. This in turn implies that $F^* F$ is invertible and

$$B^{-1} I \leqslant (F^* F)^{-1} \leqslant A^{-1} I. \tag{2.79}$$

Define now, for each $n \in \mathbb{N}$:

$$\widetilde{\psi}_n = (F^* F)^{-1} \psi_n, \tag{2.80}$$

so that $\psi_n = F^* F \widetilde{\psi}_n$. Then the following is true:

Theorem 2.4.2. *The vectors $\{\widetilde{\psi}_n\}$ constitute a frame, with frame bounds B^{-1}, A^{-1} and frame operator $\widetilde{F} = F(F^* F)^{-1}$. In addition, the expansion*

$$f(x) = \sum_{n=1}^{\infty} \langle \psi_n | f \rangle \, \widetilde{\psi}_n(x), \tag{2.81}$$

converges strongly in $\mathfrak{H}$, that is, $\widetilde{F}^* F = I$.

Proof. That $\widetilde{\mathcal{F}} \equiv \{\widetilde{\psi}_n\}$ is the frame described in the statement results from the equalities

$$\sum_n |\langle \widetilde{\psi}_n | f \rangle|^2 = \sum_n |\langle (F^* F)^{-1} \psi_n | f \rangle|^2$$

$$= \sum_n |\langle \psi_n | (F^* F)^{-1} f \rangle|^2$$

$$= \| F(F^* F)^{-1} f \|^2 \; = \; \langle f | (F^* F)^{-1} f \rangle$$

and the inequalities (2.77). Furthermore

$$\widetilde{F}^* F = (F^* F)^{-1} F^* F = I,$$

that is, (2.81) is an identity. $\square$

In other words, the duality between the two frames may be written as $\widetilde{F}^* F = F^* \widetilde{F} = I$ or explicitly,

$$\sum_n |\psi_n\rangle\langle\widetilde{\psi}_n| = \sum_n |\widetilde{\psi}_n\rangle\langle\psi_n| = I. \tag{2.82}$$

The frame $\widetilde{\mathcal{F}} = \{\widetilde{\psi}_n\}$ is called the *dual* or *reciprocal frame* of $\mathcal{F} = \{\psi_n\}$. Notice that the dual of a tight frame is again a tight frame. This notion is crucial for applications. In the case of wavelet expansions, it is the basis of the so-called biorthogonal scheme [Dau92], briefly discussed in Section 2.5.2.

In practice, orthonormal bases are not always available for representing arbitrary functions, but one may often use instead a good frame. By this, we mean that the expansion (2.81) converges sufficiently fast. How could one estimate the speed of this convergence? By (2.81), we need to compute $\widetilde{\psi}_n = (F^*F)^{-1}\psi_n$. If B and A are close to each other, F^*F is close to $\frac{1}{2}(B+A)I$, hence $(F^*F)^{-1}$ is close to $\frac{2}{B+A}I$ and thus $\widetilde{\psi}_n$ is close to $\frac{2}{B+A}\psi_n$. Hence we may write

$$f = \frac{2}{B+A}\sum_n \langle \psi_n | f \rangle \psi_n + Rf, \tag{2.83}$$

where

$$R = I - \frac{2}{B+A}F^*F. \tag{2.84}$$

Hence

$$(F^*F)^{-1} = \frac{2}{B+A}(I - R)^{-1}$$

$$= \frac{2}{B+A}\sum_{k=0}^{\infty} R^k. \tag{2.85}$$

The series converges in norm, since, by (2.84),

$$-\frac{B-A}{B+A}I \leqslant R \leqslant \frac{B-A}{B+A}I, \tag{2.86}$$

which implies

$$\|R\| \leqslant \frac{B-A}{B+A} = \frac{B/A - 1}{B/A + 1} < 1.$$

Therefore the expansion (2.81) converges essentially as a power series in $|B/A - 1|$. Thus the frame is good if $|B/A - 1| \ll 1$, in particular if it is tight. To the first order, the expansion (2.81) becomes

$$f = \frac{2}{B+A}\sum_n \langle \psi_n | f \rangle \psi_n. \tag{2.87}$$

The quantity

$$w(\mathcal{F}) = \frac{B-A}{B+A} \tag{2.88}$$

is called the *width* or the *snugness* of the frame $\mathcal{F}$. It measures the lack of tightness, since $w(\mathcal{F}) = 0$ iff the frame $\mathcal{F}$ is tight. Notice that a frame and its dual have the same width.

In practical applications, the infinite sum in (2.81) or (2.87) will be truncated and the approximate reconstruction so obtained is numerically stable. If the width of the frame is sufficiently small, a few terms will suffice. More details on frames may be found in [121,122,Dau92,220,Mal99].

Let us now come back to the notion of redundancy of a frame, which may be characterized in terms of the frame operator F. Let Ran $F \subset \ell^2$ denote the range of F, that is, the set of sequences $Ff = (\langle \psi_n | f \rangle)$, $f \in \mathfrak{H}$. First, we remark that the inclusion is strict if the frame vectors $\{\psi_n\}$ are linearly dependent. In that case, indeed, there exists a nonzero vector $y = (y_n) \in \ell^2$ such that $\sum_n y_n \psi_n = 0$. But then, for any $f \in \mathfrak{H}$, one has $\sum_n y_n \langle \psi_n | f \rangle = \langle y | F f \rangle = 0$, that is, $y \in (\text{Ran } F)^\perp \neq \{0\}$, where $(\text{Ran } F)^\perp$ is the orthogonal complement of Ran F in ℓ^2. Moreover, the more redundant the frame, the larger the orthogonal complement $(\text{Ran } F)^\perp$.

Since F is injective by the left inequality in (2.77), it may be inverted on its image, but the inverse operator is not uniquely defined, because it remains arbitrary on the complement $(\text{Ran } F)^\perp$. Among all possible inverses, the *pseudo-inverse* $\check{F}^{-1}$ is defined as the inverse operator that vanishes on $(\text{Ran } F)^\perp$. It is easy to show [Mal99] that it is also the inverse with the smallest norm, and it is given by $\check{F}^{-1} = (F^*F)^{-1}F^*$ (thus it is uniquely defined). From the discussion above, it is clear that $\check{F}^{-1} \equiv \widetilde{F}^*$. Thus, the pseudo-inverse of the frame operator is always bounded. This explains the rôle of the lower bound in (2.77) as guaranteeing numerical stability in the computation of the inverse operator F^{-1}. In addition, we see that the dual frame is built from the pseudo-inverse.

One should also notice that there exist alternative inversion techniques, for instance, the conjugate gradient method, that sometimes converge faster than the pure wavelet reconstruction formulas. For more information, see for instance [167].

Another useful notion (already met in the Definition 1.5.1 of a multiresolution analysis, in Section 1.5) is that of a *Riesz basis*, namely, a frame $\mathcal{F}$ of linearly independent vectors. Of course, this definition implies that the corresponding frame operator F maps $\mathfrak{H}$ onto ℓ^2, Ran $F = \ell^2$. It follows that the dual frame $\widetilde{\mathcal{F}}$ is also a Riesz basis, and in fact, it is biorthogonal to $\mathcal{F}$. To see this, apply (2.82) to a basis vector ψ_k:

$$\psi_k = \sum_n \psi_n \langle \widetilde{\psi}_n | \psi_k \rangle.$$

Since the vectors $\{\psi_n\}$ are linearly independent, this implies that $\langle \widetilde{\psi}_n | \psi_k \rangle = \delta_{n,k}$, i.e., the two bases are biorthogonal. Next, if the vectors $\{\widetilde{\psi}_n\}$ were not linearly independent, there would exist nonzero numbers $\{\lambda_n\}$ such that $\sum_n \lambda_n \widetilde{\psi}_n = 0$. But then, taking the inner product with any ψ_k gives

$$0 = \sum_n \lambda_n \langle \widetilde{\psi}_n | \psi_k \rangle = \lambda_k,$$

a contradiction. Further information about Riesz bases and their numerical implementation may be found in [Mal99].

2.4.2   Two-dimensional wavelet frames

Thus there remains the question: given a specific wavelet, does it generate a frame? The first problem is how to choose the sampling grid Λ in an optimal fashion. As in 1-D, one should take into account the geometry of the parameter space, that is, the lattice Λ must be invariant under discrete sets of dilations, rotations and translations.

Note, however, that in practice the sampling points are quite often fixed empirically. For the (a, θ) variables, in particular, they are mostly chosen on the basis of biological considerations or symmetry requirements [126,260,264].

Proceeding as in 1-D, one thus obtains the following natural discretization scheme.
- For the dilations, a logarithmic scale $a_j = a_0 \lambda^{-j}$, $j \in \mathbb{Z}$, for some $\lambda > 1$; here again we will put $a_0 = 1$.
- For the rotations, one subdivides the interval $[0, 2\pi)$ uniformly into L_0 pieces, for some natural number $L_0 \in \mathbb{N}$, that is, $\theta_l = l\theta_0$, $\theta_0 = \frac{2\pi}{L_0}$, $l \in \mathbb{Z}_{L_0} = \{0, ..., L_0 - 1\}$.
- For the translations, one takes into account the two previous discretizations, putting

$$\vec{b}_m \equiv \vec{b}_{jlm_0m_1} = \lambda^{-j}\, r_{l\theta_0}(\vec{u}_{m_0m_1}),$$

with

$$\vec{u}_{m_0m_1} \equiv (m_0\beta_0, m_1\beta_1), \ \ m_0, m_1 \in \mathbb{Z}, \ \beta_0, \beta_1 \geqslant 0.$$

Thus the discretization grid reads:

$$\Lambda = \Lambda(\lambda, L_0, \beta_0, \beta_1) = \left\{ (\lambda^{-j}, l\frac{2\pi}{L_0}, \vec{b}_{jlm_0m_1}), \ (j, l, m_0, m_1) \in \mathbb{Z} \times \mathbb{Z}_{L_0} \times \mathbb{Z}^2 \right\}.$$

$$(2.89)$$

The resulting discretized wavelet transform, which is a map from $L^2(\mathbb{R}^2, d^2\vec{x})$ to $l^2(\mathbb{Z} \times \mathbb{Z}_{L_0} \times \mathbb{Z}^2)$, reads now:

$$\begin{aligned}
S_{jlm_0m_1} &\equiv S(\vec{b}_m, \lambda^{-j}, l\theta_0) = \langle \psi_{jlm_0m_1} | s \rangle \\
&= \lambda^j \int_{\mathbb{R}^2} d^2\vec{x}\ \overline{\psi(\lambda^j r_{-l\theta_0}(\vec{x}) - \vec{u}_{m_0, m_1})}\, f(\vec{x}) \tag{2.90} \\
&= \lambda^{-j} \int_{\mathbb{R}^2} d^2\vec{k}\ e^{i\vec{b}_{m_0m_1}\cdot\vec{k}}\ \overline{\hat{\psi}(\lambda^{-j} r_{-l\theta_0}(\vec{k}))}\, \hat{f}(\vec{k}), \tag{2.91}
\end{aligned}$$

with wavelet coefficients

$$\left\{ S_{jlm_0m_1}, \ (j, l, m_0, m_1) \in \mathbb{Z} \times \mathbb{Z}_{L_0} \times \mathbb{Z}^2 \right\}.$$

Our task now is to find conditions on the grid $\Lambda(\lambda, L_0, \beta_0, \beta_1)$, that is, on the parameters $\lambda, L_0, \beta_0, \beta_1$, such that the family of wavelets $\{\psi_{jlm_0m_1}, \ (j, l, m_0, m_1) \in \mathbb{Z} \times \mathbb{Z}_{L_0} \times \mathbb{Z}^2\}$ is a frame. As in 1-D, the answer is that the 2-D wavelet transform obeys a sampling theorem, that gives a lower bound on the density of sampling points, like the standard Shannon theorem of signal analysis. The following result, first proven in [Mur90], is the exact counterpart of [122; Theorem 2.7], and the proof follows closely [Dau92; Section 3.3.2].

Theorem 2.4.3. *Assume the wavelet ψ satisfies the following conditions:*

$$(i) \quad s(\lambda, L_0, \psi) = \operatorname*{ess\,inf}_{\vec{k}\in\mathbb{R}^2} \sum_{j=-\infty}^{\infty} \sum_{l=0}^{L_0} |\widehat{\psi}(\lambda^{-j}\, r_{-l\theta_0}(\vec{k}))|^2 \tag{2.92}$$

$$= \operatorname*{ess\,inf}_{(|\vec{k}|,\theta)\in[0,\lambda)\times[0,2\pi)} \sum_{j=-\infty}^{\infty} \sum_{l=0}^{L_0} |\widehat{\psi}_p(\lambda^{-j}\,|\vec{k}|, \varphi - l\theta_0)|^2 > 0, \tag{2.93}$$

where $\vec{k} = |\vec{k}|(\cos\varphi, \sin\varphi)$ and $\widehat{\psi}_p$ is the Fourier transform of ψ in polar coordinates.

$$(ii) \quad S(\lambda, L_0, \psi) = \sup_{\vec{k}\in\mathbb{R}^2} \sum_{j=-\infty}^{\infty} \sum_{l=0}^{L_0} |\widehat{\psi}(\lambda^{-j}\, r_{-l\theta_0}(\vec{k}))|^2 \tag{2.94}$$

$$= \sup_{(|\vec{k}|,\theta)\in[0,\lambda)\times[0,2\pi)} \sum_{j=-\infty}^{\infty} \sum_{l=0}^{L_0} |\widehat{\psi}_p(\lambda^{-j}\,|\vec{k}|, \varphi - l\theta_0)|^2 < \infty. \tag{2.95}$$

$$(iii) \quad \sup_{\vec{u}\in\mathbb{R}^2} (1 + |\vec{u}|)^{1+\epsilon}\, \alpha(\vec{u}) < \infty, \tag{2.96}$$

where $\epsilon > 0$ and

$$\alpha(\vec{u}) = \sup_{\vec{k}\in\mathbb{R}^2} \sum_{j=-\infty}^{\infty} \sum_{l=0}^{L_0} |\widehat{\psi}(\lambda^{-j}\, r_{-l\theta_0}(\vec{k}) + \vec{u})||\widehat{\psi}(\lambda^{-j}\, r_{-l\theta_0}(\vec{k}))|. \tag{2.97}$$

Then there exist constants $\beta_0{}^c, \beta_1{}^c > 0$ such that:
(1) $\forall\, \beta_0 \in (0, \beta_0{}^c)$, $\beta_1 \in (0, \beta_1{}^c)$, the family $\{\psi_{ljm_0m_1}\}$ associated to $(\lambda, L_0, \beta_0, \beta_1)$ is a frame of $L^2(\mathbb{R}^2, d^2\vec{x})$;
(2) $\forall\, \delta > 0$, there exist $\beta_0 \in (\beta_0{}^c, \beta_0{}^c + \delta)$, $\beta_1 \in (\beta_1{}^c, \beta_1{}^c + \delta)$, such that the family $\{\psi_{ljm_0m_1}\}$ associated to $(\lambda, L_0, \beta_0, \beta_1)$ is not a frame of $L^2(\mathbb{R}^2, d^2\vec{x})$.

Proof. We want to find the conditions on $\lambda, L_0, \beta_0, \beta_1$ for which there exists $0 < A, B < \infty$, such that:

$$A \|f\|^2 \leq \sum_{l=0}^{L_0} \sum_{j,m_0,m_1=-\infty}^{\infty} |\langle \psi_{jlm_0m_1}|f\rangle|^2 \leq B \|f\|^2. \tag{2.98}$$

The central term in these inequalities reads

$$K = \sum_{j,l} \sum_{m_0,m_1} |\langle \psi_{jlm_0m_1} | f \rangle|^2$$

$$= \sum_{j,l} \sum_{m_0,m_1} \lambda^{-2j} \int_{\mathbb{R}^2} d^2\vec{k} \int_{\mathbb{R}^2} d^2\vec{k}' \, e^{i\,\vec{b}_{jlm_0m_1}\cdot(\vec{k}-\vec{k}')}$$

$$\times \widehat{\psi}(\lambda^{-j} r_{-l\theta_0}(\vec{k}')) \, \overline{\widehat{\psi}(\lambda^{-j} r_{-l\theta_0}(\vec{k}))} \, \overline{\widehat{f}(\vec{k}')} \, \widehat{f}(\vec{k})$$

$$= \sum_{j,l} \sum_{m_0,m_1} \lambda^{-2j} \int_{\mathbb{R}^2} d^2\vec{k} \int_{\mathbb{R}^2} d^2\vec{k}' \, e^{i\,\vec{u}_{m_0m_1}\cdot\lambda^{-j} r_{-l\theta_0}(\vec{k}-\vec{k}')}$$

$$\times \widehat{\psi}(\lambda^{-j} r_{-l\theta_0}(\vec{k}')) \, \overline{\widehat{\psi}(\lambda^{-j} r_{-l\theta_0}(\vec{k}))} \, \overline{\widehat{f}(\vec{k}')} \, \widehat{f}(\vec{k})$$

$$= \sum_{j,l} \sum_{m_0,m_1} \lambda^{-2j} \int_{\mathbb{R}^2} d^2\vec{k} \int_{\mathbb{R}^2} d^2\vec{k}' \, e^{i\,\vec{u}_{m_0m_1}\cdot(\vec{k}-\vec{k}')}$$

$$\times \widehat{\psi}(\vec{k}') \, \overline{\widehat{\psi}(\vec{k})} \, \overline{\widehat{f}(\lambda^j r_{l\theta_0}(\vec{k}'))} \, \widehat{f}(\lambda^{-j} r_{l\theta_0}(\vec{k})).$$

Using the Poisson formula,

$$\sum_{m_0,m_1=-\infty}^{\infty} e^{i\vec{u}_{m_0m_1}\cdot(\vec{k}-\vec{k}')} = \frac{4\pi^2}{\beta_0\beta_1} \sum_{m_0,m_1=-\infty}^{\infty} \delta(\vec{k}-\vec{k}'-\vec{\vec{u}}_{m_0m_1}), \qquad (2.99)$$

where

$$\vec{\vec{u}}_{m_0m_1} = (m_0\frac{2\pi}{\beta_0}, m_1\frac{2\pi}{\beta_1}), \qquad (2.100)$$

we obtain

$$K = \frac{4\pi^2}{\beta_0\beta_1} \sum_{j,l} \sum_{m_0,m_1} \int_{\mathbb{R}^2} d^2\vec{k} \, \overline{\widehat{\psi}(\vec{k})} \, \widehat{\psi}(\vec{k}-\vec{\vec{u}}_{m_0m_1})$$

$$\times \widehat{f}(\lambda^j r_{l\theta_0}(\vec{k})) \, \overline{\widehat{f}(\lambda^j r_{l\theta_0}(\vec{k}-\vec{\vec{u}}_{m_0m_1}))}$$

$$= \frac{4\pi^2}{\beta_0\beta_1} \sum_{j,l} \sum_{m_0,m_1} \int_{\mathbb{R}^2} d^2\vec{k} \, \overline{\widehat{\psi}(\lambda^{-j} r_{-l\theta_0}(\vec{k}))} \, \widehat{\psi}(\lambda^{-j} r_{-l\theta_0}(\vec{k}) - \vec{\vec{u}}_{m_0m_1})$$

$$\times \widehat{f}(\vec{k}) \, \overline{\widehat{f}(\vec{k} - \lambda^j r_{l\theta_0}(\vec{\vec{u}}_{m_0m_1}))}.$$

Let us split the double sum as $K = |P| + Q$, where $|P|$ denotes the term with $(m_0, m_1) = (0, 0)$ and Q the rest:

$$
K = |P| + Q
$$
$$
= \frac{4\pi^2}{\beta_0\beta_1} \sum_{j,l} \int_{\mathbb{R}^2} d^2\vec{k} \, |\widehat{\psi}(\lambda^{-j} r_{-l\theta_0}(\vec{k}))|^2 \, |\widehat{f}(\vec{k})|^2
$$
$$
+ \frac{4\pi^2}{\beta_0\beta_1} \sum_{j,l} \sum_{m_0,m_1\in\mathbb{Z}_*} \int_{\mathbb{R}^2} d^2\vec{k} \, \widehat{\psi}(\lambda^{-j} r_{-l\theta_0}(\vec{k})) \, \overline{\widehat{\psi}(\lambda^{-j} r_{-l\theta_0}(\vec{k}) - \vec{\tilde{u}}_{m_0m_1})}
$$
$$
\times \overline{\widehat{f}(\vec{k})} \, \widehat{f}(\vec{k} - \lambda^j r_{l\theta_0}(\vec{\tilde{u}}_{m_0m_1})).
$$

Then we obtain the following estimates.

(1) For the first term, we get immediately:

$$
\frac{4\pi^2}{\beta_0\beta_1} s(\lambda, L_0, \psi) \|\widehat{f}\|^2 \leqslant |P| \leqslant \frac{4\pi^2}{\beta_0\beta_1} S(\lambda, L_0, \psi) \|\widehat{f}\|^2,
$$

where $s(\lambda, L_0, \psi)$ and $S(\lambda, L_0, \psi)$ are defined in (2.92) and (2.94), respectively.

(2) For the second term, we obtain, by the Cauchy–Schwarz inequality:

$$
|Q| \leqslant \frac{4\pi^2}{\beta_0\beta_1} \sum_{j,l} \sum_{m_0,m_1\in\mathbb{Z}_*} \int_{\mathbb{R}^2} d^2\vec{k} \, |\widehat{\psi}(\lambda^{-j} r_{-l\theta_0}(\vec{k}))|
$$
$$
\times |\widehat{\psi}(\lambda^{-j} r_{-l\theta_0}(\vec{k}) - \vec{\tilde{u}}_{m_0m_1})| \, |\widehat{f}(\vec{k})| \, |\widehat{f}(\vec{k} - \lambda^j r_{l\theta_0}(\vec{\tilde{u}}_{m_0m_1}))|
$$
$$
\leqslant \frac{4\pi^2}{\beta_0\beta_1} \sum_{j,l} \sum_{m_0,m_1\in\mathbb{Z}_*}
$$
$$
\left[\int_{\mathbb{R}^2} d^2\vec{k} \, |\widehat{\psi}(\lambda^{-j} r_{-l\theta_0}(\vec{k}) - \vec{\tilde{u}}_{m_0m_1})| \, |\widehat{\psi}(\lambda^{-j} r_{-l\theta_0}(\vec{k}))| \, |\widehat{f}(\vec{k})|^2 \right]^{1/2}
$$
$$
\times \left[\int_{\mathbb{R}^2} d^2\vec{k} \, |\widehat{\psi}(\lambda^{-j} r_{-l\theta_0}(\vec{k}) - \vec{\tilde{u}}_{m_0m_1})| \, |\widehat{\psi}(\lambda^{-j} r_{-l\theta_0}(\vec{k}))| \right.
$$
$$
\left. |\widehat{f}(\vec{k} - \lambda^j r_{l\theta_0}(\vec{\tilde{u}}_{m_0m_1}))|^2 \right]^{1/2}
$$
$$
\leqslant \frac{4\pi^2}{\beta_0\beta_1} \sum_{j,l} \sum_{m_0,m_1\in\mathbb{Z}_*}
$$
$$
\left[\int_{\mathbb{R}^2} d^2\vec{k} \, |\widehat{\psi}(\lambda^{-j} r_{-l\theta_0}(\vec{k}) - \vec{\tilde{u}}_{m_0m_1})| \, |\widehat{\psi}(\lambda^{-j} r_{-l\theta_0}(\vec{k}))| \, |\widehat{f}(\vec{k})|^2 \right]^{1/2}
$$
$$
\times \left[\int_{\mathbb{R}^2} d^2\vec{k} \, |\widehat{\psi}(\lambda^{-j} r_{-l\theta_0}(\vec{k}))| \, |\widehat{\psi}(\lambda^{-j} r_{-l\theta_0}(\vec{k}) + \vec{\tilde{u}}_{m_0m_1})| \, |\widehat{f}(\vec{k})|^2 \right]^{1/2}.
$$

Applying Cauchy–Schwarz a second time, to the summation over j, l, then gives:

$$|Q| \leqslant \frac{4\pi^2}{\beta_0\beta_1} \sum_{m_0, m_1 \in \mathbb{Z}_*}$$

$$\left[\int_{\mathbb{R}^2} d^2\vec{k} \left\{ \sum_{j,l} |\widehat{\psi}(\lambda^{-j} r_{-l\theta_0}(\vec{k})) - \vec{\tilde{u}}_{m_0m_1})| \, |\widehat{\psi}(\lambda^{-j} r_{-l\theta_0}(\vec{k}))| \right\} |\widehat{f}(\vec{k})|^2 \right]^{1/2}$$

$$\times \left[\int_{\mathbb{R}^2} d^2\vec{k} \left\{ \sum_{j,l} |\widehat{\psi}(\lambda^{-j} r_{-l\theta_0}(\vec{k}))| \, |\widehat{\psi}(\lambda^{-j} r_{-l\theta_0}(\vec{k}) + \vec{\tilde{u}}_{m_0m_1})| \right\} |\widehat{f}(\vec{k})|^2 \right]^{1/2}.$$

The terms in braces in the integrals are majorized using (2.97), and we get finally

$$|Q| \leqslant \frac{4\pi^2}{\beta_0\beta_1} \left\{ \sum_{m_0, m_1 \in \mathbb{Z}_*} \alpha(\vec{\tilde{u}}_{m_0m_1}) \alpha(-\vec{\tilde{u}}_{m_0m_1}) \right\} \|\widehat{f}\|^2. \tag{2.101}$$

Define the quantity

$$E(\lambda, L_0, \beta_0, \beta_1, \psi) = \sum_{m_0, m_1 \in \mathbb{Z}_*} \alpha(\vec{\tilde{u}}_{m_0m_1}) \alpha(-\vec{\tilde{u}}_{m_0m_1}) = \sum_{m_0, m_1 \in \mathbb{Z}_*} [\alpha(\vec{\tilde{u}}_{m_0m_1})]^2. \tag{2.102}$$

In virtue of the decay condition (2.96) in assumption (iii), the sum over (m_0, m_1) converges. Then, using the inequality $|P| - |Q| \leqslant K \equiv |P| + Q \leqslant |P| + |Q|$, we obtain a lower bound for the left-hand side:

$$\frac{4\pi^2}{\beta_0\beta_1} \left\{ s(\lambda, L_0, \psi) - E(\lambda, L_0, \beta_0, \beta_1, \psi) \right\} \|\widehat{f}\|^2 \leqslant |P| - |Q| \tag{2.103}$$

and an upper bound for the right-hand side:

$$|P| + |Q| \leqslant \frac{4\pi^2}{\beta_0\beta_1} \left\{ S(\lambda, L_0, \psi) + E(\lambda, L_0, \beta_0, \beta_1, \psi) \right\} \|\widehat{f}\|^2. \tag{2.104}$$

By condition (2.92) in assumption (i), $s(\lambda, L_0, \psi)$ is strictly positive, and the decay condition (2.96) implies that

$$\lim_{(\beta_0, \beta_1) \to (0,0)} E(\lambda, L_0, \beta_0, \beta_1, \psi) = 0. \tag{2.105}$$

Therefore, there exists critical values $\beta_0{}^c, \beta_1{}^c > 0$ such that the first term in (2.103) is dominant for all $\beta_0 < \beta_0{}^c, \beta_1 < \beta_1{}^c$. Thus the left-hand side of (2.103) is strictly positive and yields a lower frame bound. This proves the statement. $\square$

Notice that, as in [Dau92; Section 3.3.2], we had to use in (2.92) the essential infimum, that is, infimum except on a set of measure zero, instead of the usual infimum, which is 0 here, since $\widehat{\psi}(\vec{0}) = 0$. This distinction is not necessary in (2.94), because in all practical cases, the function $\widehat{\psi}$ is continuous. Here also, the infimum in (2.93) has to be taken only over a ball of radius λ, since other values of $|\vec{k}|$ can be brought to this range by dilation with a suitable power λ^j, except for $\vec{k} = 0$, which is a set of measure zero.

From the inequalities (2.103) and (2.104), we obtain immediately estimates for the frame bounds:

Corollary 2.4.4. *Whenever the conditions of Theorem 2.4.3 are satisfied, and β_0, β_1 are such that*

$$s(\lambda, L_0, \psi) > E(\lambda, L_0, \beta_0, \beta_1, \psi), \tag{2.106}$$

then the bounds of the frame can be estimated by:

$$A \geqslant \frac{4\pi^2}{\beta_0\beta_1}\{s(\lambda, L_0, \psi) - E(\lambda, L_0, \beta_0, \beta_1, \psi)\}, \tag{2.107}$$

$$B \leqslant \frac{4\pi^2}{\beta_0\beta_1}\{S(\lambda, L_0, \psi) + E(\lambda, L_0, \beta_0, \beta_1, \psi)\}. \tag{2.108}$$

It is clear from this discussion that we have obtained essentially the same results as in the 1-D case – and they extend to three or more dimensions in the same way [283,Mur90]. As argued in [Dau92; Section 3.3.2], the moral is that, whenever the wavelet ψ is admissible and decays reasonably at infinity, it will yield a frame. This is true, for instance, for all types of Mexican hat or Morlet wavelets. However, the estimates of Corollary 2.4.4 imply, as in 1-D [103], the following inequalities as well, for all $\vec{k} \neq \vec{0}$,

$$A \leqslant \frac{4\pi^2}{\beta_0\beta_1} \sum_{j=-\infty}^{\infty} \sum_{l=0}^{L_0} |\widehat{\psi}(\lambda^{-j} \, r_{-l\theta_0}(\vec{k}))|^2 \leqslant B. \tag{2.109}$$

This puts a rather strong limitation on the wavelet, because the frame tends to become loose when λ is not very small, in particular for $\lambda = 2$, the preferred value. In 1-D, this defect may be corrected by introducing additional voices. This means that one further subdivides each octave, replacing in (2.109) the exponent j by $\frac{v}{N}j$, $v = 0, 1, \ldots, N-1$ (N is called the number of voices). The effect is to "densify" the lattice Λ, which improves the ration B/A and thus speeds up convergence of the discrete approximation. For further details, we refer the reader to [Dau92]. The same procedure may be applied in 2-D as well. However, if the speed is the determining criterion, one can do better by using the pseudo-QMF algorithm described in Section 2.6.

In addition, other discretization schemes may be considered. For instance, the discretization step in the angular variable θ may be made scale dependent. The idea is that one needs more directions at small scales (high frequencies) than at large scales (low frequencies). This is in the spirit of certain applications of discrete wavelet bases in solid state physics (see [30]), in which additional, smaller, scales are considered in the vicinity of the atomic crystal nodes. We will come back to this point in the next section.

2.4.3   Implementation of a 2-D wavelet frame

In general, implementing the inverse frame operator is a nontrivial task which requires using the expansion formula (2.84). In the case of a tight (or approximately tight) frame, however, the situation becomes much easier since the reconstruction is provided by the simple sum (2.87).

Let us particularize this for the concrete case of the 2-D Morlet wavelet. Using Lee's parametrization [253], the Fourier transform of this wavelet is simply:

$$\widehat{\psi}(k_1, k_2) = \frac{\kappa}{k_0}\sqrt{8\pi}\left\{ e^{-\frac{\kappa^2}{2k_0^2}\left((k_1-k_0)^2+4k_2^2\right)} - e^{-\frac{\kappa^2}{2k_0^2}\left(k_1^2+4k_2^2+k_0^2\right)} \right\}. \tag{2.110}$$

Chosing $\kappa = 3\sqrt{2\log 2}$ yields a spatial frequency bandwidth of approximately one octave, which is compatible with the characteristics of the receptive fields of human simple cells [372]. Using the general framework described before, one is able to compute frame bounds for this wavelet. Psychovisual experiments suggesting a density of about 16 to 20 orientations, one can obtain a tight frame by carefully choosing a discretization strategy. Figure 2.3 shows a reconstructed image with a tight frame of Morlet wavelets corresponding to the discretization grid

$$\Lambda = \left\{ \left(2^{-j/n}, l\frac{2\pi}{16}, \vec{b}_{j,l,m_0,m_1}\right), \ (j,l,m_0,m_1) \in \mathbb{Z} \times \mathbb{Z}_{16} \times \mathbb{Z}^2 \right\}, \ n = 2 \text{ and } 4,$$

that is, 2 and 4 voices, respectively, per octave and a uniform translation step $\beta_0 = \beta_1 = 0.8$.

2.4.4   The dyadic wavelet transform

As we saw already in 1-D (Sections 1.3 and 1.6.1), a slightly different construction consists in computing a hybrid wavelet transform in which we sample the scale variable, but leave translations untouched. If we use dyadic scales, the resulting set of coefficients is called the *Dyadic Wavelet Transform* and offers the advantage of being completely covariant with respect to translations, a very desirable feature [264,265]. In pattern recognition, for instance, translating the object as a whole should affect all wavelet coefficients in the same way, so as not to distort the transform. This is precisely the practical meaning of the so-called "shift invariance" (properly, shift covariance), and the condition is *not* satisfied in the usual discrete WT. In addition, when the translation parameter is properly discretized, such dyadic wavelet transforms may lead to genuine tight frames.

We denote, as usual, the scaled and translated wavelets by

$$\psi^\ell_{\vec{b},2^j}(\vec{x}) = 2^{-j}\psi^\ell(2^{-j}(\vec{x}-\vec{b})), \tag{2.111}$$

assume for simplicity that $c_\psi = 1$ and use the L^2-normalization. Then the dyadic WT of $f \in L^2$ is given by

(a)

(b)   (c)

Fig. 2.3. Reconstruction of the *lena* image with a tight frame of Morlet wavelets: (a) the original image; (b) and (c) the reconstructed image, with 2 and 4 voices, respectively, per octave.

$$\mathcal{W}^{\ell}(\vec{b}, 2^j) = \langle \psi_{\vec{b}, 2^j}^{\ell} | f \rangle = \psi_{2^j}^{\ell\#} \star f \,, \tag{2.112}$$

where $\psi_a^{\#}(\vec{x}) = \overline{\psi_a(-\vec{x})}$ and $\psi_a \equiv \psi_{\vec{0}, a}$. Note that we use a different notation, $\mathcal{W}(.,.)$ instead of $W(.,.)$, in order to emphasize the fact that we are dealing with the dyadic WT instead of the CWT. The superscript ℓ in equations (2.111), (2.112) refers to different wavelets, but the practical situation will be that of rotated versions of a single generating function.

The following result shows that the dyadic WT is a stable and complete representation of images provided the set of scaled wavelets appropriately pave the frequency plane.

Proposition 2.4.1 *If there exists two strictly positive constants A and B such that*

$$A \leqslant \sum_{\ell=1}^{L} \sum_{j \in \mathbb{Z}} |\widehat{\psi}^{\ell}(2^j \vec{k})|^2 \leqslant B, \quad \forall \vec{k}, \tag{2.113}$$

then,

$$A\|f\|^2 \leqslant \sum_{\ell=1}^{L} \sum_{j \in \mathbb{Z}} 2^{-2j} \|\mathcal{W}^{\ell}(.,2^j)\|^2 \leqslant B\|f\|^2. \tag{2.114}$$

Moreover, if we define dual or reconstruction wavelets χ^{ℓ} through the relation

$$\sum_{\ell=1}^{L} \sum_{j \in \mathbb{Z}} \overline{\widehat{\psi}^{\ell}(2^j \vec{k})}\, \widehat{\chi}^{\ell}(2^j \vec{k}) = 1, \quad \forall \vec{k} \in \mathbb{R}^2 \setminus \{\vec{0}\}, \tag{2.115}$$

then, the following reconstruction formula holds strongly in L^2:

$$f(\vec{x}) = \sum_{\ell=1}^{L} \sum_{j \in \mathbb{Z}} 2^{-2j} \left(\mathcal{W}^{\ell}(.,2^j) \star \chi^{\ell}_{2^j} \right)(\vec{x}). \tag{2.116}$$

Proof. Let d_j^{ℓ} stand for the Fourier transform of $\mathcal{W}^{\ell}(\vec{b}, 2^j)$ with respect to the variable $\vec{b}$:

$$d_j^{\ell}(\vec{k}) = 2^j \, \overline{\widehat{\psi}^{\ell}(2^j \vec{k})} \widehat{f}(\vec{k}).$$

Condition (2.113) yields:

$$A|\widehat{f}(\vec{k})|^2 \leqslant \sum_{\ell=1}^{L} \sum_{j \in \mathbb{Z}} 2^{-2j} |d_j^{\ell}(\vec{k})|^2 \leqslant B|\widehat{f}(\vec{k})|^2.$$

Integrating over $\vec{k}$ and using the Plancherel formula gives (2.114). Now taking the Fourier transform on both sides of (2.116) gives

$$\widehat{f}(\vec{k}) = \sum_{\ell=1}^{L} \sum_{j \in \mathbb{Z}} \widehat{f}(\vec{k}) \, \overline{\widehat{\psi}^{\ell}(2^j \vec{k})}\, \widehat{\chi}^{\ell}(2^j \vec{k}).$$

Finally, applying condition (2.115) gives the final result.   □

The norm equivalence represented by (2.113) shows that this family of wavelets behaves exactly like a frame, with frame operator

$$(Ff)(\vec{b}, 2^j, \ell) = 2^{-j} \mathcal{W}^{\ell}(\vec{b}, 2^j).$$

In particular, if the sum standing in the middle of (2.113) is a constant, we can take $A = B$ and we get the analogue of a tight frame.

Among all possible inverses of the frame operator F, the particular choice

$$\widehat{\chi}^\ell(\vec{k}) = \frac{\widehat{\psi}^\ell(\vec{k})}{\sum_\ell \sum_{j \in \mathbb{Z}} |\widehat{\psi}^\ell(2^j \vec{k})|^2}$$

leads to an attractive solution. In this case, indeed, the inverse frame operator given by (2.116) is the pseudo-inverse of F as shown by the following result:

Proposition 2.4.2 *Let F be the normalized frame operator*

$$[Ff](\vec{u}, 2^j, \ell) = 2^{-j} \mathcal{W}^\ell(\vec{u}, 2^j).$$

Then the following left inverse is the pseudo-inverse of F:

$$\widetilde{F} \mathcal{W}^\ell(\vec{u}, 2^j) = \sum_{\ell=1}^{L} \sum_{j \in \mathbb{Z}} 2^{-j} \left(\mathcal{W}^\ell(., 2^j) \star \chi_{2^j}^\ell \right)(\vec{u}). \tag{2.117}$$

Proof. Let $\ell^2(L^2)$ be the Hilbert space of square integrable sequences of functions $\{d_{\ell,j}\}_{\ell=1...L,\, j \in \mathbb{Z}}$,

$$d_{\ell,j} \in L^2 \quad \text{and} \quad \sum_\ell \sum_{j \in \mathbb{Z}} \|d_{\ell,j}\|^2 < +\infty.$$

We are going to show that, $\forall g \in L^2$ and $\forall \widetilde{d} \in \text{Ran}\{F\}^\perp$,

$$\langle g, \widetilde{F}\widetilde{d} \rangle = 0. \tag{2.118}$$

Let $d = Fg$ and $\widetilde{d} \in \text{Ran}\{F\}^\perp$. Using (2.117), we compute

$$\langle g, \widetilde{F}\widetilde{d} \rangle = \sum_\ell \sum_{j \in \mathbb{Z}} \int_{\mathbb{R}^2} d^2\vec{k}\, \overline{\widehat{g}(\vec{k})}\, \widehat{\widetilde{d}_{\ell,j}}(\vec{k}) \frac{\widehat{\psi}^\ell(2^j \vec{k})}{\sum_{\ell'} \sum_{j' \in \mathbb{Z}} |\widehat{\psi}^{\ell'}(2^{j'} \vec{k})|^2}.$$

Now, writing $d_{\ell,j} = Fg$, we also have

$$\overline{\widehat{g}(\vec{k})}\, \widehat{\psi}^\ell(2^j \vec{k}) = \overline{\widehat{d}_{\ell,j}(\vec{k})},$$

and using equation (2.113) we find

$$\frac{1}{B} \langle d, \widetilde{d} \rangle_{\ell^2(L^2)} \leqslant \langle g, \widetilde{F}\widetilde{g} \rangle \leqslant \frac{1}{A} \langle d, \widetilde{d} \rangle_{\ell^2(L^2)}. \tag{2.119}$$

Now, since $d \in \text{Ran}\{F\}$ and $\widetilde{d} \in \text{Ran}\{F\}^\perp$, (2.119) yields the result. $\qquad \square$

Another interesting case arises when we define a new set of wavelets through the relation

$$\widehat{\varphi}^\ell(\vec{k}) = \frac{\widehat{\psi}^\ell(\vec{k})}{\sqrt{\sum_\ell \sum_j |\widehat{\psi}^\ell(2^j \vec{k})|^2}}, \tag{2.120}$$

and use the same set of functions to implement the reconstruction:

$$f(\vec{x}) = \sum_{\ell=1}^{L} \sum_{j\in\mathbb{Z}} 2^{-2j} \left(\mathcal{W}^{\ell}(., 2^{j}) \star \varphi_{2^{j}}^{\ell} \right)(\vec{x}). \tag{2.121}$$

The reader can easily check that the wavelet transform operator is an isometry between L^2 and $\ell^2(L^2)$. The range of this operator is a closed subspace V of $\ell^2(L^2)$ characterized by a reproducing kernel, exactly as for the continuous transform. Indeed, any $\{d_{\ell,j}\} \in V$ satisfies

$$d_{\ell,j}(\vec{x}) = \sum_{\ell',j'} \left(d_{\ell',j'} \star \mathcal{K}_{\ell,\ell'}^{j,j'} \right)(\vec{x}),$$

where the reproducing kernel is given by

$$\mathcal{K}_{\ell,\ell'}^{j,j'}(\vec{x}) = \left(\widetilde{\varphi}_{2^{j}}^{\ell} \star \varphi_{2^{j'}}^{\ell'} \right)(\vec{x}). \tag{2.122}$$

This is easily checked by writing explicitly

$$d_{\ell,j}(\vec{u}) = \left[F F^{-1} d_{\ell',j'} \right](\vec{u}).$$

Taking the Fourier transform with respect to $\vec{u}$ on both sides and using (2.120) yields the result. This property is a sign of the redundancy of the representation, exactly as in the continuous case. Section 2.6 gives an efficient and automatic way for building such dyadic wavelet transforms starting from the continuous wavelet transform. This offers more flexibility in the design of directional dyadic wavelets.

2.5   Comparison with the 2-D discrete wavelet transform

Before analyzing the recent approach to fast algorithms (Section 2.6), we have to sketch briefly the 2-D discrete wavelet transform, in its various forms and generalizations.

    As mentioned in Chapter 1, a key step in the success of the 1-D *discrete* WT was the discovery that almost all examples of orthonormal bases of wavelets may be derived from a multiresolution analysis, and furthermore that the whole construction may be translated into the language of digital filters. In the 2-D case, the situation is exactly the same, as we shall see in this section. Further information may be found in [Dau92] or [Mey94].

2.5.1   Multiresolution analysis in 2-D and the 2-D DWT

The simplest approach consists in building a 2-D multiresolution analysis simply by taking the direct (tensor) product of two such structures in 1-D, one for the x direction, one for the y direction. If $\{V_j, j \in \mathbb{Z}\}$ is a multiresolution analysis of $L^2(\mathbb{R})$, then $\{^{(2)}V_j = V_j \otimes V_j, j \in \mathbb{Z}\}$ is a multiresolution analysis of $L^2(\mathbb{R}^2)$. Writing again

$^{(2)}V_j \oplus {}^{(2)}W_j = {}^{(2)}V_{j+1}$, it is easy to see that this 2-D analysis requires one scaling function: $\Phi(x, y) = \phi(x)\phi(y)$, but three wavelets:

$$\Psi^h(x, y) = \phi(x)\psi(y)$$
$$\Psi^v(x, y) = \psi(x)\phi(y) \qquad (2.123)$$
$$\Psi^d(x, y) = \psi(x)\psi(y).$$

As the notation suggests, Ψ^h detects preferentially horizontal edges, that is, discontinuities in the vertical direction, whereas Ψ^v and Ψ^d detect vertical and oblique edges, respectively. Indeed, for $j = 1$, the relation $V_1 = V_0 \oplus W_0$ yields:

$$
\begin{aligned}
{}^{(2)}V_1 &= V_1^{(x)} \otimes V_1^{(y)} \\
&= (V_0^{(x)} \oplus W_0^{(x)}) \otimes (V_0^{(y)} \oplus W_0^{(y)}) \\
&= (V_0^{(x)} \otimes V_0^{(y)}) \oplus (V_0^{(x)} \otimes W_0^{(y)}) \oplus (W_0^{(x)} \otimes V_0^{(y)}) \oplus (W_0^{(x)} \otimes W_0^{(y)}) \\
&= {}^{(2)}V_0 \oplus {}^{(2)}W_0,
\end{aligned}
$$

where $^{(2)}V_0 = V_0^{(x)} \otimes V_0^{(y)} \ni \phi(x)\phi(y)$ and $^{(2)}W_0$ is the direct sum of the three other products, generated by the three wavelets given in (2.123), respectively.

From these three wavelets, one gets an orthonormal basis of $^{(2)}V_j$ by defining $\{\Phi_{kl}^j(x, y) = \phi_{j,k}(x)\phi_{j,l}(y), k, l \in \mathbb{Z}\}$, and one for $^{(2)}W_j$ in the same way, namely $\{\Psi_{kl}^{\alpha,j}(x, y), \alpha = h, v, d \text{ and } k, l \in \mathbb{Z}\}$. Clearly this construction enforces a Cartesian geometry, with the horizontal and the vertical directions playing a preferential role. This is natural for certain types of images, such as in television, but is poorly adapted for detecting edges in arbitrary directions. Other solutions are possible, however (see below).

As in the 1-D case, the implementation of this construction rests on a pyramidal algorithm introduced by Mallat [259,260]. The technique consists in translating the multiresolution structure into the language of QMFs, and putting suitable constraints on the filter coefficients. For instance, ψ has compact support if only finitely many coefficients differ from zero. In the 2-D case, obviously one gets a low-pass filter h and three high-pass filters g^α, $\alpha = h, v, d$. Thus, a signal $f \in L^2(\mathbb{R}^2)$ is represented at resolution 2^j by the function $f_j = a_j + \sum_\alpha d_j^\alpha$, $\alpha = h, v, d$, where

$$c_j = \sum_{k,l \in \mathbb{Z}} c_{j,kl}\, \Phi_{kl}^j \in {}^{(2)}V_j$$
$$d_j^\alpha = \sum_{k,l \in \mathbb{Z}} d_{j,kl}^\alpha\, \Psi_{kl}^{\alpha,j} \in {}^{(2)}W_j^\alpha, \ \alpha = h, v, d.$$

In this scheme, the decomposition of an image into a low resolution approximation, plus three types of details (h, v, d), at successive finer scales, takes the familiar form of nested boxes, with the low resolution part in the upper left corner. Figure 2.4 presents a schematic three-level decomposition of an image into a low resolution approximation, with coefficients c_{-3}, plus increasingly finer details, of the three types, with coefficients

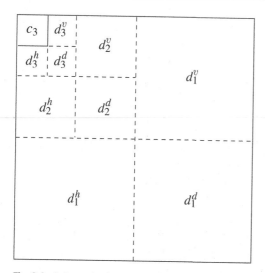

Fig. 2.4. Schematic three-level decomposition of an image into a low resolution approximation, plus increasingly finer details, of the three types (h, v, d).

d_j^α, $j = 1, 2, 3$, $\alpha = h, v, d$. Then we show a (standard) real example in Figure 2.5, which is a three-level orthonormal basis decomposition, with a Daubechies wavelet (compact support, three vanishing moments).

2.5.2   Generalizations

As in one dimension, the scheme based on orthonormal wavelet bases is too rigid for most applications and various generalizations have been proposed. We discuss some of them here, for two reasons. First, for the sake of completeness. But also in order to demonstrate that some features which are natural in the continuous transform, such as covariance, are not easy to enforce in the discrete case, however desirable they may be.

2.5.2.1   More isotropic 2-D wavelets

The tensor product scheme privileges the horizontal and the vertical directions; more isotropic wavelets may be obtained, either by superposition of wavelets with specific orientation tuning [Mar82], as we did above with the CWT, or by choosing a different way of dilating, using a nondiagonal 2-D dilation matrix, which amounts to dilating by a noninteger factor [Dau92]. Consider, for instance, the following dilation matrices:

$$D_0 = \begin{pmatrix} 2 & 0 \\ 0 & 2 \end{pmatrix}, \quad D_1 = \begin{pmatrix} 1 & 1 \\ 1 & -1 \end{pmatrix}, \quad D_2 = \begin{pmatrix} 1 & 1 \\ -1 & 1 \end{pmatrix}. \tag{2.124}$$

The matrix D_0 corresponds to the usual dilation scheme by powers of 2, whereas D_1 and D_2 lead to the so-called "quincunx" scheme [Fea90]. In the standard scheme, a unit square is dilated, in the transition $j \to j + 1$, to another square, twice bigger, with the same orientation. This means that three kinds of additional details have to be supplied,

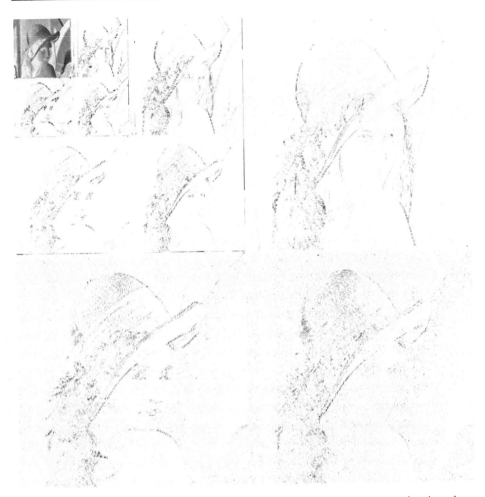

Fig. 2.5. Typical three-level decomposition of an image into a low resolution approximation, plus increasingly finer details, of the three types (h, v, d). The basic wavelet is a Daubechies wavelet with compact support and three vanishing moments.

horizontal, vertical and oblique (see Figure 2.6, left). By contrast, the same operation in the "quincunx" scheme leads to a square circumscribed to the original one, that is, rotated by 45° and larger by a factor $\sqrt{2}$, so that only one kind of additional detail is necessary (Figure 2.6, right). Indeed only one wavelet is needed in this scheme, instead of three. This is consistent with a result of Meyer, according to which the number of independent wavelets needed in a given multiresolution scheme equals $(|\det D| - 1)$, where D is the dilation matrix used.

2.5.2.2 Biorthogonal wavelet bases

In the case of the continuous transform, the wavelet used for reconstruction need not be the same as that used for decomposition, they have only to satisfy a cross-compatibility

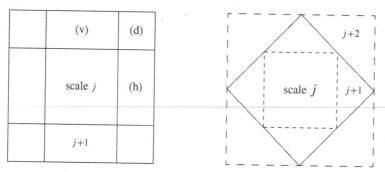

Fig. 2.6. Unit cell at successive resolutions: (left) for the "Cartesian" scheme; (right) for the "quincunx" scheme.

condition, as shown in (2.30), and this has led us to a host of different, more flexible reconstruction formulas, such as (2.31)–(2.34). The same idea in the discrete case leads to biorthogonal bases [108], i.e., one has two hierarchies of approximation spaces, $\{V_j\}$ and $\{\widetilde{V}_j\}$, with cross-orthogonality relations.

In 1-D, the construction goes as follows, and the extension to 2-D proceeds as above. Start with a scale of closed subspaces $\{V_j\}$, assuming only the existence of a scaling function $\phi \in V_0$ such that its integer translates $\{\phi_k(x) \equiv \phi(x - k), k \in \mathbb{Z}\}$ is a Riesz (or unconditional) basis of V_0 (see Section 1.5). Then, instead of orthogonalizing this basis, which would lead to the construction of an orthonormal wavelet basis, one takes the *dual* basis $\{\widetilde{\phi}_k\}$, that is, the vectors defined by the relation $\langle \phi_k | \widetilde{\phi}_l \rangle = \delta_{kl}$. Let $\widetilde{V}_0$ denote the closed subspace generated by $\{\widetilde{\phi}_k, k \in \mathbb{Z}\}$. Then the same construction is repeated for each j, using the dilation invariance of the scale $\{V_j\}$. The outcome is a second multiresolution scale $\{\widetilde{V}_j\}$, with exactly the same properties. Next, for each $j \in \mathbb{Z}$, one defines a subspace W_j by the two conditions $W_j \subset V_{j+1}$ and $W_j \perp \widetilde{V}_j$, and similarly $\widetilde{W}_j \subset \widetilde{V}_{j+1}$ and $\widetilde{W}_j \perp V_j$. In this way one obtains two sequences of subspaces $\{W_j\}$ and $\{\widetilde{W}_j\}$, with bases $\{\psi_{j,k}, j, k \in \mathbb{Z}\}$, $\{\widetilde{\psi}_{j,k}, j, k \in \mathbb{Z}\}$, respectively, which are mutually orthogonal:

$$\langle \psi_{j,k} | \widetilde{\psi}_{j',k'} \rangle = \delta_{jj'} \delta_{kk'}. \tag{2.125}$$

In terms of these bases, one gets two types of expansion formulas, for any $f \in L^2(\mathbb{R})$:

$$f = \sum_{j,k \in \mathbb{Z}} \langle \widetilde{\psi}_{j,k} | f \rangle \, \psi_{j,k}$$

$$= \sum_{j,k \in \mathbb{Z}} \langle \psi_{j,k} | f \rangle \, \widetilde{\psi}_{j,k} \tag{2.126}$$

in the case of a frame and its dual.

Translating the whole structure into the language of digital filters, a 1-D biorthogonal scheme corresponds to four filters, $(h, \widetilde{h}, g, \widetilde{g})$, where the first two are low-pass and the last two are high-pass. A corresponding characterization may be given in the 2-D case.

The resulting scheme is much more flexible and is probably the most efficient one in practical applications, both in 1-D and in 2-D (and it is widely used). For instance, it gives a better control on the regularity or decrease properties of the wavelets [108].

2.5.2.3 Wavelet packets and the Best Basis Algorithm

Electrical engineers are familiar with the notion of *subband coding scheme*. In a few words, this means that a discrete signal $\{c_n^0,\ n \in \mathbb{Z}\}$ is first subdivided into two subsignals $\{c_n^1, d_n^1,\ n \in \mathbb{Z}\}$, obtained by convolving the original signal with two filters, one low-pass and one high-pass filter, respectively. Next, each subsignal is subsampled by a factor of 2, that is, one keeps only the even or odd components, respectively (thus the total number of coefficients is unchanged). Then the operation is iterated a number of times. A reconstruction of the original signal is obtained, more or less exactly, by inverting all the operations.

In the construction of 1-D orthonormal wavelet bases, each approximation space V_j gets further decomposed into V_{j-1} and W_{j-1}, whereas the detail space W_j is left unmodified. Expanding the signals in the respective bases $\{\phi_{j-1,k}, \psi_{j-1,k}\}$ shows that the construction is in fact a subband coding scheme, but a very special one, rather asymmetrical. In order to get more flexibility, more general subband schemes have been considered, called *wavelet packets*, where both subspaces V_j and W_j are decomposed at each step [Mey94,Wic94,110,111]. Such a scheme provides rich collections ("libraries") of orthonormal bases, each one corresponding to a given decomposition of $L^2(\mathbb{R})$ into a sequence of mutually orthogonal subspaces, chosen at successive scales $j \in \mathbb{Z}$ (an example is given below). Enumerating all possible such decompositions or orthonormal bases is a nontrivial combinatorial problem, whose solution stems from graph theory [Wic94,111]. Facing such a plethora of orthonormal bases, one needs a strategy for determining the best basis to use in a given situation. An efficient solution, based on entropic criteria, has been proposed by Coifman *et al.* [Wic94,111], under the name of the Best Basis Algorithm, and it has become a standard tool in signal analysis.

Of course, all this applies verbatim in two dimensions, although the labeling of basis vectors becomes even more intricate. This holds, in particular, for the comments made at the end of Section 1.5 concerning the numerical implementation of finite reconstruction formulas.

In order to fix ideas, we show in Figure 2.7 the subband subdivision scheme of the standard wavelet 1-D multiresolution analysis, in the case of a three-level decomposition. This corresponds to the following decomposition into orthogonal subspaces:

$$V_0 = W_{-1} \oplus W_{-2} \oplus W_{-3} \oplus V_{-3},$$

or, in the notation of Section 1.5 [compare (1.58)], the representation of a signal $s \equiv s_0 \in V_0$ by its wavelet coefficients $s_0 = (d_{-1}, d_{-2}, d_{-3}, c_{-3})$. By comparison, Figure 2.8 shows the modified subdivision scheme used in the wavelet packets formalism, together

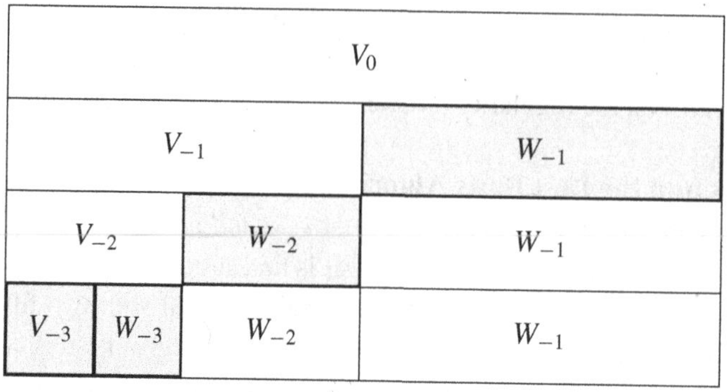

Fig. 2.7. The 1-D wavelet subband scheme, with a three-level decomposition.

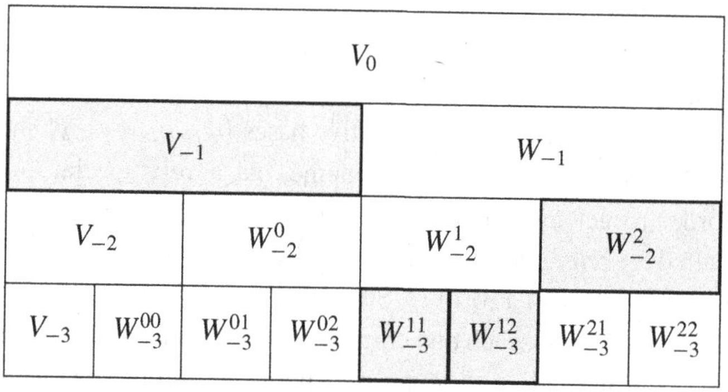

Fig. 2.8. The 1-D wavelet packet subband scheme, with a particular choice of three-level decomposition.

with a particular choice of three-level decomposition, namely [compare Figure 1.11 (b) and (c)]:

$$V_0 = V_{-1} \oplus W^2_{-2} \oplus W^{11}_{-3} \oplus W^{12}_{-3}.$$

In the corresponding orthonormal basis, the signal is represented as $s \equiv s_0 = (c_{-1}, dd_{-2}, ccd_{-3}, dcd_{-3})$. The notation proceeds as follows. For $j = -1$, the coefficients are (c_{-1}, d_{-1}). Then at each step, the coefficient x_j is replaced by the pair (cx_{j-1}, dx_{j-1}). Thus, for $j = -2$, one has, from left to right in Figure 2.7, $(cc_{-2}, dc_{-2}, cd_{-2}, dd_{-2})$; for $j = -3$, $(ccc_{-3}, dcc_{-3}, cdc_{-3}, ddc_{-3}, ccd_{-3}, dcd_{-3}, cdd_{-3}, ddd_{-3})$; and so on. For more details on the basis labeling, and wavelet packets in general, we refer to [Wic94] or [110,111].

2.5.2.4   The lifting scheme: second-generation wavelets

One can go one step beyond, and abandon the regular dyadic scheme and the Fourier transform altogether. Using the "lifting scheme" leads to the so-called *second-generation wavelets* [349,350], which are essentially custom-designed for any given problem. The starting point is that, in a biorthogonal scheme, one scale $\{V_j\}$ does not determine its counterpart $\{\widetilde{V}_j\}$ uniquely, but the freedom left in the generating wavelet is known explicitly, and reduces to an arbitrary trigonometric polynomial. Thus the idea is to start from a given biorthogonal scheme with filters $(h, \widetilde{h}, g, \widetilde{g})$, then tranform it using that freedom into a new one $(h^{(1)}, \widetilde{h}^{(1)}, g^{(1)}, \widetilde{g}^{(1)})$, and so on, by a succession of "lifting steps."

But one must generalize first the very notion of biorthogonal scheme, in order to get a more flexible scheme. To that effect, one weakens Definition 1.5.1 of a multiresolution analysis $\{V_j, j \in \mathbb{Z}\}$ by replacing the two conditions (1) and (2) (which enforce the scale invariance and thus the dyadic scheme) by the single condition (3) for each $j \in \mathbb{Z}$, V_j has a Riesz basis $\{\varphi_{j,k}, k \in \mathcal{K}(j)\}$, the elements of which are called *scaling functions*.

Here $\mathcal{K}(j)$ is a general index set, which allows irregular sampling (no translation invariance!). One assumes only that $\mathcal{K}(j) \subset \mathcal{K}(j+1)$, without the dilation relation given by condition (1). In the same way, one considers a dual scale $\{\widetilde{V}_j\}$, with dual scaling functions $\{\widetilde{\varphi}_{j,k}, k \in \mathcal{K}(j)\}$, biorthogonal to the previous ones:

$$\langle \varphi_{j,k} | \widetilde{\varphi}_{j,k'} \rangle = \delta_{kk'}, \ k, k' \in \mathcal{K}(j). \tag{2.127}$$

Then the filter $h \equiv h_{j,k,l}$ enters through a refinement equation

$$\varphi_{j,k} = \sum_{l \in \mathcal{K}(j+1)} h_{j,k,l} \, \varphi_{j+1,l}, \tag{2.128}$$

and similarly for $\widetilde{h} \equiv \widetilde{h}_{j,k,l}$ (all filters are assumed to be finite). The two filters $h, \widetilde{h}$ are then biorthogonal (see (2.131) below).

Next one defines wavelets in the usual way. A family of functions $\{\psi_{j,m}, m \in \mathcal{M}(j)\}$, where $\mathcal{M}(j) = \mathcal{K}(j+1) \setminus \mathcal{K}(j)$, is a set of wavelet functions if: (i) the space $W_j = \overline{\text{span}} \{\psi_{j,m}, m \in \mathcal{M}(j)\}$ is a complement of V_j in V_{j+1}, and $W_j \perp \widetilde{V}_j$; (ii) the set $\{\psi_{j,m}/\|\psi_{j,m}\|, j \in \mathbb{Z}, m \in \mathcal{M}(j), \}$ is a Riesz basis for $L^2(\mathbb{R})$. Dual wavelets are vectors of the biorthogonal basis,

$$\langle \psi_{j,m} | \widetilde{\psi}_{j',m'} \rangle = \delta_{jj'} \delta_{mm'}. \tag{2.129}$$

They span spaces $\widetilde{W}_j$, which complement $\widetilde{V}_j$ in $\widetilde{V}_{j+1}$, and $\widetilde{W}_j \perp V_j$. By construction, the wavelets satisfy refinement relations:

$$\psi_{j,m} = \sum_{l \in \mathcal{K}(j+1)} g_{j,m,l} \, \varphi_{j+1,l}, \tag{2.130}$$

which thus define the filter $g \equiv \{g_{j,m,l}\}$, and similarly for the dual filter $\widetilde{g}$.

Altogether, the four filters $h, \widetilde{h}, g, \widetilde{g}$ satisfy biorthogonality relations:

$$\sum_l g_{j,m,l} \, \widetilde{g}_{j,m',l} = \delta_{mm'}, \qquad \sum_l h_{j,k,l} \, \widetilde{g}_{j,m,l} = 0$$

$$\sum_l h_{j,k,l} \, \widetilde{h}_{j,k',l} = \delta_{kk'}, \qquad \sum_l g_{j,m,l} \, \widetilde{h}_{j,k,l} = 0. \tag{2.131}$$

Similar relations may be written for scaling and wavelet functions. Now, of course, since the index sets $\mathcal{K}(j)$ are general, some extra care is required to guarantee the convergence of all the expansions (hence the finiteness condition on the filters), and also the rapidity of the algorithm. We refer to [350] for technical details.

This scheme becomes simpler if one introduces an operator notation (familiar in the signal processing literature), as follows.

- The filter $h_{j,k,l}$ is embodied in the operator $H_j : \ell^2(\mathcal{K}(j+1)) \to \ell^2(\mathcal{K}(j))$, defined by $b = H_j a$, where $a \equiv (a_l) \in \ell^2(\mathcal{K}(j+1)), b \equiv (b_k) \in \ell^2(\mathcal{K}(j))$, and

$$b_k = \sum_{l \in \mathcal{K}(j+1)} h_{j,k,l} \, a_l.$$

- The filter $g_{j,m,l}$ is embodied in the operator $G_j : \ell^2(\mathcal{K}(j+1)) \to \ell^2(\mathcal{M}(j))$, and similarly for the operators $\widetilde{H}_j, \widetilde{G}_j$.

In this notation, the conditions for exact reconstruction can be written in matrix form as:

$$\begin{pmatrix} \widetilde{H}_j \\ \widetilde{G}_j \end{pmatrix} \begin{pmatrix} H_j^* & G_j^* \end{pmatrix} = \begin{pmatrix} 1 & 0 \\ 0 & 1 \end{pmatrix} \quad \text{and} \quad \begin{pmatrix} H_j^* & G_j^* \end{pmatrix} \begin{pmatrix} \widetilde{H}_j \\ \widetilde{G}_j \end{pmatrix} = 1.$$

$$(2.132)$$

Now we may describe the lifting scheme. As already mentioned, the idea is to exploit the freedom in designing a set of filters $\widetilde{H}_j, \widetilde{G}_j$ biorthogonal to a given one H_j, G_j. The freedom is that of an arbitrary operator $S_j : \ell^2(\mathcal{M}(j)) \to \ell^2(\mathcal{K}(j))$. In components, this operator is represented by a set of coefficients, $S_j \equiv s_{j,k,m}$. In the usual, first generation wavelet scheme, this in turn would be given by a trigonometric polynomial $s(\omega)$.

The technique proceeds in two steps.

(i) First a *lifting* step, which consist in passing from a given biorthogonal filter set $\{H_j, \widetilde{H}_j, G_j, \widetilde{G}_j\}$ to a new one, $\{H_j, \widetilde{H}_j^{(1)}, G_j^{(1)}, \widetilde{G}_j\}$, where

$$\widetilde{H}_j^{(1)} = \widetilde{H}_j + S_j \widetilde{G}_j, \qquad G_j^{(1)} = G_j - S_j^* H_j,$$

the two filters $H_j, \widetilde{G}_j$ being unchanged. Hence, the original scaling functions $\varphi_{j,k}$ do not change, but all the other functions $\widetilde{\varphi}_{j,k}, \psi_{j,m}, \widetilde{\psi}_{j,m}$ are modified.

(ii) A *dual lifting* step, leading from the set $\{H_j, \widetilde{H}_j^{(1)}, G_j^{(1)}, \widetilde{G}_j,\}$ to the set $\{H_j^{(1)}, \widetilde{H}_j^{(1)}, G_j^{(1)}, \widetilde{G}_j^{(1)},\}$, where

$$H_j^{(1)} = H_j + \widetilde{S}_j G_j^{(1)}, \qquad \widetilde{G}_j^{(1)} = \widetilde{G}_j - \widetilde{S}_j^* \widetilde{H}_j^{(1)}.$$

and $\widetilde{S}_j : \ell^2(\mathcal{M}(j)) \to \ell^2(\mathcal{K}(j))$ is another operator. Here the two filters $\widetilde{H}_j^{(1)}, G_j^{(1)}$ remain unchanged.

Each of the two steps preserves the biorthogonality of the filter sets, as can be checked easily on the conditions (2.132). Of course, one has to verify along the way that the new scaling and wavelet functions belong correctly to the appropriate spaces (in particular, that they are square integrable).

Then it may be shown that *any* biorthogonal filter set may be obtained by this procedure after a finite number of steps, starting from a trivial set, called the *Lazy wavelet*, because it does nothing but split the sequences into two subsets of components. More precisely, the Lazy wavelet corresponds to the filter set $H_j = \widetilde{H}_j = E$, $G_j = \widetilde{G}_j = D$, where $E : \ell^2(\mathcal{K}(j+1)) \to \ell^2(\mathcal{K}(j))$ and $D : \ell^2(\mathcal{K}(j+1)) \to \ell^2(\mathcal{M}(j))$ are simply the restriction or subsampling operators. [In the engineering literature, the lazy wavelet is also called a polyphase filter (of size 2) [Vet95].]

The resulting scheme is fast, and independent from translation invariance and from the Fourier transform. Thus it applies to wavelets on intervals or on curves, and in higher dimensions as well, for instance, to wavelets on two-dimensional manifolds [Mal99,349,350]. Here the idea is to start from a succession of finer and finer grids. Consider a 2-D manifold (such as the 2-sphere) and choose a certain grid $\mathcal{G}(j)$ on it. Then refine the latter to a grid $\mathcal{G}(j+1)$, of the form

$$\mathcal{G}(j+1) = \mathcal{G}(j) \cup \mathcal{C}(j),$$

where $\mathcal{C}(j)$ denotes the complement. A typical example is to start from a triangulation of the manifold and refine it by bisecting each side, as illustrated in Figure 2.9 in the case of the sphere. Then the multiresolution spaces are defined as $V_{j+1} = \ell^2(\mathcal{G}(j+1))$, $V_j = \ell^2(\mathcal{G}(j))$, $W_j = \ell^2(\mathcal{C}(j))$, with appropriate bases $\{\varphi_{j,k}\} \in V_j$, $\{\psi_{j,m}\} \in W_j$, and the whole machinery is put into operation. The resulting tool has proven to be extremely versatile and efficient. For instance, Schröder and Sweldens [336] have applied it to the design of wavelets on the sphere, with a very convincing application to the reproduction of coastlines on a terrestrial globe (we will see in Chapter 9, Section 9.2, another approach to the same problem, this one directly based on the CWT).

As a final remark, we may point out that the lifting scheme opens the door to nonlinear multiresolution decompositions, such as the median transform of Bijaoui [Sta98] or the

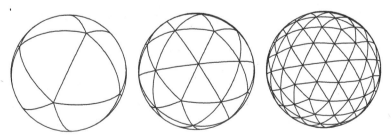

Fig. 2.9. Typical grid refining for applying the 2-D lifting scheme: the geodesic sphere construction, starting with the icosahedron on the left (subdivision level 0) and the next two subdivision levels (from [336]).

morphological wavelets introduced by Goutsias and Heijmans [198,219]. The latter are explicitly based on the lifting scheme and make the connection with the standard field of mathematical morphology. We note finally that the lifting scheme by itself offers a very pedagogical entrty into the wavelet world, as examplified in the little volume of Jensen and la Cour-Harbo [Jen01].

2.5.2.5 Integer wavelet transforms

In their standard numerical implementation, the classical (discrete) WT converts floating point numbers into floating point numbers. However, in many applications (data transmision from satellites, multimedia), the input data consists of integer values only and one cannot afford to lose information: only lossless compression schemes are allowed. Recent developments, based on the lifting scheme, have produced new methods that allow one to perform all calculations in integer arithmetic [2,92]. In addition, such methods also improve the performances of lossy compression techniques [324].

2.6   Bridging the gap: continuous wavelet packets and fast algorithms

2.6.1   Custom design of dyadic frames

Besides the full discretization described in Section 1.3, and the discrete WT just discussed, there is an intermediate procedure, introduced in [159], under the name of *infinitesimal multiresolution analysis*. It consists in discretizing the scale variable alone, on an arbitrary sequence of values (not necessarily powers of a fixed ratio). This leads to fast algorithms that could put the CWT on the same footing as the DWT in terms of speed and efficiency, by extending the advantages of the latter to cases where no exact QMF is available. We describe the method in 2-D, the 1-D case (already sketched in Section 1.6.1) being easily derived on this basis. Interested readers should refer to [Tor95] for further details.

Instead of the standard L^2-normalization used in (2.13), it is more convenient to choose the L^1-normalization and use $\psi_{(\vec{b},a)} = a^{-2}\psi(a^{-1}(\vec{x} - \vec{b}))$. Note that, for simplicity, we consider here only isotropic wavelets, but the extension to the general case is straightforward (see Section 2.6.3).

Let us start with the L^1-reconstruction formula associated to the CWT in two dimensions:

$$f(\vec{x}) = \int_0^\infty \frac{da}{a} \int_{\mathbb{R}^2} d^2\vec{b} \; \check{S}(\vec{b}, a) \, \psi_{(\vec{b},a)}(\vec{x}). \tag{2.133}$$

The basic idea behind the proposed construction is now to segment the integral over scales in (2.133) and replace it by a sum over dyadic intervals. This is done first by rewriting the reconstruction formula as

$$f(\vec{x}) = \int_0^\infty \frac{da}{a} \, d_a(\vec{x}),$$

where we have defined the *infinitesimal detail*

$$d_a(\vec{x}) = \int_{\mathbb{R}^2} d^2\vec{b} \; \check{S}(\vec{b}, a) \, \psi_{(\vec{b},a)}(\vec{x}).$$

By virtue of Young's convolution inequality, $d_a \in L^2$ and, taking its Fourier transform, we obtain

$$\widehat{d_a}(\vec{k}) = |\widehat{\psi}(a\vec{k})|^2 \, \widehat{f}(\vec{k}). \qquad (2.134)$$

These equations show that d_a represents the amount of information captured by the wavelet between scales a and $a + da$, hence the name "infinitesimal details." Summing all these details, that is, integrating over the scale variable, reproduces the original signal. In the same vein, we can synthesize a low resolution approximation of f by integrating up to a given resolution, say a_0:

$$f_{a_0}(\vec{x}) = \int_{a_0}^{\infty} \frac{da}{a} \, d_a(\vec{x}).$$

Taking Fourier transforms on both sides suggests to introduce the following Fourier multiplier:

$$|\widehat{\phi}(\vec{k})|^2 = \int_{1}^{\infty} \frac{da}{a} \, |\widehat{\psi}(a\vec{k})|^2. \qquad (2.135)$$

It is then shown in [159] that the approximation f_a can be written

$$f_a(\vec{x}) = \int_{\mathbb{R}^2} d^2\vec{b} \; \langle \phi_{(\vec{b},a)} | f \rangle \, \phi_{(\vec{b},a)}(\vec{x}), \qquad (2.136)$$

and that the following limit holds in the strong sense in L^2:

$$\lim_{a \to 0} f_a = f. \qquad (2.137)$$

Thus, following [Tor95], we speak of the *bilinear formalism*. Remark also that (2.135) implies

$$\lim_{|\vec{k}| \to \infty} |\widehat{\phi}(\vec{k})|^2 = 0$$

and thus defines a scaling function.

Now, starting from an approximation of f at scale $a_0 = 2^{-j_0}$, we can refine up to an arbitrary resolution by adding up details. For this purpose, we introduce slices of details

$$D_j(\vec{x}) = \int_{2^{-(j+1)}}^{2^{-j}} \frac{da}{a} \, d_a(\vec{x}).$$

Taking Fourier transforms on both sides, we have

$$\widehat{D_j}(\vec{k}) = \widehat{f}(\vec{k}) \int_{2^{-(j+1)}}^{2^{-j}} \frac{da}{a} \, |\widehat{\psi}(a\vec{k})|^2$$

and this leads us to define the *integrated wavelet packets*:

$$|\widehat{\Gamma}(\vec{k})|^2 = \int_{1/2}^{1} \frac{da}{a} \, |\widehat{\psi}(a\vec{k})|^2 .$$ (2.138)

This function satisfies a two-scale relation, the analog of (1.62):

$$|\widehat{\Gamma}(\vec{k})|^2 = |\widehat{\phi}(\tfrac{1}{2}\vec{k})|^2 - |\widehat{\phi}(\vec{k})|^2.$$ (2.139)

Finally, putting equations (2.136) and (2.138) together, we obtain the following dyadic decomposition:

$$f = \int_{\mathbb{R}^2} d^2\vec{b} \, \langle \phi_{(\vec{b},2^{-j_o})} | f \rangle \, \phi_{(\vec{b},2^{-j_o})} + \sum_{j=j_o}^{\infty} \int_{\mathbb{R}^2} d^2\vec{b} \, \langle \Gamma_{(\vec{b},2^{-j})} | f \rangle \, \Gamma_{(\vec{b},2^{-j})},$$ (2.140)

which holds in L^2 norm. In order to simplify our notations, we will write the wavelet coefficients of f as $\mathcal{W}_f(\vec{b}, 2^{-j}) = \langle \Gamma_{(\vec{b},2^{-j})} | f \rangle$ and introduce the approximation coefficients $\mathcal{S}_f(\vec{b}, 2^{-j}) = \langle \phi_{(\vec{b},2^{-j})} | f \rangle$. Equation (2.140) now reads

$$f(\vec{x}) = \langle \mathcal{S}_f(\cdot, 2^{-j_o}) | \phi_{(\cdot,2^{-j_o})} \rangle(\vec{x}) + \sum_{j=j_o}^{\infty} \langle \mathcal{W}_f(\cdot, 2^{-j}) | \Gamma_{(\cdot,2^{-j})} \rangle(\vec{x})$$ (2.141)

(scalar product over $\vec{b}$ in each term). We have thus built a dyadic wavelet transform starting from the CWT. It is important to realize that the scaling function ϕ and the integrated wavelet packet Γ inherit the localization and smoothness properties of ψ. In view of the considerable freedom we have in the choice of ψ, we are now able to easily design custom, translation invariant dyadic frames.

Remark: More flexibility is obtained if one subdivides the scale interval $[1/2, 1]$ into n subbands, by $a_0 = 1/2 < a_1 < \ldots < a_n = 1$. In that case, one ends with one scaling function $\Phi(\vec{x})$ and n integrated wavelets $\Psi_i(\vec{x})$, $i = 0, \ldots n - 1$, corresponding to integration from a_{i-1} to a_i. This more efficient version allows one to compute explicitly the characteristics of the wavelet packet, such as its central frequency, its standard deviation, etc., following (1.14)–(1.15), and these in turn may be expressed in terms of the corresponding quantities of the mother wavelet ψ. An application to sound analysis is given in [233].

A simpler decomposition formula, called the *linear scheme* in [Tor95], arises when one starts from the so-called Morlet reconstruction formula (2.34) (this is the scheme we have sketched in the 1-D case in Section 1.5).

The same reasoning as before leads us to introduce, as in (1.61), a scaling function

$$\widehat{\phi}(\vec{k}) = \int_{1}^{\infty} \frac{da}{a} \, \widehat{\psi}(a\vec{k}),$$ (2.142)

and integrated wavelet packets

$$\widehat{\Gamma}(\vec{k}) = \int_{1/2}^{1} \frac{da}{a} \, \widehat{\psi}(a\vec{k}) \tag{2.143}$$

$$= \widehat{\phi}(\tfrac{1}{2}\vec{k}) - \widehat{\phi}(\vec{k}). \tag{2.144}$$

One notices that these wavelets are simply expressed as a difference of smoothing functions, as in 1-D, (1.62). In this case, the reconstruction formula is much simpler, since it involves a straight sum of approximation and wavelet coefficients:

$$f(\vec{x}) = \mathcal{S}_f(\vec{x}, 2^{-j_o}) + \sum_{j=j_o}^{\infty} \mathcal{W}_f(\vec{x}, 2^{-j}). \tag{2.145}$$

One of the main advantages of this construction is that it allows to build wavelets and scaling functions that have fast decay both in the spatial and frequency domains. This is very useful in applications where one wants to use wavelets that have sharp prescribed localization properties in the Fourier domain and are also of fast decay in the spatial domain, as it is the case with Gabor functions. This is very difficult to achieve in practice. For example, if one wants to use spline-based wavelet frames, it appears that, although the spatial localization is very good, splines are not sharply localized in Fourier variables (they have an algebraic decay, see [14] for a review) and can even show disturbing sidelobes. A concrete example is given by texture analysis where the latter are distinguished on the basis of the statistics of frequency subbands as measured using Gabor wavelets [234]. If one wants to use a dyadic frame, special care has to be given to the frequency localization of the wavelets and this can be easily done using the technique described above.

2.6.2 Example of a typical design

We will now apply the previous formalism to a concrete example. Our aim is to construct isotropic dyadic wavelets and scaling functions belonging to the class of C^{∞} functions with fast decay. Our starting point is a family of wavelets associated to pseudo-differential operators defined by multiplication by $k^n = \left(k_1^2 + k_2^2\right)^{n/2}$ in the Fourier domain:

$$\widehat{\psi}(\vec{k}) = k^n e^{-k^2}. \tag{2.146}$$

The associated scaling functions are computed using (2.135):

$$\widehat{\phi}_n(k) = \int_k^{\infty} \frac{da}{a} \, a^n e^{-a^2}, \quad (n \geqslant 2)$$

and it satisfies the following recurrence relation:

$$\widehat{\phi}_n(k) = \frac{1}{2} k^{n-2} e^{-k^2} + \frac{(n-2)}{2} \widehat{\phi}_{n-2}(k),$$

with

$$\widehat{\phi}_2(k) = \frac{1}{2}e^{-k^2} , \quad \widehat{\phi}_3(k) = \frac{1}{2}ke^{-k^2} + \sqrt{\frac{\pi}{4}}\,\mathrm{erfc}(k).$$

Here the error function is defined by

$$\mathrm{erfc}(k) = \frac{4}{\sqrt{\pi}}\int_k^\infty dk\, e^{-k^2}.$$

Normalizing this family of functions by $\int_{\mathbb{R}} d^2\vec{x}\,\phi_n(\vec{x}) = 1$ leads us to define

$$\widehat{\phi}_n(k) = \alpha_n^{-1}\widehat{\phi}(k),$$

where the constant α_n is defined by

$$\alpha_n = \widehat{\phi}_2(0)\prod_{i=1}^{\frac{n-2}{2}} \frac{n - 2i}{2} , \quad n \text{ even}, \quad n > 2,$$

$$\alpha_n = \widehat{\phi}_3(0)\prod_{i=1}^{\frac{n-3}{2}} \frac{2i + 1}{2} , \quad n \text{ odd}, \quad n > 3.$$

The recursion starts with $\alpha_2 = \widehat{\phi}_2(0)$ and $\alpha_3 = \widehat{\phi}_3(0)$. Scaling functions of the lowest orders are listed in Table 2.1. Using (2.138) or (2.144), we obtain the desired family of isotropic wavelets. An example of such a scaling function of order 4 is given in Figure 2.12, in the next section. Note that the parameter n also controls the number of vanishing moments of the associated wavelet. As an example of this technique, Figure 2.10 shows a three-level decomposition of the *lena* image into isotropic wavelet packets.

2.6.3 Designing directional dyadic frames

The previous construction applied only to isotropic wavelets and yielded frames of isotropic elements. In the following we will add to this scheme sensitivity to local orientation. A simple and straightforward way to achieve this is to start from an isotropic integrated wavelet Γ and segment it into directional ones (a precise definition of this

Table 2.1. *Scaling functions of lowest orders*

Order	Scaling function
$n = 2$	$\widehat{\phi}_2(k) = e^{-k^2}$
$n = 3$	$\widehat{\phi}_3(k) = \frac{1}{2}ke^{-k^2} + \frac{\sqrt{\pi}}{4}\mathrm{erfc}(k)$
$n = 4$	$\widehat{\phi}_4(k) = e^{-k^2}(1 + k^2)$
$n = 5$	$\widehat{\phi}_5(k) = \alpha_5^{-1}\left(\frac{1}{2}k^3e^{-k^2} + \frac{3}{4}ke^{-k^2} + \frac{3\sqrt{\pi}}{8}\mathrm{erfc}(k)\right)$

Fig. 2.10. Three levels of decomposition of the *lena* image using an isotropic wavelet packet frame. The lower right image is the low resolution approximation.

notion will be given in Section 3.3). This can be done by introducing an angular window $\eta(\varphi)$, $\varphi \in [0, 2\pi)$, in the Fourier domain and then defining a new wavelet

$$\widehat{\Psi}(k, \varphi) = \widehat{\Gamma}(k)\eta(\varphi).$$

Note that this construction amounts to work with wavelets that are separable in *polar* coordinates. The choice of the angular window is restricted by the need for an exact, linear or bilinear, reconstruction formula. More precisely, if one makes use of (2.141), η has to satisfy

$$\sum_{\ell=0}^{L-1} \left| \eta\left(\varphi - \frac{2\pi\ell}{L}\right) \right|^2 = 1, \qquad (2.147)$$

while the simpler formula (2.145) requires

$$\sum_{\ell=0}^{L-1} \eta\left(\varphi - \frac{2\pi\ell}{L}\right) = 1, \tag{2.148}$$

where we have assumed L orientations. An additional requirement, further explored in Chapter 3, is that the support of η be strictly less than π in order to have some directional sensitivity. In order to preserve the frequency localization of $\widehat{\Gamma}$, it is also important that η be regular enough. The optimal choice is thus to build a partition of the circle using a suitable compactly supported C^{∞} function.

Altogether, we introduce the L angular windows $\eta_{\ell}(\varphi) \equiv \eta\left(\varphi - \frac{2\pi\ell}{L}\right)$, $\ell = 0, 1, \ldots, L-1$, and the corresponding directional wavelets

$$\widehat{\Psi}^{\ell}(k, \varphi) = \widehat{\Gamma}(k)\,\eta_{\ell}(\varphi), \ \ell = 0, 1, \ldots, L-1. \tag{2.149}$$

An example of a dyadic directional wavelet built using this technique is depicted on Figures 2.11 and 2.12.

2.6.4   Implementation using approximate QMFs

One of the main drawbacks of these oriented frames is that they are not designed to be implemented using a fast pyramidal algorithm. Nevertheless we will now show that one can design special QMF pairs that allow for a very good approximation of the discrete WT and provide a substantial gain in computational speed. This technique is mainly an extension to 2-D of the original work of Muschietti and Torrésani [291].

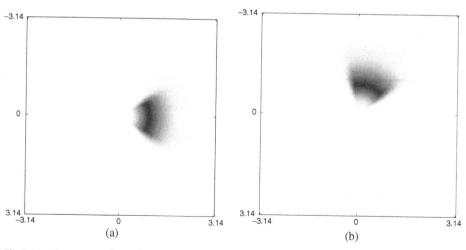

Fig. 2.11. Fourier transform of a directional dyadic wavelet of order 4 and angular resolution of $\pi/5$ for two values of the rotation parameter: (a) $\ell = 0$; and (b) $\ell = 1$.

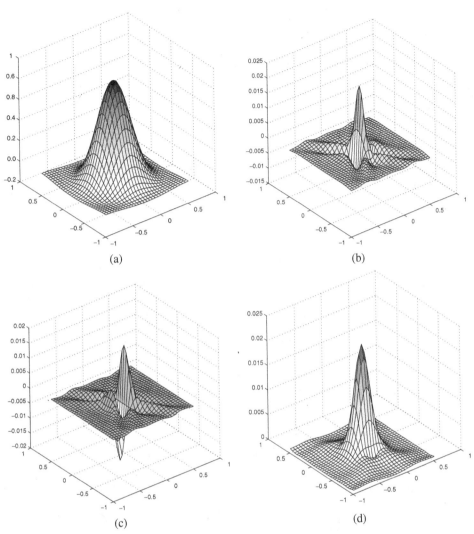

Fig. 2.12. (a) Scaling function of order 4; (b–d) the associated wavelet with angular resolution of $\pi/5$: (b) real part; (c) imaginary part; and (d) modulus.

Since we work with translation invariant frames, we will now sample the position parameter of the DWT over a regular grid, that is, we will consider wavelets and scaling functions indexed by the integer grid (we use here the L^1-normalization):

$$\Psi^\ell_{j;m,n}(x, y) = 2^{2j}\Psi^\ell(2^j(x-m, y-n)), \quad j, m, n \in \mathbb{Z}.$$

We then obtain a *discrete* dyadic wavelet transform (compare Section 2.4.4) by just restricting the DWT to this particular grid:

$$\mathcal{W}^\ell_j f(m, n) = \langle \Psi^\ell_{j;m,n} | f \rangle, \quad \mathcal{S}_j f(m, n) = \langle \phi_{j;m,n} | f \rangle. \tag{2.150}$$

Let us assume there exists 2-D discrete filters h and g^ℓ, $\ell = 0 \ldots L - 1$, such that one can compute these quantities using a pyramidal algorithm [260], that is,

$$\mathcal{S}_j f(m, n) = \sum_{p,q \in \mathbb{Z}} \overline{h_{p,q}}\, \mathcal{S}_{j+1} f(m - 2^{j+1} p, n - 2^{j+1} q) \tag{2.151}$$

$$\mathcal{W}_j^\ell f(m, n) = \sum_{p,q \in \mathbb{Z}} \overline{g_{p,q}^\ell}\, \mathcal{S}_{j+1} f(m - 2^{j+1} p, n - 2^{j+1} q). \tag{2.152}$$

This is equivalent to asking that the related wavelets and scaling function satisfy a two-scale equation of the form:

$$\widehat{\phi}(2\vec{k}) = h(\vec{k}) \widehat{\phi}(\vec{k}), \tag{2.153}$$

$$\widehat{\Psi}^\ell(2\vec{k}) = g^\ell(\vec{k}) \widehat{\Psi}^\ell(\vec{k}), \tag{2.154}$$

where $h(\vec{k}) \equiv h(k_x, k_y)$ and $g^\ell(\vec{k}) \equiv g^\ell(k_x, k_y)$, $\ell = 0 \ldots L - 1$, are the Fourier series of the filters h and g^ℓ respectively. As already stressed in Chapter 1, we know that the same filters can be used to reconstruct the signal provided they satisfy a QMF relation:

$$|h(\vec{k})|^2 + \sum_{\ell=0}^{L-1} |g^\ell(\vec{k})|^2 = 1.$$

This is the exact discrete equivalent of (2.141). Similarly, one can design filters that implement the weaker formula (2.145), provided they satisfy a simpler constraint:

$$h(\vec{k}) + \sum_{\ell=0}^{L-1} g^\ell(\vec{k}) = 1.$$

The main problem at this point is that the integrated directional wavelets we have designed in Section 2.6.1 do *not* in general satisfy any two-scale equation. Another way to formulate the problem is to remark that, although there usually exists regular multipliers $\check{h}$ and $\check{g}^\ell$ satisfying

$$\widehat{\phi}(2\vec{k}) = \check{h}(\vec{k}) \widehat{\phi}(\vec{k}), \quad \widehat{\Psi}^\ell(2\vec{k}) = \check{g}^\ell(\vec{k}) \widehat{\phi}(\vec{k}),$$

these are in general not $2\pi \times 2\pi$ periodic and hence cannot be used to compute successive approximations and details as in (2.151) and (2.152). Nevertheless it is crucial to notice that these multipliers, exactly as the filters h and g^ℓ, are always multiplied by $\widehat{\phi}$. Now, if the latter is very well localized in $[-\pi, \pi] \times [-\pi, \pi]$, the lack of periodicity of $\check{h}$ and $\check{g}^\ell$ is compensated by the localization of $\widehat{\phi}$ and it seems then reasonable to look for good approximations of $\check{h}$, $\check{g}^\ell$ using periodic filters. For this purpose, let us assume that our signal lives in $L^2(\mathbb{R}^2)$ and introduce the subspace of $L^2(\mathbb{R}^2)$ spanned by the integer translates of the scaling function defined in (2.135):

$$\mathcal{V}_0 = \{f \in L^2 \mid f = \sum_{m,n \in \mathbb{Z}} c_{m,n} \phi(x - m, y - n),\ c_{m,n} \in \ell^2\}$$

and suppose that the family $\{\phi(x - m, y - n), \ m, n \in \mathbb{Z}\}$ forms a Riesz basis of $\mathcal{V}_0$. Finding a best approximant for $\check{h}$ can be formulated as finding an element of $\mathcal{V}_0$ whose distance to $\frac{1}{4}\phi(x/2, y/2)$ is minimal in L^2. Similarly, finding best approximants for the $\check{g}^\ell$ is equivalent to finding those elements of $\mathcal{V}_0$ that minimize the distance to $\frac{1}{4}\Psi^\ell(x/2, y/2)$. In other words, the problem is to minimize the L^2 distances

$$v(\check{h}, h^a) = \left[\iint_{\mathbb{R}^2} d^2\vec{k} \ |\check{h}(\vec{k}) - h^a(\vec{k})| \ |\widehat{\phi}(\vec{k})|^2 \right]^{1/2},$$

$$v(\check{g}^\ell, g^{\ell,a}) = \left[\iint_{\mathbb{R}^2} d^2\vec{k} \ |\check{g}^\ell(\vec{k}) - g^{\ell,a}(\vec{k})| \ |\widehat{\phi}(\vec{k})|^2 \right]^{1/2}.$$

Now since $\mathcal{V}_0$ is a vector subspace of a Hilbert space, the projection theorem applies and guarantees the existence and uniqueness of such solutions. Suppose that $h^a \in \mathcal{V}_0$ is the solution for $\check{h}$. It can be expanded as

$$h^a = \sum_{m,n \in \mathbb{Z}} h^a_{m,n} \phi_{m,n},$$

where the $h^a_{m,n}$ are the Fourier coefficients of the approximate filter

$$h^a(\vec{k}) \equiv h^a(k_x, k_y) = \frac{1}{4\pi^2} \sum_{m,n \in \mathbb{Z}} h^a_{m,n} e^{-i(mk_x + nk_y)}.$$

The projection theorem gives also the following characterization of these coefficients:

$$\iint_{\mathbb{R}^2} dx\, dy \ \overline{\phi(x + m, y + n)} \left[\sum_{p,q \in \mathbb{Z}} h^a_{p,q} \phi(x + p, y + q) - \frac{1}{4}\phi(\frac{x}{2}, \frac{y}{2}) \right] = 0,$$

$$\forall m, n \in \mathbb{Z}. \qquad (2.155)$$

Similarly, for the directional wavelets, one obtains L approximate filters $g^{\ell,a}_{m,n}$ characterized by

$$\iint_{\mathbb{R}^2} dx\, dy \ \overline{\phi(x + m, y + n)} \left[\sum_{p,q \in \mathbb{Z}} g^{\ell,a}_{p,q} \phi(x + p, y + q) - \frac{1}{4}\Psi^\ell(\frac{x}{2}, \frac{y}{2}) \right] = 0,$$

$$\forall m, n \in \mathbb{Z}. \qquad (2.156)$$

Finally, for $j \leqslant -1$, the approximate low resolution and detail coefficients read

$$\mathcal{S}^a_j f(m, n) = \sum_{p,q \in \mathbb{Z}} \overline{h^a_{p,q}} \ \mathcal{S}^a_{j+1} f(m - 2^{j+1}p, n - 2^{j+1}q) \qquad (2.157)$$

and

$$\mathcal{W}^{\ell,a}_j f(m, n) = \sum_{p,q \in \mathbb{Z}} \overline{g^{\ell,a}_{p,q}} \ \mathcal{S}^a_{j+1} f(m - 2^{j+1}p, n - 2^{j+1}q). \qquad (2.158)$$

The following theorem, due to Gobbers and Vandergheynst [360], extends to 2-D the result obtained in 1-D by Muschietti and B. Torrésani [291]. It gives explicit formulas for the best approximants h^a and $g^{\ell,a}$, as well as an estimation of the error with respect to the original coefficients $\mathcal{S}_j f(m, n)$ and $\mathcal{W}_j^\ell f(m, n)$:

Theorem 2.6.1 *(i) The optimal filters h^a and $g^{\ell,a}$, solutions of (2.155) and (2.156), are given by*

$$h^a(k_x, k_y) = \frac{\sum_{p,q} \widehat{\phi}\left(2(k_x + 2\pi p), 2(k_y + 2\pi q)\right) \overline{\widehat{\phi}(k_x + 2\pi p, k_y + 2\pi q)}}{\sum_{p,q} |\widehat{\phi}(k_x + 2\pi p, k_y + 2\pi q)|^2},$$

$$(2.159)$$

$$g^{\ell,a}(k_x, k_y) = \frac{\sum_{p,q} \widehat{\Psi}^\ell\left(2(k_x + 2\pi p), 2(k_y + 2\pi q)\right) \overline{\widehat{\phi}(k_x + 2\pi p, k_y + 2\pi q)}}{\sum_{p,q} |\widehat{\phi}(k_x + 2\pi p, k_y + 2\pi q)|^2}.$$

$$(2.160)$$

(ii) If we write[†] $C_0 = 2 \operatorname*{ess\,sup}\limits_{\vec{k} \in \mathbb{R}^2} h^a(\vec{k})$ and $C_\ell = 2 \operatorname*{ess\,sup}\limits_{\vec{k} \in \mathbb{R}^2} g^{\ell,a}(\vec{k})$, then, for all $j \leqslant -1$,

$$\|\mathcal{S}_j^a f - \mathcal{S}_j f\|_\infty \leqslant 2\, v(\check{h}, h^a)\, \frac{1 - C_0^{|j|}}{1 - C_0}\, \|f\|_2,$$

and

$$\|\mathcal{W}_j^{\ell,a} f - \mathcal{W}_j^\ell f\|_\infty \leqslant 2 \left(v(\check{g}^\ell, g^{\ell,a}) + C_\ell\, v(\check{h}, h^a)\, \frac{1 - C_0^{|j|-1}}{1 - C_0} \right) \|f\|_2. \quad (2.161)$$

Proof. We will essentially follow the proof given in [291]. The first part is obtained by taking the Fourier transform of (2.155) and (2.156) and using the periodicity of h^a and $g^{\ell,a}$. We then have

$$\int_0^{2\pi} dk_x \int_0^{2\pi} dk_y\, e^{ik_x m + ik_y n} \left\{ 2\, h^a(k_x, k_y) \sum_{k,l \in \mathbb{Z}} |\widehat{\phi}(k_x + 2\pi k, k_y + 2\pi l)|^2 \right.$$

$$\left. - \sum_{k,l \in \mathbb{Z}} 2\, \overline{\widehat{\phi}(k_x + 2\pi k, k_y + 2\pi l)}\, \widehat{\phi}\left(2(k_x + 2\pi k), 2(k_y + 2\pi l)\right) \right\} = 0,$$

$$(2.162)$$

which gives the result for h^a and similarly for $g^{\ell,a}$. As for the second part of the theorem, the inequality $\|f\|_\infty \leqslant \|\widehat{f}\|_1 / 4\pi^2$ allows us to work directly in the Fourier domain. For the purpose of the calculation, let us introduce the intermediate quantities

---

[†] Since this quantity is orientation independent, we drop the corresponding superscript for ease of notation.

$$\tilde{\mathcal{S}}_j(m, n) = \sum_{k,l \in \mathbb{Z}} \overline{h_{k,l}^a}\, \mathcal{S}_{j+1}\big(m - 2^{j+1}k, n - 2^{j+1}l\big),$$

$$\tilde{\mathcal{W}}_j^\ell(m, n) = \sum_{p,q \in \mathbb{Z}} \overline{g_{p,q}^{k,a}}\, \mathcal{S}_{j+1}\big(m - 2^{j+1}p, n - 2^{j+1}q\big).$$

We have the inequality

$$\left\| \widehat{\mathcal{S}_j^a} - \widehat{\mathcal{S}_j} \right\|_1 \leqslant \left\| \widehat{\mathcal{S}_j^a} - \widehat{\tilde{\mathcal{S}}_j} \right\|_1 + \left\| \widehat{\tilde{\mathcal{S}}_j} - \widehat{\mathcal{S}_j} \right\|_1, \tag{2.163}$$

and similarly for $\left\| \widehat{\mathcal{W}_j^{\ell,a}} - \widehat{\mathcal{W}_j^\ell} \right\|_1$. Let us now compute the second term in the right-hand side of (2.163):

$$\left\| \widehat{\tilde{\mathcal{S}}_j} - \widehat{\mathcal{S}_j} \right\|_1 = \iint_{S^1 \times S^1} dk_x\, dk_y\, 2^j$$

$$\left| \sum_{k,l \in \mathbb{Z}} \hat{f}(k_x + 2\pi k, k_y + 2\pi l)\hat{\phi}\big(2^j(k_x + 2\pi k), 2^j(k_y + 2\pi l)\big) \right.$$

$$- \sum_{k,l \in \mathbb{Z}} \hat{f}(k_x + 2\pi k, k_y + 2\pi l)\, h^a\big(2^{j-1}k_x, 2^{j-1}k_y\big)$$

$$\left. \times \hat{\phi}\big(2^{j-1}(k_x + 2\pi k), 2^{j-1}(k_y + 2\pi l)\big) \right|.$$

Using the Cauchy–Schwarz inequality and the periodicity of h^a, we find

$$\left\| \widehat{\tilde{\mathcal{S}}_j} - \widehat{\mathcal{S}_j} \right\|_1 \leqslant 8\pi^2 \|f\|_2 \sqrt{\int_{\mathbb{R}^2} d^2\vec{k} \left| h^a(\vec{k})\, \hat{\phi}(\vec{k}) - \hat{\phi}(2\vec{k}) \right|^2}$$

$$\leqslant 8\pi^2 \|f\|_2\, \nu(\check{h}, h^a).$$

For the first term on the right-hand side of (2.163), we find

$$\left\| \widehat{\mathcal{S}_j^a} - \widehat{\tilde{\mathcal{S}}_j} \right\|_1 \leqslant 2 \int_{\mathbb{R}^2} d^2\vec{k} \left| h^a\big(2^{j+1}\vec{k}\big) \right| \left| \widehat{\mathcal{S}_{j+1}^a}(\vec{k}) - \widehat{\mathcal{S}_{j+1}}(\vec{k}) \right|$$

$$\leqslant 2 \operatorname*{ess\,sup}_{\vec{k} \in \mathbb{R}^2} |h^a| \left\| \widehat{\mathcal{S}_{j+1}^a} - \widehat{\mathcal{S}_{j+1}} \right\|_1.$$

Combining these estimations, we have

$$\left\| \widehat{\mathcal{S}_j^a} - \widehat{\tilde{\mathcal{S}}_j} \right\|_1 \leqslant 8\pi^2 \|f\|_2\, \nu(\check{h}, h^a) + C_0 \left\| \widehat{\mathcal{S}_{j+1}^a} - \widehat{\mathcal{S}_{j+1}} \right\|_1. \tag{2.164}$$

By iteratively bounding the last term of (2.164) in the same way, we finally obtain

$$\left\| \widehat{\mathcal{S}_j^a} - \widehat{\tilde{\mathcal{S}}_j} \right\|_1 \leqslant 8\pi^2 \|f\|_2\, \nu(\check{h}, h^a) \left(1 + C_0 + C_0^2 + \ldots + C_0^{|j|-1}\right)$$

$$= 8\pi^2 \|f\|_2\, \nu(\check{h}, h^a)\, \frac{1 - C_0^{|j|}}{1 - C_0}. \tag{2.165}$$

An equivalent processing of the wavelet coefficients concludes the proof. $\square$

There remain several open questions related to the design of these approximate QMFs. In particular, in contradiction to the orthogonal case discussed in the previous chapter, the convergence of the cascade algorithm is not ensured. It is thus impossible to tell precisely if these filters really correspond to a frame and what would be the properties of such a frame. This scheme is thus really a handy numerical shortcut that allows one to efficiently compute an *approximation* of the original frame expansion.

2.6.5   Some implementation issues

When implementing (2.157) and (2.158), the first fact to consider is that, generally, we possess only a finite number $M \times N$ of samples $f(m, n)$ (with $(m, n) \in [0 \ldots M - 1, 0 \ldots N - 1]$) of the signal to be analyzed. A decision has thus to be made about the nature of the signal outside this range.

Furthermore, the filters will practically be computed on a finite grid. That is, we will only use a finite number $P \times Q$ of filter coefficients $h_{p,q}^a$ and $g_{p,q}^{k,a}$, with $(p, q) \in [-P/2 \ldots P/2 - 1, -Q/2 \ldots Q/2 - 1]$.

In a first approach, the signal is considered to be zero outside this range. Unfortunately, this decision leads to impractical and inefficient algorithms, as one has to compute $(M + 2^{|j|-1}(P - 1)) \times (N + 2^{|j|-1}(Q - 1))$ coefficients for each $j < 0$ to avoid side effects.

A second approach is to consider the signal as being periodic of period $M \times N$. With this approach, fast circular convolution algorithms can be used and we are led to the following algorithm structure:

   (i) compute h^a and $g^{\ell,a}$ and their associated impulse responses using Theorem 2.6.1;
  (ii) compute a first approximation of the analyzed signal f using (2.150); this step is traditionally skipped and the signal is considered as its first approximation;
 (iii) for each $j < 0$, iteratively compute details and approximations using (2.157) and (2.158).

The cost of this algorithm is:

$$C(M, N) = c \cdot M \cdot N$$

where the constant c depends on the size of the impulse responses of the filters. This dependency strongly limits the power of this algorithm as experiments show that, even for small sizes, this algorithm is always slower than the FFT-based algorithm used to compute the CWT. As those sizes may not be arbitrarily chosen, another algorithm should be used in order to get valuable results with small computation times.

The obvious way to handle this problem is to replace convolutions in direct space by products of Fourier transforms in frequency space. The main advantage of this technique is that we no longer need to restrict ourselves to use small filters, as we may now use impulse responses that are the same size as the signal, thus giving much more precise results than with the above algorithm. For the sake of simplicity, we will develop this new algorithm in the 1-D case, its extension to the 2-D directional case being straightforward.

For the first step, we have to compute two periodic convolution products:

$$c_{-1,k} = \sum_{n=0}^{N-1} c_{0,k-n} h_n$$

$$d_{-1,k} = \sum_{n=0}^{N-1} c_{0,k-n} g_n.$$

By introducing $\{C_{j,n}\} = FFT_N (\{c_{j,k}\})$, $\{D_{j,n}\} = FFT_N (\{d_{j,k}\})$, $\{H_n\} = FFT_N$ $(\{h_k\})$ and $\{G_n\} = FFT_N (\{g_k\})$ where the notation FFT_N means the FFT algorithm applied to a sequence of length N, we get: $C_{-1,n} = \sqrt{N} C_{0,n} H_n$ and $D_{-1,n} = \sqrt{N} C_{0,n} G_n$. We may then get the details back in the real space with $\{d_{j,k}\} = IFFT_N (\{D_{j,n}\})$.

For the second step, we have to compute:

$$c_{-2,k} = \sum_{n=0}^{N-1} c_{-1,k-2n} h_n$$

$$d_{-2,k} = \sum_{n=0}^{N-1} c_{-1,k-2n} g_n.$$

Given the periodicity of the signal, we may rewrite this as:

$$c_{-2,k}^0 \equiv c_{-2,2k} = \sum_{n=0}^{N/2-1} c_{-1,2(k-n)}(h_n + h_{n+N/2})$$

$$c_{-2,k}^1 \equiv c_{-2,2k+1} = \sum_{n=0}^{N/2-1} c_{-1,2(k-n)+1}(h_n + h_{n+N/2})$$

$$d_{-2,k}^0 \equiv d_{-2,2k} = \sum_{n=0}^{N/2-1} c_{-1,2(k-n)}(g_n + g_{n+N/2})$$

$$d_{-2,k}^1 \equiv d_{-2,2k+1} = \sum_{n=0}^{N/2-1} c_{-1,2(k-n)+1}(g_n + g_{n+N/2})$$

for $k = 0, \ldots, N/2 - 1$. Let us now define $h_n^1 = h_n + h_{n+N/2}$ for $n = 0, \ldots, N/2 - 1$, we get:

$$c_{-2,k}^0 = \sum_{n=0}^{N/2-1} c_{-1,k-n}^0 h_n^1 \qquad c_{-2,k}^1 = \sum_{n=0}^{N/2-1} c_{-1,k-n}^1 h_n^1 \qquad (2.166)$$

$$d_{-2,k}^0 = \sum_{n=0}^{N/2-1} c_{-1,k-n}^0 g_n^1 \qquad d_{-2,k}^1 = \sum_{n=0}^{N/2-1} c_{-1,k-n}^1 g_n^1 \qquad (2.167)$$

where $c^0_{-1,k} = c_{-1,2k}$, $c^1_{-1,k} = c_{-1,2k+1}$, $d^0_{-1,k} = d_{-1,2k}$ and $d^1_{-1,k} = d_{-1,2k+1}$. In the Fourier space, (2.166) and (2.167) may be rewritten as:

$$C^0_{-2,n} = \sqrt{N/2} C^0_{-1,n} H^1_n \qquad C^1_{-2,n} = \sqrt{N/2} C^1_{-1,n} H^1_n$$
$$D^0_{-2,n} = \sqrt{N/2} C^0_{-1,n} G^1_n \qquad D^1_{-2,n} = \sqrt{N/2} C^1_{-1,n} G^1_n$$

where $\{C^0_{-2,n}\} = FFT_{N/2}\left(\{c^0_{-2,k}\}\right)$, $\{C^1_{-2,n}\} = FFT_{N/2}\left(\{c^1_{-2,k}\}\right)$, $\{D^0_{-2,n}\} = FFT_{N/2}\left(\{d^0_{-2,k}\}\right)$ and $\{D^1_{-2,n}\} = FFT_{N/2}\left(\{d^1_{-2,k}\}\right)$. A straightforward calculation gives:

$$H^1_n = \sqrt{2} H_{2n}$$

for $n = 0, \ldots, N/2 - 1$. Furthermore:

$$C^0_{-1,n} = \frac{1}{\sqrt{2}}(C_{-1,n} + C_{-1,n+N/2})$$

$$C^1_{-1,n} = \frac{e^{2i\pi \frac{n}{N}}}{\sqrt{2}}(C_{-1,n} - C_{-1,n+N/2})$$

for $n = 0, \ldots, N/2 - 1$. The last equation is particularly interesting as it introduces the same twiddle factors as those already present in the traditional FFT implementations.

Extending the above results to the following steps, one comes to the conclusion that the complexity of this new algorithm is also of order $N \log_2(N)$, but with a hidden constant that is exactly half the one encountered in the FFT-based algorithm, this constant being associated with the pyramidal structure. The same ideas apply to the 2-D case, thus giving an algorithm of complexity

$$C(M, N) = M \cdot N \cdot \log_2(M.N),$$

with a hidden constant also exactly half that of the FFT-based algorithm. As a matter of comparison, Figure 2.13 shows timings of this algorithm for different image sizes. Timings of the usual pyramidal algorithm and standard implementation in the Fourier domain are also displayed.

Putting it all together, we have a new fast algorithm, perfectly suited to compute the 2-D CWT, faster than the traditional "pseudo-pyramidal" algorithm, and sharper. Furthermore, it is essential to note that the whole construction is equivalent to that leading to the FFT algorithm in the Fourier transform theory. It should be noted that a Fourier implementation of the pyramidal algorithm is quite natural when one addresses the problem of designing maximally regular wavelets. That is why the algorithm described above shares common features with the Fourier implementation of the Meyer wavelet decomposition [Kol97]. We refer the interested reader to the work of Rioul and Duhamel [327] for more general considerations on implementing pyramidal algorithms in the frequency domain.

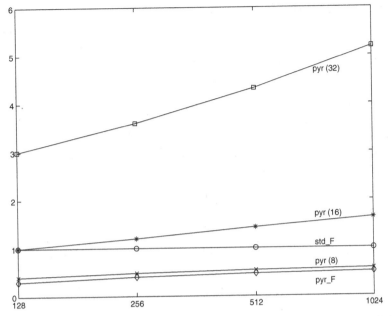

Fig. 2.13. Computation time for different implementations of the 2-D pyramidal algorithm as a function of the image width (square images assumed): pyr(n) is the standard pyramidal algorithm with n by n filters, pyr_F is the modified Fourier pyramidal algorithm explained in the text. The graph is normalized with respect to the standard algorithm with convolution in the Fourier domain (std_F).

2.7   Steerable filters

While looking for a flexible tool for processing oriented data, Freeman and Adelson [170] introduced some time ago the concept of steerable filters. These were further developed by Perona [311] and Simoncelli *et al.* [342]. Here again one obtains a multiscale pyramid decomposition, which is quite efficient in a number of problems, mostly related to machine vision. Similar techniques have been used with the Gabor transform [234]. We will briefly describe this scheme and compare it to the directional wavelet packets of Section 2.6.

The basic idea is quite simple, and best illustrated on the example of a Gaussian kernel $G(x, y)$. From the partial derivatives $G_x(x, y) = -2x\, e^{x^2+y^2}$, $G_y(x, y) = -2y\, e^{x^2+y^2}$, one computes the derivative in the direction θ:

$$G_\theta(x, y) = \cos\theta\, G_x(x, y) + \sin\theta\, G_y(x, y).$$

Since convolution is a linear operation, one may use G_θ for filtering an image f in the direction θ by superposing the filterings in directions x and y:

$$F_\theta(f)(\vec{b}) = (G_\theta \star f)(\vec{b}) = \cos\theta\, F_x(\vec{b}) + \sin\theta\, F_y(\vec{b}).$$

This is the property of orientability. More generally, a filter f is *orientable* or *steerable* if any oriented version of it may be obtained from a finite number of basic orientations:

$$f(r_\theta(\vec{x})) = \sum_{m=1}^{M} k_m(\theta)\, f(r_{\theta_m}(\vec{x})). \tag{2.168}$$

The weights $\{k_m(\theta),\, m = 1, \dots, M\}$ are called interpolation functions. (The notion of orientability may be extended to other transformations, such as scaling [311], but we will not consider these generalizations here.)

When the filter f admits a finite Fourier series (i.e., it is a real trigonometric polynomial),

$$f(r, \varphi) = \sum_{n=-N}^{N} a_n(r)\, e^{in\varphi}, \quad \vec{x} \equiv (r, \varphi), \tag{2.169}$$

Freeman and Adelson have shown that the interpolation functions must satisfy the relation

$$\begin{pmatrix} 1 \\ e^{i\theta} \\ \vdots \\ e^{iN\theta} \end{pmatrix} = \begin{pmatrix} 1 & \cdots & 1 \\ e^{i\theta_1} & \cdots & e^{i\theta_M} \\ \vdots & & \vdots \\ e^{iN\theta_1} & \cdots & e^{iN\theta_M} \end{pmatrix} \begin{pmatrix} k_1(\theta) \\ \vdots \\ k_M(\theta) \end{pmatrix} \tag{2.170}$$

and that the minimal number of interpolation functions is always larger than the number of nonzero coefficients in the angular Fourier series (2.169) of the filter f. These steerable filters obviously generalize the interpolation properties associated to the partial derivatives G_x, G_y, which are thus prototypes of such filters. Moreover, the steerability property (2.168) is independent of the radial part of the filter, so that one may use functions that generate a dyadic pyramid.

The main virtue of steerable filters is to allow the computation of filtering in *any* direction from the interpolation functions and the basic filters. This explains their intensive use in vision for studying oriented features in images [311,342]. From the algorithmic point of view, the complexity of steerable filters is that of the associated dyadic pyramid, thus comparable to that of directional wavelet packets. Note, however, that separable filters may sometimes be obtained, which is an additional bonus.

So the natural question arises, can directional wavelets be made steerable? Unfortunately, the answer is no. Indeed, the Fourier transform of a steerable filter is also steerable. On the other hand, the angular support of a directional wavelet (2.149) necessarily has a width smaller than π, so that the relation (2.169) can never be satisfied. More intuitively, the steerability condition (2.168) requires that the basic filters overlap in a fixed way, in order to ensure the existence of interpolation functions, whereas the angular overlap of directional wavelet packets may be taken to be arbitrarily small. What about angular resolution? That of steerable filters may be made as good as one wishes, simply by taking more basic filters. However, because of the substantial overlap

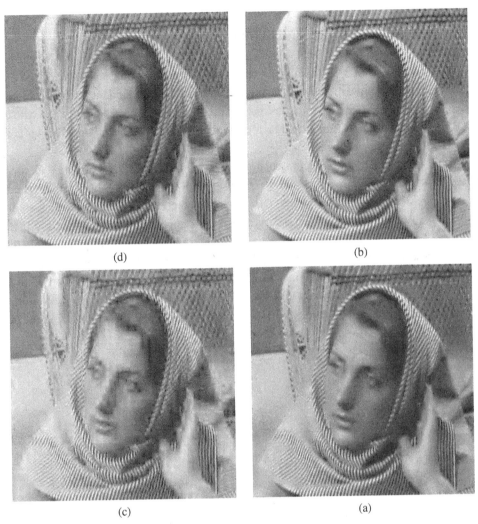

Fig. 2.14. Four reconstructions of the image *barbara*: (a) 2-D wavelet orthonormal basis; (b) redundant frame (3 bits/coefficient); (c) and (d) the same with 2 bits/coefficient.

between the latter, the steerable scheme will always require more basic filters than the directional wavelet packets in order to achieve a high angular resolution, so that, in the end, the computing cost may become prohibitive.

In conclusion, steerable filters and directional wavelet packets are two comparable, yet incompatible, tools for decomposing an image into an oriented pyramid. In the former case, however, a large number of basic filters, with fixed overlap, is required for achieving a high angular resolution. This precludes their use in applications that require a maximal decorrelation of orientations, such as watermarking of images or texture segmentation. Here directional wavelet packets are the best choice. We will explore these applications in the next chapters (Sections 4.7.2 and 5.5, respectively).

2.8   Redundancy: plus and minus

Exactly as in 1-D, redundancy has many advantages, that more than compensate the higher computational cost it implies. In a nutshell, redundant decompositions lead to better quality reconstructions and are more robust to noise. Actually, the whole discussion of Section 1.6.2 could be repeated here almost verbatim, in particular concerning the robustness issue.

We highlight the reconstruction aspect with one striking example. We consider the standard *barbara* image and decompose it in two ways, first with an orthonormal wavelet basis (using 2-D Daubechies DB4 wavelets), then with a redundant frame of directional wavelets. The images reconstructed by the two methods are presented in Figure 2.14. Panels (a) and (b) show the reconstruction using 3 bits per coefficient, while (c) and (d) show the result obtained with 2 bits per coefficient. In either case, the resulting image is visually better when the redundant frame is used, (b) or (d). The orthonormal basis gives more artifacts and distortions. Of course, the effect is more marked in the 2 bit case. Although the two results are poor, we show them for emphasizing the point.

3   Some 2-D wavelets and their performance

3.1   Which wavelets?

The next step is to choose an analyzing wavelet ψ. At this point, there are two possibilities, depending whether one is interested or not in detecting oriented features in an image, i.e., regions where the amplitude is regular along one direction and has a sharp variation along the perpendicular direction.

(i) *Isotropic wavelets*

If one wants to perform a pointwise analysis, i.e., when no oriented features are present or relevant in the signal, one may choose an analyzing wavelet ψ which is invariant under rotation. Then the θ dependence drops out, for instance, in the reconstruction formula (2.26). The most familiar example is the isotropic 2-D Mexican hat wavelet (2.21).

(ii) *Anisotropic wavelets*

When the aim is to detect oriented features in an image (for instance, in the classical problem of edge detection or in directional filtering), one has to use a wavelet which is *not* rotation invariant. The best angular selectivity will be obtained if ψ is *directional*, which means that the (essential) support of $\widehat{\psi}$ in spatial frequency space is contained in a convex cone with apex at the origin (by which we mean that the wavelet is numerically negligible outside the cone). Typical directional wavelets are the 2-D Morlet wavelet (2.22) or the conical wavelets.

There are many ways of designing wavelets of either kind, but in fact almost all of those available on the market may be obtained by a general procedure, outlined in the proposition below. The starting point is a scaling function, that is, a function $\phi(\vec{x})$, whose integral over the plane does not vanish. Then wavelets can be built by taking derivatives of the scaling function or the difference of two scaling functions. The most obvious example of a scaling function is a Gaussian, which is very easy to use and essentially localized in a disk.

Another method, that we shall describe in Chapter 6, is to impose the saturation of the uncertainty product of two or more operators corresponding to infinitesimal generators of elementary operators, namely the dilation, orientation and translation operators, as given in Section 2.1 above.

Proposition 3.1.1 *Let ϕ be a (sufficiently smooth) scaling function, that is, a function satisfying:*

$$\int_{\mathbb{R}^2} d^2\vec{x}\ \phi(\vec{x}) \neq 0. \tag{3.1}$$

Then the functions ψ_1 and ψ_2 defined below are wavelets:

$$\psi_1(\vec{x}) = \sum_{n=1}^{N}\sum_{m=1}^{M} c_{nm} \frac{\partial^n}{\partial x^n} \frac{\partial^m}{\partial y^m} \phi(\vec{x}), \quad \text{where } \vec{x} \equiv (x, y),\ N, M \geqslant 1, \tag{3.2}$$

$$\psi_2(\vec{x}) = \frac{1}{a_1}\left[U(\vec{b}_1, a_1, \theta_1)\phi\right](\vec{x}) - \frac{1}{a_2}\left[U(\vec{b}_2, a_2, \theta_2)\phi\right](\vec{x}),$$

$$(\vec{b}_1, a_1, \theta_1) \neq (\vec{b}_2, a_2, \theta_2), \tag{3.3}$$

where $U(\vec{b}, a, \theta)$ is the unitary operator already defined in (2.13):

$$\left[U(\vec{b}, a, \theta)s\right](\vec{x}) = a^{-1}s(a^{-1}r_{-\theta}(\vec{x} - \vec{b})).$$

The proof is straightforward. We shall now examine in detail several examples of wavelets of each kind. As will be seen in the sequel, almost all of them fall into the general types described in the proposition, possibly combining the two operations of derivation and difference. It turns out also that in many cases, the basic scaling function is the Gaussian $\exp(-|\vec{x}|^2)$. This is not a coincidence. In order to understand this, assume the wavelet ψ to be, as in (3.2), some derivative of a scaling function ϕ,

$$\psi(\vec{x}) = \frac{\partial^n}{\partial x^n} \frac{\partial^m}{\partial y^m} \phi(\vec{x}).$$

Then we may rewrite the basic formula (2.19) as a genuine convolution:

$$S(\vec{b}, a, \theta) = \left(\phi^{\#}_{a,\theta} * \frac{\partial^n}{\partial x^n} \frac{\partial^m}{\partial y^m} s\right)(\vec{b}) \tag{3.4}$$

$$= \left(\frac{\partial}{\partial b_x}\right)^n \left(\frac{\partial}{\partial b_y}\right)^m \left(\phi^{\#}_{a,\theta} * s\right)(\vec{b}), \tag{3.5}$$

where

$$\phi^{\#}_{a,\theta}(\vec{x}) = a^{-1}\overline{\phi(-r_{-\theta}(\vec{x})/a)}.$$

Equation (3.5) shows that $S(\vec{b}, a, \theta)$ is simply the derivative of the signal rotated and blurred at resolution (scale) a. Thus we expect large wavelet coefficients to occur at locations of sharp variation of s. Since the Gaussian is a very common kernel for blurring, it is not a surprise that many wavelets will be related to derivatives of a Gaussian. As for the popularity of the latter, it is due to several factors: it is very well localized, both in position and in Fourier space, it is indeed optimal for the joint space–frequency localization, as argued by Gabor [180], and finally it is rotation invariant, thus it does not privilege any particular direction.

3.2 Isotropic wavelets

3.2.1 The 2-D Mexican hat and its generalizations

In its isotropic version, this is simply the Laplacian of a Gaussian:

$$\psi_{H}(\vec{x}) = -\Delta \, \exp(-\tfrac{1}{2}|\vec{x}|^2) = (2 - |\vec{x}|^2) \, \exp(-\tfrac{1}{2}|\vec{x}|^2). \tag{3.6}$$

This is a real, rotation invariant wavelet, with vanishing moments of order up to 1, also known in the literature as the LOG wavelet. It is shown, in position domain and in spatial frequency space, in Figures 3.1 and 3.2, respectively. It was originally introduced by Marr and Hildreth [Mar82,266], in their pioneering work on vision, precisely because it

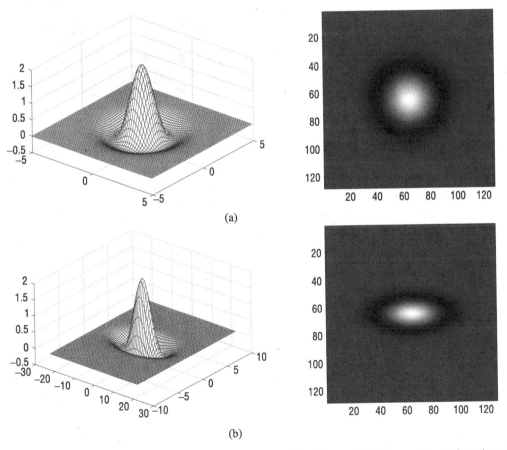

Fig. 3.1. The 2-D Mexican hat wavelet in position domain, seen in 3-D perspective (left) and gray levels (right): (a) the isotropic wavelet; (b) the anisotropic wavelet with $\epsilon = 2$.

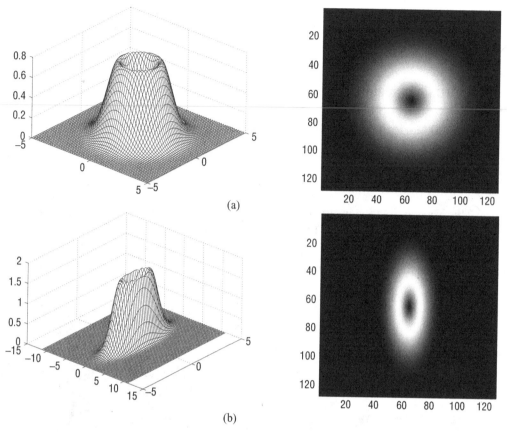

Fig. 3.2. The same wavelets as in Figure 3.1, seen in spatial frequency space.

is obtained by applying an isotropic differential operator of second order to the Gaussian (this was in fact the original motivation of [266]). The Mexican hat will be efficient for a fine pointwise analysis, but *not* for detecting directions. This will be confirmed by quantitative calibration tests in Section 3.4 below.

In addition, and for the same reasons, one may also use higher order Laplacians of the Gaussian,

$$\psi_{\mathrm{H}}^{(n)}(\vec{x}) = (-\Delta)^n \, \exp(-\tfrac{1}{2}|\vec{x}|^2).$$

(3.7)

For increasing n, these wavelets have more and more vanishing moments, and are thus sensitive to increasingly sharper details. An interesting technique, pioneered in 1-D by A. Arnéodo [49], is to analyze the same signal with several wavelets $\psi_{\mathrm{H}}^{(n)}$, for different n. The features common to all the transforms surely belong to the signal, they are not artifacts of the analysis.

In several applications, it is useful to introduce an additional parameter, namely, the width σ of the Gaussian (in $\vec{k}$-space). Although σ is redundant, since the Gaussian can be dilated to an arbitrary width, nevertheless it helps to fix explicitly the central frequency. Thus one uses, instead of (3.6), the wavelet

$$\psi_H(\vec{x}) = -\Delta \, \exp(-\frac{\sigma^2|\vec{x}|^2}{2}), \quad \widehat{\psi_H}(\vec{k}) = \frac{|\vec{k}|^2}{\sigma^2} \, \exp(-\frac{|\vec{k}|^2}{2\sigma^2}) \, . \tag{3.8}$$

An approximate version of the Mexican hat has been introduced by Arnéodo *et al.* [Arn95,44,171]; under the name of *Bessel filter*, namely:

$$\widehat{\psi_B}(\vec{k}) = 1 \, , \text{ for } R/\gamma \leqslant |\vec{k}| \leqslant R \, , \tag{3.9}$$
$$= 0 \, , \text{ otherwise,}$$

where R and R/γ are the radii of the external and internal disks, respectively. The inverse Fourier transform $\psi_B(\vec{x})$ is a Bessel function (hence the name):

$$\psi_B(\vec{x}) \simeq \frac{J_1(r)}{r} - \frac{1}{\gamma^2} \frac{J_1(r/\gamma)}{r/\gamma}, \quad r = |\vec{x}|. \tag{3.10}$$

This filter was designed and used systematically for the so-called *optical wavelet transform*, which consists in a hardware (optical) realization of the CWT. We will come back to this application in Section 5.4.1 [Arn95,46]. Another isotropic wavelet, very similar to the previous one, has been introduced in [175] under the rather funny name of *Pet hat*. It is defined in Fourier space as

$$\widehat{\psi}(\vec{k}) = \cos^2\left(\frac{\pi}{2} \log_2 \frac{|\vec{k}|}{2\pi}\right), \text{ for } \pi < |\vec{k}| < 4\pi, \tag{3.11}$$
$$= 0 \, , \text{ for } |\vec{k}| < \pi \text{ and } |\vec{k}| > 4\pi.$$

This wavelet has a better resolving power in scale than the Mexican hat, hence it is more efficient in sorting objects in astrophysical images according to their characteristic scale, which is precisely the aim of the authors of [175]. Yet another one, which has the advantage of being both isotropic and continuous, is the Halo wavelet [117], defined as

$$\widehat{\psi}(\vec{k}) = c \, e^{-(|\vec{k}|^2 - |\vec{k}_o|^2)}. \tag{3.12}$$

This is a real wavelet, that selects precisely the annular region $|\vec{k}| \simeq |\vec{k}_o|$.

3.2.2   Difference wavelets

Among the many wavelets (or filters) proposed in the literature, an interesting class consists of wavelets obtained as the difference of two positive functions, according to (3.3) in Proposition 3.1.1. In order to get an isotropic wavelet in this way, the only possibility is to take the difference between a single isotropic function ϕ and a contracted version of the latter, that is, the particular case where only the scale factors differ in (3.3). Indeed, if ϕ is a smooth non-negative function, integrable and square integrable, with all moments of order one vanishing at the origin, then the function ψ given by the relations

$$\psi(\vec{x}) = \alpha^{-2}\,\phi(\alpha^{-1}\vec{x}) - \phi(\vec{x}), \quad \widehat{\psi}(\vec{k}) = \widehat{\phi}(\alpha\vec{k}) - \widehat{\phi}(\vec{k}) \quad (0 < \alpha < 1) \tag{3.13}$$

is easily seen to be a wavelet satisfying the admissibility condition (2.17). Since ϕ is typically a smoothing function, the wavelet ψ is called the "Difference-of-Smoothings" or DOS wavelet [Duv91].

A typical example is the "Difference-of-Gaussians" or DOG wavelet, obtained by taking for ϕ an isotropic Gaussian:

$$\psi_{\mathrm{D}}(\vec{x}) = \tfrac{1}{2\alpha^2}\,\exp(-\tfrac{1}{2\alpha^2}|\vec{x}|^2) - \exp(-\tfrac{1}{2}|\vec{x}|^2) \quad (0 < \alpha < 1). \tag{3.14}$$

The DOG filter is a good substitute for the isotropic Mexican hat (for $\alpha^{-1} = 1.6$, their shapes are extremely similar), frequently used in psychophysics works [DeV88,Duv91,124]. It was also considered by Grossmann [209] for signal analysis, together with more general linear combinations of Gaussians. An immediate application is the construction of wavelets on the 2-sphere, simply by lifting a DOG wavelet in the tangent plane to the sphere by inverse stereographic projection (Section 9.2).

Another example is the "Difference-of-Mesas" or DOM filter, corresponding to a function ϕ which is a smoothed version of the characteristic function of a disk (a 'mesa' function) [369,370]. The resulting annular wavelet has been used, together with the Halo wavelet (3.12), in the detection of Einstein gravitational arcs in cosmological pictures [82] (see Section 5.1). The principle behind this application is again the filtering property: the wavelet detects preferentially objects that resemble it.

More generally, the concept of the difference wavelet is useful for reducing noise in images. Take an image, consisting of an object to be identified, embedded in noise or clutter. Let $\phi(\vec{x})$ be an averaged version of the image. Then the corresponding difference function $\psi(\vec{x})$ given by (3.13) is a wavelet ideally suited for the analysis of the object in question [16]. Indeed the difference operation substantially reduces the background noise, and ψ incorporates a maximal amount of resemblance with the object (*a priori* information). A similar subtraction technique, known in astronomy as "unsharp masking," is commonly used in the treatment of astrophysical images for enhancing the relevant information (such as galaxies or stars), while reducing the noise. Here one computes first the high-frequency content of the image as the difference $I_{\mathrm{h}} = I_{\mathrm{o}} - I_{\mathrm{s}}$ between the original image I_{o} and a smoothed version I_{s} of it, and adds it to the original image, with a multiplicative factor λ. Then, in the corrected image $I_{\mathrm{corr}} = I_{\mathrm{o}} + \lambda I_{\mathrm{h}} = I_{\mathrm{o}} + \lambda(I_{\mathrm{o}} - I_{\mathrm{s}})$, the high-frequency details are enhanced over the background (but severe distortions may result if λ is chosen too big) [258]. An example of unsharp masking may be found at the address `http://www.chapman.edu/oca/gallery2/demo.htm`. A similar procedure has been introduced for the problem of automatic target recognition (ATR). We will discuss the corresponding algorithm in Section 4.2.1 and the general problem of image denoising in Section 4.6.

An additional advantage of these difference wavelets is that they lead to interesting and fast algorithms, for instance, in the formalism of continuous wavelet packets [159]. Indeed, in 1-D, we have seen in (1.62), that the integrated wavelet $\Psi(x)$ associated to the scaling function $\Phi(x)$ is precisely $\Psi(x) = 2\Phi(2x) - \Phi(tx)$, and this property extends to 2-D, equation (2.144) (the L^1 normalization is used here, contrary to (3.13)). Notice, finally, that ϕ, and thus also ψ, need not be isotropic. The directional wavelet packets constructed in Section 2.6 are a striking example.

3.3   Directional wavelets

3.3.1   Oriented wavelets and edge detection

Detecting oriented features (segments, edges, vector field, . . .) poses a major challenge in computer vision, and many techniques have been proposed in the literature to meet it. In the context of wavelet analysis, one needs a wavelet which is directionally selective. A natural way of designing such a wavelet is to modify an isotropic one, such as the Mexican hat, simply by stretching it. Mathematically, this amounts to replacing in (3.6) $\vec{x}$ by $A\vec{x}$, where $A = \mathrm{diag}[\epsilon^{-1/2}, 1]$, $\epsilon \geqslant 1$, is a 2×2 anisotropy matrix. However, such a wavelet still acts as a second-order operator and detects singularities in all directions and it is of little use in practice. Indeed, on one hand, the calibration tests that we will discuss in Section 3.4 below show that it performs poorly [13]. On the other hand, there is a theorem due to Daugman [125], according to which no real wavelet rendered anisotropic by a mere stretching in one direction can have a good directional selectivity, no matter how large the anisotropy ϵ is taken.

3.3.1.1   Some precursors
Stretching an isotropic wavelet being inefficient for inducing directional selectivity, the next step is to take a directional derivative. As we have seen in Section 3.2.1, Marr and Hildreth [266] apply the Laplacian to the Gaussian, thus getting the Mexican hat wavelet, precisely because it is an *isotropic* differential operator of second order. As a consequence, the Mexican hat is inefficient at detecting directions. In order to get a good edge detector, Canny [98] designed a filter which is optimal for several criteria (detection, localization, uniqueness of answer). However, this filter is numerically heavy to implement and it is advantageously replaced, to a very good approximation, by the first derivative of a Gaussian. The wavelet

$$\psi^{(1)}(\vec{x}) = \frac{d}{dx} e^{-|\vec{x}|^2/\sigma^2} \qquad (3.15)$$

detects edges oriented in the y-direction, and it suffices to rotate it to get an edge detector that is sensitive to an arbitrary direction. In addition, Canny considered many different values of the width σ of the Gaussian $G(x)$, which amounts to varying the scale. Furthermore, given an image s, his technique consists in locating the maxima of $\psi^{(1)} \star s$, which are the zero-crossings of $\partial_x^2 G \star s$, i.e., precisely the technique of Mallat [261,264]. In other words, Canny was very close to a primitive version of wavelet analysis!

However, for computational reasons, one is forced to sample the orientation angles, keeping only a few values. But, as remarked by Perona in 1992 [311],

... this practice has the strong drawback of introducing anisotropies and algorithmic difficulties in the computational implementations. It would be preferable to keep thinking in terms of a continuum, of angles for example, and to be able to localize the orientation of an edge with the maximum accuracy allowed by the filter one has chosen.

Perona's answer to this objection is to advocate the use of *steerable filters*, described in Section 2.7. However, taking together Canny's approach and Perona's remark leads directly to the continuous wavelet transform, which thus appears as a very natural tool for edge detection, and more generally in computer vision.

Coming back to Canny's work, it is instructive to compare the first, second and third derivatives of the Gaussian (see [Bha99]; Chapter 4) and test their performance, exactly as in 1-D [49]. The Canny edge detector was later improved by Deriche [135,136] and Bourennane *et al.* [83], still keeping the same philosophy. The former gets for the optimal filter the function

$$f(x) = -c\, e^{-\alpha|x|} \sin \omega x \simeq -c\, x\, e^{-\alpha|x|} \quad \text{for} \quad \alpha/\omega \gg 1. \tag{3.16}$$

An alternative is to consider a mixed derivative, such as $\partial_x \partial_y \exp(-|\vec{x}|^2)$. This wavelet has good capabilities for detecting *corners* in a contour, but we will describe in Section 3.3.3 another one that performs even better, the so-called end-stopped wavelet of [Bha99,76].

3.3.1.2   The concept of directional wavelets

Although the directional derivative wavelets just described do have some capabilities of directional filtering, they are by far not sufficient, because their angular selectivity is rather poor. In order to go beyond, we introduce the concept of directional wavelets [18,19,24]:

Definition 3.3.1. *A wavelet ψ is said to be* directional *if the effective support of its Fourier transform $\widehat{\psi}$ is contained in a convex cone in spatial frequency space $\{\vec{k}\}$, with apex at the origin, or a finite union of disjoint such cones (in that case, one will usually call ψ multidirectional).*

Since it may sound counter-intuitive, this definition requires a word of justification. According to (2.20), the wavelet acts as a filter in $\vec{k}$-space (multiplication by the function $\widehat{\psi}$). Suppose the signal $s(\vec{x})$ is strongly oriented, for instance a long segment along the

x-axis. Then its Fourier transform $\widehat{s}(\vec{k})$ is a long segment along the k_y-axis. In order to detect such a signal, with a good directional selectivity, one needs a wavelet ψ supported in a narrow cone in $\vec{k}$-space. Then the WT is negligible unless $\widehat{\psi}(\vec{k})$ is essentially aligned onto $\widehat{s}(\vec{k})$: directional selectivity demands restriction of the support of $\widehat{\psi}$, not ψ. The corresponding standard practice in signal processing is to design an adequate filter in the *frequency* domain (high pass, band pass, ...). In addition, there are cases (magnetic resonance imaging, for instance) where data are acquired in $\vec{k}$-space (then called the measurement space) and the image space is obtained after a Fourier transform: here again directional filtering takes place in $\vec{k}$-space.

According to this definition, the anisotropic Mexican hat is not directional, since the support of $\widehat{\psi}_{\rm H}$ is centered at the origin, no matter how big its anisotropy is. Indeed, the detailed tests described in [13] confirm its poor performances in selecting directions. We will come back to this point, with quantitative results, in Section 3.4. A review of directional wavelets and their use may be found in [19].

3.3.2   The 2-D Morlet wavelet

This is the prototype of a directional wavelet:

$$\psi_{\rm M}(\vec{x}) = \exp(i\vec{k}_o \cdot \vec{x})\,\exp(-\tfrac{1}{2}|A\vec{x}|^2) - \exp(-\tfrac{1}{2}|A^{-1}\vec{k}_o|^2)\,\exp(-\tfrac{1}{2}|A\vec{x}|^2), \qquad (3.17)$$

$$\widehat{\psi}_{\rm M}(\vec{k}) = \sqrt{\epsilon}\,(\exp(-\tfrac{1}{2}|A^{-1}(\vec{k} - \vec{k}_o)|^2) - \exp(-\tfrac{1}{2}|A^{-1}\vec{k}_o|^2)\,\exp(-\tfrac{1}{2}|A^{-1}\vec{k}|^2)). \quad (3.18)$$

The parameter $\vec{k}_o$ is the wave vector, and $A = \text{diag}[\epsilon^{-1/2}, 1], \epsilon \geqslant 1$, is a 2×2 anisotropy matrix. The correction term in (3.17) and (3.18) enforces the admissibility condition $\widehat{\psi}_{\rm M}(\vec{0}) = 0$. However, since it is numerically negligible for $|\vec{k}_o| \geqslant 5.6$, one usually drops it altogether (but not always, see Section 3.4). In that case, putting $\epsilon = 1$, we obtain the function

$$\psi_{\rm G}(\vec{x}) = \exp(i\vec{k}_o \cdot \vec{x})\,\exp(-\tfrac{1}{2}|\vec{x}|^2). \qquad (3.19)$$

This function is well-known in the image processing literature under the name of Gabor function [126]. One reason of its popularity is its computational simplicity. Another one, in particular in the modeling of human vision, is that a large fraction of cells in the primary visual cortex of primates (including man, presumably) have a receptive field that resembles a Gabor function [DeV88,369,370,372]. An example is shown, for $\epsilon = 2, \vec{k}_o = (0, 6)$, in Figures 3.3 and 3.4.

The Gabor function (3.19) has the qualitative behavior expected from a wavelet. It is well localized, both in position space, around the origin, and in spatial frequency space, around $\vec{k} = \vec{k}_o \neq 0$, but, strictly speaking, it is not admissible. On the other hand, the full Morlet function given in (3.17)–(3.18) is always admissible but, for small $|\vec{k}_o|$, it is useless as a wavelet, since $\widehat{\psi}_{\rm M}$ consists then essentially of two disjoint pieces. Thus the Morlet function is interesting for practical wavelet analysis only for $|\vec{k}_o|$ large enough.

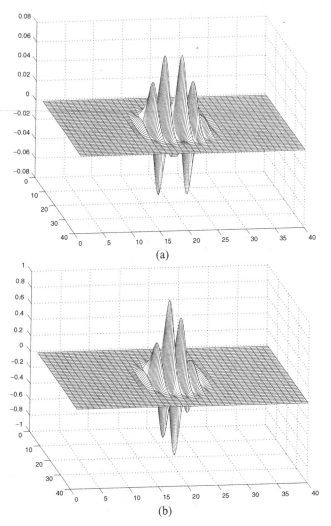

Fig. 3.3. The 2-D Morlet wavelet $\psi_M(\vec{x})$ (or rather $\psi_G(\vec{x})$), with $\epsilon = 2, \vec{k}_o = (0, 6)$ in 3-D perspective, in position domain: (a) real part; (b) imaginary part.

The Morlet wavelet is complex. The modulus of the truncated wavelet ψ_G is a Gaussian, elongated in the x direction if $\epsilon > 1$, and its phase is constant along the direction orthogonal to $\vec{k}_o$ and linear in $\vec{x}$, $\mathrm{mod}(2\pi/|\vec{k}_o|)$, along the direction of $\vec{k}_o$. Thus, plotting the phase of $\psi_G(\vec{x})$ as a function of $\vec{x}$, we get a succession of straight lines, perpendicular to $\vec{k}_o$, and with intensity varying periodically and linearly from 0 to 2π. As compared to the 1-D case, the additional feature here is the inherent directivity of the wavelet ψ_G, entirely contained in its phase. This turns out to be a crucial ingredient in the study of directional features of objects (Chapters 4 and 5). Indeed, from the fact that the WT is a convolution of the signal with the dilated wavelet, we see that the wavelet ψ_G smoothes the signal in all directions, but detects the sharp transitions in the direction

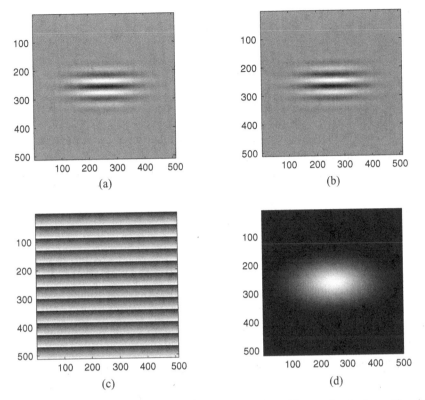

Fig. 3.4. The 2-D Morlet wavelet $\psi_M(\vec{x})$ with $\epsilon = 2$, $k_0 = 6$ in gray levels, in position domain: (a) real part; (b) imaginary part; (c) phase; and (d) modulus.

perpendicular to $\vec{k}_o$. In Fourier space, the effective support ("footprint") of the function $\widehat{\psi}_M$ is an ellipse centered at $\vec{k}_o$ and elongated in the k_y direction, thus contained in a convex cone. Since the ratio of the axes is equal to $\sqrt{\epsilon}$, the cone becomes narrower as ϵ increases. Clearly this wavelet will detect preferentially singularities (edges) in the x direction, and its angular selectivity increases with $|\vec{k}_o|$ and with the anisotropy ϵ. The best selectivity will be obtained by taking $\vec{k}_o$ parallel to the long axis of the ellipse in $\vec{k}$-space, that is, $\vec{k}_o = (0, k_o)$. The Morlet wavelet ψ_M (or rather ψ_G) then becomes (see Figures 3.3 and 3.4):

$$\psi_G(\vec{x}) = \exp(ik_o y) \, \exp[-\tfrac{1}{2}(\tfrac{1}{\epsilon}x^2 + y^2)], \quad \vec{x} = (x, y). \tag{3.20}$$

Many variants of the basic wavelets may be designed for specific problems. For instance, we know that the Mexican hat is very good at detecting discontinuities (e.g., edges) in an image, but it is not directional. On the other hand, the Morlet wavelet is directional, but mostly selective in spatial frequency (these statements will be proved in Section 3.4). Both properties may be combined in a single wavelet, the *Gabor* (or modulated) *Mexican hat* wavelet, defined as follows:

$$\psi_{GM}(\vec{x}) = -(\epsilon \frac{\partial^2}{\partial x^2} + \frac{\partial^2}{\partial y^2})\{\exp(ik_o y) \exp[-\frac{1}{2}(\frac{1}{\epsilon}x^2 + y^2)]\}, \quad \epsilon \geqslant 1, \qquad (3.21)$$

$$= (2 - \frac{1}{\epsilon}x^2 - (y - ik_o)^2)\exp(ik_o y) \exp[-\frac{1}{2}(\frac{1}{\epsilon}x^2 + y^2)], \qquad (3.22)$$

$$\widehat{\psi}_{GM}(\vec{k}) = \sqrt{\epsilon}(\epsilon k_x^2 + k_y^2) \exp[-\frac{1}{2}(\epsilon k_x^2 + (k_y - k_o)^2)]. \qquad (3.23)$$

Notice that no correction term is needed here, the function ψ_{GM} is admissible as it stands. This wavelet, introduced in [289], is extremely efficient in detecting edges, even in the presence of heavy noise. One of its possible applications may be character recognition (see Section 4.1).

Going back to the Gabor or (truncated) Morlet wavelet (3.20), we notice that it has the general form

$$\psi_D(\vec{x}) = \phi(x)\psi(y), \quad \vec{x} = (x, y), \qquad (3.24)$$

where ϕ is a 1-D scaling function (a bump, typically a Gaussian) and ψ is a 1-D wavelet. Functions of this type provide an easy way to design a separable, yet directional wavelet; in the present case, a horizontal one. This technique is due to Bournay Bouchereau [Bou97] (see Section 5.2.2).

A related example is the Gabor-like wavelet of Unser [357], obtained by replacing the Gaussian in (3.20) by another window function, typically a B-spline.

3.3.3   End-stopped wavelets

A basic problem in the characterization of an image, for instance, in the comparison of two images, is the identification of specific features. In human vision, this is achieved by the so-called *saccadic* movement of the eyes: the eyes scan the scene freely, momentarily focus on some point of interest, and quickly move on to the next target point in the scene [Yar67]. The visual jump from one target to another is called a saccade, the target points of consecutive saccades being points of interest which stand out against the general background of the scene. This process has been analyzed in detail and modeled with wavelets by Bhattacharjee [Bha99], that we now quote.

... There is evidence that such features of interest are identified by the lower levels of the visual system, and are not a result of conscious reasoning. Psychovisual studies on several mammals show the presence of certain cells, called *hypercomplex* cells or *end-stopped* cells, in the primary visual cortex. End-stopping behavior is related to oriented linear stimuli, that is, end-stopped cells are activated, under certain conditions, by linear stimuli having a particular orientation. Two kinds of end-stopping behavior have been identified. The *single* end-stopped cells respond strongly if a line in a particular orientation ends within the receptive field of the cell. For real-world scenes, these cells respond strongly to corners, or points of high curvature in general. Some other cells respond strongly only to short, oriented, linear stimuli. Such cells are called *double* or *complete* end-stopped cells. For cells of this kind, the response is strong as long as the stimulus has a specific orientation (the characteristic orientation varies from cell to cell), and both ends of the stimulus lie within the receptive field of the cell. Thus, double end-stopped cells respond to short linear segments in images.

In order to model faithfully this physiological behavior, Bhattacharjee [76,Bha99] introduces two specific wavelets, called *end-stopped wavelets*, that we now describe. According to the discussion above, the responses of end-stopped cells are related to end points of linear structures lying in a specific orientation. Thus, end-stopping can be simulated by isolating linear structures in the image that have a particular orientation, and then processing these structures further to locate their end points, or to determine their lengths.

The first stage, to detect lines having a specific orientation, can be achieved with a Morlet (more properly, a Gabor) wavelet. Then, the end-points of a line can be detected by applying the first derivative of a Gaussian filter along the line. Combining the two operations yields the end-stopped wavelet ψ_{EI}, namely

$$\psi_{\text{EI}}(\vec{x}) = \tfrac{1}{4}\, x\, \exp[-\tfrac{1}{4}\{(x^2 + y^2) + k_o(k_o - 2iy)\}], \tag{3.25}$$

in position space, and

$$\widehat{\psi}_{\text{EI}}(\vec{k}) = \exp[-\tfrac{1}{2}(k_x^2 + (k_y - k_o)^2)]\, \left\{ -ik_x \exp[-\tfrac{1}{2}(k_x^2 + k_y^2)] \right\} \tag{3.26}$$

in spatial frequency space. Equation (3.26) clearly shows that $\widehat{\psi}_{\text{EI}}$ is the product of two components. The first factor is a Morlet wavelet, with wave vector $\vec{k}_o = (0, k_o)$ oriented along the k_y-axis. The second factor is a first derivative of a Gaussian oriented along the frequency axis k_x, that is, in the direction perpendicular to the orientation of the Morlet wavelet. On the other hand, if one regroups the two Gaussian factors in (3.26), the result is then simply the derivative in the k_x-direction of a Morlet wavelet of width $\sigma = 1/\sqrt{2}$ and wave vector $\tfrac{1}{2}\vec{k}_o$ (up to a multiplicative constant). An example is shown, in spatial frequency space, in Figure 3.5.

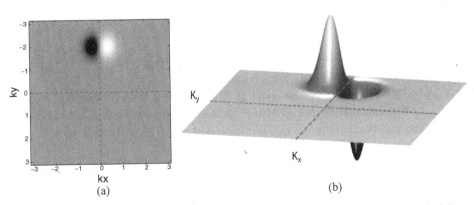

(a)   (b)

Fig. 3.5. The end-stopped wavelet $\widehat{\psi}_{\text{EI}}(\vec{k})$ in spatial frequency space: (a) in gray levels; (b) in 3-D perspective.

As we will see in Section 4.1, the end-stopped wavelet ψ_{E1} performs extremely well for detecting corners in an image. This can be understood intuitively as follows. According to the figure, $\widehat{\psi}_{E1}$ may be interpreted as the difference between two Morlet wavelets. Thus an infinite rod oriented along the k_y-axis will not be seen, but if we take a finite segment, the end-points will appear. On the other hand, if the rod or the segment are slightly misaligned, they will be seen. Thus the end-stopped wavelet ψ_{E1} detects direction by a zero-crossing, hence it has a better resolution compared to the plain Morlet wavelet.

If one wants to detect, in addition, the size l of a linear structure, oriented in a certain direction θ, one may use the double end-stopped wavelet, ψ_{E2}. This wavelet indeed produces a strong response at the center of any linear structure that is oriented at θ and has a length close to l. Compared with ψ_{E1}, the new wavelet is obtained by replacing the first derivative of a Gaussian by a second derivative, that is, a Mexican hat. The latter is used to determine the range of lengths to which the wavelet should respond. Its efficiency is easy to understand. Quoting [Bha99] again,

... The Mexican hat is characterized by an excitatory central region surrounded by an inhibitory region. In 1-D, consider a linear stimulus of length $l = l_1$, and a Mexican hat filter which has an excitatory region of length $l' > l_1$. The filter will produce the strongest response when it is exactly centered over the stimulus. Moreover, as l increases, the maximum response of the filter will initially increase. This response will increase as long as the length $l \leqslant l'$. After l exceeds l', a part of the stimulus will fall in the inhibitory region of the filter, thus damping the response of the filter. Thus, given several linear stimuli of different lengths, the Mexican hat filter responds most strongly to the stimulus which is closest to (but not greater than) the width of the excitatory portion of the filter.

Thus we obtain the required wavelet by taking, as before, a Morlet wavelet of unit width, with wave-vector $\vec{k}_o = [0, k_o]$ oriented along the k_y-axis, multiplied by a Mexican hat of width σ, oriented along the k_x-axis, i.e., perpendicular to the orientation of the Morlet factor. This yields for the double end-stopped wavelet ψ_{E2} the function

$$\psi_{E2}(\vec{x}) = \frac{2\sigma^2(x^2 + \sigma^2 + 1)}{(\sigma^2 + 1)^3} \exp\left(-\frac{\sigma^2(x^2 + y^2) + k_o(k_o - 2i\sigma^2 y)}{2(\sigma^2 + 1)}\right), \qquad (3.27)$$

$$\widehat{\psi}_{E2}(\vec{k}) = -\frac{1}{8\sigma^4}(k_x^2 - \sigma^2)\exp[-\tfrac{1}{2\sigma^2}(k_x^2 + k_y^2)]\exp[-\tfrac{1}{2}(k_x^2 + (k_y - k_o)^2)].$$

As for the pure Mexican hat, the width parameter σ allows us to control the resolving power of the wavelet.

The two end-stopped wavelets, ψ_{E1} and ψ_{E2}, have been used successfully in [Bha99] for the detection of characteristic features in images, in the general context of image retrieval. We will discuss this application in more details in Section 4.3.

3.3.4   Conical wavelets

In order to achieve a genuinely oriented wavelet, it suffices to consider a smooth function $\widehat{\psi}^{(C)}(k)$ with support in a strictly convex cone $\mathcal{C}$ in spatial frequency space and behaving inside $\mathcal{C}$ as $P(k_1, \ldots, k_n)e^{-\zeta \cdot k}$, with $\zeta \in \mathcal{C}$ and $P(.)$ denotes a polynomial in n variables.

Alternatively one may replace the exponential by a Gaussian, which gives a better localization in spatial frequency. In both cases, the resulting wavelets will be called *conical*.

We begin with the former case, thus obtaining the class of *Cauchy* wavelets [18,19,24]. For simplicity, we consider a strictly convex cone, symmetric with respect to the positive k_x-axis, namely

$$\mathcal{C} \equiv \mathcal{C}(-\alpha, \alpha) = \{\vec{k} \in \mathbb{R}^2 \mid -\alpha \leqslant \arg \vec{k} \leqslant \alpha, \, \alpha < \pi/2\},$$

that is, the convex cone determined by the unit vectors $\vec{e}_{-\alpha}, \vec{e}_{\alpha}$. The dual cone, with sides perpendicular to those of the first one, is also convex and reads:

$$\widetilde{\mathcal{C}} = \mathcal{C}(-\tilde{\alpha}, \tilde{\alpha}) = \{\vec{k} \in \mathbb{R}^2 \mid \vec{k} \cdot \vec{k}' > 0, \, \forall \vec{k}' \in \mathcal{C}(-\alpha, \alpha)\},$$

where $\tilde{\alpha} = -\alpha + \pi/2$. Therefore $\vec{e}_{-\alpha} \cdot \vec{e}_{\tilde{\alpha}} = \vec{e}_{\alpha} \cdot \vec{e}_{-\tilde{\alpha}} = 0$.

Given the fixed vector $\vec{\eta} = (\eta, 0), \eta > 0$, we first define the (symmetric) *Cauchy wavelet* in spatial frequency variables:

$$\widehat{\psi}_m^{(\alpha)}(\vec{k}) = \begin{cases} (\vec{k} \cdot \vec{e}_{\tilde{\alpha}})^m \, (\vec{k} \cdot \vec{e}_{-\tilde{\alpha}})^m \, e^{-\vec{k} \cdot \vec{\eta}}, & \vec{k} \in \mathcal{C}(-\alpha, \alpha) \\ 0, & \text{otherwise.} \end{cases} \tag{3.28}$$

The Cauchy wavelet $\widehat{\psi}_m^{(\mathcal{C})}(\vec{k})$ is strictly supported in the cone $\mathcal{C}(-\alpha, \alpha)$ and the parameter $m \in \mathbb{N}^*, m \geqslant 1$, gives the number of vanishing moments of $\widehat{\psi}$ on the edges of the cone, and thus controls the regularity of the wavelet.

An interesting aspect of this wavelet is that its inverse Fourier transform may be calculated exactly [24], with the result:

$$\psi_m^{(\alpha)}(\vec{x}) = \frac{(-1)^{m+1}}{2\pi} \, (m!)^2 \, \frac{(\sin 2\alpha)^{2m+1}}{[\vec{z} \cdot \sigma(\alpha)\vec{z}]^{m+1}}, \tag{3.29}$$

where we have introduced the complex variable $\vec{z} = \vec{x} + i\vec{\eta} \in \mathbb{R}^2 + i\widetilde{\mathcal{C}}$ and the 2×2 matrix

$$\sigma(\alpha) = \begin{pmatrix} \cos^2 \alpha & 0 \\ 0 & -\sin^2 \alpha \end{pmatrix}.$$

Indeed, from the definition (3.28), we get:

$$\psi_m^{(\alpha)}(\vec{x}) = \frac{1}{2\pi} \int_{\mathcal{C}(-\alpha,\alpha)} d^2\vec{k} \, e^{i\vec{k}\cdot\vec{x}} \, (\vec{k} \cdot \vec{e}_{\tilde{\alpha}})^m \, (\vec{k} \cdot \vec{e}_{-\tilde{\alpha}})^m \, e^{-\vec{k}\cdot\vec{\eta}}$$

$$= \frac{(-1)^m}{2\pi} \, [\vec{e}_{\tilde{\alpha}} \cdot \vec{\nabla}_{\vec{x}}]^m \, [\vec{e}_{-\tilde{\alpha}} \cdot \vec{\nabla}_{\vec{x}}]^m \int_{\mathcal{C}(-\alpha,\alpha)} d^2\vec{k} \, e^{-\vec{k}\cdot(\vec{\eta} - i\vec{x})}.$$

The integral on the right-hand side is convergent, since $\vec{k} \cdot \vec{\eta} > 0$. Write $\vec{\xi} = \vec{\eta} - i\vec{x} = -i(\vec{x} + i\vec{\eta})$ and let A be the matrix that maps the unit vectors $\vec{e}_1, \vec{e}_2$ onto $\vec{e}_{-\alpha}, \vec{e}_{\alpha}$, respectively: $(\vec{e}_\nu)^i = A_\nu{}^j \, (\vec{e}_j)^i, \, \nu = \pm\alpha$ (where we use the usual summation convention), so that $k^j = A_\nu{}^j \, k^\nu$ (contravariant coordinates) and $\xi_\nu = A_\nu{}^j \, \xi_j$ (covariant coordinates). Explicitly, we have:

$$A = \begin{pmatrix} \cos\alpha & \cos\alpha \\ -\sin\alpha & \sin\alpha \end{pmatrix}, \quad \text{so that} \quad \det A = \sin 2\alpha.$$

In the new (nonorthogonal) coordinates, the cone becomes

$$\mathcal{C}(-\alpha,\alpha) = \{\vec{k} \in \mathbb{R}^2 : k^\nu \geqslant 0, \, \nu = \pm\alpha\},$$

and the integral may be evaluated immediately:

$$\int_{\mathcal{C}(-\alpha,\alpha)} d^2\vec{k} \, e^{-\vec{k}\cdot\vec{\xi}} = \int_{\mathcal{C}(-\alpha,\alpha)} dk^1 \, dk^2 \, \exp(-A_\nu^{\,j} k^\nu \xi_j)$$

$$= \det A \int_0^\infty dk^\alpha \int_0^\infty dk^{-\alpha} \, \exp(-k^\nu \xi_\nu),$$

$$= \frac{\det A}{\xi_\alpha \, \xi_{-\alpha}}$$

$$= \frac{\sin 2\alpha}{(\vec{e}_\alpha \cdot \vec{\xi})(\vec{e}_{-\alpha} \cdot \vec{\xi})}$$

$$= \frac{-\sin 2\alpha}{[(\vec{x} + i\vec{\eta}) \cdot \vec{e}_\alpha][(\vec{x} + i\vec{\eta}) \cdot \vec{e}_{-\alpha}]}.$$

Then the result follows by differentiation, if one remembers that $\vec{e}_{\tilde{\alpha}} \cdot \vec{e}_{-\alpha} = \vec{e}_{-\tilde{\alpha}} \cdot \vec{e}_\alpha = 0$, $\vec{e}_{\tilde{\alpha}} \cdot \vec{e}_\alpha = \vec{e}_{-\tilde{\alpha}} \cdot \vec{e}_{-\alpha} = \sin 2\alpha$. Indeed:

$$(\vec{e}_{\tilde{\alpha}} \cdot \vec{\nabla}_{\vec{x}}) \frac{1}{[(\vec{x} + i\vec{\eta}) \cdot \vec{e}_{-\alpha}]} = \frac{\vec{e}_{\tilde{\alpha}} \cdot \vec{e}_{-\alpha}}{[(\vec{x} + i\vec{\eta}) \cdot \vec{e}_{-\alpha}]^2} = 0,$$

$$(\vec{e}_{\tilde{\alpha}} \cdot \vec{\nabla}_{\vec{x}})^m \frac{1}{[(\vec{x} + i\vec{\eta}) \cdot \vec{e}_\alpha]} = \frac{(-1)^m \, m! \, (\vec{e}_{\tilde{\alpha}} \cdot \vec{e}_\alpha)^m}{[(\vec{x} + i\vec{\eta}) \cdot \vec{e}_\alpha]^{m+1}}$$

$$= \frac{(-1)^m \, m! \, (\sin 2\alpha)^m}{[(\vec{x} + i\vec{\eta}) \cdot \vec{e}_\alpha]^{m+1}},$$

and similarly for the other factor. Thus one obtains as the final result the function $\psi_m^{(\alpha)}$ given in (3.29). Clearly, this function is square integrable, and admissible in the sense of (6.15), in other words, it is a wavelet. We show in Figures 3.6 and 3.7 various aspects of the Cauchy wavelet $\psi_4^{(10)}$. This is manifestly a highly directional filter, strictly supported in the cone $\mathcal{C} = \mathcal{C}(-10°, 10°)$.

Notice the slow decay in $\vec{x}$-space of the conical wavelet $\psi_m^{(\alpha)}$, independent of α. This is the price to pay for forcing the wavelet to be strictly supported in a cone, and is to be expected in the light of standard results on the localization properties of wavelets (theorems of Balian–Low and Battle) [Fei98,36]. However, as we have already stressed in Section 3.3.1.2, this is irrelevant for the analysis, only the behavior in $\vec{k}$-space counts.

The construction generalizes in a straightforward way [24] to an arbitrary convex cone $\mathcal{C} \equiv \mathcal{C}(\alpha, \beta) = \{k \in \mathbb{R}^2 \,|\, \alpha \leqslant \arg k \leqslant \beta\}$, with dual $\tilde{\mathcal{C}} \equiv \mathcal{C}(\tilde{\alpha}, \tilde{\beta}) = \{k \in \mathbb{R}^2, k \cdot k' > 0, \, \forall k' \in \mathcal{C}(\alpha, \beta)\}$, where $\tilde{\beta} = \alpha + \pi/2$, $\tilde{\alpha} = \beta - \pi/2$, and arbitrary vanishing

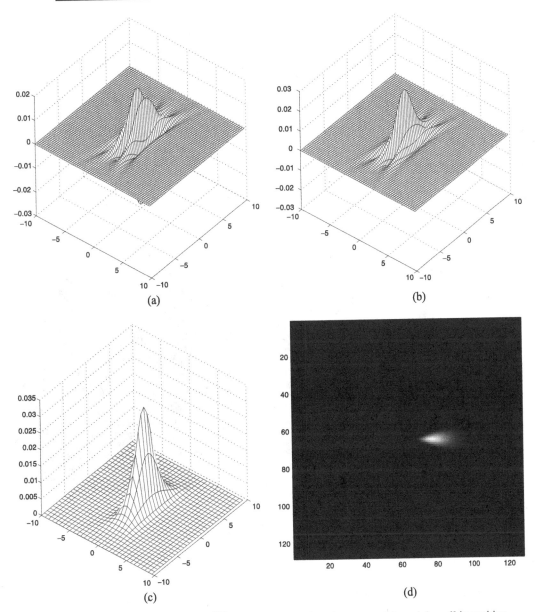

Fig. 3.6. The Cauchy wavelet $\psi_4^{(10)}$: (a) real part; (b) imaginary part; (c) modulus, all in position domain; and (d) modulus in frequency domain.

moments $l, m > 0$ on the cone edges. For any fixed $\vec{\eta} \in \widetilde{\mathcal{C}}$, the resulting wavelet reads in spatial frequency space as

$$\widehat{\psi}_{lm}^{(\mathcal{C})}(k) = \begin{cases} (k \cdot e_{\tilde{\alpha}})^l \, (k \cdot e_{\tilde{\beta}})^m \, e^{-k \cdot \eta}, & k \in \mathcal{C}(\alpha, \beta); \\ 0, & \text{otherwise,} \end{cases} \tag{3.30}$$

and in position space,

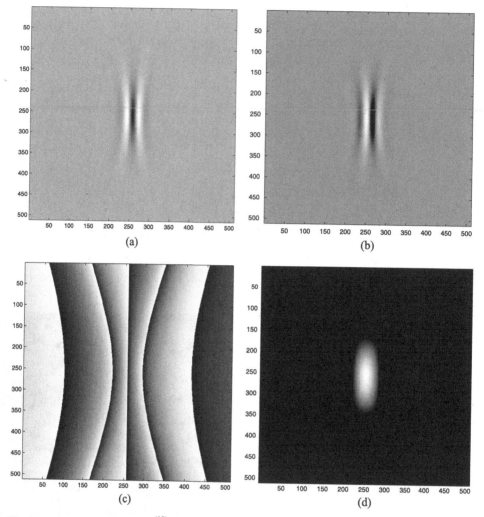

Fig. 3.7. The Cauchy wavelet $\psi_4^{(10)}$, in gray levels, in position domain: (a) real part; (b) imaginary part; (c) phase; and (d) modulus.

$$\psi_{lm}^{(C)}(x) = \text{const.} \ (\vec{z} \cdot \vec{e}_\alpha)^{-l-1} \ (\vec{z} \cdot \vec{e}_\beta)^{-m-1}, \tag{3.31}$$

where again $\vec{z}$ denotes the complex variable $\vec{z} = \vec{x} + i\vec{\eta} \in \mathbb{R}^2 + i\widetilde{C}$.

Actually the origin of the name "Cauchy" is the following example. For $\alpha = 0, \beta = \pi/2, \eta = e_{\pi/4}$ and $m = 1$, one gets:

$$\psi_1^{(C)}(x) = \frac{1}{2\pi} \ (1 - ix)^{-2}(1 - iy)^{-2}, \tag{3.32}$$

i.e., the product of two 1-D Cauchy wavelets [Hol95]; that is, derivatives of the Cauchy kernel $(z - t)^{-1}$. Of course, this example is of little use in practice. Indeed, the main interest of Cauchy wavelets is their good angular selectivity, which requires a narrow

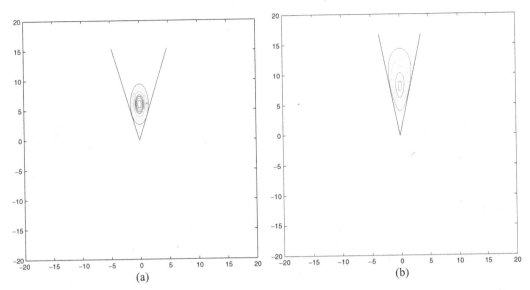

Fig. 3.8. Two directional 2-D wavelets, in $\vec{k}$ space, seen in level curves: (a) the Morlet wavelet $\widehat{\psi}_\mathrm{M}$ ($\vec{k}_o = (0, 6)$, $\epsilon = 5$); (b) the Cauchy wavelet $\widehat{\psi}_4^{(10°)}$ with supporting cone $\mathcal{C}_{10} = \mathcal{C}(-10°, 10°)$, rotated by $90°$ for the sake of comparison.

cone. For applications, it turns out that the wavelet $\psi_4^{(10)}$, with support in the cone $\mathcal{C}_{10} = \mathcal{C}(-10°, 10°)$ has properties very similar to those of the Morlet wavelet (3.17) with $|\vec{k}_o| = 5.6$, except that here the opening angle of the cone is totally controllable. For a Morlet wavelet, on the contrary, the cone gets narrower for increasing $|\vec{k}_o|$, but then the amplitude decreases as $\exp(-|\vec{k}_o|^2)$. In that sense, Cauchy wavelets are better adapted. We show side by side in Figure 3.8 the Morlet wavelet $\widehat{\psi}_\mathrm{M}$ with $\vec{k}_o = (0, 6)$, $\epsilon = 5$ (left) and the Cauchy wavelet $\widehat{\psi}_4^{(10)}(k)$, rotated by $90°$ for the sake of comparison (right). Quantitative comparisons will be made in Section 3.4.

In 1-D, a wavelet ψ is called *progressive* or a Hardy function [Hol95,205], if $\widehat{\psi}(\omega) = 0$ for $\omega < 0$. This in turn may be expressed in terms of the Hilbert transform, defined by $\widehat{Hf}(\omega) = -i \operatorname{sign} \omega\, f(\omega)$, namely

$$\psi = (1 + iH)\phi, \ \phi \in L^2(\mathbb{R}, dt)$$

(that is, ψ is the analytic signal associated to ϕ). Equivalently, ψ belongs to the Hardy space $H_+^2(\mathbb{R})$ of square integrable functions which extend analytically into the upper half-plane. We claim that the conical wavelets are the 2-D analogs of this concept; that is, the genuine 2-D progressive wavelets.

In order to prove that statement, we first notice that the convex cone $\mathcal{C}(\alpha, \beta)$ may also be expressed in terms of the covariant coordinates $k_{\bar{\nu}} = (\vec{e}_{\bar{\nu}} \cdot \vec{k})$, $\nu = \alpha, \beta$:

$$\mathcal{C}(\alpha, \beta) = \{\vec{k} \in \mathbb{R}^2 : k_{\bar{\alpha}} \geqslant 0, k_{\bar{\beta}} \geqslant 0\}. \tag{3.33}$$

Consider the directional Hilbert transforms:

$$\widehat{H_{\tilde{\nu}}f}(\vec{k}) = -i \operatorname{sign} k_{\tilde{\nu}} f(\vec{k}). \tag{3.34}$$

Given $\phi \in L^2(\mathbb{R}^2, d^2\vec{x})$, define the function

$$\psi = (1 + iH_{\tilde{\alpha}} + iH_{\tilde{\beta}} - H_{\tilde{\alpha}}H_{\tilde{\beta}})\phi$$
$$= (1 + iH_{\tilde{\alpha}})(1 + iH_{\tilde{\beta}})\phi. \tag{3.35}$$

Then it is easy to see, as in [366], that $\widehat{\psi}(\vec{k})$ vanishes outside the cone $\mathcal{C}(\alpha, \beta)$, and indeed:

$$\widehat{\psi}(\vec{k}) = \begin{cases} 4\widehat{\phi}(\vec{k}), & \vec{k} \in \mathcal{C}(\alpha, \beta), \\ 0, & \text{otherwise.} \end{cases} \tag{3.36}$$

Therefore the inverse Fourier transform $\psi(\vec{x})$ is the boundary value of a function $\psi(\vec{z})$ holomorphic in the tube $\mathbb{R}^2 + i\widetilde{\mathcal{C}}$; i.e., a 2-D Hardy function. For a fixed convex cone $\mathcal{C}(\alpha, \beta)$, the set of all such functions constitutes a Hilbert space, naturally denoted $H^2_{(\alpha, \beta)}$, which is unitary equivalent, via the complex Fourier transform, to the space $L^2(\widetilde{\mathcal{C}}(\alpha, \beta), d^2\vec{k})$ [Ste71; Theorem VI.3.1]. In that sense, conical wavelets are a genuine multidimensional generalization of the 1-D Hardy functions, much more so than the so-called 2-D Hardy functions defined by Dallard and Spedding (in particular, their "Arc" wavelet) [117]. Among them Cauchy wavelets are particularly simple (they occupy a special niche, as we will see in Section 8.2). In conclusion, notice that we are talking here of a Hardy space $H^2_{(\alpha, \beta)}$, but similar considerations may be made for Hardy spaces $H^1_{(\alpha, \beta)}$, in terms of the Riesz operators, which are a natural multidimensional generalization of the Hilbert transform [366].

Cauchy wavelets have a good angular selectivity, provided one chooses a narrow cone. However their radial selectivity is not terribly good, because the exponential decays rather slowly as $|\vec{k}| \to \infty$. In order to obtain a better radial localization, one may replace the exponential by a Gaussian in k_x [Vdg98]. This has the effect to concentrate the wavelet on its central frequency $(\sqrt{2m}, 0)$. The resulting wavelet is called the *Gaussian conical* wavelet. The *angular* selectivity of ψ is specified by the angular aperture of the cone and is well controlled by the parameter α. However, the radial selectivity is only roughly fixed by the moment number m. The resulting wavelet is very similar to the Cauchy wavelet, except that it is more concentrated in spatial frequency space, since it is also localized in scale, around the central scale a_o. However, although the pure Gaussian is well peaked, the addition of a large number of vanishing moments tends to spread it. Thus, one can achieve an even better scale localization by using an appropriate width $\sigma > 0$ for the Gaussian. We may still improve on this by adding another parameter $\chi(\sigma) > 0$, whose sole rôle is to control the radial support of ψ [26,27]. This leads to the following formula for our conical wavelet:

$$\widehat{\psi}_{\text{c}}(\vec{k}) = \begin{cases} (\vec{k} \cdot \vec{e}_{-\tilde{\alpha}})^m (\vec{k} \cdot \vec{e}_{\tilde{\alpha}})^m \, e^{-\frac{\sigma}{2}(k_x - \chi(\sigma))^2}, & \vec{k} \in C(-\alpha, \alpha), \\ 0, & \text{otherwise,} \end{cases} \tag{3.37}$$

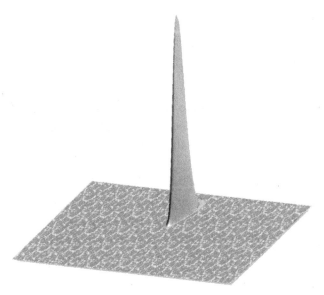

Fig. 3.9. The Gaussian conical wavelet (3.37), in frequency space, for $m = 4$, $\alpha = 10°$, $\sigma = 1$.

where $\chi(\sigma) = \sqrt{2m}\,\frac{\sigma - 1}{\sigma}$ is called the *center correction* term. Notice that the central frequency is the point $(\sqrt{2m}, 0)$ for any σ, and for $\sigma = 1$, one recovers the pure Gaussian conical wavelet. This is the wavelet we will mostly use in the sequel. It is shown in frequency domain, in Figure 3.9. It is clear on this picture why this wavelet is sometimes called the shark wavelet!

Another alternative is the *conical Mexican hat*, introduced by Murenzi *et al.* [289]. The wavelet with support in $C = C(\beta)$ is defined, for any $\vec{\eta} \in \widetilde{C}$ and $l, m \in N^*$, as

$$\widehat{\psi}_{lm}^{(C)}(\vec{k}) = \begin{cases} (\vec{k} \cdot \vec{e}_{-\tilde{\beta}})^m \, (\vec{k} \cdot \vec{e}_{\tilde{\beta}})^l \, (\epsilon k_x^2 + k_y^2) e^{-\frac{1}{2}(\epsilon k_x^2 + k_y^2)}, & \vec{k} \in C(\beta) \\ 0, & \text{otherwise.} \end{cases} \tag{3.38}$$

Besides these directional wavelets, there exist two other tools especially designed for the detection of lines or curves, called *ridgelets* and *curvelets*, respectively. These actually define new transforms, that we will discuss later, in Section 11.1.

3.3.5   Multidirectional wavelets

Given a directional wavelet ψ, as above, it is easy to build a multidirectional one, with n-fold symmetry simply by superposing n suitably rotated copies of ψ:

$$\psi_n(\vec{x}) = \frac{1}{n} \sum_{k=0}^{n-1} \psi(r_{-\theta_k}(\vec{x})), \ \theta_k = k\frac{2\pi}{n}, \ k = 0, 1, \ldots, n - 1. \tag{3.39}$$

Taking, for instance, $n = 4$ and for ψ a Gabor (truncated Morlet) wavelet, we get the following real wavelet with four-fold symmetry:

$$\psi_{4M}(x, y) = \frac{1}{2}(\cos k_o x + \cos k_o y)\, e^{-\frac{1}{2}(x^2 + y^2)}. \tag{3.40}$$

This wavelet filters out all features which are not primarily horizontal or vertical. In the same way, one gets wavelets with symmetry 6 or 10, which may find applications, respectively, in biological problems or the analysis of quasicrystals. In general, multidirectional wavelets should be useful for pattern recognition.

Notice that a similar construction was proposed by Watson [369,370]. His *fan filters* are obtained by taking first the difference between two "mesa" functions, which yields an annular wavelet, and then repeatedly bisecting the spatial frequency space and taking only one side (i.e., the associated analytic signal). The allowed directions θ are thus restricted to a fan-shaped region:

$$0 \leqslant 2\theta \leqslant \frac{2\pi}{2^{n-1}} \quad (n = 2, 3, \dots). \tag{3.41}$$

This construction may then be generalized to arbitrary angles [306]. Actually, a comparison with Section 2.6 immediately shows that the construction of directional wavelet packets is based on the very same idea.

These fan filters have all the properties of directional wavelets, including admissibility in the form (2.17). Applying on these filters discrete rotations and scaling, Watson builds a pyramid of oriented filters as a tool for data compression and signal reconstruction after coding, in a model of human vision. This is in fact a discretized version (in polar geometry) of the CWT. Another example, very similar to the previous one, is that of the steerable filters, described in Section 2.7.

3.4   Wavelet calibration: evaluating the performances of the CWT

Given a wavelet, what is its angular and scale selectivity (resolving power)? What is the minimal sampling grid for the reconstruction formula (2.26) that guarantees that no information is lost? The answer to both questions resides in a *quantitative* knowledge of the properties of the wavelet, that is, the tool must be *calibrated*.

To that effect, one takes the WT of particular, standard signals. Three such tests are useful, and in each case the outcome may be viewed either at fixed (a, θ) (position representation) or at fixed $\vec{b}$ (scale-angle representation).

- *Point signal:* for a snapshot at the wavelet itself, one takes as the signal a delta function, i.e., one evaluates the impulse response of the filter:

$$\langle \psi_{a,\theta,\vec{b}} | \delta \rangle = a^{-1}\, \overline{\psi(a^{-1} r_{-\theta}(-\vec{b}))}. \tag{3.42}$$

- *Reproducing kernel:* taking as the signal the wavelet ψ itself, one obtains the reproducing kernel K, which measures the *correlation length* in each variable $a, \theta, \vec{b}$:

$$c_\psi \, K(a, \theta, \vec{b} | 1, 0, \vec{0}) = \langle \psi_{a,\theta,\vec{b}} | \psi \rangle = a^{-1} \int \overline{\psi(a^{-1} r_{-\theta}(\vec{x} - \vec{b}))} \, \psi(\vec{x}) \, d^2\vec{x}. \quad (3.43)$$

- *Benchmark signals:* for testing particular properties of the wavelet, such as its ability to detect a discontinuity or its angular selectivity in detecting a particular direction, one may use appropriate "benchmark" signals.

3.4.1   The scale and angle resolving power

Suppose the wavelet ψ has its effective support in spatial frequency in a vertical cone of aperture $\Delta\varphi$, corresponding to $\vec{k}_o = (0, k_o)$. The width of $\widehat{\psi}$ in the x and y directions is given by $2w_x$ and $2w_y$, respectively:

$$w_x = \frac{1}{\|\widehat{\psi}\|} \left[\int d^2\vec{k} \, k_x^2 |\widehat{\psi}(\vec{k})|^2 \right]^{1/2}, \quad w_y = \frac{1}{\|\widehat{\psi}\|} \left[\int d^2\vec{k} \, (k_y - k_o)^2 |\widehat{\psi}(\vec{k})|^2 \right]^{1/2}. \tag{3.44}$$

Then the wavelet $\widehat{\psi}$ is concentrated in an ellipse of semi-axes w_x, w_y, and its radial support is $k_o - w_y \leqslant |\vec{k}| \leqslant k_o + w_y$. Thus the scale width or scale resolving power (SRP) of ψ is defined as:

$$SRP(\psi) = \frac{k_o + w_y}{k_o - w_y}. \tag{3.45}$$

In the same way, one defines the angular width or angular resolving power (ARP) by considering the tangents to that ellipse. Then a straightforward calculation yields:

$$ARP(\psi) = 2 \cot^{-1} \frac{\sqrt{k_o^2 - w_y^2}}{w_x} \simeq \Delta\varphi. \tag{3.46}$$

For instance, if ψ is the (truncated) Morlet wavelet (3.17), one obtains:

$$SRP(\psi_{\mathrm{M}}) = \frac{k_o\sqrt{2} + 1}{k_o\sqrt{2} - 1}, \quad ARP(\psi_{\mathrm{M}}) = 2 \cot^{-1} \sqrt{\epsilon(k_o^2 - 1)}, \tag{3.47}$$

and, for $k_o \gg 1$:

$$ARP(\psi_{\mathrm{M}}) = 2 \cot^{-1}(k_o\sqrt{\epsilon}). \tag{3.48}$$

This last expression coincides with the empirical result of [13]: the angular sensitivity of ψ_{M} depends only on the product $k_o\sqrt{\epsilon}$. Notice also that the SRP is independent of the anisotropy factor ϵ.

    If ψ is the Cauchy wavelet (3.28) with support in the cone $\mathcal{C}(-\alpha, \alpha)$, the ARP is simply the opening angle 2α of the supporting cone.

Fig. 3.10. Filter bank obtained with the Morlet wavelet (5 scales, 16 orientations).

3.4.2   The reproducing kernel and the resolving power of the wavelet

A natural way of testing the correlation length of the wavelet is to analyze systematically its reproducing kernel. Let the effective support of the wavelet ψ in spatial frequency be, in polar coordinates, $\Delta\rho$ and $\Delta\varphi$. Then an easy calculation [18] shows that the effective support of K is given by $a_{min} = (\Delta\rho)^{-1} \leqslant a \leqslant a_{max} = \Delta\rho$ for the scale variable, and $-\Delta\varphi \leqslant \theta \leqslant \Delta\varphi$ for the angular variable. Thus we may define the wavelet parameters (or resolving power) $\Delta\rho$, $\Delta\varphi$ in terms of the parameters Δa, $\Delta\theta$ of K, as:

- scale resolving power (SRP): $\Delta\rho = \sqrt{\Delta a} = \sqrt{a_{max}/a_{min}}$;
- angular resolving power (ARP): $\Delta\varphi = \frac{1}{2}\Delta\theta$.

In this way, one may design a wavelet filter bank $\{\widehat{\psi}_{a_j,\theta_\ell}(\vec{k})\}$, which yields a complete tiling of the spatial frequency plane, in polar coordinates [15,18]. An example is given in Figure 3.10, for the case of the Morlet wavelet. Clearly this analysis is only possible within the scale-angle representation. Thus it requires the use of the CWT, and it is outside of the scope of the DWT, which is essentially limited to a Cartesian geometry (see Section 2.6).

3.4.3   Calibration of a wavelet with benchmark signals

The capacity of the wavelet at detecting a discontinuity may be measured on the (benchmark) signal consisting of an infinite rod (see [13] for the full discussion). The result is that both the Mexican hat and the Morlet wavelet are efficient in this respect.

For testing the angular selectivity of a wavelet, one computes the WT of a semi-infinite rod, sitting along the positive x-axis, and modeled as usual with a delta function:

$$s(\vec{x}) = \vartheta(x)\,\delta(y), \tag{3.49}$$

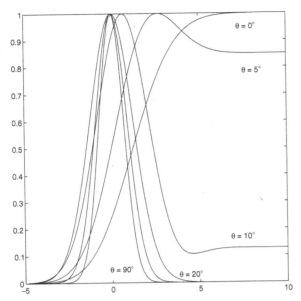

Fig. 3.11. Angular selectivity of the Morlet wavelet for different values of the orientation angle: $\theta = 0°, 5°, 10°, 20°, 45°$ and $90°$. The graph shows the modulus $|S((b_x, 0), 1, \theta)|$ as a function of b_x.

where $\vartheta(x)$ is the step function. Plugging this expression into the definition of the transform yields (we take $a = 1$ for simplicity):

$$S(\vec{b}, 1, \theta) = \int_0^{+\infty} dx\, \overline{\psi\left(r_{-\theta}(x - b_x, -b_y)\right)}. \tag{3.50}$$

Let us take first a Morlet wavelet with $\epsilon = 5$, oriented at an angle θ, and compute the CWT of s as a function of b_x. As illustrated by Figure 3.11, the result is that ψ_M detects the orientation of the rod with a precision of the order of $5°$. Indeed, for $\theta < 5°$, the WT is a "wall," increasing smoothly from 0, for $x \leqslant -5$, to its asymptotic value (normalized to 1) for $x \geqslant 5$. Then, for increasing misorientation θ, the wall gradually collapses, and essentially disappears for $\theta > 10°$. Only the tip of the rod remains visible, and for large θ ($\theta > 45°$), it gives a sharp peak.

Essentially the same result is obtained with a Cauchy wavelet supported in the cone $\mathcal{C}(-10°, 10°)$, of opening angle $ARP = 20°$, as shown in Figure 3.12(a). Conversely, one sees in panel (b) that, for a fixed misorientation angle $\theta = 20°$, the Cauchy wavelet yields the same selectivity for $ARP \leqslant 20°$.

On the contrary, as observed on Figure 3.13, the same test performed with an anisotropic Mexican hat gives a result almost independent of θ. Even varying the anisotropy factor ϵ doesn't really change the result: the discontinuity is detected by a sharper variation, but the sensitivity to its orientation is not greatly improved.

The conclusion is that the Morlet and the Cauchy wavelets are highly sensitive to orientation, but the anisotropic Mexican hat is not.

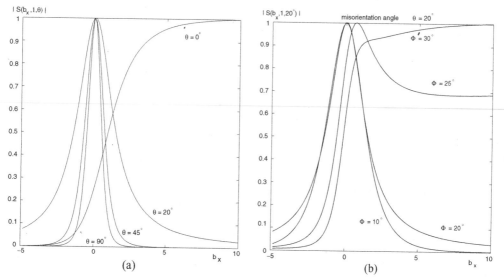

Fig. 3.12. Angular selectivity of the Cauchy wavelet $\psi_{1,1}$: (a) for a cone of fixed width $\Phi = 2\alpha = 20°$ and for different values of the orientation angle: $\theta = 0°, 5°, 10°, 20°, 45°$ and $90°$; (b) for a fixed value of the misorientation angle $\theta = 20°$ and various values of the ARP $\Phi = 2\alpha$. The graph shows the modulus $|S((b_x, 0), 1, \theta)|$, respectively $|S((b_x, 0), 1, 20°)|$, as a function of b_x.

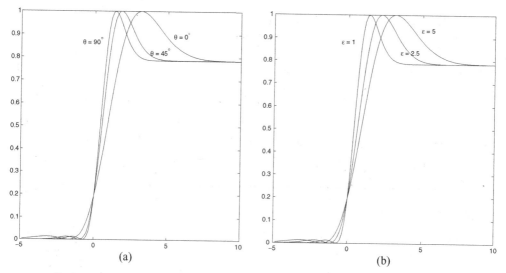

Fig. 3.13. Angular selectivity of the anisotropic Mexican hat wavelet with: (a) anisotropy $\epsilon = 5$ and for different values of the orientation angle ($\theta = 0°, 45°$ and $90°$) and (b) for a fixed orientation ($\theta = 0°$) but various anisotropy factors ($\epsilon = 1, 2.5$ and 5).

Let now the signal be a segment. If one uses a Morlet or a Cauchy wavelet as above, the WT reproduces the segment if the misorientation $\Delta\phi$ between the signal and the wavelet is smaller than $5°$, but the segment becomes essentially invisible for $\Delta\phi > 15°$, except for the tips (these are point singularities). In the end, the image of the segment reduces to two peaks corresponding to the two endpoints (see Figure 3.14).

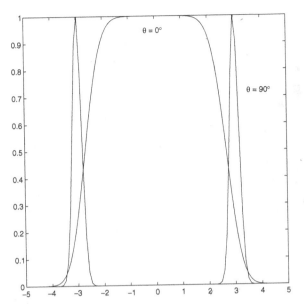

Fig. 3.14. Modulus of the wavelet transform of a segment positioned at $\theta = 0°$, using a Morlet wavelet for several values of the orientation angle θ.

This is exactly the property used crucially in the measurement of the velocity field of a turbulent fluid (see Section 5.3.3 below).

One may remark that the precision mentioned here is obtained with the modulus of the WT. In fact, if the wavelet is complex (like ψ_M), one may also exploit the phase of the WT, and it gives a higher precision yet [13]. But this is practical only on academic signals, real data are in general too noisy and only the modulus is useful.

Another way of comparing the angular selectivity of the two wavelets is to analyze a directional signal in the angle–angle representation (α, θ) described above. The result confirms the previous one [17].

In order to illustrate the difference in angular selectivity between the anisotropic Mexican hat and the Morlet wavelet, we analyze a directional signal with both of them and view the transform in the angle–angle representation described in Section 2.3.3. The result is shown in Figure 3.15. The signal is a rectangular slab of size 3×2, positioned radially at $\pi/2$ and it is analyzed with an anisotropic Mexican hat with $\epsilon = 2$ (left) and a Morlet wavelet (right). The figures show the modulus of the CWT, at range $|\vec{b}| = 3$ and scale $a = 1$, in the angle–angle representation (α, θ). For the Mexican hat, the transform exhibits a maximum located at $\alpha = \pi/2, \theta = \pi/2$, as it should, and the graph is periodic both in α and in θ. For the Morlet wavelet, several maxima can be distinguished, all located around $\alpha = \pi/2$. A careful inspection shows that the strongest peaks at $\theta = \pi/2$ and $3\pi/2$ correspond to the two longest edges of the rectangle while the smaller peaks at $\theta = 0, \pi, 2\pi$ correspond to the smaller edges. Notice however that, for the Mexican hat, the peak is sharp in α and quite broad in θ, as expected, since this wavelet has a very good resolution in position (α), but a rather poor one in directional

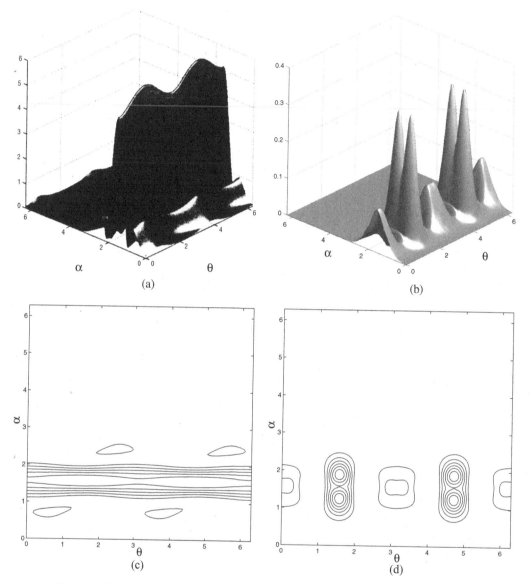

Fig. 3.15. CWT of a slab of size 3×2 positioned radially at $90°$, analyzed with an anisotropic Mexican hat (left) with $\epsilon = 2$ and a Morlet wavelet (right), both at scale $a = 1$. The figures in the top row give a 3-D perspective view of the transform, in the angle–angle representation (α, θ), at range $|\vec{b}| = 3$. The same result is shown in level curves in the figures of the bottom row.

selectivity (θ). In addition, the contour of the slab can be precisely detected by tracking the zero crossings of this representation. On the opposite, the Morlet wavelet gives a sharp peak in both variables, since it has good selectivity in both [13], but the contour is less visible. This example illustrates Daugman's theorem [126], and at the same time the usefulness of the angle–angle representation.

4 Applications of the 2-D CWT. I: image processing

The 2-D CWT has been used by a number of authors, in a wide variety of problems [Com89,Mey91,Mey93]. In all cases, its main use is for the *analysis* of images, since image synthesis or compression problems are rather treated with the DWT. In particular, the CWT can be used for the detection or determination of specific features, such as a hierarchical structure, edges, filaments, contours, boundaries between areas of different luminosity, etc. Of course, the type of wavelet chosen depends on the precise aim. An isotropic wavelet (e.g. a Mexican hat) often suffices for pointwise analysis, but a directional wavelet (e.g. a Morlet or a conical wavelet) is necessary for the detection of oriented features in the signal. Somewhat surprisingly, a directional wavelet is often more efficient in the presence of noise.

In the next two chapters, we will review a number of such applications, including some nonlinear extensions of the CWT. First, in the present chapter, we consider various aspects of image processing. Then, in Chapter 5, we will turn to several fields of physics where the CWT has made an impact. Some of the applications are rather technical and use specific jargon. We apologize for that and refer the reader to the original papers for additional information.

4.1 Contour detection, character recognition

4.1.1 The detection principle

Exactly as in the 1-D case, the WT is especially useful to detect *discontinuities* in images, for instance the *contour* [Mur90,13] or the *edges* of an object (which are discontinuities in the luminosity), and in particular its *corners* [204,266]. For that purpose, one may first ignore the directions and perform a pointwise analysis. Then, the simplest choice is an isotropic wavelet, such as the radial Mexican hat ψ_H given in (3.6) or (3.8). In that particular case, the effect of the WT consists in smoothing the signal with a Gaussian and taking the Laplacian of the result. Thus large values of the amplitude will appear at the location of the discontinuities, in particular the contour of objects. In particular, the corners of the contour, which are point singularities, will be highlighted. In addition, if the wavelet is real, the CWT detects the convexity of each corner: a convex corner gives rise to a sharp positive peak, whereas a concave one yields a negative peak. However, it turns out that other wavelets are useful too. For instance, in the presence of heavy noise, directional wavelets outperform the Mexican hat. On the other hand, if only the corners of the contour are needed, as in character recognition, mixed derivatives of the Gaussian or, even better, the end-stopped wavelets become the first choice.

In order to test these properties on a concrete example, we begin by analyzing a simple geometric object, namely, a set with the shape of the letter **L**, represented by its characteristic function. Thus our test image is the white **L**-shaped region against a black background, the signal presented in Figure 4.1(a) or, equivalently, in Figure 4.2(a). First, we analyze the effect of scaling on the WT, i.e., going to finer and finer scales. For a pointwise analysis, we choose the isotropic Mexican hat ψ_{H} given in (3.6) [13]. The CWT is plotted in Figure 4.1 for three values of the scale parameter a (conveniently taken as powers of 2), $a = 2^j$, $j = 3, 2, 1$. Each WT is plotted both in 3-D perspective and in level curves. From these pictures, the following observations can be made.

For a large value of a, the WT sees only the object as a whole, thus allowing the determination of its position in the plane. When a decreases, increasingly finer details appear. In this simple case, the WT vanishes both inside and outside the contour, since the signal is constant there. Eventually, only the contour remains and it is perfectly seen at $a = 2$. This is the analog of the precise localization of discontinuities in 1-D [204]. Of course, if one takes values of a that are too small, numerical artifacts (aliasing) appear and spoil the result. This is only a numerical limitation, however, that could be improved by a finer discretization (but with a longer computing time). We notice that the exterior contour is a sharp negative "wall," whereas the interior contour is a positive one. The same effect would appear in 1-D if one would consider, for instance, the full WT of a square pulse. The jump from 0 to 1 gives a negative minimum followed by a sharp positive maximum, and the jump from 1 to 0 gives the opposite pattern.

Note also that the corners of the figure are highlighted in the WT by sharp peaks. The amplitude is larger at these points, since the signal is singular there in *two* directions, as opposed to the edges. In addition the WT detects the *convexity* of each corner. The six convex corners give rise to positive peaks, whereas the concave one yields a negative peak. Here we see again the advantage of using a real wavelet and plotting the WT itself, *not* its modulus, which is a frequent practice.

The conclusion is that the CWT is an efficient edge detector, provided it is evaluated at a scale that is sufficiently small (i.e., high spatial frequency), but still avoiding numerical artifacts (aliasing).

The next step is to compare the performances of various wavelets on the same signal, and the **L**-shape is an ideal benchmark for making the comparison. The result is shown in Figure 4.2 (taken from [Bha99]). Panel (a) gives the signal. Panel (b) is the response of the isotropic Mexican hat wavelet (this is in fact the negative of Figure 4.1(d)!). As noted there already, the response of the wavelet is strongest at the corners, but is also

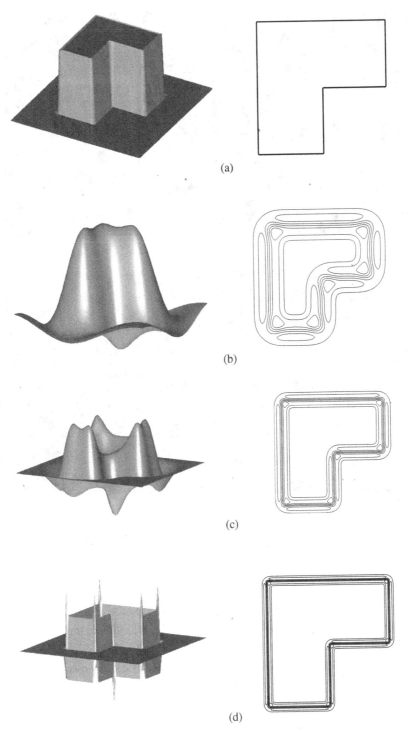

Fig. 4.1. The **L**-shape and its CWT, obtained with an isotropic Mexican hat, presented in 3-D perspective (left column) and in level curves (right column), at three successive scales: (a) the signal; (b) $a = 8$; (c) $a = 4$; (d) $a = 2$.

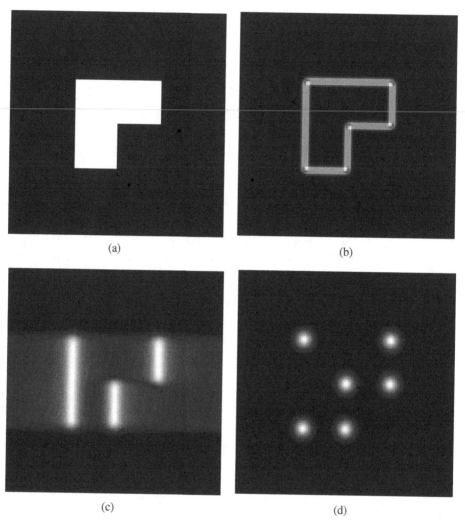

Fig. 4.2. Comparison between the different wavelets applied to the **L**, all at scale $a = 2$: (a) the test image; (b) response of the isotropic Mexican hat wavelet; (c) response of the 2-D Morlet wavelet with $|\vec{k}_0| = 6$, $\theta = 0°$; (d) response of the end-stopped wavelet $\psi_{\rm EI}$ at $\theta = 0°$ (from [Bha99]).

fairly strong at other points along the edges. Panel (c) is the result of applying a 2-D Morlet wavelet, oriented horizontally ($\theta = 0°$), which detects only the vertical edges, as it should (it is an efficient directional filter, as we will see below). Notice the horizontal "leaking" of the response: this is an artifact due to the use of a scale that is too small for the signal (so that the wavelet gets too wide in spatial frequency space). Finally, panel (d) is the response of the end-stopped wavelet $\psi_{\rm EI}$, again oriented horizontally. This wavelet responds only to endpoints of vertical edges, which in this case are all the corners of the contour. Unlike the previous wavelets, however, it shows no response to other points along the edges.

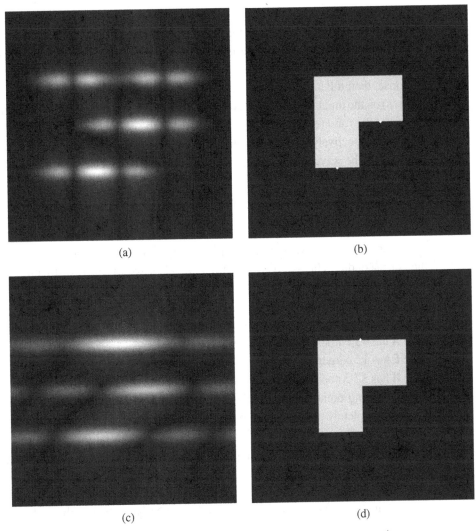

(a) (b)

(c) (d)

Fig. 4.3. Analysis of the **L**-shape with the double end-stopped wavelet ψ_{E2} with $|\vec{k}_0| = 6, \theta = 90°$: (a) response of the wavelet of width $\sigma = 12$; (b) local maxima of (a), thresholded at 50%, superimposed (white crosses) on the input image. The wavelet identifies the short horizontal edges; (c) and (d) the same analysis for the same wavelet of width $\sigma = 36$. Now long horizontal edges are selected (from [Bha99]).

In a last case, we analyze the **L**-shape with the double end-stopped wavelet ψ_{E2}, and the result is shown in Figure 4.3, again taken from [Bha99]. This wavelet is sensitive both to the orientation and to the size of the edges, and its behavior may be understood as follows. The Morlet component of the filter selects linear structures which are perpendicular to the orientation of the wavelet, i.e., here horizontal segments. Then the Mexican hat component, which operates parallel to the linear structures detected by the Morlet component, produces strong responses only for those structures that approach the characteristic length of the Mexican hat (which is determined by σ). Thus, the ψ_{E2} wavelet detects linear structures of given size and orientation.

In our example, the ψ_{E2} wavelet is oriented vertically, using two different values of σ. Figures 4.3(a) and (b) show the response of the wavelet with $\sigma = 12$, first the raw response, then the local maxima of (a), thresholded at 50% and superimposed (white crosses) on the input image. The wavelet identifies the short horizontal edges. Similarly, Figures 4.3(c) and (d) show the raw response and the corresponding thresholded local maxima, respectively, for the wavelet with $\sigma = 36$. For this value of σ, the strongest response of the wavelet corresponds to the longer horizontal edge in the input image.

4.1.2   Application to character recognition

We will now apply the technique developed in the previous section to the problem of character recognition. Our signal will be a set of simplified characters, modeled by the corresponding characteristic function. At this stage of our investigation, we only consider characters composed of segments or union of rectangles. Let us consider for instance a few simple characters (Figure 4.4). We see that
- the **L** has six corners: five convex and one concave,
- the **A** has 12 corners: six convex and six concave,
- the **E** has 12 corners: eight convex and four concave,
- the **H** has 12 corners: eight convex and four concave.

The interesting point is that, in this case, the number of concave corners and convex corners completely characterizes these letters. Therefore, an automatic recognition of these characters requires a fast algorithm for extracting this particular information and encoding it. In what follows we will propose a first step for designing such an algorithm, based on the 2-D CWT.

In order to follow the construction in detail, we focus on a thick letter **A**, represented by its characteristic function. Of course, this object behaves exactly as the academic

Fig. 4.4. A set of simplified characters: the letters **L, A, E, H**.

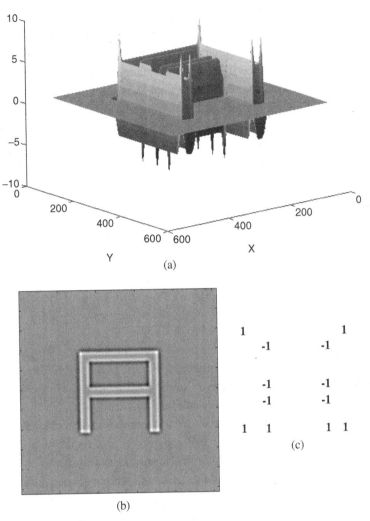

Fig. 4.5. Detecting the contour of the letter A with the radial Mexican hat: (a) the CWT at $a = 0.075$, in 3-D perspective; (b) the same, in level curves; (c) coding of the same by the signs of the respective corners.

image of the preceding section. If one works again with a radial Mexican hat, and goes down to a sufficiently small scale, the CWT reveals the contour of the letter. Moreover, the corners of the figure are highlighted in the WT by sharp peaks, the sign of which is determined by the convexity of the corresponding corner, since the wavelet is real. The result is shown in Figure 4.5. In panel (a), we show the WT of the letter at scale $a = 0.075$, in 3-D perspective. Here we see clearly the twelve peaks corresponding to each corner, some positive (for the six convex corners), some negative (for the six concave corners). Panel (b) presents the same result in level curves, and panel (c) shows the coding of the corners by a logical flag (± 1 for concavity or convexity).

This exercise leads to an algorithm for automatic character recognition [17]. The basic idea of the method is to treat only the significant parts of the signal, focusing on the information needed for unambiguous recognition. In the case of simple letters, this information is entirely contained in the high-frequency components, namely corners and edges. Take, for example, the letter A of Figure 4.5. It can be entirely characterized by the succession of its 12 corners and the additional information that consists in deciding whether a corner is concave or convex. That is, twelve points and a logical flag (concavity or convexity) for each point.

The following simple algorithm achieves this treatment. It consists in locating the local maxima of the CWT and eliminating everything else by thresholding, and it is able to detect an A unambiguously.

- Compute the CWT $S(\vec{b}, a_f)$ in position representation with the Mexican hat wavelet at the finest relevant scale $a = a_f$. The transform exhibits local extrema at the corners and is positive (negative) for a convex (concave) corner. Compute the absolute extrema of the transform

$$ m(a_f) = \min_{\vec{b}}\{S(\vec{b}, a_f)\}, \qquad M(a_f) = \max_{\vec{b}}\{S(\vec{b}, a_f)\}. $$

- To get rid of the other high-frequency components, threshold the transform using a negative value $T_- > m(a_f)$ and a positive value $T_+ < M(a_f)$, both directly computed from the CWT. All the values between T_- and T_+ are set to zero (in the terminology of Donoho [146,147], this is a hard thresholding, see Section 4.6).
- We are left with an image, denoted by $T S(\vec{b}, a_f)$, composed of positive and negative peaks at the position of the corners, which we encode as a vector with components $+1$ or -1 depending on the local sign of the (thresholded) transform.

Using this simple technique we are able to deal with simple shapes, especially with characters that are not corrupted with noise. In the case where we have additive noise or if we need to be more accurate, we use the same treatment at a different small scale, that is $a \in \left[a_{\min}, (a_{\max}/a_{\min})^{1/2}\right]$. We obtain a sequence of images $\{T S(\vec{b}, a_j)\}_{(j=1,\dots,N)}$. Adding these images together gives an image from which one encodes again the local maxima and minima, which are now enhanced against the background noise. The scheme of this algorithm is displayed in Figure 4.6 and applied for recognition in Figure 4.7, for the (noiseless) case of the four letters of Figure 4.4. Panel (a) shows the CWT of the signal with a Mexican hat wavelet, at scale $a = 1.5$. Panel (b) gives the skeleton of that CWT, thresholded at 90% of the maximal values; as expected, only the top of the peaks survive, and they are shown with crosses for the positive peaks (maxima, convex corners) and circles for negative ones (minima, concave corners). Needless to say, this algorithm works only for a *real* wavelet and with the values of the CWT itself, not its absolute value.

Actually, since only the corners are needed, we may as well use a wavelet that sees *only* the corners, not the edges. Typically, a directional wavelet (when it is misaligned),

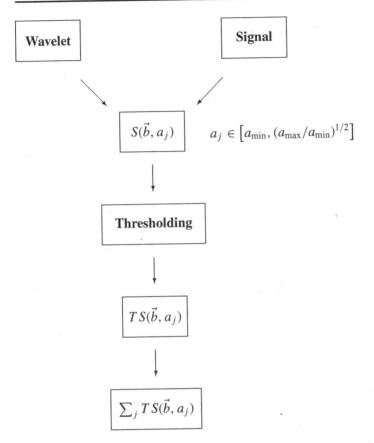

Fig. 4.6. Strategy for character recognition.

or a real wavelet such as the gradient wavelets $\partial_x \exp(-|\vec{x}|^2)$ or $\partial_x \partial_y \exp(-|\vec{x}|^2)$, or even better, the end-stopped wavelet ψ_{EI} described in the previous section. The latter has indeed been designed specifically for that purpose.

This simple technique may be further improved by adding some denoising and inclusion of a second wavelet capable of dealing with letters of arbitrary shape (for instance, a ring-shaped wavelet sensitive to circular shapes). In addition, the automatic recognition device will need some *training*. An elegant solution would then be to use the simple wavelet treatment as a preprocessing for some sort of "intelligent" device, such as a neural network.

However, when noise is present, a different strategy works better, namely, to use a directional wavelet instead of an isotropic one. The reason is that the detection capability of the former is more robust to noise. This feature may be understood as follows: to specify a direction is an additional element of information, that is present in the signal, but not in the noise, in general, so that the SNR ratio improves. In order to show this, we will analyze below another set of simple letters, namely, **A**, **B** and **C**, in the presence of increasingly strong noise (additive Gaussian noise). Now the technique used here is

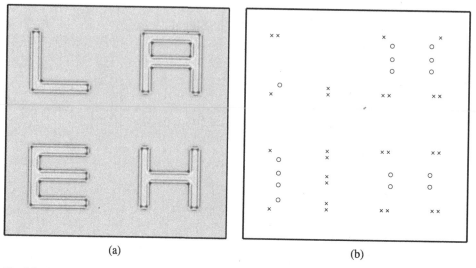

(a)   (b)

Fig. 4.7. Application of the algorithm to character recognition: (a) wavelet tranform of the signal in Figure 4.4; (b) the same after thresholding: maxima (convex corners) are indicated with crosses, minima (concave corners) with circles.

a particular instance of the general problem of feature detection and recognition, so we postpone the details of the algorithm to the next section.

We may also notice that directional wavelets, namely directional derivatives of a smoothing function, have been applied to the same problem of character recognition by Hwang and Chang [230], implementing the wavelet maxima technique of Mallat and Zhong [264,265].

4.2   Object detection and recognition in noisy images

Suppose we have an image containing a certain number of targets, embedded in a cluttered environment: how can one detect and identify the various targets in an automated way? Stated explicitly [16], the purpose of automatic target detection and recognition (ATR) is the use of computer processing to detect and recognize signatures in sensor data, especially targets embedded in a cluttered environment, with the aim of neutralizing potential threats to military and civilian populations while minimizing the required resources and the risk to human life (this problem has an obvious military connotation, which explains the jargon used!). Such targets can be tanks, planes, other vehicles, missiles, ground troops, etc. Clutter can be grass, trees, topographical features, atmospheric phenomena (i.e., clouds, smoke, etc.). In general, the situation can be modeled using the following equation:

$$s(\vec{x}) = n(\vec{x}) + \sum_{l=1}^{L} T_l(\vec{x}). \tag{4.1}$$

where $n(\vec{x})$ represents an additive noise (clutter plus measurement noise), $T_l(\vec{x})$ are targets to be detected and recognized, and $s(\vec{x})$ represents the accessible measured signal.

Automatic or assisted target detection and identification requires the ability to extract the essential features of an object from (usually) cluttered environments. However, detection and identification lead to different requirements. Typically, to provide detection and identification of difficult targets while maintaining full surveillance coverage, a coarse resolution sensor is required for detection, while a fine resolution sensor is necessary for recognition (identification). This suggests the use of multiscale techniques, which provide the flexibility to utilize only the resolution required at each level and, perhaps, allow optimal processing for each of the required operations.

Many methods are used for the ATR problem: classical pattern matching, model-based schemes, dyadic wavelets, subband coding, even some attempts using neural networks [155,247]. Here we will describe an approach based on the 2-D continuous wavelet transform, which could offer a great improvement over traditional pattern matching methods.

The rationale for using the CWT for ATR is the following. Typical features to be extracted from the image of a target are its position, its spatial extent, and its shape, including its orientation and symmetry. Thus the relevant parameters are position, scale and orientation, that is, exactly those considered in the 2-D continuous wavelet transform. The scale dependence allows sensitivity to variations in sensor resolution, as well as determination of target size, or equivalently the target distance, for instance in optical or infrared imagery. Rotation dependence leads to robust behavior in identifying the orientation of the target. So, unlike other methods, the 2-D wavelet transform incorporates several parameters directly relevant to the essential features of an object. Projection of the transform can thus provide a useful set of image representations for fully automated discrimination. In addition, wavelet methods yield a consistent and efficient image reconstruction algorithm. Indeed, the CWT has been used successfully in a number of situations, notably in infrared imagery. We will discuss this application in the next two sections.

4.2.1   Principle of the ATR wavelet algorithm

A simple ATR algorithm based on the 2-D CWT has been proposed in [16]. It consists in a two-stage strategy and relies in an essential way on the successive use of the two basic representations described in Section 2.3.3, the position and the scale-angle representations. A similar technique has been used previously in the analysis of acoustic wave trains in water, see Section 5.3.4. The algorithm reads as follows (Figure 4.8).

At the first stage, we compute the CWT in the *position representation* at all relevant scales $a = a_j$ and angles $\theta = \theta_j$, that is, $S(\vec{b}, a_j, \theta_j)$, $j = 1, 2, \ldots$. For the detection, we take the image obtained for each fixed pair (a_j, θ_j), threshold it, and add all the images together. Thresholding is performed in an adaptive way, becoming more severe

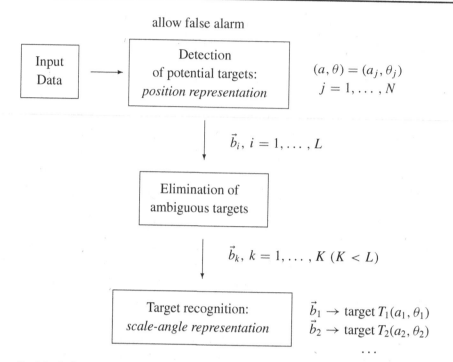

Fig. 4.8. A diagram illustrating the two-step strategy for ATR.

for smaller a. Note that other nonlinear transformations (e.g., enhancement, morphological operators) may also be applied. The effect of this procedure is to suppress the clutter information, while preserving the target information. As a result, the latter is reinforced and becomes visually enhanced.

Next, we compute the centroids $\vec{b} = \vec{b}_i$, $i = 1, \ldots, L$ in the resulting composite image. These centroids correspond to the positions of potential targets. False alarms are of course possible, but one may control the false-alarm rate by adjusting the thresholds in order to eliminate spurious false detection.

Then, at the second stage, one switches to the *scale-angle representation* and computes the wavelet transform of the composite image at each remaining centroid $\vec{b} = \vec{b}_k$, $k = 1, \ldots, K$ ($K \leqslant L$). If the centroid $\vec{b}_k$ corresponds to a genuine target T_k, the corresponding wavelet transform will exhibit a unique maximum (a_k, θ_k), which gives the size and the orientation of the target T_k. Moreover, the signature of each target in the scale-angle representation allows the discrimination between different targets.

An academic example of application of the algorithm just described, more precisely the first stage of it, is presented in Figure 4.9. We take our favorite **L**-shape (a), embedded in a Gaussian white noise (b), with a signal-to-noise ratio of 18. The wavelet transform, with a Mexican hat, is taken at six different scales, $a = 1, 2, 4, 8, 16, 32$, each image is properly thresholded (at $90\%/a$), and the six images are added together. The resulting composite image (c) shows the reconstructed object, the noise has largely been suppressed.

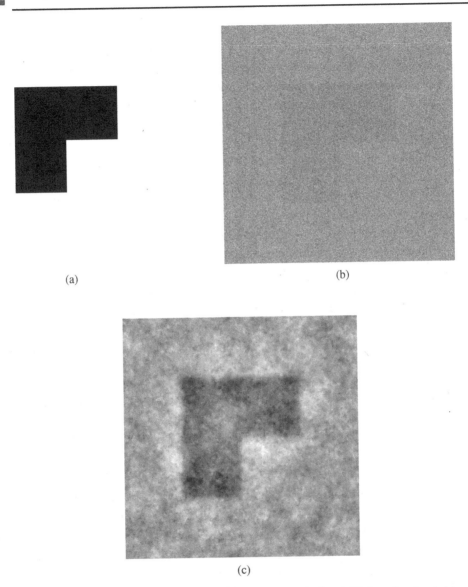

(a)

(b)

(c)

Fig. 4.9. Reconstruction of a signal embedded in noise. (a) The signal; (b) the noisy signal; (c) reconstruction of the signal with six scales.

Of course, this method is rather primitive, although it is fast and robust. An alternative technique for image denoising relies on the use of directional wavelet packets (Section 2.6). We will make a detailed comparison between the two methods in Section 4.6.

4.2.2   Application to infrared radar imagery: position features

The real power of this approach is, of course, better appreciated in real life situations, preferably difficult. A prime example is that of infrared imagery (FLIR or Forward Looking Infrared Radar imagery). Automatic target detection and identification for

FLIR imagery requires the ability to extract the essential features of an object from cluttered environments under the condition that the range to the target is unknown. Moreover, the gray-scale of a target in FLIR imagery displays a great variability: a succession of dark and bright areas depending on the temperature radiating from various parts of the target (for example, a target can be hot or cold). Most of the time, the target does not appear in ideal conditions and it is very difficult to estimate the sensor output probability density function (PDF). One therefore uses algorithms that first extract features of interest (such as structural, spatial and frequency features) and then classifies the objects based on those features (a review of these issues may be found in [134]).

Multiscale techniques, such as the CWT, are highly desirable, because they can extract and normalize both the unknown scale and orientation of the target. As indicated above, typical features to be extracted from the image of a target in a cluttered environment are the position, the spatial extent, and the shape of the target, including its orientation and symmetry. This justifies the use of the 2-D CWT in the ATR problem, and in particular for infrared imagery. An additional issue is to determine which wavelet will perform best, an isotropic one (Mexican hat) or a directional one (Morlet or Cauchy).

Now, in the presence of heavy noise, it is better to apply the ATR algorithm described above not to the CWT amplitude itself, but rather to one of the partial energy densities discussed in Section 2.3.4. We will then speak of *CWT features* to be extracted from the image. This opens a choice between two solutions. The most obvious one is to take the position energy density of the signal $s(\vec{x})$, given in (2.56), namely,

$$P[s](\vec{b}) = \int_0^\infty \frac{da}{a^3} \int_0^{2\pi} d\theta \ |S(\vec{b}, a, \theta)|^2, \tag{4.2}$$

and this we shall do in this section. We test the algorithm on FLIR data from the TRIM2 database, namely, a set of images each of which contains four targets (various types of tanks), seen under 21 different aspect angles (we recall that the aspect angle α is the polar angle in the position representation, $\vec{b} = (|\vec{b}|, \alpha)$). Figures 4.10 and 4.11 present two such images, together with the corresponding receiver operator characteristics (ROC) curves. These curves plot the probability of detection versus false alarm rate, for the whole set of images containing a given type of tank at 21 aspect angles, analyzed in turn with the Mexican hat, the Morlet, and the Cauchy wavelet. The first stage of the ATR algorithm is then applied to the image in Figure 4.11(a) and the results are presented in Figure 4.12. The upper row shows the output of the first step of the algorithm, i.e., the CWT position features, evaluated with the three wavelets successively. Clearly, the targets stand out from their background with more definition than the original. The bottom row then gives the output of the second step of the algorithm, that is, the use of thresholding, morphological transformations, and other nonlinear transformations. The particular image presented corresponds to thresholding at gray-scale value 40. Finally, Figure 4.13 gives the Receiver operator characteristics (ROC) curves obtained

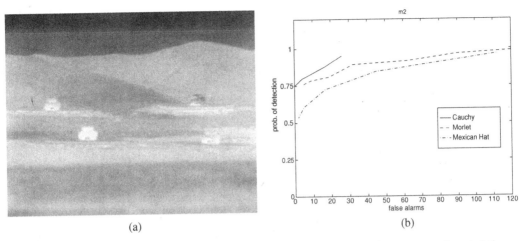

(a)   (b)

Fig. 4.10. (a) Original image containing four M2 tanks (left); (b) Receiver operator characteristics (ROC) curves for the set of images containing M2 tanks at 21 aspect angles.

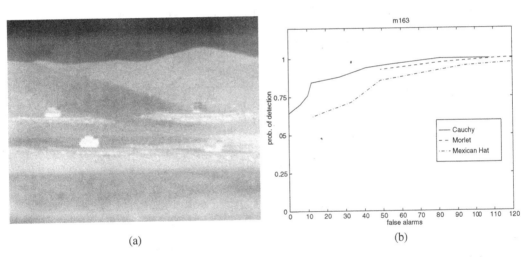

(a)   (b)

Fig. 4.11. (a) Original image containing four M163 tanks; (b) Receiver operator characteristics (ROC) curves for the set of images containing M163 tanks at 21 aspect angles.

by combining images with all types of tanks, again for the three wavelets, Mexican hat, Morlet, and Cauchy wavelets.

The results show that directional wavelets such as the Morlet and the Cauchy wavelets perform better than an isotropic one such as the Mexican hat. Moreover, the Cauchy wavelet performs better than the Morlet wavelet. This can be understood as follows. The selected images contain objects plunged in clutter noise. Detection of an object of interest supposes, for example, the ability to capture the internal structure (for example, wheels, doors, etc . . .) of the object, the boundaries (edges, corners) between the

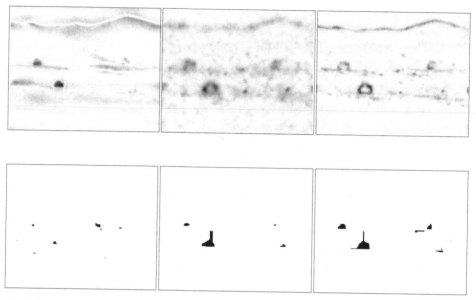

Fig. 4.12. CWT position densities (top row) and detection results (bottom row) of the original image in Figure 4.11 with the Mexican hat (first column), the Morlet (second column), and the Cauchy wavelet (third column), respectively.

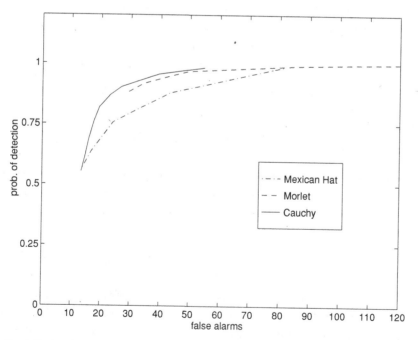

Fig. 4.13. Receiver operator curves for the detection algorithms on data containing M1, M2, M163, and M60 tanks.

object and its surrounding. This justifies the use of directional wavelets in this context, and explains their better performance over nondirectional filters. This technique might be applied with different wavelets, for example, target-adapted wavelets. It might also be compared to or combined with the CWT-based classification algorithm developed in [287].

4.2.3   Scale-angle features and object recognition

The detection algorithm described in the previous section was based on CWT position features extracted from the image. An alternative solution is to use the scale-angle energy density (scale-angle spectrum) of the signal $s(\vec{x})$, given in (2.57), namely,

$$\mathsf{M}[s](a, \theta) = \int_{\mathbb{R}^2} d^2\vec{b} \, |S(\vec{b}, a, \theta)|^2. \tag{4.3}$$

In this section, we shall present an algorithm based on scale-angle CWT features, then apply it again to automatic character recognition and target recognition in FLIR images. As we will see, this algorithm allows significant reduction of the data needed for an efficient recognition and it is robust against noise. As in the previous case, we will compare the performance of two wavelets, this time two directional wavelets derived from the Mexican hat, the modulated or Gabor Mexican hat (3.22) and the conical Mexican hat (3.38). Detailed comparative tests show that the conical Mexican hat wavelet outperforms both the Gabor Mexican hat and the usual Morlet wavelet, and all of them outperform traditional methods of character recognition such as template matching [289].

As a first application, we analyze another set of simple letters, namely, **A**, **B**, and **C**, in the presence of increasingly strong noise (additive Gaussian noise). The characters are shown in Figure 4.14(a), at various SNR levels, 20, 25, 30, 35, and 40 dB. Then we compute the continuous wavelet transform of these letters, using the conical Mexican hat wavelet (3.38). The result is presented in Figure 4.14(b) in the scale-angle representation, evaluated at the center of each character. In this way, the directional features of the object are enhanced, and indeed are detected despite the noise.

Given any letter in Figure 4.14(a), the problem is to recognize it (in an automated way), that is, to determine which of the 26 letters of the alphabet it resembles most – or actually coincides with. This is an instance of object recognition in a noisy environment, more precisely, identification of a noisy object within a preassigned collection of test objects – here the 26 letters of the alphabet. Here again, the standard technique consists of choosing a certain number of characteristic features that suffice to discriminate unambiguously among the test objects. Then one arranges them in a feature vector and measures the distance between the feature vector of the unknown object and those of each test object. The one that yields the smallest distance gives the answer.

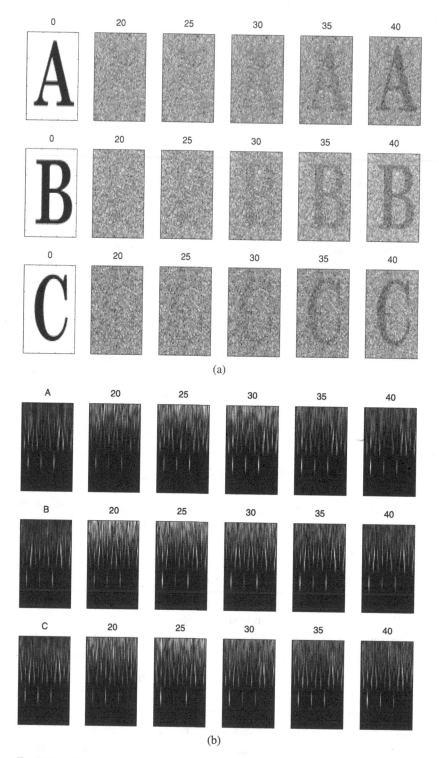

Fig. 4.14. (a) Test data for the recognition algorithm: three characters **A**, **B**, and **C** at various SNR levels, 20, 25, 30, 35, and 40 dB. (b) Continuous wavelet transform of the same, using the conical Mexican hat wavelet. The CWT is presented in the scale-angle representation, evaluated at the center of the character.

4.2.3.1   The recognition algorithm

We will first describe the algorithm in complete generality. Applications to specific problems will be given in the following sections.

Consider a training set, consisting of objects divided into q classes, X^j, $j = 1, 2, \ldots, q$, the class X^j having p_j prototypes, $X^j = (X_1^j, \ldots, X_l^j, \ldots, X_{p_j}^j)$, and an unknown object, Y, that we wish to classify as object X_j, for some j. The algorithm for classification is described as follows:

- For every $j = 1, 2, \ldots, q$, compute the 2-D CWT scale-angle energy density $\mathsf{M}[X_l^j](a, \theta)$ for each element X_l^j of the class X^j.
- For every $j = 1, 2, \ldots, q$, compute the mean and the standard deviation of the elements of the class X^j,

$$\mu_{X^j}(a, \theta) = \frac{1}{p_j} \sum_{l=1}^{p_j} \mathsf{M}[X_l^j](a, \theta) \tag{4.4}$$

$$\sigma_{X^j}(a, \theta) = \left[\frac{1}{p_j} \sum_{l=1}^{p_j} \left\{ \mathsf{M}[X_l^j](a, \theta) - \mu_{X^j}(a, \theta) \right\}^2 \right]^{1/2}. \tag{4.5}$$

- Compute the scale-angle energy density $\mathsf{M}[Y](a, \theta)$ of the test object.
- Select the scale-angle feature vectors $\mathsf{V}[Y]$, $\mathsf{V}[X^j]$ for the unknown object Y and for each class X^j, i.e., $\mathsf{M}[Y](m, n)$ and $\mu_{X^j}(m, n)$, $\sigma_{X^j}(m, n)$, $m = 1, 2, \ldots, N_1$, $n = 1, 2, \ldots, N_2$, where N_1, N_2 are the chosen number of scales and angles, respectively.
- Compute the distance between them,

$$d_j = d(Y, X^j) \equiv d(\mathsf{V}[Y], \mathsf{V}[X^j]), \ j = 1, \ldots, q. \tag{4.6}$$

- Then, if d_k is the minimum of the set $\{d_j, j = 1, \ldots, q\}$, the object Y is classified as belonging to the class X^k.

In principle, one can use any distance, such as the Euclidean distance

$$d(Y, X^j) = \frac{1}{N_1 N_2} \sum_{m=1}^{N_1} \sum_{n=1}^{N_2} \left\{ \mathsf{M}[Y](m, n) - \mu_{X^j}(m, n) \right\}^2,$$

or the maximum likelihood distance, which is given by

$$d(Y, X^j) = \frac{1}{N_1 N_2} \sum_{m=1}^{N_1} \sum_{n=1}^{N_2} \left[2 \log \sigma_{X^j}(m, n) + \frac{\left\{ \mathsf{M}[Y](m, n) - \mu_{X^j}(m, n) \right\}^2}{\sigma_{X^j}(m, n)} \right].$$

Clearly this algorithm may lead to false recognition, if two or more distances d_j are very close to each other.

4.2.3.2   Application to character recognition

The algorithm has been tested [289] on the 26 characters of the alphabet (A,B,C, ...) at various noise levels (additive Gaussian noise). In this case, one has $q = 26$, that is, each X^j is one letter of the alphabet, whereas the p_j prototypes X_l^j correspond to different copies of the same letter, distorted, rotated, embedded in different noises. For

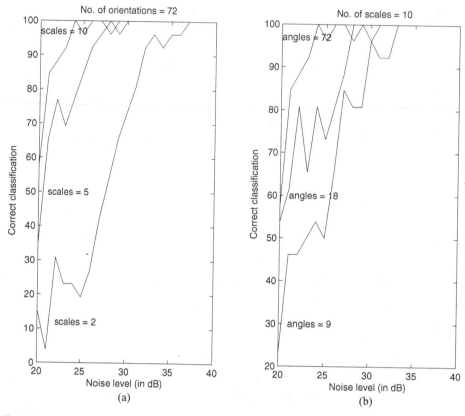

Fig. 4.15. Results of recognition as a function of noise level for various lengths of the feature vector in the scale-angle plane, using the modulated (Gabor) Mexican hat wavelet with $k_0 = 6$ and eccentricity $\epsilon = 2$; (a) fixed number (72) of orientations, different numbers of scales; (b) fixed number (10) of scales, different numbers of orientations.

instance, the five letters **A** on the top row of Figure 4.14(a) are five prototypes of an A, with different levels of noise.

The results of the analysis are summarized in Figures 4.15 and 4.16. Both give the performance of the recognition algorithm for various lengths of the feature vector as a function of the noise level (SNR), using the modulated (Gabor) Mexican hat and the conical Mexican hat, respectively. In Figure 4.16 the result of the traditional template matching is added for comparison. In this experiment, it turns out that the conical Mexican hat performs better than the modulated Mexican hat, which in turn performs better than the traditional Morlet wavelet. The same behavior is observed in other cases, for instance in the recognition and classification of targets in FLIR imagery [285–289].

4.2.3.3 Application to radar imagery

Other angular features of images have been used successfully in recent work by Kaplan and Murenzi on Synthetic Aperture Radar (SAR) images [241]. The problem here is to have a good estimate of the position of a target (in jargon, pose estimation as part of

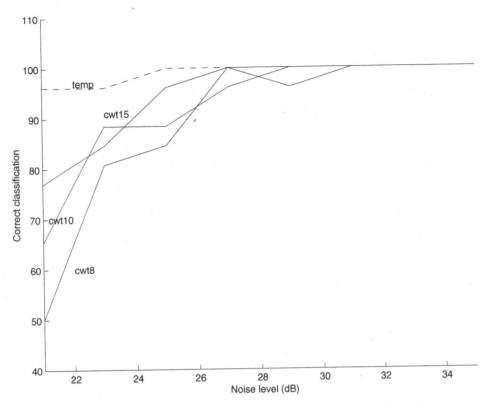

Fig. 4.16. Results of recognition as a function of noise level for various lengths of the feature vector in the scale-angle plane; the wavelet now is the conical Mexican hat.

ATR). Since the latter is assumed to be at the center of the image, one fixes the position at $\vec{b} = \vec{0}$ in the CWT and integrates the energy density over scale only, thus defining the angular energy density

$$\mathsf{E}[s](\theta) = \int_{a_{\min}}^{a_{\max}} \frac{da}{a^3} |S(\vec{0}, a, \theta)|^2. \tag{4.7}$$

The pose estimate is then simply the orientation that maximizes the angular energy density, $\widehat{\theta} = \arg\max_\theta \mathsf{E}[s](\theta)$. This technique yields very good results, using the fast circular convolution algorithm described in Section 2.3.1 with an isotropic Mexican hat wavelet (but, surprisingly, *not* with a Morlet wavelet).

4.3   Image retrieval

4.3.1   The problem of content-based image retrieval

Digital images are everywhere, and they often come in huge quantities: remote-sensing agencies, medical imaging facilities, art museums, travel agencies, law-enforcement agencies, all have to deal with large collections of archived digital images. Depending on the particular situation, these collections may be homogeneous (as in medical imaging or fingerprint archives), or completely heterogeneous. Thus there arises the problem of *image retrieval*: given an image, one has to identify those images in the collection that resemble it most (a typical example is the identification of a suspect's fingerprint among those archived in the police records). Beyond a few dozen images, manual browsing is practically impossible, a purely computerized solution is necessary. Traditional methods of object classification, based on external descriptors, are inoperant for images, especially in heterogeneous collections. One has to resort to a description derived automatically from the image itself, and this leads to a fairly new field of research, called *content-based image retrieval*.

Although several methods have been proposed, we will consider here only the wavelet-based scheme designed by Bhattacharjee in his thesis [Bha99], where an overview of other methods may be found. The roots of this approach lie in the theory of vision initiated by Marr [Mar82], namely the process of comparing images must necessarily be based on low-level information extracted from the images, especially in the case of a heterogeneous collection of images, where no assumption is made about the content of an image. Indeed, methods based on segmentation do not make sense in such a situation. Subdividing the image into blocks, and performing block-wise comparisons of images does not work either, because this approach is not invariant to rotation and cannot support arbitrary subimage-queries.

Thus image comparison should be based on visually significant structural features detected in the images. Indeed, this seems to be the case when human subjects view a scene. Psychophysical experiments show that even when analyzing a static scene, our eyes do not remain continually focused on a single retinal image, but rather perform the so-called saccadic movements described in Section 3.3.3. As explained there, the target points of consecutive saccades are points of interest which stand out against the general background of the scene. Thus it remains to identify those key points, and we are facing again the problem of identifying relevant features in an image. According to experiments, the crucial ones are low-level features, which can be classified in various ways (see [Bha99] for a complete discussion). For instance, one may distinguish patches of uniform intensity; edges or lines; and corners or line ends. Among these, the last kind seems to encode the maximum amount of image information. For this reason, it is natural to design wavelets that respond precisely to these features, and the end-stopped wavelets described in Section 3.3.3 are Bhattacharjee's answer to that question. Before going further into their actual implementation, we shall now briefly describe the whole recognition scheme.

The first stage is the feature point detection scheme, which forms the core of the image comparison process. However, image comparison based simply on the positions or color attributes of the feature points would not be very robust. Thus, for each feature point, one constructs a description based on the texture of the immediate neighborhood of the point. This is obtained by using suitable filters, namely directional derivatives of a Gaussian, of order one, two and three. The responses of these filters are organized in an ordered set and may be represented as a vector. Thus each feature point detected in the image produces one vector of responses, called a *token*. These tokens carry information about a small region in the image, including directional information. The last step of the process is the actual measure of similarity between images, based on tokens. This requires a rather complex image indexing strategy. The outcome is an efficient technique for content-based image retrieval, which proceeds iteratively. To quote the author [Bha99]:

... To use the system, the user presents the query in the form of an image. The system then sorts all images in the collection in order of decreasing similarity to the query, and returns the top few images as the answer-set. The size of the answer-set is specified by the user. From the answer-set, the user may mark the images that are relevant to the user's needs, and provide this information to the image retrieval system as feedback. The system then *refines* the query automatically, based on the relevance information provided, and subsequently returns another answer-set of images to the user. This process may be iterated till the user is satisfied.

This image recognition scheme turns out to be both efficient and reasonably economical, as attested by several explicit examples given in [Bha99].

4.3.2   Feature point detection using an end-stopped wavelet

To conclude this section, we discuss now in more detail the wavelet-based algorithm for feature point detection [Bha99,76,77] (we quote freely from these works). Both end-stopped wavelets ψ_{E1} and ψ_{E2} can be used to select meaningful points in images, and this is why we have discussed them both in Section 3.3.3. However, the author chooses to consider the former only, because, as explained above, he is essentially interested in detecting corner-like features, to which ψ_{E1} responds best. In addition, he works at a fixed scale, since an exhaustive search of the scale space is computationally prohibitive. Furthermore, for a heterogeneous collection of images, to which new images may be added as and when they become available, it is impossible to identify a specific set of scales that will be appropriate for analyzing all images in the collection. Thus, the algorithm applies the ψ_{E1} at the same scale to all images in the collection. There are two negative consequences of this compromise.

- The feature point detection scheme is not scale invariant. Consider two images I_1 and I_2, where I_2 is a subsampled version of I_1 by a factor of two along each dimension. If the ψ_{E1} wavelet is applied at the same scale to both images, we are not guaranteed that the set of feature points detected in I_2 will have a one-to-one correspondence with the set of feature points detected in I_1. That is, the system is not invariant to scale changes even by a factor of four. However, as the experimental results demonstrate, the final system is quite robust to small variations of scale, presumably because of the redundancy of information between responses at nearby scales.

- Some of the feature points detected in an image may not be well localized, that is, the local maxima of the wavelet response may not coincide with the corresponding image feature. This happens because the scale of analysis is coarser than the most appropriate local scale, so the local maximum of the wavelet response does not fall on the feature in question, due to the interaction with other neighboring features. Such seemingly spurious points would pose serious problems for recognition systems. However, the performance of the proposed image retrieval system is not appreciably impaired by the use of a fixed scale of analysis. This is because the comparison of images is not based on a recognition process, but rather, is based on a comparison of small image-patches surrounding the feature points.

  This being said, the algorithm reduces essentially to the standard one described in Section 4.1.2.

  (i) Transform the input image, at a preselected scale, s_0, with the ψ_{EI} wavelet at N different orientations $\theta = \theta_0, \theta_1, \cdots, \theta_{N-1}$. The result is a set of N *response images*, each showing strong responses near the end-points of linear structures oriented perpendicular to θ.

  (ii) For each pixel position, retain only the strongest response value among all the orientations. This produces the so-called *maxima image*.

  (iii) Detect peaks of significant local maxima in the maxima image. The coordinates of these peaks give the feature points.

In practice, the author uses $N = 18$, which corresponds to an orientation resolution of $10°$, to cover the entire semicircle evenly. In fact, the angular resolving power of the ψ_{EI} wavelet is $19.2°$, as measured by the standard benchmarking technique described in Section 3.4.1, but the response is quite weak away from the axis, so that a significant overlap is recommended. Finally, the scale of analysis chosen is $a = 8$.

We conclude by showing some experimental results obtained with this feature point detection scheme. In particular they demonstrate the robustness of the scheme in the face of rotation and cropping.

Figure 4.17 shows the feature points detected for two images of very different kinds. In both cases, the detected points are marked by bright crosses superimposed on the input image. Figure 4.17(c) shows the result for a face image, and Figure 4.17(d) shows the points detected in an ornament image. The ornament images depict very complex hand-drawn artwork, which were designed to be trademarks of publishers in the nineteenth century (these images have been scanned from photocopies of old books, as gray level images).

Figure 4.18 demonstrates the robustness of the proposed feature detection scheme towards rotation and cropping. The image in Figure 4.18(a) shows the feature points detected for a subimage of the ornament image shown in Figure 4.17(b). Note that most of the feature points marked in Figure 4.18(a) have corresponding feature points in Figure 4.17(d). Figure 4.18(b) shows the feature points detected for a subimage extracted from a rotated version of the image shown in Figure 4.17(b). The image has

(a)   (b)

(c)   (d)

Fig. 4.17. Results of feature point detection: (a) face test image; (b) ornament test image; (c) feature points detected in (a); (d) feature points detected in (b). The ornament image shown in (b) is a black-on-white pattern that has been scanned as a gray level image. In both (c) and (d), the white crosses mark the positions of the detected feature points (from [Bha99]).

(a)   (b)

Fig. 4.18. Robustness to cropping and rotation: (a) feature points detected in a subimage of Figure 4.17(b); (b) feature points marking detected in a rotated, cropped version of Figure 4.17(b). Most feature points are detected in both images (from [Bha99]).

been rotated by $17°$. Again, most of the feature points in the corresponding section of the original image [Figure 4.17(d)] are also detected in the rotated, cropped version.

4.4   Medical imaging

Wavelet applications abound in medicine and biology, both in 1-D and in 2-D, often with the discrete WT [Ald96]. For the 2-D case, one may quote image segmentation, mammography, tomography [68,315], and magnetic resonance imaging (MRI) [217]. Here we will mention only one potential application, still under development, namely the technique of spiral reconstruction in MRI. The problem at hand is the so-called spiral acquisition method in MRI, that is, the Fourier transform of the image to be analyzed is sampled along several interleaving spirals [84]. This raises the question of the completeness of the reconstruction of the original image from such data. This is clearly a case of nonuniform sampling, and completion means that the set of sampling points generate a frame. The question was analyzed and solved by Benedetto [67], using powerful mathematical tools, such as Beurling's theorem. Actually, 2-D wavelets may also be used in this context, the hope being that they may allow to bypass the infamous "gridding" problem, namely the necessity of adjusting the sampling points on a Cartesian grid in order to apply the FFT algorithm.

4.5   Detection of symmetries in patterns

4.5.1   The tools for symmetry detection

Wavelets may be used for evaluating the symmetry of a given pattern under discrete rotations and dilations, as was demonstrated in detail in [24], on which this section is based. The method presented here allows one to determine, in a straightforward and economical way, all the (possibly hidden) symmetries of a given pattern. Of course, invariance under separate rotations or dilation is easy and there are various methods for determining it. But the determination of combined dilation–rotation invariances (helicoidal symmetries) of a given pattern is much more delicate and, in fact, we do not know any other method for doing it.

The technique uses, in an essential way, the angular selectivity of the directional wavelets. In order to achieve good precision, one needs a directional wavelet with a very good selectivity in the scale-angle variables. The Gaussian conical wavelet (3.37) is an extremely efficient tool in that respect and we will use it systematically.

However, it is a fact that most patterns of interest possess only *local* or *approximate* symmetries of this type, i.e., without true periodicity or discrete translational invariance. They are thus quasiperiodic sets, such as quasilattices, planar self-similar (Penrose) tilings or diffraction patterns of quasicrystals. These objects have indeed a local symmetry only, but the method applies as well, because the local character of the wavelet transform allows us to treat exact and local symmetries precisely on the same footing.

For quasiperiodic objects, the useful information is contained, to a first approximation, only in the scale and angle variables, since there is no spatial periodicity as for regular (crystallographic) tilings [Bar94,61,322]. In such a case, one may ignore the dependence of the CWT on the translation degrees of freedom, represented by the parameter $\vec{b}$. One possibility is to use the scale-angle representation, which consists of fixing the position parameter $\vec{b}$ (Section 2.3.3). However, this may lead to ambiguities, because the result, including its symmetries, depends sensitively on the value of $\vec{b}$ that has been chosen [24] (an example is given in Section 4.5.2.3). The alternative is to average over all values of $\vec{b}$ and consider the scale-angle spectrum, as defined in (2.57).

$$
\begin{aligned}
M[s](a, \theta) &= \int_{\mathbb{R}^2} d^2\vec{b} \, |S(\vec{b}, a, \theta)|^2 \\
&= (2\pi a)^2 \int_{\mathbb{R}^2} |\widehat{s}(\vec{k})|^2 |\widehat{\psi}(a \, r_\theta^{-1}\vec{k})|^2 \, d^2\vec{k} ,
\end{aligned}
\tag{4.8}
$$

where $S(\vec{b}, a, \theta)$ is the wavelet transform of the signal $s(\vec{x})$ with respect to a directional wavelet.

Clearly, $M[s]$ gives the intensity of the spectrum of s, namely, the contents of $|\hat{s}|^2$ is averaged locally around $a r_\theta^{-1}(\vec{k})$, according to the shape of $|\psi|^2$. Furthermore, if ψ is directional, $\widehat{\psi}$ is supported in a narrow cone, and then (4.8) "probes" the behavior of the signal in the direction θ, as the beam of a torchlight exploring a target. This intuitively explains all the results that follow. Positions are not considered in the analysis, because only the modulus of $\widehat{s}$ is used. This is why the method may be interesting in a (quasi)crystallographic context, where only amplitudes of the diffraction spectrum are recorded in experiments. In practice, however, the scale-angle spectrum is often not very readable, because some of the (approximate) symmetries may be rather weak. Therefore, exactly as for the wavelet transform itself, one plots instead the skeleton of the scale-angle spectrum, which in this case reduces to the set of local maxima (isolated peaks in the scale-angle spectrum), and then the possible periodicity properties become clearly visible.

In addition, the scale-angle measure is well adapted for analyzing a *statistical symmetry* [321]. This is a weaker concept of symmetry, which corresponds to the invariance under rotation, or dilation, of the frequency of appearance of any given local configuration inside of the pattern. This is clearly the relevant concept when one is dealing with an approximate symmetry.

Let us thus assume that a certain substructure of s interferes positively with the wavelet $\psi_{\vec{b}_o, a_o, \theta_o}$; that is, the corresponding wavelet coefficient is large. Assume further that this configuration occurs a certain number of times in s, giving to $M[s]$ a local maximum at a point (a_0, θ_0). Then, if s is statistically symmetric under a certain rotation

R_ρ and a certain dilation D_λ, the local maximum of $\mathsf{M}[s]$ has a mirror image at $(a_0\lambda, \theta_0 + \rho)$ by the covariance of the CWT (Proposition 2.2.3). The important fact is that not only the maxima are relevant, but all the points of the scale-angle spectrum, because, in the definition of statistical symmetry, any local configuration has importance and gives its contribution to $\mathsf{M}[s]$.

Of course, it is nice to detect (statistical) symmetries of the image, but one also wants to know whether one has found *all* of them. An answer to that question is given by the *voting algorithm* introduced in [Vdg98], extending a technique of Hwang and Mallat [228].

The idea is to make a vote on the more significant symmetries of the image s under consideration. First, one computes the correlation $\mathsf{P}[s](\tau_0, \alpha_0)$ between the scale-angle spectrum $\mathsf{M}[s]$ and the version obtained under dilation by a factor λ_0 and rotation by an angle α_0, that is, $\mathsf{M}[s](a\lambda_0^{-1}, \theta - \alpha_0)$. Introducing logarithmic coordinates $a = e^t$, $\lambda_0 = e^{\tau_0}$, and defining $\mathsf{M}'[s](t, \theta) = \mathsf{M}[s](e^t, \theta)$, one has $\mathsf{M}[s](a\lambda_0^{-1}, \theta - \alpha_0) = \mathsf{M}'[s](t - \tau_0, \theta - \alpha_0)$. The correlation $\mathsf{P}[s]$ is thus given by:

$$\mathsf{P}[s](\tau_0, \alpha_0) = \|s\|_2^{-2} \int_{t_{\min}}^{t_{\max}} dt \int_0^{2\pi} d\theta \, \mathsf{M}'[s](t - \tau_0, \theta - \alpha_0) \, \mathsf{M}'[s](t, \theta), \qquad (4.9)$$

where τ_0 ranges from 0 to the width of the logarithmic scale interval. In practice, of course, we take a bounded interval $[t_{\min}, t_{\max}]$ for t and we work with discrete steps and thus the integration is approximated by a summation over a linear grid $\Gamma \subset [t_{\min}, t_{\max}] \times [0, 2\pi]$.

Then the algorithm allocates a vote to the point (τ_0, α_0) if $\mathsf{P}[s](\tau_0, \alpha_0)$ exceeds a given constant $K > 0$ (which specifies the error that is tolerated). Once a vote has been cast for a point (τ_0, α_0), one identifies all its integer multiples $(n\tau_0, n\alpha_0)$ that lie within Γ, and give all their votes to (τ_0, α_0) [Vdg98,24].

4.5.2   Detecting symmetries in 2-D patterns

We shall now apply the method developed in Section 4.5.1 to the detection of rotation–dilation symmetries of certain classes of 2-D patterns, following essentially [24,26].

4.5.2.1   Geometric patterns

We begin with a simplified version and eliminate the scale dependence by integrating over a, thus ending with the angular spectrum $\alpha[s](\theta)$ of the object, defined in (2.60). In general, $\alpha[s](\theta)$ is 2π-periodic. However, when the analyzed object has rotational symmetry n, i.e., it is invariant under a rotation of angle $2\pi/n$, then $\alpha[s]$ is in fact $2\pi/n$-periodic.

To give a simple example, consider three geometrical figures, a square, a regular hexagon and a rectangle [24]. The square and the hexagon have symmetry $n = 4$ and $n = 6$, respectively, and thus their angular spectrum show four and six equal peaks, respectively (Figure 4.19). The width of these peaks is simply the aperture of the support

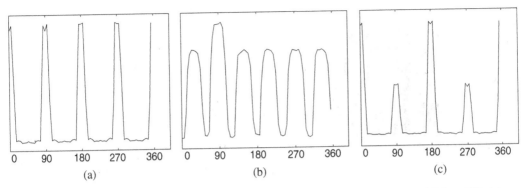

Fig. 4.19. Angular measure of regular figures obtained with a Cauchy wavelet ($ARP = 20°$): (a) a square; (b) a regular hexagon; and (c) a rectangle, with side ratio $2:1$ (from [24]).

cone (i.e., the ARP) of the wavelet. The case of the rectangle is more interesting. It has symmetry $n = 2 \times 2$ (two mirror symmetries, or rotations by π around both Ox or Oy), and that is reflected on the graph of its angular spectrum: there are two large peaks corresponding to the longer edges and two smaller peaks corresponding to the shorter ones, and the ratio $2:1$ between the two equals that of the lengths of the corresponding edges. Indeed, the wavelet catches the direction of the *edges*, not that of the corners, so that indeed the maxima of $\alpha[s]$ are again at $\theta = 0°, 90°, 180°, 270°$, just as for the square, but now the amplitudes are different. This also explains why the peak at 90° in the case of the hexagon, panel (b), is slightly higher: the vertical sides in the original figure are sharp, the oblique ones are ragged, for numerical reasons, so that the former give a larger response to the wavelet. This explains why one needs a highly directional wavelet in this case.

It is remarkable that the scale-angle spectrum technique works in the presence of severe noise. Let us take again a square pattern and compute its CWT with a directional wavelet, first without noise (Figure 4.20), then with moderate additive Gaussian noise (Figure 4.21, top panels), finally with severe additive Gaussian noise (Figure 4.21, bottom panels). In each figure, we show successively: (c) the angular spectrum (2.60), which reveals the fourfold symmetry of the pattern; and (d) the scale spectrum (2.61), which measures the size of the object (from [289]).

Next we proceed to patterns with a genuine combined rotation–dilation symmetry. In this case we need the full scale-angle spectrum $M[s](a, \theta)$, which will again be $2\pi/n$-periodic in θ if the pattern has rotational symmetry n. In addition, if the object is invariant under dilation by a factor a_o, then $M[s]$ is $(\log a_o)$-periodic in $\log a$. Thus in the case of an inflation invariance, $M[s](a, \theta)$ is a doubly periodic function in $\log a$ and θ (note that, like wavelet transforms themselves, a scale-angle spectrum is usually plotted as a function of $\log a$ and θ).

The first object we analyze is a "twisted snowflake," that is, a mathematical snowflake [43,44] with the following modified construction rule: upon each downscaling by a

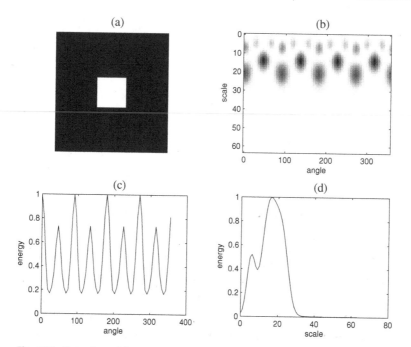

Fig. 4.20. Detection of the symmetries of a square pattern: (a) the signal; (b) the scale-angle spectrum; (c) the angular spectrum; and (d) the scale spectrum (from [289]).

factor of 3, the figure is turned by $36°$. The scale-angle spectrum of this object, given in Figure 4.22(b), shows precisely the combined symmetry. The set of four maxima at a given scale a_o is reproduced, at scale $a_o/3$, but translated in θ by $36°$.

4.5.2.2 Quasiperiodic point sets

An interesting class of point sets is that of the quasilattices based on algebraic numbers (see [Bar94,61] for a systematic analysis). All of them possess a rotational symmetry of order n, where n may be crystallographically allowed ($n = 1, 2, 3, 4, 6$) or not (e.g. $n = 5, 8, 10, 12$). Moreover, each pattern is invariant under dilation by a characteristic factor, which is an integer equal to $2 \cos(2\pi /n)$ in the first case, and an irrational number β_n in the second case (thus called quasicrystallographic). But in fact there is more. In many cases, there is in addition a combined rotation–dilation symmetry, typically a rotation by π/n together with a dilation by a factor δ_n related to β_n.

As an example, we consider the octagonal pattern shown in Figure 4.23 [24]. It has a global symmetry $n = 8$ and is invariant, by construction, under dilation by a factor $1 + \sqrt{2}$. But the scale-angle spectrum $\mathsf{M}[s](a, \theta)$ (calculated with a Gaussian conical wavelet) reveals *two* combined rotation–dilation symmetries, namely a rotation of $\pi/8$ together with a dilation by a factor $\delta_1 = \sqrt{2} \cos(\pi/8)$, or $\delta_2 = 2 \cos(\pi/8)$, respectively. The remarkable fact is that these two additional symmetries were discovered on the graph of the scale-angle spectrum, *not* on the tiling itself! (Actually both symmetries

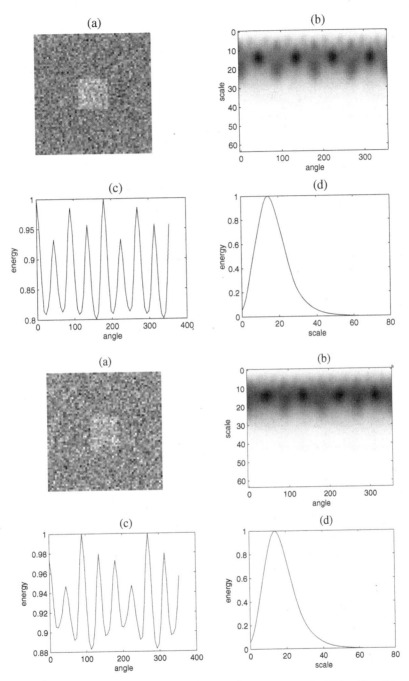

Fig. 4.21. The same analysis as in Figure 4.20, in the presence of additive Gaussian noise: (top) moderate noise [signal-to-noise ratio (SNR) = 26 dB]; (bottom) severe noise (SNR = 22 dB).

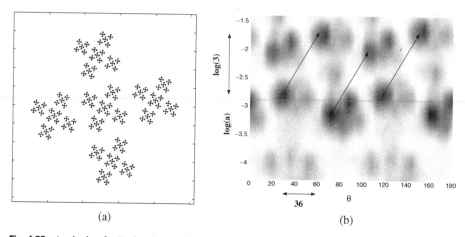

(a)   (b)

Fig. 4.22. Analysis of a "twisted snowflake": (a) the pattern; (b) the scale-angle spectrum
$\mathsf{M}[s](a, \theta)$, computed with a Cauchy wavelet ($m = 4$, $\gamma = 10°$). Corresponding local maxima are
shifted by 36° and a scaling ratio of 3 (from [24]).

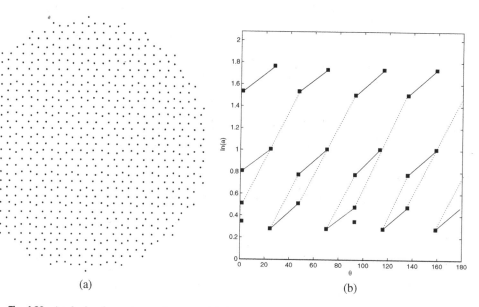

(a)   (b)

Fig. 4.23. Analysis of an octagonal pattern: (a) the pattern; (b) the local maxima of its scale-angle
spectrum $\mathsf{M}[s](a, \theta)$; this pattern has a rotation symmetry by $\pi/4$, and two distinct mixed
symmetries, consisting of a rotation by $\pi/8$ combined with a dilation by $\delta_1 = \sqrt{2}\cos(\pi/8)$,
$\delta_2 = 2\cos(\pi/8)$, respectively. Homologous maxima are linked by a line segment, continuous for δ_1
and dashed for δ_2 (from [24]).

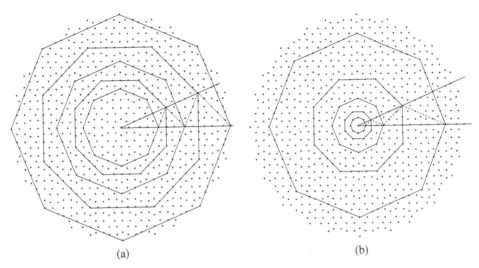

(a)   (b)

Fig. 4.24. Two sets of octagons on the octagonal tiling obtained by successive applications of a rotation by $\pi/8$ combined with a dilation by (a), $\delta_1 = \sqrt{2}\cos(\pi/8)$; (b), $\delta_2 = 2\cos(\pi/8)$ (from [24]).

were later derived by a geometrical argument.) A nice way of visualizing the symmetries is to draw successive octagons, representing the orbits of successive points under a rotation by $\pi/4$ (Figure 4.24). Note that, for a better visualization of these orbits, we have brought back all successive summits into the first sector $0 \leqslant \theta \leqslant \pi/8$. This operation reveals an additional difference between the two symmetries. Indeed the pattern on the right is invariant under the combined operation δ_2-dilation + rotation by $\pi/8$, and this operation generates a semigroup (every point has a successor, not necessarily a predecessor, i.e., the inverse operation is not a symmetry). This semigroup has apparently infinitely many different orbits (on the portion of the tiling visible on the figure, we have detected 10 different orbits). However, the other combined operation, δ_1-dilation + rotation, is *not* an exact symmetry, it is only approximate. For instance, some orbits stop after a few iterations, or have gaps.

Now comes the question, did we detect *all* symmetries of the octagonal pattern? The answer is in fact yes, as shown by the result of the voting algorithm described in Section 4.5.1, presented in Figure 4.25. The graph indeed shows the two pure operations, rotation by $\pi/4$ and dilation (ρ_0) by a factor $1 + \sqrt{2}$, and the points ρ_1 and ρ_2 corresponding to the combined rotation–dilation operations with dilation ratio δ_1 and δ_2, respectively. Note that the pure dilation ρ_0 is equivalent under the $\pi/4$-periodicity to the product $\rho_1 \cdot \rho_2$.

4.5.2.3   Other examples of aperiodic patterns

There are many more examples of patterns that exhibit this kind of combined symmetries. A whole class is that of tilings of the plane, some of which are commonly known

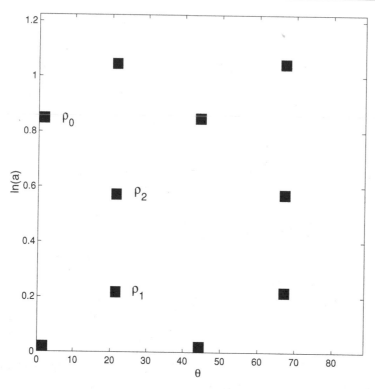

Fig. 4.25. The set of renormalization parameters $\rho = (\delta, \theta)$ obtained by the voting algorithm on the scale-angle spectrum of the octagonal tiling given in Figure 4.23. The point on the horizontal axis corresponds to the $\pi/4$-periodicity. The points ρ_1 and ρ_2 correspond to the combined rotation–dilation operations with dilation ratio δ_1 and δ_2, respectively, whereas ρ_0 is a pure dilation, equivalent to the product $\rho_1 \cdot \rho_2$ under the $\pi/4$-periodicity. The other, unmarked, points are translations of the previous ones under both periodicities, in a and θ (from [24]).

under the name of Penrose tilings (these are dual to the preceding type, in the sense that they are obtained by drawing the Voronoi cells of the point set). We show a typical example in Figure 4.26. From the scale-angle spectrum, obtained with a Gaussian conical wavelet (3.37), with parameters $m = n = 4$, $\sigma = 16$, we conclude that this pattern has a rotation symmetry by $\pi/5$, a dilation symmetry by $\tau = 2\cos(\pi/5) = \frac{1}{2}(1 + \sqrt{5})$, the golden mean, and a mixed symmetry, consisting of a rotation by $\pi/10$ combined with a dilation by $\lambda = 1.36$. Incidentally, these examples show why it is safer to integrate over all scales in order to isolate the angular behavior, rather than to fix a certain scale $a = a_o$ and consider $\mathsf{M}[s](a_o, \theta)$. If a_o coincides with one of the characteristic scales, $a_1, a_2, \ldots$, the result is correct, but if a_o falls in between, no maximum will be seen, and the symmetry is not detected. The effect is shown in Figure 4.27.

Another interesting class of examples may be found in various pattern-forming phenomena in fluids [194]. Typically nonlinear waves at the surface of a fluid generate a regular pattern, via an unstability and a bifurcation. Most of these patterns have a

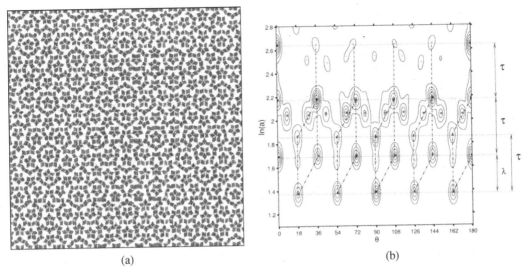

(a)   (b)

Fig. 4.26. Symmetry detection with the CWT: (a) a Penrose tiling; (b) the corresponding scale-angle spectrum $M[s](a, \theta)$, obtained with a Gaussian conical wavelet ($\gamma = 10°$, $m = 4$, $\sigma = 16$). Homologous maxima are linked by a line segment (from [24]).

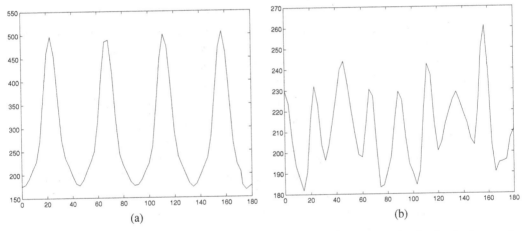

(a)   (b)

Fig. 4.27. The scale-angle measure of the octagonal pattern from Figure 4.23, for fixed values a_o of the scale: (a) for $\ln a_o = 1.20$, the periodicity is obvious; (b) for $\ln a_o = 1.13$, between two lines of maxima, the symmetry is not seen (from [24]).

rotational invariance of some order, and some of them are quasicrystalline. As a result, they lend themselves quite naturally to a wavelet analysis [86]. The best known case is the instability demonstrated by Faraday in 1831. The resulting pattern was known to have a rotational symmetry of order $n = 12$. Our standard analysis indeed yields this symmetry, together with the corresponding invariance under dilation by the corresponding factor $\beta_{12} = 2 + \sqrt{3} \simeq 3.73$. In addition, we find, as before, a combined symmetry of a rotation by $2\pi/24 = 15°$ together with a dilation by $\delta = 1.89 \simeq \sqrt{\beta_{12}}$, which was unexpected [26]. It can actually be proved that such a combined rotation–dilation invariance, with half the rotation angle, is always present in all these "algebraic quasicrystals" [Bar94].

This technique permits one to determine, in a straightforward way, the (possibly hidden) symmetries of a given pattern. This applies to a genuine lattice, but also to a quasilattice, for which the symmetry is only local, for instance, the diffraction spectrum of a quasicrystal.

Let us remind the reader that quasicrystals are those remarkable alloys discovered in 1984 [338], whose X-ray diffraction patterns show local n-fold point symmetry for $n = 5, 8, 10$, or 12. The latter are crystallographically forbidden for being incompatible with translational invariance (the only rotational symmetries compatible with lattice periodicity are of order $n = 1, 2, 3, 4$, or 6). These diffraction patterns display bright Bragg peaks of unequal intensity and they are self-similar with irrational scaling factors. More precisely, the involved irrationals are the following algebraic numbers:

$$\tau = \tfrac{1}{2}(1 + \sqrt{5}) = 2\cos(\pi/5) \quad \text{(pentagonal or decagonal quasilattices)}$$
$$\beta_8 = 1 + \sqrt{2} = 1 + 2\cos(\pi/4) \quad \text{(octagonal case)}$$
$$\beta_{12} = 2 + \sqrt{3} = 2 + 2\cos(\pi/12) \quad \text{(dodecagonal case)},$$

that is, precisely the dilation factors discussed above. Similarly, some of the wave functions for transport electrons in quasicrystals are critical: they are neither localized (as would be the case in a random amorphous structure), nor spread out (as for perfect periodic crystals). Moreover, they display self-similarity too. For example, in the five-fold case, self-similarity ratios are typically powers of the golden mean $\tau = \tfrac{1}{2}(1 + \sqrt{5})$.

Thus we expect that the present wavelet-based method will yield interesting physical applications in the field of crystallography, in three possible directions. The first one concerns the diffraction patterns, where, at a given resolution level, it is necessary to classify and label the Bragg peaks according to their position and intensity. Secondly, two-dimensional wavelets based on the number τ seem particularly appropriate for the scanning analysis of patterns obtained through tunneling or atomic force microscopy of quasicrystalline surfaces. The third application concerns the determination of electronic wave functions in quasicrystals (explicit construction by using a discrete wavelet basis, adapted to the given symmetry type). Actually, these applications may be made easier if one uses a set of wavelets directly adapted to the symmetry. We will describe some of these in Section 11.4.2.

4.5.2.4 Point sets generated from noncrystallographic Coxeter groups

A completely different kind of example is given by quasilattices derived from infinite dimensional extensions of noncrystallographic Coxeter groups. The idea is to generate an aperiodic point set by applying successive reflections and translations to the root system. The example we have analyzed is based on an affine extension of the Coxeter group H_2 [27,303]. This group is isomorphic to the dihedral group of order 10 and it has a noncrystallographic root system (that contains explicitly the golden mean τ). Then,

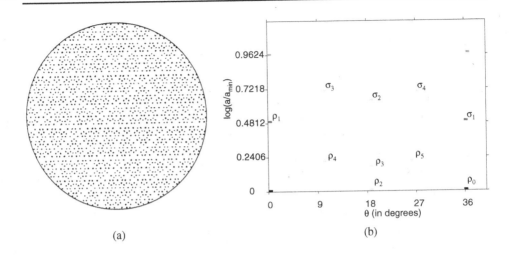

(a)   (b)

Fig. 4.28. (a) A H_2^{aff}-induced quasicrystal centered on $(0, 0)$; the picture shows the part of the quasicrystal located within a radius of 2; (b) result of the voting algorithm with a Gaussian conical wavelet ($\gamma = 3.5°$, $m = 4$, $\sigma = 10$) (from [27]).

applying a number of reflections and translations in a two-dimensional subspace of the root space, one generates planar aperiodic point set, called an H_2^{aff}-induced quasicrystal and shown in Figure 4.28(a). Next we apply the voting algorithm of Section 4.5.1. The result is shown in Figure 4.28(b). We observe six basic symmetries, labeled by ρ_j with $j = 0, \ldots, 5$. Among these, the pure operations may be considered as major symmetries, because the intensity of their maxima is dominant, namely the rotation by $\pi/5$, denoted ρ_0, and the dilation by τ, denoted ρ_1. All others are weaker symmetries. One of them, denoted ρ_2, is again a pure rotation symmetry, with angle $\pi/10$, but since the intensity observed is much weaker than the intensity of ρ_0, it is not possible to view ρ_2 as the generator of the latter. In addition, we have combined symmetries, ρ_3, ρ_4, and ρ_5. The remaining symmetries, labeled by σ_j, with $j = 1, \ldots, 4$, are compositions of the basic ones. Indeed, $\sigma_3, \sigma_2, \sigma_4, \sigma_1$ may be obtained by combining the pure dilation ρ_1 with ρ_4, ρ_3, ρ_5, and ρ_0, respectively.

It remains to provide a geometrical interpretation of all these symmetries. As building blocks of the quasiperiodic tiling, one may take a decagon contained in Figure 4.28(a), just like the octagons of Figure 4.24, and its successive translations. Each translated decagon is reproduced several times during the successive rotations of an angle of $k\pi/5$ with $k = 1, \ldots, 9$. Thus any symmetry found in this construction will be important for the whole tiling. Furthermore, any symmetry which is present already in the nontranslated decagon will appear with an even stronger intensity, because it appears more frequently and thus leads to a more dominant statistical symmetry. Then one finds that each of the symmetries σ_2 and ρ_3 corresponds to the ratio of two specific segments in the basic decagon, with the angle between them. In addition, the weaker symmetries σ_3, σ_4, ρ_4, and ρ_5 can be traced to similar geometric relations in the figure [27].

4.6   Image denoising

A very successful application of wavelets in image processing is to denoising. Most of the authors so far have used the DWT for that purpose, and indeed orthogonal or biorthogonal wavelet bases. In some cases, wavelet frames have also been exploited and also the lifting scheme of Section 2.5.2.4 [Jan00,Jan01]. A notable improvement was obtained by E. Candès with a new tool, better adapted to images, called *curvelets* [Can98,96,97,346] (we will describe these in Section 11.1.4). It turns out, however, that the techniques described in the present book may be useful too. We have already seen an example in Section 4.2, namely, the first stage of the ATR algorithm. Although it is rather primitive, this technique has the advantage of being very simple. A much more elaborate approach is to take directional dyadic wavelet frames, introduced in Section 2.6.3. Without going into technical details, it is instructive to compare the two methods, and this we shall do here.

The key notion in every denoising method is *thresholding*. The idea is that only the relevant features of an image yield significant wavelet coefficients, whereas the noise gives many small coefficients, spread more or less everywhere in the parameter space. Thus it suffices to put to zero all the coefficients that lie below a fixed threshold Λ. The problem, however, is how to choose the latter. A sophisticated and highly successful technique was introduced by Donoho and Johnstone [146,147]. In addition, one has to choose between two versions:

- *hard thresholding:* here the small coefficients, i.e. $|c_{jk}| < \Lambda$, are replaced by 0 and the rest remains untouched. As a consequence, artificial discontinuities are created.
- *soft thresholding* or *wavelet shrinkage:* in order to remove these discontinuities, all the remaining coefficients are shifted by $\pm\Lambda$, so as to make them continuous.

This thresholding technique, initially developed in 1-D, extends to 2-D in an obvious way.

The ATR method has been described in Section 4.2, and one observes that it uses a hard thresholding, with a threshold level Λ fixed arbitrarily. As for the dyadic wavelet packets, we start with a dyadic tight frame generated by a wavelet that satisfies the conditions (compare Proposition 2.4.1, in particular, (2.114))

$$\sum_{\ell=0}^{L-1} |\widehat{\Psi}(r, \varphi - 2\pi\ell/L)|^2 = |\widehat{\psi}(r)|^2, \tag{4.10}$$

where L is the number of orientations, and $\widehat{\psi}(r)$ is the Fourier transform of an isotropic dyadic wavelet, i.e.,

$$\sum_{j=-\infty}^{+\infty} |\widehat{\psi}(2^j r)|^2 = 1 \tag{4.11}$$

and

$$\sum_{j=-\infty}^{J} |\widehat{\psi}(2^j r)|^2 = |\widehat{\phi}(2^J r)|^2, \tag{4.12}$$

where ϕ is the associated 2-D scaling function [265]. Using a directional frame allows us to put more redundancy in the technique and to benefit from the fact that directional wavelets will emphasize oriented features, such as edges, etc. Even though (2.140) [or (2.141)] is a continuous formula, all computations may be carried out in a discrete setting, either by means of the sampling theorem, or by using the approximate QMFs introduced in Section 2.6.4. It suffices then to compute the wavelet coefficients and to threshold them appropriately. The choice of the threshold is a crucial matter and usually requires an estimation of the standard deviation of the contaminating noise. Finally, reconstructing using the thresholded coefficients yields an estimated, denoised image.

In order to illustrate the efficiency of the method, we present in Figure 4.29 a comparison between the ATR technique and the present one. We choose again our familiar L-shape, embedded in increasingly severe Gaussian noise, with standard deviation $\sigma = \frac{1}{4}255, \frac{1}{2}255$, and 255, and corresponding PSNR 12.07, 6.04, and 0.01 dB, respectively. In the middle column, we show the image denoised with the ATR method. The CWT is computed with an isotropic Mexican hat over five dyadic scales, 2, 4, 8, 16, and 32. Then, a hard thresholding is applied at each scale layer with a relative (with respect to the maximum) threshold decreasing with scale: 50, 25, 12.5, 6.25, and 3.125%, respectively. Finally, all these modified scale layers are summed, giving an image denoised, in the sense that the object has been considerably enhanced over the noise. In the right-hand column of the figure, we show the result obtained with a directional dyadic wavelet packet, using five scales and eight orientations. The denoised images have PSNRs equal to 25.97, 21.63, and 18.16 dB, respectively. The result is obviously better than the one obtained with the ATR method.

As a second example, we show in Figure 4.30 the denoising of the `lena` image with the directional dyadic wavelet packet method, using five scales and 16 orientations. Panel (a) shows the original image, panel (b) the noisy version, obtained by adding to the original picture a white noise of $\sigma = 25.5$ (PSNR: 20.02 dB). Panel (c) shows the denoised image, which now has a PSNR of 31.07 dB.

4.7   Nonlinear extensions of the CWT

4.7.1   Local contrast

The intensity of light around us varies considerably, in fact by several orders of magnitude. Our visual system is well adapted to this situation. Indeed it analyzes the spatial organization of the luminous field by relying on the *contrast* of objects and figures contained in the images. Intuitively, contrast is defined as the ratio between a variation

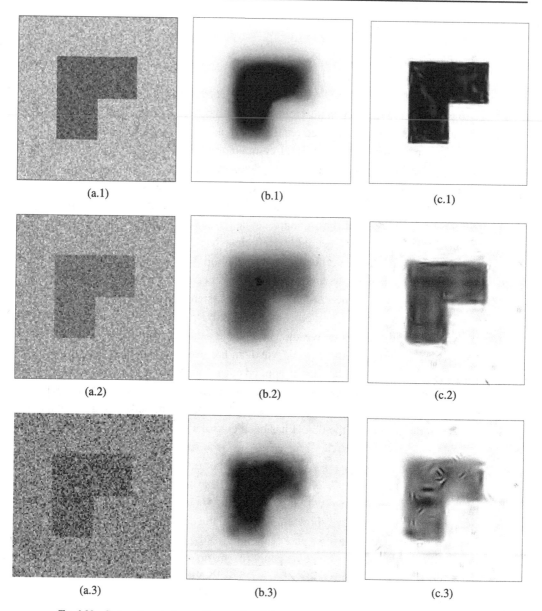

Fig. 4.29. Comparison between the two denoising methods. The left column (a) shows the signal, with noise increasing from top to bottom, the middle one (b) is the result of the ATR algorithm, the right one (c) that obtained with a directional dyadic wavelet packet.

of luminance and a reference level of luminance. It is mathematically expressed using *Weber's law*:

$$C^{\mathrm{W}} = \frac{\Delta L}{L}. \tag{4.13}$$

(a)

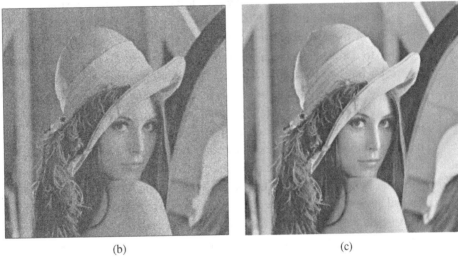

(b)   (c)

Fig. 4.30. Denoising of `lena` with a directional dyadic wavelet packet.

This definition is often used for small patches with a luminance offset ΔL on a uniform background of luminance L. In the case of sinusoids or other periodic patterns of symmetrical deviations ranging from L_{min} to L_{max}, which are also very popular in vision experiments, one generally uses the *Michelson contrast* [Mic27], namely,

$$C^{M} = \frac{L_{max} - L_{min}}{L_{max} + L_{min}}. \tag{4.14}$$

While these two definitions are good predictors of perceived contrast for the above-mentioned classes of simple stimuli, they fail when the stimuli become more complex and cover a wider frequency range, for example, Gabor patches [307]. It is also evident that neither of these simple global definitions is appropriate for measuring contrast in natural images, because the brightest and darkest points would determine the contrast of the entire image, whereas actual human contrast perception varies with the *local* average luminance.

In order to address these issues and provide a quantitative definition of contrast, Peli [308] proposed a *local band-limited contrast*:

$$C_j^P(x, y) = \frac{\psi_j * I(x, y)}{\phi_j * I(x, y)}, \tag{4.15}$$

where $I(x, y)$ is the input image, ψ_j is a band-pass filter at level j of a filter bank, and ϕ_j is the corresponding low-pass filter. The normalization by the low-pass signal takes into account the local luminance variations. Modifications of this contrast definition have been used in a number of vision models [118,257] and are in good agreement with psychophysical experiments on Gabor patches [307].

The particular form of (4.15) suggests use of the wavelet transform for describing the variations of luminance. Now the WT is a space-scale analysis, and the spatial extension of the wavelets is characterized explicitly by their scale factor. Thus it is possible to define at each scale a different normalization, similar to a local average. So, following Duval–Destin [Duv91,12], one is led to the notion of *local contrast*, defined by combining the wavelet transform with an *adaptive* normalization. The latter will be obtained by projecting the signal, at a given scale, on a local weight function, chosen with the same localization properties as the wavelets. This local mean value will be called *luminous level*. This is the background against which luminance variations are measured, and the WT may be interpreted as a representation of these luminance variations within an image. The resulting contrast analysis is *nonlinear*, but it presents several advantages. It is particularly well adapted to the processing of positive signals. It also yields a multiplicative reconstruction process, which preserves positivity. Let us give some details and an example of application.

Let $h \in L^1(\mathbb{R}^2) \cap L^2(\mathbb{R}^2)$ be a non-negative, rotation invariant, weight function, normalized to $\|h\|_{L^1} = 1$. Given an image, represented by a non-negative function f, the luminous level with respect to the weight function h is defined as

$$M_a[f](\vec{b}) = \langle h_{(\vec{b},a)} | f \rangle, \quad h_{(\vec{b},a)}(\vec{x}) = a^{-2} h\left(a^{-1}(\vec{x} - \vec{b})\right). \tag{4.16}$$

Note that we use throughout the L^1-normalization, that is, $h_{(\vec{b},a)}$ instead of the usual $h_{\vec{b},a}$ (see Section 2.2). This is more natural in this context, since all the functions $h_{(\vec{b},a)}$ have the same L^1-norm.

Then we define the *local contrast* as the ratio of the CWT to the corresponding luminous level (the wavelet ψ is assumed to be also rotation invariant):

$$C_a[f](\vec{b}) = \frac{F_a(\vec{b})}{M_a[f](\vec{b})} = \frac{\langle \psi_{(\vec{b},a)} | f \rangle}{\langle h_{(\vec{b},a)} | f \rangle} = \frac{\langle \psi_{\vec{b},a} | f \rangle}{\langle h_{\vec{b},a} | f \rangle}, \tag{4.17}$$

where $F_a(\vec{b}) \equiv \check{F}(\vec{b}, a) = \langle \psi_{(\vec{b},a)} | f \rangle$ is the CWT of f with the L^1-normalization (but the local contrast is independent of the normalization). In order to make sense, this definition requires that the support of ψ be contained in the support of h. The local contrast is nonlinear, but its behavior is controlled by an integral condition. Large absolute values of contrast imply the existence of a region where the luminance signal is very small. A typical example, very natural in the study of vision, is to take for h a Gaussian and for ψ a Mexican hat.

But one can do better and take for ψ the difference wavelet associated to h, as given in (3.13). Then the local contrast becomes

$$C_a[f](\vec{b}) = \frac{\langle h_{(\vec{b},a\alpha)} | f \rangle}{\langle h_{(\vec{b},a)} | f \rangle} - 1, \tag{4.18}$$

and the existence condition is simply that the support of h be star-shaped.

This formula in turn leads to a multiplicative reconstruction scheme. Indeed, estimates of the luminous level at smaller and smaller scale factors a may be considered as smoothened versions of the image with progressively contracted weight functions h. Then, as for the WT, the approximation of a function at a given scale may be written in terms of the approximation at a larger scale and the complementary signal:

$$\begin{aligned}
M_{a\alpha}[f] &= M_a[f] \cdot (C_a[f] + 1), \\
M_{a\alpha^2}[f] &= M_{a\alpha}[f] \cdot (C_{a\alpha}[f] + 1) \\
&= M_{a\alpha}[f] \cdot (C_a[f] + 1) \cdot (C_{a\alpha}[f] + 1),
\end{aligned} \tag{4.19}$$

and, by recurrence:

$$M_{a\alpha^n}[f] = M_{a\alpha}[f] \cdot (C_a[f] + 1) \dots (C_{a\alpha^{n-1}}[f] + 1). \tag{4.20}$$

$M_{a\alpha^n}[f]$ is the nth resolution approximation of f; it is the image as seen through the smoothing function h contracted by a factor $a\alpha^n$ ($a < 1$). One notices the obvious analogy with the usual multiresolution analysis (Section 1.5). The formalism may be generalized further to the so-called *infinitesimal contrast analysis* developed in [159].

An interesting application of the notion of local contrast is the design of an algorithm for the matching of stereoscopic images [313,314]. The technique consists in using a localized correlation function for comparing the two images, by means of the 2-D CWT. The locality of the latter provides a good estimate of the disparity between images, at each scale. However, in order to prevent the occurrence of artifacts due to the high sensitivity of the localized correlation function to the local mean value of the signals, one must normalize the latter in an adaptive way, and this leads precisely to local contrast described above.

Peli and Duval–Destin's definition of local contrast as defined above measures contrast only as incremental or decremental changes from the local background, which is analogous to the symmetric (in-phase) responses of vision mechanisms. However, a

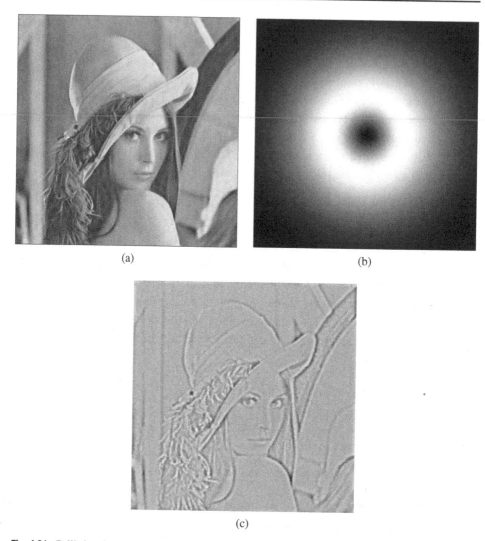

Fig. 4.31. Peli's local contrast (c) from equation (4.15) computed for the *lena* image (a) using an isotropic band-pass filter (b).

complete description of contrast for complex stimuli has to include the antisymmetric (quadrature) responses as well [347].

The problem is illustrated in Figure 4.31, which shows the contrast C^P computed with an isotropic band-pass filter for the *lena* image. It can be observed that C^P does not predict correctly the perceived contrast, as it varies between positive and negative values of similar amplitude at the border between bright and dark regions and exhibits zero-crossings right where the perceived contrast is actually highest.

This behavior can be understood when C^P is computed for sinusoids with a constant C^M. The contrast computed using only a symmetric filter actually oscillates between $\pm C^M$ with the same frequency as the underlying sinusoid, which complicates establishing a correspondence between such a local contrast measure and data from psychophysical experiments.

These examples underline the need for taking into account both the in-phase and the quadrature component in order to be able to relate a generalized definition of contrast to the Michelson contrast of a sinusoid grating. Analytic filters represent an elegant way to achieve this. The magnitude of the analytic filter response, which is the sum of the energy responses of in-phase and quadrature components, exhibits the desired behavior, i.e., it gives a constant response to sinusoid gratings.

Unfortunately, extending the Hilbert transform to 2-D is not a straightforward task. However, as already stressed in Chapter 3, directional wavelets offer a pleasant alternative, since the Fourier transform of the wavelet is included in a convex cone with apex at the origin and of aperture strictly smaller than π. This means that at least three such wavelets are required to cover all possible orientations uniformly, but otherwise there is no restriction on the number of filters.

There are many applications where isotropy is required. In these cases, it is important to combine the analytic responses defined above into an isotropic contrast measure. Working in polar coordinates (r, φ) in the Fourier domain, we choose, as in Section 2.6.3, a directional wavelet $\widehat{\Psi}(r, \varphi)$ satisfying the above requirements and the conditions (4.10)–(4.12) above. Note that the function ϕ in (4.12) need not be a scaling function associated to ψ, but it should at least have the same localization properties in order to provide for a meaningful normalization of the luminance level.

Now it is possible to construct an isotropic contrast measure from the energy sum of directional filter responses [374]:

$$C_j^{\mathrm{I}}(x, y) = \frac{\sqrt{2 \sum_\ell |\Psi_{j\ell} * I(x, y)|^2}}{\phi_j * I(x, y)}, \tag{4.21}$$

where $\Psi_{j\ell}$ denotes the wavelet dilated by 2^{-j} and rotated by $2\pi\ell/L$. If the directional wavelet Ψ belongs to $L^1(\mathbb{R}^2) \cap L^2(\mathbb{R}^2)$, the convolution in the numerator of (4.21) is again a square integrable function, and (4.10) shows that its L^2-norm is exactly what would have been obtained using the isotropic wavelet ψ. C_j^{I} is thus an orientation- and phase-independent quantity, but being defined by means of analytic filters, it behaves as prescribed with respect to sinusoidal gratings (i.e., $C_j^{\mathrm{I}}(x, y) \equiv C^{\mathrm{M}}$ in this case).

Examples for this isotropic contrast are shown in Figure 4.32. It can be seen that the contrast features obtained with C_j^{I} correspond very well to perceived contrast. The combination of the directional analytic filter responses produces a naturally meaningful phase-independent measure of isotropic contrast.

This technique may be applied for improving the contrast in any kind of image. An example of application to a photograph was given in [12]. Here we show one with a medical image (Figure 4.33). The image f is decomposed over N contrast levels, as in (4.20), using the couple Gaussian–DOG. For each level j, one defines the contrast chart as the modulus of the local contrast,

(a)   (b)   (c)

Fig. 4.32. Isotropic contrast of the *lena* image as described by equation (4.21) at three different levels, 0, 1, and 2.

$$M_j(\vec{b}) = |C_{2^j}[f](\vec{b})|, \ \ j = 1, \ldots, N. \tag{4.22}$$

Then one interprets the product of the N charts, $S(\vec{b}) = \prod_{j=1}^{N} M_j(\vec{b})$ as a measure of the correlation between the successive scales of the image at the point $\vec{b}$. After thresholding, one obtains a binary image or mask. The latter is used in medical imagery, for instance, as a preprocessing to more sophisticated algorithms. It is taken as *a priori* knowledge and helps to reduce the amount of computation.

4.7.2 Watermarking of images

Digital image watermarking consists in embedding a digital signature in an image. This operation is usually performed by slightly modifying the visual information in such a way that the perturbation is invisible to human eyes, but can still be recovered by using an appropriate algorithm. This embedding can be performed directly in the spatial domain, but also in the frequency domain (using DCT[†] coefficients) or, as we shall see now, in the wavelet domain. Finally, one often asks that the watermark should be *robust*, that is, it should survive common image alterations: geometrical image transformations, addition of noise, lossy compression or even print-scan procedure. We refer to [214] for further details.

The generic picture of an image watermarking application is depicted in Figure 4.34. The inputs of the system consist in the original image, the watermark and an optional public or secret key. The watermark, or digital signature, can be of various nature: number, image or text. The key is used to encrypt the watermark and prevents it being read by unauthorized parties. In the sequel, we will mainly focus on the embedding part of the system and, more precisely, we will see how the wavelet transform can be used in conjunction with a vision model for robust and imperceptible image watermarking.

[†] DCT = discrete cosine transform (see [Mal99]).

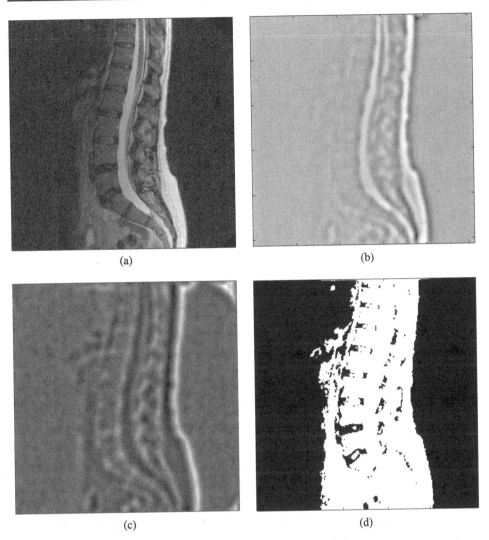

(a)

(b)

(c)

(d)

Fig. 4.33. Contrast analysis of a medical image: (a) the original image; (b) the CWT with a Mexican hat ($j = -1$); (c) the contrast chart $M_j(\vec{b})$, $j = -1$; (d) the resulting binary image. Many more details are seen on the two bottom images than on the ordinary CWT.

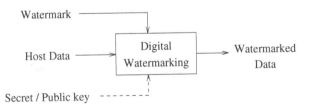

Fig. 4.34. Typical image watermarking system.

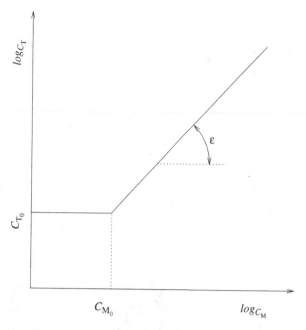

Fig. 4.35. Contrast masking model describing the relationship between the masker contrast and the target contrast at detection threshold.

In order to model the visibility of the watermark in the original image, there are mainly two effects that need to be taken into account, namely, contrast sensitivity and masking. Contrast sensitivity, as we have seen previously, describes the response of the human visual system to the contrast of a stimulus. Masking, on the other hand, describes the phenomenon in which a signal, the masker, is capable of "hiding" a second signal, the target. In other words, the target visibility depends on the presence of a masker. It is possible to combine contrast sensitivity and masking in a model that describes the relation between the masker contrast and the target contrast at detection threshold. Figure 4.35 shows such a model where we have represented on the horizontal axis the logarithm of the masker contrast C_M, and on the vertical axis we have the logarithm of the target contrast C_T. The curve is divided into a threshold range, where the target detection threshold is independent of the masker contrast, and a masking range, where it grows as a power of the masker contrast. The mathematical description of this model is given by:

$$C_T(C_M) = \begin{cases} C_{T_0} & \text{if } C_M < C_{M_0}, \\ C_{T_0} \left(C_M / C_{M_0} \right)^{\varepsilon} & \text{otherwise.} \end{cases} \qquad (4.23)$$

The model contains three parameters, ε, C_{T_0} and C_{M_0}, which specify the size of the threshold and the masking range as well as the slope of the transducer function. They have to be determined by means of subjective experiments.

This model is applied in [359] to a watermarking scheme based on spatial spread-spectrum modulation, as proposed by Kutter [Kut99]. Each bit to be embedded in the image is represented by a two-dimensional pseudo-random pattern. The statistics of the pattern are bimodal with equal probabilities for -1 and 1. The random patterns of all bits are superimposed as follows:

$$w(x, y) = \alpha(x, y) \sum_i p_i(x, y), \tag{4.24}$$

where $p_i(x, y)$ are the pseudo-random modulation functions for bit i, $\alpha(x, y)$ is the watermark weighting function, and $w(x, y)$ is the resulting watermark which is added to the image. In this watermarking scheme, the pseudo-random patterns p_i are sparse, which means that the superposition of all patterns does not necessarily modify all pixels in the image. To quantify the sparseness, we introduce the density D of the watermark, which is given by the modified number of pixels divided by the total number of pixels in the image.

The watermark weighting function $\alpha(x, y)$ is computed using the introduced masking model and the local isotropic contrast measure presented in Section 4.7. For computing the local contrast according to (4.21), we use directional wavelet frames as described in Section 2.6, based on the scaling functions of Table 2.1. The minimum number of orientations required by the analytic filter constraint, i.e., an angular support smaller than π, is three. The human visual system emphasizes horizontal and vertical directions, so four orientations should be used as a practical minimum. To give additional weight to diagonal structures, eight orientations are preferred. We only use the highest frequency band of the pyramidal decomposition, because masking is strongest when masker and target have similar frequencies. Furthermore, higher levels tend to smear the local contrast and are thus not suitable for this kind of application. The watermark weighting function α is now computed as follows:

$$\alpha(x, y) = C_T(C_0^I)(x, y) \cdot \phi_0 * I(x, y), \tag{4.25}$$

where C_0^I is the local isotropic contrast of the masker image at level 0, C_T is the corresponding target contrast threshold as given by our masking model, and ϕ_0 is a low-pass filter. The local amplitude of the watermark at the threshold of visibility is thus determined by the multiplication of the isotropic contrast values with the corresponding low-pass filtered image. Finally, the parameters of our vision model (C_{T_0}, C_{M_0} and ε) have been determined by performing subjective tests [359].

Figure 4.36 shows weighting masks for the *lena* image at watermark densities of 0.4 and 1, respectively. For illustrative purposes, the figures to the right visualize the segmentation into threshold and masking ranges. The dark areas correspond to regions where only contrast sensitivity is exploited, and the bright areas show regions where the masking effect is dominant.

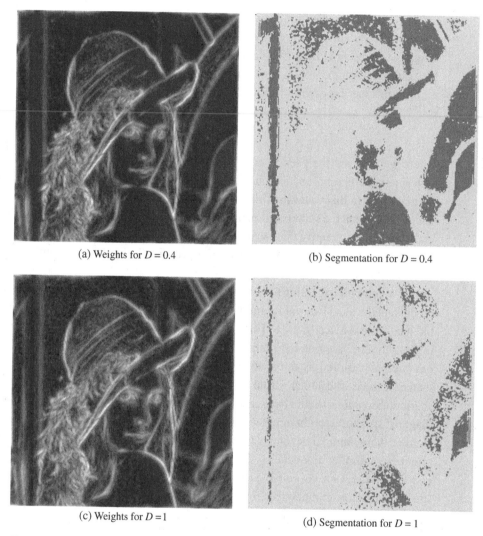

(a) Weights for $D = 0.4$ (b) Segmentation for $D = 0.4$

(c) Weights for $D = 1$ (d) Segmentation for $D = 1$

Fig. 4.36. Watermark weighting function of the *lena* image for two different density values (left column). The segmented images (right column) illustrate the threshold and masking ranges of the contrast masking model, represented by dark and bright areas, respectively.

In comparison with other watermarking schemes, this weighting mask based on the simple masking model presented above facilitates the insertion of a watermark with higher energy while preserving the visual quality of the image, leading to a watermark that is more robust. It has also been applied successfully to watermarking the blue channel of color images [251].

5 Applications of the 2-D CWT. II: physical applications

In the previous chapter, we have discussed a number of applications of the 2-D CWT that belong essentially to the realm of image processing. Besides these, however, there are plenty of applications to genuine physical problems, in such diverse fields as astrophysics, geophysics, fluid dynamics or fractal analysis. Here the CWT appears as a new analysis tool, that often proves more efficient than traditional methods, which in fact rarely go beyond standard Fourier analysis. We will review some of these applications in the present chapter, without pretention of exhaustivity, of course. Our treatment will often be sketchy, but we have tried to provide always full references to the original papers.

5.1 Astronomy and astrophysics

5.1.1 Wavelets and astronomical images

Astronomical imaging has distinct characteristics. First, the Universe has a marked hierarchical structure, almost fractal. Nearby stars, galaxies, quasars, galaxy clusters and superclusters have very different sizes and live at very different distances, which means that the scale variable is essential and a multiscale analysis is in order, instead of the usual Fourier methods. This suggests wavelet analysis. Now, the main problem is that of detecting particular features, relations, groupings, etc., in images, which leads us to prefer the *continuous* WT over the discrete WT. Finally, there is in general no privileged direction, nor particular oriented features, in astrophysical images. All this leads us to use the CWT with an isotropic 2-D wavelet. In addition, astrophysical images are very noisy. In particular the bright sky and our own galaxy (the Milky Way) represent noise, which must be removed, with a technique similar to that used in 1-D for the subtraction of unwanted lines or noise in spectra [210]. Here statistical techniques play an essential rôle. All these considerations characterize the type of wavelet applications that have been developed in astronomy and astrophysics.

The first attempt to apply the CWT to astrophysical images is due to the group of A. Bijaoui in Nice, in 1990 (see [80] for a review). In their pioneering paper [343], they used the CWT for the analysis of galaxy clusters, with the 2-D Mexican hat. Similar techniques were exploited by a large number of authors, especially in the last few years. In this section, we will review some of this work, and also present two novel applications, one to solar physics, the other one to the detection of gamma-ray sources in the Universe.

5.1.2   Structure of the Universe, cosmic microwave background (CMB) radiation

In several papers [161,162,193,343,344], the authors exploit galaxy counts to identify galaxy groupings, from compact groups (0.5 degrees or tens of kpc[†] in extent) to clusters (down to 1 degree, from hundreds of kpc to some Mpc), to large-scale structures or superclusters (5 degrees or tens of Mpc or more), including the determination of a possible hierarchy between them. The same technique allows the detection of voids, that is, large regions (up to 60 Mpc) with very few galaxies, and also to a neat definition of each of these notions. The results of such work leads to the analysis of the large-scale structure of the Universe, thus to cosmological considerations. For instance, the distribution of groups of various size and of voids points to a possible fractal structure of the Universe. On the other hand, the multiscale approach yields much valuable insight into the inner structure of individual clusters [193]. Here, as in all papers analyzing galaxy maps, the basic data is a bidimensional distribution of Dirac delta functions, possibly weighted according to some statistical criterion. The same type of data will be used in the next two sections.

A byproduct of such hierarchical analyses is the multiscale vision model developed by Bijaoui and his group [78,79] in order to detect and characterize structures of different sizes (for numerical reasons, also linked to the necessity of denoising the images, they later switched to a discrete WT, based on spline wavelets). For instance, they propose in [254] a morphological indicator allowing a comparison between various cosmological models (for instance, cold versus hot dark matter).

In the same vein, a group from Santander, Spain, has undertaken a systematic analysis, by wavelet methods, of the COBE data on the cosmic microwave background (CMB) radiation. As a first step, they study the local (i.e., in small sky patches) temperature anisotropies in the CMB, including denoising the images [332,333]. In these papers, the authors use both the CWT and the DWT (the latter especially for denoising). As for the former choice, they first consider a 2-D CWT without a rotation parameter, but with two independent scalings in the x and y directions, then the usual isotropic Haar and Mexican hat wavelets. Next [100,361], they use isotropic wavelets to detect and determine the flux of point sources superimposed on the CMB, in conditions simulating the Planck Surveyor mission. As they point out, the advantage of the wavelet method is that no assumption has to be made regarding the statistical properties of the point

---

[†] 1 pc = 1 parsec = 3.26 light years.

source population or the underlying emission of the CMB. Since the CMB observations are performed with antennas that are best modeled by a Gaussian beam, it turns out that the isotropic Mexican hat wavelet is in fact optimal for detecting point sources. In a further work [362], a detailed comparison is made of the wavelet method with the standard maximum-entropy method. The conclusion is that the two methods are in fact complementary and can be combined to improve the accuracy of the detection.

More recently, the Santander group has turned to a global analysis of the CMB, trying to detect potential non-Gaussian CMB temperature fluctuations. This is an important observation for cosmology, for any non-Gaussianity would be evidence for a departure from standard inflationary theories. Since the data used in these experiments is the full sky COBE-DMR data, it is clear that the sphericity of the data has to be taken into account. As a consequence, one has to resort to spherical wavelets. A first attempt was made by Barreiro *et al.* [62], using discrete spherical Haar wavelets, constructed with the lifting scheme of Schröder and Sweldens [336], described in Section 2.5.2.4. Then, following the same logic that recommends the use of the CWT with an isotropic Mexican hat, the Santander group introduced the *spherical Mexican hat* (see below), establishing the superior capability of the latter over the Haar wavelets [267]. The net result of these investigations is that the CMB temperature fluctuations are consistent with a Gaussian distribution, thus vindicating the standard theories [101]. Finally [363], the same group has used the same spherical Mexican hat wavelet for extending their previous work [361] on simulated Planck maps, thus achieving a large catalog of potential point sources.

Coming back to the CWT on the 2-sphere S^2, a mathematically precise transform was constructed in [29] and will be discussed at length in Chapter 9, Section 9.2. The idea is simply to take the plane $\mathbb{R}^2$ as the tangent plane at the North Pole of S^2 and lift functions on $\mathbb{R}^2$ to functions on S^2 by inverse stereographic projection. Introducing polar coordinates both on the plane and on the sphere, the correspondence reads:

$$S^2 \ni (\theta, \varphi) \Longleftrightarrow (r, \varphi) \equiv (2 \tan \frac{\theta}{2}, \varphi).$$

For square integrable functions, this leads to a unitary map between the respective Hilbert spaces, $I^{-1} : L^2(\mathbb{R}^2, d\vec{x}) \to L^2(S^2, \sin \theta \, d\theta \, d\varphi)$, namely,

$$(I^{-1}f)(\theta, \varphi) = \frac{2}{1 + \cos \theta} f(2 \tan \frac{\theta}{2}, \varphi). \tag{5.1}$$

In the case of an isotropic wavelet $\psi(r)$, with $r = |\vec{x}|$, the correspondence is simply

$$(I^{-1}\psi)(\theta) = \frac{2}{1 + \cos \theta} \psi(2 \tan \frac{\theta}{2}). \tag{5.2}$$

Then, choosing for ψ the isotropic Mexican hat wavelet ψ_{H}, one gets the *spherical Mexican hat wavelet*:

$$\psi_{\text{H.sph}}(\theta) = 4 \frac{1 - 2\tan^2\frac{\theta}{2}}{1 + \cos\theta} \, \exp(-2\tan^2\tfrac{\theta}{2}). \tag{5.3}$$

It should be noted that, in this parametrization, the scaling $\vec{x} \mapsto a^{-1}\vec{x}$ in the tangent plane becomes on the sphere: $\tan\frac{\theta}{2} \mapsto a^{-1}\tan\frac{\theta}{2}$. We refer to Chapter 9 for the full analysis.

Another research area where the CWT has been used is the detailed analysis of individual galaxies, notably in the group of P. Frick [173,174]. Of particular importance is the cross-correlation between images obtained at different wavelengths. To that effect, the authors of [174] consider the (normalized) *wavelet cross-correlation function* or *wavelet correlation coefficient*, obtained by polarization from the wavelet spectrum (2.61) $W[s](a)$:

$$C[s_1, s_2](a) = \frac{\int d^2\vec{b} \, S_1(\vec{b}, a) \, \overline{S_2(\vec{b}, a)}}{(W[s_1](a) \, W[s_2](a))^{1/2}}, \tag{5.4}$$

and originally introduced by Hudgins *et al.* [227]. Using this tool, the authors study a particular spiral galaxy, called NGC 6946, comparing the images of total radio emission, red light and mid-infrared dust emission on all scales. Note that in their treatment they use both the Mexican hat wavelet and, for a better separation of scales, their own isotropic wavelet, called Pet hat and defined in (3.11). In a later work from Frick's group [175], spherical wavelets (in a somewhat primitive form) are used for isolating coherent structures in the distribution of the Faraday rotation measure of extragalactic radio sources, that is, a weighted integral of the longitudinal magnetic field along the line of sight. In addition, since these sources are given as irregularly distributed points in the sky, they adapt to the 2-D spherical situation the technique of *gapped wavelets* introduced previously in 1-D [172].

A final application in astrophysics, still under development, is to *gravitational lensing*, namely, the detection of Einstein arcs in cosmological pictures [82]. Whenever the light from a distant bright object (a quasar) is seen through a galaxy, the latter behaves as a gravitational lens, so that the point source appears as a ring, or a portion of a ring ("arclet"), if the alignment is not exact. By measuring the radius of that ring, one may infer the distance of the source. This may be done in two steps. The center of the arc is obtained with an annular-shaped wavelet, such as the Bessel filter (3.9), Frick's wavelet (3.11), or the annular Halo wavelet (3.12), used at a rather large scale (e.g. $a = 2$). This determination is quite robust to noise, in particular, to spurious bright points, that mimic nearby stars. The arc itself is obtained with an isotropic Mexican hat, at a smaller scale (e.g. $a = 0.5$). By superposing the two transforms and applying a severe thresholding (up to 95%) for eliminating the noise, one obtains an image with three bright spots: two points of the arc, around the endpoints, and the center of the corresponding circle. From this, one can reconstruct the arc unambiguously, and thus one obtains a tool for measuring in a simple way the distance of quasars, for instance.

Before turning to more specific 2-D applications, we hasten to add that 1-D wavelet analysis has been used in various current problems is astrophysics. A case in point is the analysis of the solar neutrino capture rate data from the Homestake experiment [215], a crucial ingredient in the resolution of the celebrated solar neutrino problem.

5.1.3   Application of the CWT in solar astronomy

Since 1996, the *Extreme-ultraviolet Imaging Telescope* (EIT) on board the *Solar and Heliospheric Observatory* (SoHO) satellite observes the Sun in four wavelengths: 171, 195, 284 and 304 Å. These correspond respectively to particular emission lines of Iron (IX-X, XII, XV) and Helium (II), and thus to temperatures typical of those of the *Sun corona* in the first three wavelengths, and of the *transition region* in the fourth one [130].

The Sun corona is physically very complex and contains a huge amount of different events appearing at different locations and scales. Solar astronomers are interested in the physics which can be deduced from them in order to improve our knowledge of the global Sun. One way to achieve this is to make time statistics on features of *special* solar objects. In addition, because of the large number of EIT pictures (currently greater than 100 000), astronomers aim at an automatic analysis.

However, many conceptual problems arise due to the difference between the human description of things and the true (logical) computer vision. These can be summarized into two main questions.

- How to define a Sun corona object in simple terms, that is, in sufficiently simple concepts which can be managed by a computer program?
- How to determine the relevant characteristics of such an object and how to translate them as simply as possible?

After a short description of the common Sun corona objects, we will show that the continuous wavelet transform (CWT) offers tools to answer these fundamental questions (we basically follow [37]). Notice also that Bijaoui and his group have applied their vision model to the analysis of EIT images of the solar corona [316].

5.1.3.1   Special coronal objects

The physical objects of the Sun corona result in general from convective motions in the solar mantle and/or of magnetic interactions with hot material. Here is a list of the principal objects, ordered by size, from the smallest to the largest (for more information, see [130,278]).

Magnetic network: In the red 304 Å images, the *magnetic network* constitutes a textured solar background resulting of the advection of small magnetic flux by the convective motion in the solar mantle.

Brightenings: Brightenings are visible in all EIT images and are related to magnetic topology changes to a lower energy state. Their typical scale in an image is close of the pixel size, but they brighten and fade away on a time scale ranging from several minutes to hours.

Flares: A sudden and energetic local *brightening* in an *active region* (see below).

Bright points: Bright points (BP) are small regions with enhanced emission. They are located above pairs of magnetic features of opposite polarity in the photosphere. We can see them in the quiet corona and in *coronal holes*. They present a lifetime ranging between two hours and two days.

Magnetic loops (or Loops): These objects result from the filling of magnetic field lines with plasma. Because the temperature of this material varies along the loop, the footpoints of the loop are more precise in the 171 Å, because they are cooler than the loop summit, which is better seen in the 195 Å. The magnetic loops may be part of the same active region, connecting two regions of opposite flux, or even join different ARs.

Active regions: They show up as a region of large increase in the ultraviolet flux on the image. Their typical size is about 10% of the solar radius. Physically, these active regions (AR) contain hot material in smaller and larger *loops* around and inside a region of enhanced magnetic flux. Because active regions are deeply related to the well-known *Sun spots*, they appear in two bands of latitude according to the evolution of the main solar cycle of 11 years: They live at high latitudes at the solar minimum (beginning of the cycle) and move towards the equator at the solar maximum.

Coronal holes: Coronal holes (CH) are large regions where the magnetic field lines are open and are advected by the solar wind into interplanetary space. Because the energy is advected away, the CH are colder than the closed magnetic field regions and they appear effectively like dark holes in the EIT images. Their morphology evolves with time and they become very small during the Sun maximum.

A visual summary of the solar objects defined above is presented in Figure 5.1. There are also features which are not related to the Sun physics, but either to defects of the SoHO satellite due to its aging, or to its interaction with some external events. The main ones are the *cosmic ray hits*, which, by the interactions of cosmic rays with the EIT CCD camera, plague the images with many bright pixels or bright straight lines, depending on the cosmic orientation relatively to the CCD surface.

We should mention finally that all these images have a noise component, namely, a readout noise coming from the CCD camera, the solar noise and the photon-shot noise (Poisson noise). The global noise is well approximated by Gaussian statistics, because of the high counting effect (central limit theorem).

5.1.3.2 Distribution of small features

To start with, let us define the specific wavelet tools needed for the present application, which aims at selecting some of the solar corona phenomena. We restrict ourselves to

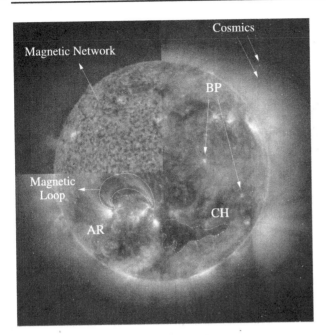

Fig. 5.1. The main Sun corona objects. The top left quadrant is the 304 Å wavelength image, the rest corresponds to 171 Å.

isotropic wavelets, since directions are irrelevant in the present context. Given an image $s(\vec{x})$, we consider its CWT $S(\vec{b}, a)$ with respect to a Mexican hat and the corresponding energy density $E[s](\vec{b}, a) = |S(\vec{b}, a)|^2$. Then, as in Section 2.3.5, we define ridges $\mathcal{R}_j$, and the corresponding amplitude $\mathcal{A}_j$ (2.63) and slope $\mathcal{S}_j$ (2.64). These two parameters are sufficient for the detection and discrimination of small features contained in the image s. A precious tool to that effect is the *histogram* of the amplitude as a function of the slope, or the *slope–amplitude histogram*.

Let a_0 the smallest relevant scale. Choose a sequence $\{\vec{b}_j, 0 \leqslant j \leqslant K - 1\}$ of maxima of $E[s](\vec{b}, a_0)$, belonging to ridges $\{\mathcal{R}_j, 0 \leqslant j \leqslant K - 1\}$. Then, given the set of all corresponding couples $(\mathcal{S}_j, \mathcal{A}_j)_{0 \leqslant j \leqslant K-1}$, the histogram is built by the following simple algorithm:

- determine the desired size of the histogram $\mathcal{H}$, say $M \times N$, and initialize $\mathcal{H}$ as the zero $M \times N$ matrix;
- compute $\mathcal{S}_{\min}$ and $\mathcal{S}_{\max}$, the minimum and the maximum of all the slopes $(\mathcal{S}_j)_{0 \leqslant j \leqslant K-1}$, respectively;
- compute $\mathcal{A}_{\min}$ and $\mathcal{A}_{\max}$, the minimum and the maximum of all the amplitudes $(\mathcal{A}_j)_{0 \leqslant j \leqslant K-1}$, respectively;
- form the discretized slope $\widetilde{\mathcal{S}}_m = \mathcal{S}_{\min} + \frac{m}{M-1}(\mathcal{S}_{\max} - \mathcal{S}_{\min})$ for $0 \leqslant m \leqslant (M - 1)$, and the discretized amplitude $\widetilde{\mathcal{A}}_n = \mathcal{A}_{\min} + \frac{n}{N-1}(\mathcal{A}_{\max} - \mathcal{A}_{\min})$ for $0 \leqslant n \leqslant (N - 1)$;

- then, for $k = 0, \ldots, K - 1$:
  - take the slope $\mathcal{S}_k$ and compute the index m such that $\widetilde{\mathcal{S}}_m$ is the nearest discretized slope from $\mathcal{S}_k$, that is, the index m such that

$$-0.5 < (M - 1)\frac{\widetilde{\mathcal{S}}_m - \mathcal{S}_k}{\mathcal{S}_{max} - \mathcal{S}_{min}} \leqslant 0.5; \tag{5.5}$$

  - do the same with the amplitude and determine the index n such that

$$-0.5 < (N - 1)\frac{\widetilde{\mathcal{A}}_n - \mathcal{A}_k}{\mathcal{A}_{max} - \mathcal{A}_{min}} \leqslant 0.5; \tag{5.6}$$

  - increment the (m, n) entry of $\mathcal{H}$ by one,

$$\mathcal{H}_{mn} := \mathcal{H}_{mn} + 1.$$

The histogram $\mathcal{H}$ reflects the 2-D distribution of the slope–amplitude couples. Identifying distinct areas inside $\mathcal{H}$ is equivalent to detecting different classes of small objects contained in the image s.

Before computing any histogram, a practical remark must be made about the difference between the continuous theoretical world and the discretized view of the programming. Indeed, actual computation requires an adequate sampling of the image s and of the wavelet ψ. Therefore, the scale a cannot effectively go to zero in (2.63) and (2.64). Indeed, the wavelet ψ must be sampled sufficiently on the grid determining the image. Thus, there will be a minimal scale a_0 for which $\widehat{\psi}_{\vec{b} a_0}$ is essentially contained in the frequency domain $[-\pi, \pi) \times [-\pi, \pi)$ (assuming the sampling period T is equal to 1).

5.1.3.3 Analysis of academic objects

To test our method, we begin by analyzing two types of objects that will model small features in EIT images. Take first the smallest possible object, a singularity of height c localized on a point $\vec{u}$, represented by a Dirac distribution $s(\vec{x}) = c\,\delta^{(2)}(\vec{x} - \vec{u})$. One readily computes the CWT of s and the corresponding energy density

$$\mathsf{E}[s](\vec{b}, a) = \frac{c^2}{a^4}\left|\psi(a^{-1}(\vec{u} - \vec{b}))\right|^2. \tag{5.7}$$

It is easy to see that, if ψ has a maximum in $\vec{x} = \vec{0}$, then $\mathsf{E}[s]$ is maximum in $\vec{b} = \vec{u}$ for all scales. The equation of the associated ridge is simply $(\vec{u}, a)$ for all $a \in \mathbb{R}^+$. The amplitude of this ridge is given by $\ln \mathcal{A}_{\vec{u}} = -4 \ln a_0 + \ln c$ (thus it tends to ∞ as $a_0 \to 0$) and the corresponding slope has the value -4.

The second object is a simple Gaussian localized in $\vec{w}$, of width σ and height D, $s(\vec{x}) = D \exp(-\frac{1}{2}|\vec{x} - \vec{w}|^2/\sigma^2)$. A detailed calculation shows that $\mathsf{E}[s]$ has also a vertical ridge localized in $\vec{b} = \vec{w}$ with a maximum in $a = \sigma$. The amplitude of this ridge is proportional to D and the slope is now positive.

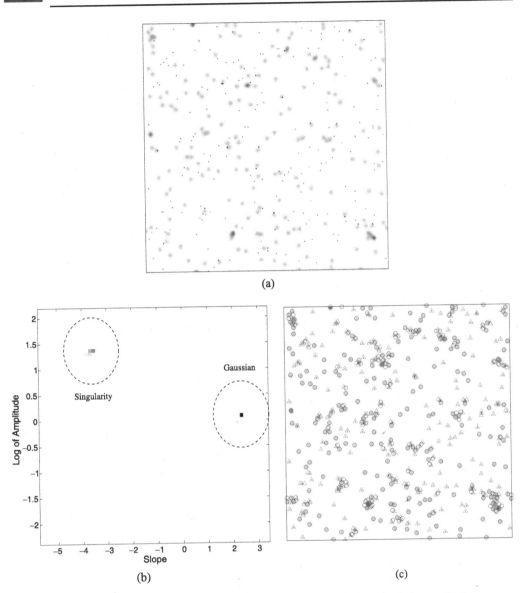

Fig. 5.2. Analysis of singularities and Gaussians. (a) The original academic image; (b) the slope–amplitude histogram (the logarithm of the amplitude is plotted to reduce the range); (c) the selection of points in the singularity population (triangles), or in the Gaussian population (circles).

These two examples show that the amplitude yields a criterion for selecting small objects according to their intensity. Then the slope decides between a singularity or a larger object modeled by a Gaussian.

The procedure is illustrated in Figure 5.2. We analyze an academic image s of size 256×256, shown in panel (a), and consisting of a collection of randomly placed singularities and Gaussians of small size, and compute the corresponding slope–amplitude histogram, knowing that, in this case, the minimal scale a_0 of the Mexican hat is close to 0.9.

In Figure 5.2(b), we clearly see two distinct populations in the slope–amplitude histogram. The population on the left-hand side (left dashed circle) corresponds to singularities of s with a slope centered around -4. The right area (right circle) corresponds to the Gaussians. A rough selection of points according to the sign of the slope is made in Figure 5.2(c). Negative slopes are represented by triangles, and positive ones by circles; singularities and Gaussians are effectively selected separately.

5.1.3.4   Application to EIT images

We can now apply the preceding technique to the selection of *cosmic ray hits* and of *bright points* in the EIT images. The former are well described by singularities, because cosmics burn only a few pixels on the CCD camera of the satellite, and the latter can be modeled by Gaussians of small size.

The analyzed EIT image, shown in Figure 5.3(a), is the top-left quadrant of a 284 Å wavelength image of the Sun. The slope–amplitude histogram, computed for $a_0 = 1$, is presented in Figure 5.3(b). We notice that the distinction between populations is not as neat as in the academic example. The reason is that the white noise present in the picture recording has a main effect of spreading the well-defined areas of Figure 5.2(b).

Next, we impose on the histogram of Figure 5.3(b) an additional selection criterion. For cosmics, we choose the maxima $\vec{b}_j$ of $E[s](., a_0)$ such that $\ln \mathcal{A}_j > 2$ and $\mathcal{S}_j < 0$ and, for bright points, those with $\mathcal{S}_j > 0$. The amplitude thresholding prevents us from taking singularities that are too faint coming from quantization and Gaussian noise. The result is shown in Figure 5.3(c). The cosmics are detected everywhere in the image (because they are not related to solar physics), while the bright points appear mainly on the solar disk (*on-disk*).

In Figure 5.4, we make a zoom on a particular *on-disk* area of the Sun. The selection effect is now clearer than in the global image.

5.1.3.5   Conclusion and open questions

We have presented in the previous section a simple method based on the CWT to discriminate two kinds of simple events in the Sun corona pictures, the cosmic ray hits and the bright points. However, there is ample room for improving the range and efficiency of the method. First, noise has been suppressed in the slope–amplitude histogram by a hard thresholding. However, a precise statistical study remains to be made on these selections according to the SNR of the analyzed images and their CWTs.

Then, one may try to characterize more complex solar phenomena, such as the active regions, the magnetic loops or the textured magnetic network. A possibility is to exploit the full information carried by the vertical ridges of the CWT. Indeed, the method described above uses only the first relevant scale of the latter, that is, only their

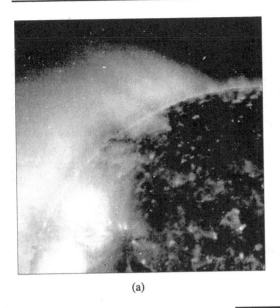

(a)

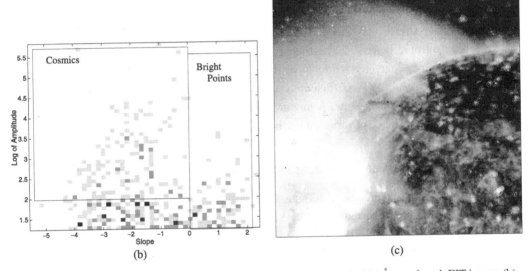

(b)   (c)

Fig. 5.3. Analysis of an EIT image. (a) The top-left quadrant of a 284 Å wavelength EIT image; (b) the slope–amplitude histogram; (c) the selected cosmics (triangles) and bright points (circles).

beginning. Information about possible maxima of $E[s]$ along these ridges is interesting too, for instance for determining the typical scale which defines each type object, as in the 1-D analysis of impact experiments [358].

Several hierarchical criteria based on the CWT may also help us to detect the inclusion of small events into larger ones, such as the brightenings within the active regions. The shape of strong response areas in the CWT at different scales could be useful in this context, as found in [344], for instance.

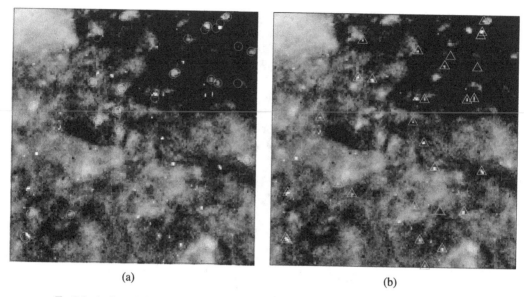

Fig. 5.4. A closer look on a small on-disk region of the Sun: (a) Bright points selection; (b) cosmics selection.

Finally, directional wavelets may be useful, too, since many solar events present an anisotropic behavior. The magnetic loops, for instance, are locally equivalent to straight lines characterized by a particular width. At a scale proportional to this width, the anisotropic CWT coefficients should vary for different angles θ relatively to the main direction of this line. A magnetic loop signature could perhaps be found inside this variation (this is a variation on the problem of detection of oriented contours).

5.1.4   Detection of gamma-ray sources in the Universe

Another topic where the CWT has been applied successfully is the analysis of the X-ray structure of various objects, such as clusters of galaxies, following a suggestion by Grebenev *et al.* [201]. This leads to a different class of problems. Indeed, such sources are frequently at the limit of detection, so that statistical considerations become crucial. In particular, we are here often in the photon-counting regime, the photon per pixel statistics is significantly different from Gaussian and most sources are extended. The analysis of such images by wavelet methods was further developed by Damiani *et al.* [115]. The work reported here uses a similar approach, but goes significantly deeper in the analysis. In particular, considerable care is devoted to the presence of Poisson noise in the photon flux. We again follow [37].

When it comes to the detection and analysis of gamma-ray sources in the Universe, the way data analysis is carried out depends on the energy range explored by the telescope. This is not only due to the nature of emitting objects, and to the difficulty of designing appropriate detectors, but also to the gradually *lower photon counting rate* as the energy increases. For example, in average 100 ultraviolet photons from the Sun are expected to be detected in one second by each pixel of the SoHO CCD camera (Section 5.1.3), whereas about 1 gamma photon is recorded by the whole gamma-ray space telescope EGRET during the same period! This correspondingly decreases accuracy and significance of any statistical decision, like event detection. Equally important for the data analyst, the nature of photon-counting processes induces an intrinsic "noise," called Poisson noise, requiring more statistical care than the usual Gaussian noise.

The problem we address in this section is the detection of sources in the raw data of the above-mentioned telescope EGRET (20 MeV – 30 GeV photons). Sources are point-like objects like pulsars or active galactic nuclei and appear in the data as a few detected photons coming from the same direction in the sky. The whole issue is to give a meaning to the coincidence of finding these photons together, hence to conclude (or not) that they were produced by chance from the diffuse background (interaction of cosmic rays with interstellar clouds). In addition to the detection significance, the position, magnitude and spectral characteristics of a source are other desirable quantities determined from the data. This may seem a very humble problem to solve, but, as outlined above, the scope of questions one can answer at 1 GeV is considerably restricted compared to the wealth of the analysis in the previous section.

5.1.4.1 Sample data and the classical solutions

Every dot on Figure 5.5 is a detected photon, of energy 100 MeV or above, during the viewing period 21.0 of EGRET. A "position" on this counting map refers to a direction in the sky, and the map is modeled as approximately flat. Such counting maps are very broadly modeled as counting (Poisson) processes from two contributions, the background flux and the flux from the sources. That is, we do not directly observe light

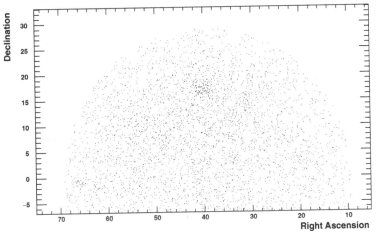

Fig. 5.5. Detected photons above 100 MeV during EGRET viewing period 21.0.

intensities, but rather photons that are randomly created from the corresponding physical objects (point-like objects or extended objects like interstellar gases). Moreover, the detector is far from being perfect: direction and energy of an incoming photon are recorded with an error that translates into the convolution of the above-mentioned fluxes by a bell-shaped function, the PSF (point-spread function), which is here more heavy-tailed than a Gaussian [268].

As most recognition tasks in data analysis, gamma-ray source detection is often carried out by a *maximum likelihood* (ML) method. That is, a parametrized source model is fitted to data through maximization of the probability that the data arose as a realization of the suggested model. This involves a heavy nonlinear optimization procedure and an initial guess from the user to set the parameter values (height, width and position, say) to their optimal values. But one eventually ends up with a very faithful account of the physical properties of each source. There are statistical reasons to think that it is hard to beat the quality of ML estimation, like the minimum variance property (see [Ead71]). The reference for ML source detection in EGRET data is [268].

Our concern, however, is to develop a simpler method of source detection based on the continuous wavelet transform. Roughly speaking, the idea is to group the events in a chosen interval of energies into a single 2-D counting map, as in Figure 5.5, and to take its CWT with an isotropic wavelet, typically a Mexican hat. The source candidates are the maxima of the wavelet transform. Then, based on some statistical criterion, a detection significance will be given to each maximum. The higher the significance, the more likely the candidate to be a true source.

In order to give the status of source candidate to the maxima of the wavelet transform, we must make sure that relevant information (the sources) is properly decorrelated from noise (the background). For this purpose, the Mexican hat wavelet is a good choice, because: (i) its isotropic bell shape responds mostly to bell-shaped sources; (ii) its good localization in space allows to discriminate events according to their position, more efficiently than the Laplacian of the heavy-tailed PSF; and (iii) its good localization in the frequency plane permits us to discriminate events according to their relative scale.

The statistical performance of a wavelet method is presumably poorer than what can be achieved by ML. However, the latter has two drawbacks, namely, high computational complexity of the implementation and supervision of the optimization process. These issues now simply disappear, since wavelets allow a real-time automatic processing of the same job. To stress again the difference between the two tools, we can think of wavelet methods as providing an initial guess for more thorough ML identification of the source parameters, or as an on-board data processing module to warn the astronomers in case of a sudden gamma-ray burst.

Before giving the details of our procedure, let us mention that, as in the analysis of the distant Universe [78,79], related problems like the identification of extended objects, that is, multiscale structures, have also been attacked with the discrete WT

[Sta98,345]. As usual, this allows a faster implementation, but the choice of filter is severely restricted, in particular, the Mexican hat is not admissible. Thus a balance must be made between quality of the analysis and speed of implementation.

We can also note at this point that the "almost flat" approximation used above, namely that a direction in the sky (a point on the sphere) can locally be represented by two planar coordinates, is not necessary. If a more global data analysis is required, we can always switch to the genuine spherical wavelet transform, mentioned in Section 5.1.2, and discussed at length in Section 9.2. This makes no conceptual difference in what follows, only the algorithms will be more CPU-time consuming.

5.1.4.2 Decision criteria and results

It remains to describe the procedure itself. In what follows, we will concentrate on the wavelet aspects and skip most of the statistical arguments, which may be found in detail in [37]. The problem may be subdivided in a series of questions, each of which requires to choose a decision criterion, following more specific questions.

(i) *What is the detection criterion?*

Or in other words, how big should the values of the wavelet transform be to conclude that a peak is indeed a source? Our criterion is based on a physical model of the background interstellar gamma-ray emission, related to the distribution of hydrogen in the galaxy. The idea is to measure the discrepancy between this model and the data in the wavelet domain. Peaks will be considered sources if they significantly overshoot the model.

(ii) *How to estimate the total photon flux from a source?*

Intuitively, the bigger the value of the wavelet transform at the position of a source, the larger the flux of the source; and, since the wavelet transform is linear, this relation should be linear too. That is correct, modulo complications due to the presence of an unknown background of magnitude comparable to that of the source. We can however use our coarse *a priori* background model to remove part of this bias.

Our estimator of the flux of a source detected to be at position x is not the value $y_{obs}(x)$ of the wavelet transform of the observed data, at scale a and at position x, but rather

$$\Phi = \frac{y_{obs}(x) - W[\mu_B](x)}{W[\mu_S](x)},$$

where $W[\mu_B](x)$ and $W[\mu_S](x)$ are the wavelet transforms at x (and at a given scale, as always) of the modeled background flux and of the modeled source flux, respectively. This estimator can be shown to be asymptotically unbiased (i.e., when the exposure time or all the fluxes tend to infinity), provided the models are correct. Confidence intervals on this statistic can also be derived.

(iii) *How to estimate the position of a source?*

Intuitively, the source candidates should be located at the maxima of the wavelet transform. As above, this is true only for flat backgrounds. We account for its nonuniformity in a correction to the quantity to maximize in order to get the position of the source candidate. The position estimator is thus not $\mathrm{argmax}_x\, y_{\mathrm{obs}}(x)$ but rather

$$x^* = \underset{x}{\mathrm{argmax}}\left[y_{\mathrm{obs}}(x) - W[\mu_B](x) \right].$$

Again, assuming the models are correct, this subtraction restores asymptotic unbiasedness. Confidence regions corrected in this way may be seen on the example presented in Figure 5.7 below.

(iv) *How to choose the scale of the wavelet for best performance?*

Since the physical source is point-like, recorded sources look like the impulse response of the detector, i.e., the PSF. Hence, the best choice for the scale parameter is that leading to a wavelet with a width comparable to that of the PSF. This width does not vary much from source to source, it only depends on the energy of the incoming photons. A source emitting proportionally more at high energies than low energies is said to be "hard," or to have a low *spectral index* and has a rather peaked PSF: it is best detected at small scale. On the contrary, a "soft" source has a rather flat PSF and is best detected at a larger scale. This dependence of the optimal scale parameter on the spectral index is illustrated on Figure 5.6.

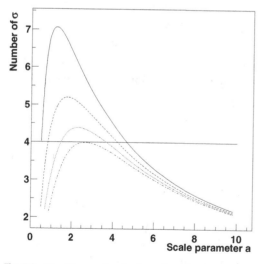

Fig. 5.6. Significance for the detection of a source, expressed in number of sigmas, as a function of the scale parameter of the wavelet. The wavelet is centered at the position of the source. Each curve refers to a different spectral index, from lower (peaked curve) to higher (flat curve). The position of the maxima of these curves changes as the spectral index changes.

Most of the 270 EGRET sources are not identified. The intrinsic resolution of high energy gamma-ray detectors is not good enough to provide strong constraints on the source position; it is therefore difficult to find counterparts at other wavelengths. Their nature is still a mystery, that the next generation telescope GLAST will help to solve. A large fraction of identified sources consists in active galaxies whose nucleus is a massive black hole (up to $10^9 M_{Sun}$) surrounded by an accretion disk of matter falling in the gravitational well. In addition, strong jets of ultra-relativistic matter and radiation are emitted perpendicularly to the disk. Active galactic nuclei (AGN) detected in gamma-rays above 100 MeV have a jet pointed towards the Earth. Their emission is very variable, so that they are often undetected when they are in a quiescent state and then, in a short time, they become very bright.

To give an example, EGRET viewing period 21.0 is a high latitude observation in which an AGN is in flaring state. 3EG J0237+1635 was detected at $10\,\sigma$ in [213] and is $16\,\sigma$ here. Several other sources are also present above $4\,\sigma$. The procedure described above has been applied and results are shown in Figure 5.7. One can see that all but one of the sources are seen. One should also note that the bright AGN position is slightly wrong, because of the presence of a faint source in the vicinity that is not detected. To detect this kind of source, the algorithm must be applied a second time after the addition of the significant sources to the background.

This attempt at developing an alternative method to the usual maximum likelihood estimation will probably prove to be relevant in years to come. Indeed, the next generation gamma-ray telescope, GLAST, is to be launched in March 2006 and its complexity

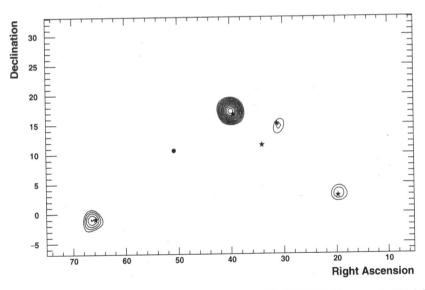

Fig. 5.7. Detected sources using the wavelet analysis of the EGRET viewing period 21.0 for E>100 MeV, the contours give the significance level from 4 to 16 σ. Superimposed stars give the positions of the third EGRET catalog sources detected over 4 σ (from [213]).

(parameters to take into account, volume of the data stream) will make it impossible to build a source catalog following EGRET's old-school procedure. The algorithms must be made more efficient in some way. Wavelets will not be the key to the whole problem, of course, but will hopefully help develop alternative viewpoints.

5.2   Geophysics

5.2.1   Geology: fault detection

As we have seen in Section 1.7, wavelet analysis was born in geophysics, as the empirical method designed by J. Morlet for analyzing the recordings of microseisms used in oil prospection. Thus it was to be expected that wavelets would soon find applications in other geophysical problems. It was indeed the case, as can be seen from the reviews [Fou94] or [250], where mostly 1-D applications are discussed, however.

Then an interesting application of 2-D directional wavelets to geology was initiated in 1995 by Ouillon [Oui95,298]. The object to be analyzed is a system of geological faults covering a large area in the Arabian peninsula, which shows a self-similar behavior over scales from a few meters to hundreds of kilometers. Standard methods for analyzing such a system are based on renormalization group techniques or on the multifractal formalism (see Section 5.4). What the authors propose here is a continuous wavelet analysis, with directional wavelets, combined with a multifractal analysis. The motivation for this choice is that the relevant information to be measured is the anisotropy of the fault field, and the variation of this anisotropy with scale.

In order to understand the idea, let us consider a synthetic so-called *en échelons* fracture [298], depicted in Figure 5.8(a). At a small scale, the dominant orientation of this object is vertical, but at large scale, one sees only an oblique line pointing NE at 45°. Analyzing it with an isotropic wavelet would reveal these details, without focusing on directions, whereas an anisotropic one will enhance the direction response. Thus the authors of [298] chose the latter, namely an anisotropic Mexican hat (see Section 3.3.1).

The originality of the method is an optimized local filtering of the WT, as follows. The wavelet used is an anisotropic Mexican hat with anisotropy factor ϵ. One computes the WT of the image for a number of couples (θ, ϵ). Then, for each point in the signal, one selects the pair (θ, ϵ) that gives the largest value of the CWT among all those computed, the *Optimum Anisotropic Wavelet Coefficient (OAWC)*. Thus, for each point, the OAWC selects the local filter which best matches the signal at the chosen resolution (scale). Then one thresholds the OAWC map in order to keep only the most significant features. The ridges of the remaining map correspond to the dominant structures detected. In the case of a fault array, these ridges (called virtual rupture lines or VLR) correspond to the faults as seen on a map. Finally, one draws a histogram of

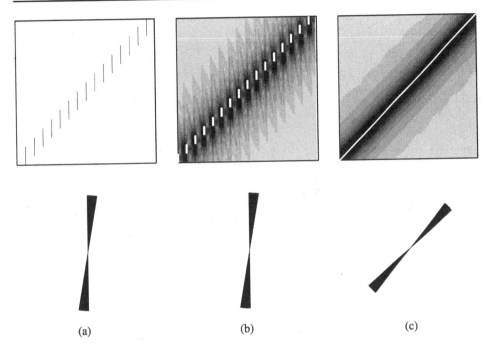

Fig. 5.8. NOAWC analysis of the synthetic *en échelons* fracture. Top row: (a) the signal; (b) the NOAWC map at the resolution of 2 pixels; (c) the same at the resolution of 4 pixels. The bottom row shows the corresponding orientation roses (from [Gai00]).

the azimuths θ of the optimal wavelets associated with the points of the VLRs. This histogram, called a *rose* by geologists, depicts clearly the anisotropy of the object and its variation with scale. In further papers [120,181], this OAWC method was further improved by adding an adaptive normalization, in the sense that each OAWC is divided by its theoretical maximum corresponding to a perfect match between the wavelet and the object. The so-called NOAWC so obtained is a local indicator of the quality of the match.

To give an example, we present in Figure 5.8 the NOAWC analysis of the *en échelons* fracture signal (from [Gai00]). Panels (b) and (c) show the NOAWC map at the resolution of 2 and 4 pixels, respectively, with the VLRs enhanced in white, and on the bottom row, the corresponding orientation roses. Whereas the 2 pixels rose points at 90° (vertical orientation), that at 4 pixels resolution points at 60°. The interesting information is then the critical scale corresponding to a brutal shift in the orientation of the rose (see [Gai00] and [298,299] for more details).

The NOAWC method has been applied successfully to the analysis of geological fault arrays and to the so-called rock fabric analysis, where "fabric" means the "complete spatial and geometrical configuration of all those components that make up a deformed rock" [Gai00,182]. As an example, we present in Figure 5.9 the NOAWC multiscale analysis of a 150-km-wide fault field, taken from [Gai00]. As in the previous

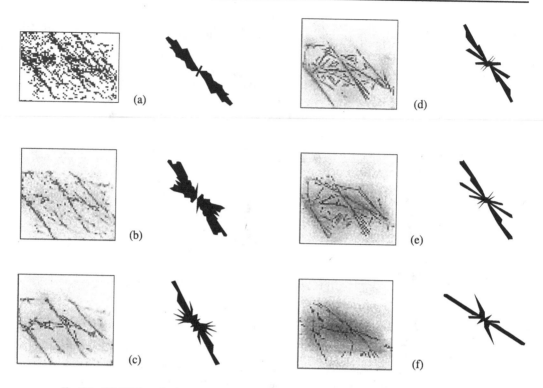

Fig. 5.9. NOAWC analysis of a real map of a fault field. (a) The original map; (b)–(f) NOAWC maps at scale $a = 2, 4, 8, 16, 32$. On the right, we show the corresponding orientation roses (from [Gai00]).

case, we show in the successive panels the NOAWC maps at smaller and smaller resolutions, together with the corresponding orientation roses. The dominant direction of the latter clearly varies with scale. The critical scales where transitions take place are then determined by a multifractal analysis. We refer to [299] and [Gai00] for further details.

As already mentioned, the authors of all these papers use for the NOAWC method an anisotropic Mexican hat, which has a rather poor directional selectivity. However, the elliptical shape of the "footprint" of the wavelet plays an essential role in the method. This suggests the use of a Morlet wavelet instead of a Mexican hat. We should expect a much better precision, but the experiment has yet to be done.

5.2.2   Seismology

As it is well-known [Bur98], the wavelet saga started with Jean Morlet, a French geophysicist working in oil prospection for Elf-Aquitaine. The technique consists of the sending of an impulse into the ground (by an explosion or any other means) and analyzing the signal reflected by the various discontinuities in the underground, down

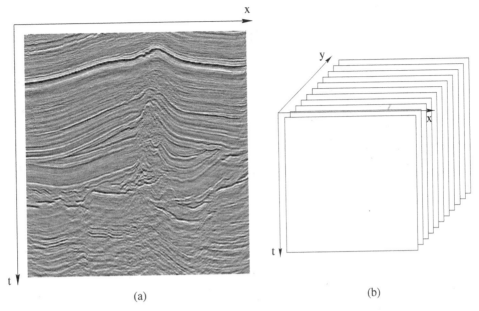

Fig. 5.10. (a) A seismic section; (b) a seismic block (from [Bou97]).

to 8000 m. These correspond to abrupt changes in density and composition of the rock, a necessary (but by far not sufficient!) condition for the presence of oil or gas. Clearly the resulting signal will be extremely noisy. Then, on a purely empirical basis, Morlet had the idea of representing this signal by a linear superposition of contributions obtained by dilating/contracting a fixed mother function (the analyzing wavelet). The method worked reasonably well, but it took a year of work between J. Morlet and the theoretical physicist A. Grossmann to understand its exact mathematical structure – namely, the content of Chapter 1 [199,205].

From there the theory of wavelets expanded in all directions, as we have seen in the preceding pages, until the loop was completed in the Ph.D. thesis of E. Bournay Bouchereau [Bou97], where she looked again at the very problem of seismic exploration treated by Morlet. The raw data are the so-called *seismic sections*, that is, 2-D plots where the vertical t axis represents twice the time needed by the wave to reach the corresponding rock layer (thus depth) and the x axis the horizontal distance between two successive receptors. A typical seismic section is shown in Figure 5.10(a). For a comprehensive study, one groups together a collection of parallel sections, thus getting a 3-D *seismic block*, as shown in Figure 5.10(b). Now, given a section, or a portion thereof, the goal is to detect the geological faults it contains. The technique developed in [Bou97] consists in taking the 2-D continuous WT of the section with a directional wavelet, either a Morlet wavelet or a separable directional one, on the model of (3.24). (The CWT is used here, instead of a discrete WT, in order to maintain translation covariance, which is essential for pattern identification.) This CWT easily detects faults,

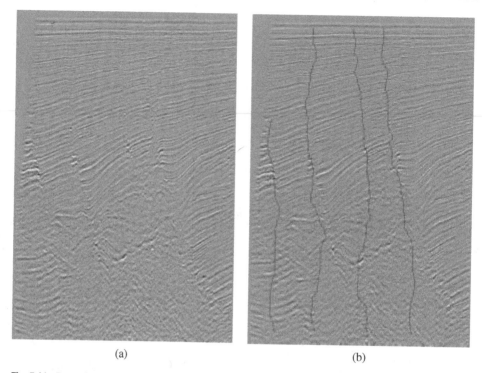

(a)   (b)

Fig. 5.11. Detection of faults in a seismic section: (a) a section; (b) faults detected with a Morlet wavelet analysis (from [Bou97]).

which can be considered as lines of singularities. In the presence of a high noise (deep layers), a preliminary adaptive filtering improves the efficiency of the detection. An example of fault detection is given in Figure 5.11.

In addition to the work just described, we ought to mention other applications of directional wavelets in seismology. For instance, the NOAWC method of the previous section has been used for the study of natural seismological events (spatial distribution of hypocenters of an earthquake sequence) [Gai00]. Even 3-D wavelets (see Section 9.1), namely a 3-D Mexican hat, have been used for the description of seismicity of a large area in the western Alps [75]; and, needless to say, 1-D wavelet analysis has also been applied to seismic time series, in particular, the arrival time of the various components, S-wave, P-wave, etc. (see, for instance, [104,297,Oon00]).

Finally, it is amusing to note that essentially the same technique based on 2-D directional wavelets has been used for certain problems in metallurgy [236], owing to the similarity between a metallurgical image (inner structure of a piece of metal) and a seismic image.

5.2.3   Climatology

Before concluding this section, we have to say a few words on the use of wavelet analysis in climatology. Most of the applications are in 1-D, typically time series analysis, using often the discrete WT. We refer to [Fou94] or [250] for a review. An exception is the work of Hudgins *et al.* [227] on atmospheric turbulence, which uses the CWT in an essential way (this is the paper in which they introduced the wavelet cross-correlation function or wavelet cross-spectrum). Another one in the same spirit is that of van Milligen [273] on turbulence in fusion plamas. In both cases, the wavelet method turns out to be superior to the standard Fourier techniques.

As for 2-D examples, the prime domain is again that of turbulence in fluids, that we will discuss in Section 5.3.1. In addition to the latter, an interesting application was made by Kumar [249], namely, to determine the so-called scale space anisotropy of geophysical fields. By this, one means that such fields are not only highly anisotropic over a wide range of scales, for dynamical reasons, but, in addition, different scale features are oriented in different directions (exactly as for the geological fault arrays described above). A typical example is provided by hurricanes, where the scale anisotropy is obvious. In many other cases, however, the anisotropy is present in a subtle way, that cannot be properly detected by classical techniques, such as spectrum- or correlation-based techniques. As an alternative, Kumar uses a 2-D Morlet wavelet analysis to characterize scale space anisotropy in radar-depicted spatial rainfall, by studying the fraction of energy in different directions at different scales. For that purpose, he introduces the relative scale-angle spectrum (2.58), which is sufficiently sensitive to reveal the subtle presence of scale space anisotropy in random fields. In the particular example treated here, the author is able to conclude that " . . . a rainfall field might show an anisotropic structure that might not be obvious from a typical spectral analysis and may have wider implications in modeling and sampling problems."

Another application, closely related to the previous one, is the use of 2-D wavelets for enhancing thin-line features in meteorological radar reflectivity images [212]. Thin-line features in reflectivity correspond to surface wind convergence lines that can potentially lead to the initiation of thunderstorms. Thus the detection, preferably automatic, of such features is an important ingredient in the short time forecasting of thunderstorms. It turns out that a directional wavelet is required, namely a 2-D Morlet wavelet or a separable substitute built on the model (3.24). Once again, we see the superior discrimination power of directional wavelets in physical applications!

In addition to the directional aspects analyzed in the previous applications, it is a fact that many meteorological phenomena have a distinctly fractal behavior. Clouds are a good example, but several artificial examples, such as random surfaces, share the same property. Thus, it is not surprising to find several applications of wavelets to such fractal structures. We will discuss some of them in Section 5.4.1.

5.3   Applications in fluid dynamics

The wavelet transform, both continuous and discrete, has been successfully applied to the analysis of 2-D developed turbulence in fluids, especially for localization of coherent structures in the distribution of energy or enstrophy. This topic is briefly described in Section 5.3.1. In addition, we will describe here two other applications of 2-D wavelets in fluid dynamics, which rely on the possibility of local filtering, both in direction and in position, with directional wavelets.

5.3.1   Detecting coherent structures in turbulent fluids

Turbulence in fluids is a phenomenon that has resisted analysis until now. After more than a century of research, no real theoretical understanding of the dynamics of a turbulent flow has been achieved. There only exist various statistical or phenomenological models, which are widely used in practical applications, but lack a genuine justification. Even the very definition of the terms used does not always achieve a consensus among physicists. On the other hand, there is a huge amount of experimental data. In order to understand the rôle of wavelets in this context, we have to go back to the basics. For a general review, we refer to [165].

The starting point is the system of Navier–Stokes (NS) equations governing the evolution of an incompressible Newtonian fluid:

$$\frac{\partial \vec{v}}{\partial t} + (\vec{v} \cdot \vec{\nabla})\vec{v} + \frac{1}{\rho}\vec{\nabla}p = \nu \Delta \vec{v} + \vec{F} \tag{5.8}$$

$$\vec{\nabla} \cdot \vec{v} = 0, \tag{5.9}$$

supplemented by adequate initial and boundary conditions. In these equations, $\vec{v} \equiv \vec{v}(\vec{x}, t)$ is the velocity, $p \equiv p(\vec{x}, t)$ the pressure, $\vec{F}$ the external force per unit mass, ρ a constant density, and ν the constant viscosity. The NS equations (5.8)–(5.9) are often expressed in terms of *vorticity*, namely $\vec{\omega} = \vec{\nabla} \times \vec{v}$, which measures the local rotation rate of the fluid. In dimension 2, the NS equations are formally the same, but the velocity field reads $\vec{v}(\vec{x}) = (u(x, y), v(x, y))$ and the vorticity reduces to the pseudoscalar $\omega = \partial_x v - \partial_y u$. Then the fundamental quantities are the total *energy* and the total *enstrophy*, defined as, respectively

$$E(t) = \frac{1}{2}\int_{\Omega} d^2\vec{x}\, |\vec{v}(\vec{x}, t)|^2, \qquad Z(t) = \frac{1}{2}\int_{\Omega} d^2\vec{x}\, |\omega(\vec{x}, t)|^2,$$

and their Fourier transforms, the energy and enstrophy spectra (Ω is the volume occupied by the fluid).

Fully developed turbulence is the regime of very large Reynolds numbers Re $\sim 1/\nu$, $\nu \to 0$ (in practice, in aeronautics, meteorology or combustion, for instance, Re varies between 10^6 and 10^{12}). In this regime, the nonlinear advection term $(\vec{v} \cdot \vec{\nabla})\vec{v}$ becomes dominant, by several orders of magnitude. As with the semiclassical limit $\hbar \to 0$ in quantum mechanics, this changes the character of the equation. As a result, not only is there no analytical solution known, but the NS equations cannot be solved numerically in this regime with present day computers, unless some drastic simplifications are made. Instead, since turbulent flows are highly unpredictable, one has to use statistical models, requiring some basic assumptions, such as statistical homogeneity and isotropy, or ergodicity, which allows one to replace ensemble averages by space averages. All this led, for example, to the celebrated 1941 cascade model of Kolmogorov [165].

Yet turbulent fluids often exhibit *coherent structures*, that is, structures in the energy or the enstrophy spectrum that persist through a large range of scales (vorticity tubes, often called filaments), but are highly unstable. Clearly, the mere existence of these invalidates the statistical assumptions. In addition, the averaging processes, while satisfactory at low Reynolds numbers, ignore the coherent structures, since they have a small extent in space and in time. These are, however, an essential aspect of fully developed turbulence, and thus statistical models are inadequate for $Re \gg 1$. This situation led Marie Farge to introduce, back in 1988 [163], wavelet methods for detecting and analyzing the time evolution of such coherent structures. Since then, she and her collaborators have devoted a huge amount of research work in this direction. Many different techniques have been used, wavelets (CWT and DWT), wavelet packets, multifractal techniques. In retrospect, the basic idea is always to separate the coherent structures, which are analyzed with wavelets, from the background flow, treated by statistical methods. This is, of course, not the place to go into the details of this considerable, but rather specialized work, which represents one of the most spectacular applications of wavelet analysis in physics. We refer the interested reader to the extensive review papers by M. Farge *et al.*, which contain references to the original work [164,165,335]. In addition, a 1-D application to intermittent turbulence in atmospheric data is given in [211].

5.3.2   Directional filtering

We will turn instead to applications that use specifically directional wavelets. As a consequence of their good directional selectivity, the Morlet and Cauchy wavelets are quite efficient for directional filtering. In order to illustrate the point, we analyze in Figure 5.12 a pattern made of rods in many different directions (a). Applying the CWT, with a Cauchy wavelet in a fixed direction (here horizontal), selects all those rods with roughly the same direction (b), whereas the other ones, which are misaligned, yield only a faint signal corresponding to their tips, in agreement with the behavior discussed above. Since this is in fact noise, one performs a thresholding to remove it, thus getting a clear picture (c). In this way, one can count the number of objects that lie in any particular direction. Note that the same pattern was analyzed with a Morlet wavelet in [18,19], and the result is slightly less neat.

Figure 5.13 presents another example of directional filtering, this one with the Gaussian conical wavelet (3.37). The picture represents bacteria, seemingly at random. However, after filtering successively at $-10°$, $45°$, and $135°$, one realizes this latter orientation is significantly more populated.

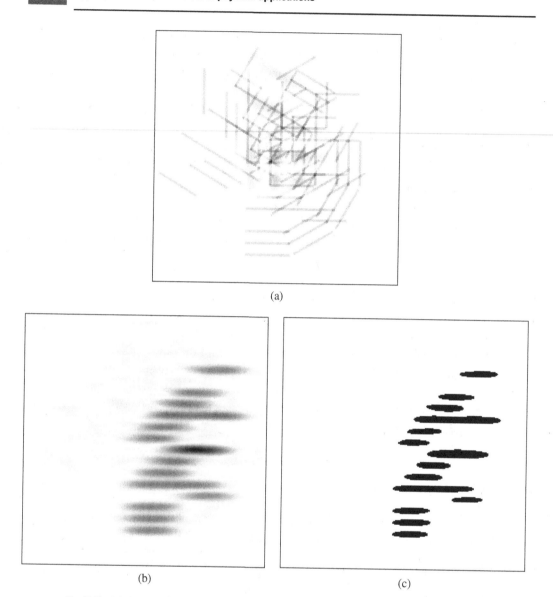

Fig. 5.12. Directional filtering with a Cauchy wavelet ($ARP = 20°$) oriented at $\theta = 0°$: (a) the pattern; (b) the CWT; (c) the same after thresholding at 25%.

5.3.3 Measuring a velocity field

In the first example [Wis93,375], the aim is to measure the velocity field of a 2-D turbulent flow around an obstacle. Velocity vectors are materialized by small segments, by the technique of discontinuous tracers. Tiny plastic balls are seeded into the flow and illuminated by a "plane of light," in order to get a 2-D image. Then two successive photos are taken with a fast CCD camera, with exposure times of 700 and 6000 μs, respectively. In this way one gets a "dot-bar" signature for each tracer, which materializes

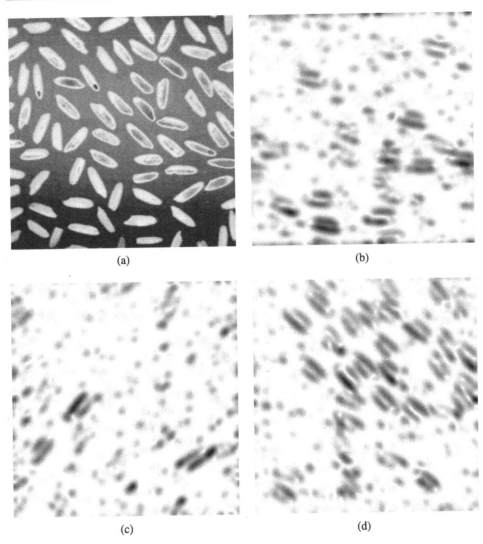

Fig. 5.13. Another example of directional filtering, with a Gaussian conical wavelet: (a) the original image, representing bacteria; (b) filtering at $-10°$; (c) the same at $45°$; (d) the same at $135°$.

the direction and the length of the local velocity (see an example in Figure 5.14, taken from [Wis93]). In order to get sufficiently many data points, one superposes several such pictures, typically 16. First one computes the WT of the resulting image with a Morlet wavelet, which selects those vectors that are closely aligned with the wavelet. Then a second analysis is performed with a wavelet oriented in the orthogonal direction, thus completely misoriented with respect to the selected vectors. Now the WT sees only the tips of the vectors and their length may be easily measured. The same two operations are then repeated with various successive orientations of the wavelet. Using appropriate thresholdings, the complete velocity field may thus be obtained, in a

Fig. 5.14. The dot-bar signature of tracers in the fluid flowing from left to right (from [Wis93]).

totally automated fashion, with an efficiency sensibly better than with more traditional methods. Two examples of reconstructed velocity fields from [Wis93] are given in Figure 5.15, corresponding to a quasi-laminar flow and a turbulent flow around an obstacle (again the flow comes from the left; units are normalized to the size of the experimental area). Notice that the analysis gives in principle both the modulus and the phase of the WT. But here, contrary to the simple applications like contour detection [13], the phase cannot be exploited, the data are too noisy. Thus one loses some precision on the orientation. Nevertheless, the method is remarkably efficient.

In the same vein, the 2-D CWT (again Cartesian only) has been proposed for improving the method of holographic particle velocimetry, which consists in measuring the velocity of particles in a fluid by exploiting holograms of fluid volumes [10].

5.3.4   Disentangling of a wave train

A second example originates from underwater acoustics. When a point source emits a sound wave above the surface of water, the wave hitting the surface splits into several components of very different characteristics (called respectively "direct," "lateral," and "transient"). The resulting wave train is represented by a linear superposition of damped plane waves, and the goal is to measure the parameters of all components. This phenomenon has been analyzed successfully with the WT both in 1-D [334] and in 2-D [18], and the extension to a 3-D version is straightforward.

Let us give some details of the method in the 2-D case. The signal representing the underwater wave train is taken as a linear superposition of damped plane waves:

$$f(\vec{x}) = \sum_{n=1}^{N} c_n \, e^{i\vec{k}_n \cdot \vec{x}} \, e^{-\vec{l}_n \cdot \vec{x}}, \qquad (5.10)$$

where, for each component, $\vec{k}_n$ is the wave vector, $\vec{l}_n$ is the damping vector, and c_n a complex amplitude. Then, using successively the scale-angle and the position

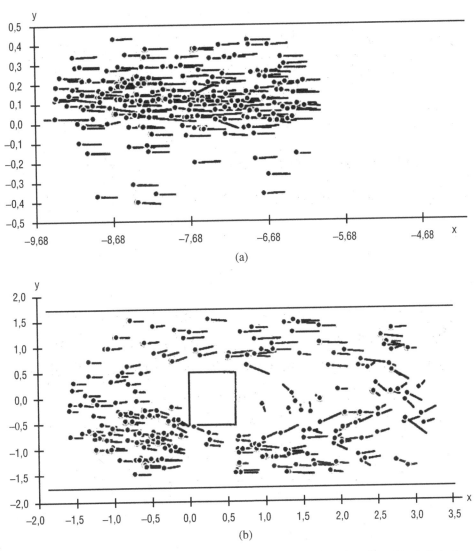

Fig. 5.15. Two examples of reconstructed velocity fields: (a) a quasi-laminar flow; (b) a turbulent flow around an obstacle (from [Wis93]).

representations described in Section 2.2.3, one is able to measure all the $6N$ parameters of this signal with remarkable ease and precision.

The method proceeds in three steps and uses explicitly the phase space interpretation. First one computes the CWT of the signal (5.10) with a Morlet wavelet. By linearity, the result is the linear superposition of the contributions of the various components. Moreover, each component is the product of two factors, where the first one depends on $\vec{b}$ only and the second one on (a, θ) only:

$$F(\vec{b}, a, \theta) = \sum_{n=1}^{N} c_{\vec{b}, n} \, \check{F}_n(a, \theta). \tag{5.11}$$

Actually, the resulting function may be written explicitly in terms of the phase space variables introduced in Section 2.3.2, for instance, $\check{F}_n(a, \theta) \equiv \check{F}_n(\vec{v})$.

Now we go to the scale-angle representation and consider the WT (5.11) for fixed $\vec{b}$. Then a straightforward calculation shows that, for each term in this superposition, $a^{-1}\check{F}_n(a, \theta)$ admits a unique local maximum. Now, in the full transform (5.11), each term has its own local maximum, but these need not be well separated: one maximum may hide another one, totally or partially. This masking effect will happen, for instance, when:

- one component has a much bigger amplitude, $|c_{\vec{b},n}| \gg |c_{\vec{b},m}|$, for all m not equal to n (total masking);
- two wave vectors are close to each other, $\vec{k}_n \simeq \vec{k}_m$, but with different amplitudes, $|c_{\vec{b},n}| > |c_{\vec{b},m}|$ (partial masking).

In both cases, the two waves can be separated, by increasing the selectivity of the wavelet (for instance, using a Morlet wavelet with a more anisotropic modulus). If the two waves have close wave vectors ($\vec{k}_n \simeq \vec{k}_m$) with similar amplitudes ($|c_{\vec{b},n}| \simeq |c_{\vec{b},m}|$), but different damping vectors ($\vec{k}_n \neq \vec{k}_m$), then they can still be separated, by changing the observation point $\vec{b}$. Otherwise the method will fail, the two waves interfere inextricably, none of them dominates the other one.

When the masking effect is not too important, the maxima will be sufficiently prominent that the interferences between the different components will become negligible (in the modulus) and one may write:

$$|F(\vec{b}, a, \theta)| \simeq \sum_{n=1}^{N} |c_{\vec{b},n}| \, |\check{F}_n(a, \theta)|, \tag{5.12}$$

One then reverts to the position representation, choosing for (a, θ) each maximum successively. In each case, the filtering effect of the CWT essentially eliminates all components except one, which is then easy to treat. In this way, one is able to measure all the $6N$ parameters of the signal easily.

A striking example is given in [18], illustrating the power of the method as well as the rôle of the anisotropy factor ϵ of the Morlet wavelet. The signal is the superposition of four damped plane waves, with different wave vectors $\vec{k}_n$, except that the directions of $\vec{k}_1$ and $\vec{k}_4$ differ by 20° only. As a result, wave # 1 partially masks wave # 4: when the analysis is performed with a Morlet wavelet with $\epsilon = 1$, the corresponding maxima in the scale-angle representation are not well separated [Figure 5.16(a)]. When one uses instead a wavelet with $\epsilon = 5$, the two maxima are clearly identifiable and and can be localized precisely (b). Notice that the "footprint" of the wavelet is not an ellipse, because the radial coordinate used is a^{-1}, not a. Now the procedure allows to reconstruct each of the four components almost perfectly [18]. Only wave # 4 keeps some trace of interference with wave # 1, the others are indeed pure waves. In order to remove this effect, one should first subtract wave # 1 from the signal and redo the analysis.

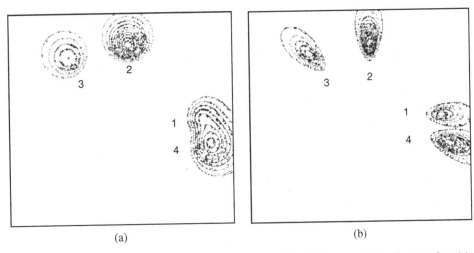

Fig. 5.16. Disentangling of a four component wave train with a Morlet wavelet: the four maxima (a) with $\epsilon = 1$; (b) with $\epsilon = 5$ (from [18]).

5.4   Fractals and the thermodynamical formalism

5.4.1   Analysis of 2-D fractals: the WTMM method

Many physical phenomena require a wide range of scales for a complete description of their properties. The paradigm, of course, are fractals, which are complex mathematical objects that have no natural length scale. More precisely, a fractal, be it in 1-D or in 2-D, is by definition self-similar under dilation, either globally (genuine fractal) or locally (multifractal). Physical examples abound. For instance, all kinds of random walks used to mimic various noisy dynamical behaviors, financial time series, geologic shapes (such as the fault systems descibed in Section 5.2.1), interfaces that develop in growth processes far from equilibrium, fractal growth processes (such as the so-called diffusion-limited aggregates), electrodeposition clusters, etc. A fractal is in general a very irregular object (for instance, its support may be a Cantor-like set), hence it should be represented by a *singular measure*, rather than a function. In order to cope with such situations, the so-called *multifractal formalism* has been developed. Now, central concepts of the theory, such as generalized fractal dimensions or spectrum of singularities of the measure, are closely related to ideas from statistical mechanics (as a matter of fact, the standard "box counting" method is already of a statistical nature). Thus one speaks also of a *thermodynamical formalism* of fractal analysis. A review of this formalism may be found in [47] or [293].

   Since scaling is the most significant operation in the context of fractals, the CWT is a natural tool for analyzing them. Clearly the *continuous* version of the WT is essential here, since the characteristic scaling ratio is unknown *a priori*. The first step is to extend the CWT to singular measures. This was achieved in 1-D by Holschneider [221] and Arnéodo *et al.* [45], then extended to 2-D by Arnéodo and his group in Bordeaux, including one of us (R.M.) [43]. We briefly describe the key steps of the formalism.

We take the most general case, namely, an object described by a fractal measure μ on $\mathbb{R}^2$. The standard thermodynamical formalism (spectrum of generalized fractal dimensions) yields only statistical information about the object as a whole. To get precise local information requires a wavelet transform. The CWT of the measure μ with respect to the wavelet ψ is defined as

$$T[\mu](\vec{b}, a, \theta) = \int d\mu(\vec{x}) \,\overline{\psi(a^{-1}r_{-\theta}(\vec{x} - \vec{b}))}. \tag{5.13}$$

In the isotropic case, we write simply

$$T[\mu](\vec{b}, a) = \int d\mu(\vec{x}) \,\overline{\psi(a^{-1}(\vec{x} - \vec{b}))}. \tag{5.14}$$

Assume now that the measure has the following scaling behavior (self-similarity) around the point $\vec{x}_o$:

$$\mu(\mathcal{B}(\vec{x}, \lambda\epsilon)) \sim \lambda^{\alpha(\vec{x}_o)} \mu(\mathcal{B}(\vec{x}, \epsilon)), \quad \lambda > 0, \tag{5.15}$$

where $\mathcal{B}(\vec{x}, \epsilon)$ is a ball of radius ϵ around $\vec{x}_o$ and $\alpha(\vec{x}_o)$ is the local scaling exponent. Using the covariance property of the CWT (Proposition 2.2.3), it is easily shown that the WT of the measure μ scales in the same way:

$$T[\mu](\vec{x}_o + \lambda\vec{b}, \lambda a, \theta) \sim \lambda^{\alpha(\vec{x}_o)} T[\mu](\vec{x}_o + \vec{b}, a, \theta), \quad \lambda \to 0^+. \tag{5.16}$$

This relation is the key to the wavelet analysis of fractals. For instance, the local exponent $\alpha(\vec{x}_o)$ may be obtained by plotting $\log |T[\mu](a, \theta, \vec{b})|$ versus $\log a$, for a small enough. This would suffice for an exact (global) fractal, such as a numerical snowflake, for which α is constant over the whole object. For a genuine multifractal, $\alpha(\vec{x}_o)$ varies from point to point, and then (5.16) allows one to compute the generalized fractal dimensions and the singularity spectrum of the object. This wavelet approach to the thermodynamical formalism of fractal analysis has been developed systematically by Arnéodo and his collaborators in Bordeaux. It is instructive to give some hints about the basic ideas of the formalism.

We first go back to the standard multifractal formalism. Given a singular measure μ, it can be described in terms of its q-moments as follows. Suppose we cover the support of μ with $N(\epsilon)$ boxes $\mathcal{B}_i(\epsilon)$ of size ϵ. Then the scaling behavior of μ will be deduced from the *partition function*

$$\mathcal{Z}(q, \epsilon) = \sum_{i=1}^{N(\epsilon)} \mu_i^q(\epsilon), \quad \text{where } \mu_i \equiv \mu(\mathcal{B}_i(\epsilon)), \; q \in \mathbb{R}. \tag{5.17}$$

In the limit $\epsilon \to 0^+$, $\mathcal{Z}(q, \epsilon)$ behaves as a power law:

$$\mathcal{Z}(q, \epsilon) \sim \epsilon^{\tau(q)}, \quad \epsilon \to 0^+.$$

Finally, the spectrum of generalized fractal dimensions is obtained from the exponents $\tau(q)$ by the relation

$$D_q = \tau(q)/(q-1),$$

and these in turn completely characterize the singular behavior of the original measure μ, in particular, the local Hölder exponents that describe it (see [47,292]).

Now the rationale for reinterpreting this formalism in terms of wavelets is that the WT tends to "forget" the regular part of the signal (because of the vanishing moments) and focus on the singular part. A naive way of achieving this would be to take as partition function, instead of (5.17),

$$\mathcal{Z}(q, a) = \int d\vec{x} \, |\vec{T}[\mu](\vec{x}, a)|^q, \quad q \in \mathbb{R} \tag{5.18}$$

(in 1-D, one uses simply $|T(x, a)|$ as WT modulus, see (2.66)). This is a bad choice, however, since $\mathcal{Z}(q, a)$ might diverge for $q < 0$. Instead one replaces the integral over $\vec{x}$ by a discrete sum over the local maxima of $T[\mu](\vec{x}, a)$, for fixed a, that is, precisely the WTMM introduced in Section 2.3.5. The justification of this choice is that [262] (i) the maxima lines (ridges) have the same scaling behavior as the WT itself, and (ii) each maxima line $l = (\vec{b}_l(a), a)$ points, as $a \to 0$, to a point $\vec{b}_l(0)$ which corresponds to a singularity of μ and, in addition, the WT modulus scales along the line as

$$|\vec{T}[\mu](\vec{b}_l(a), a)| \sim a^{\alpha(\vec{b}_l(0))}.$$

Thus the wavelet plays the role of a generalized "oscillating box" and the scale a defines its size. Actually, the definition of the partition function can be further refined by using explicitly the WTMMM, as follows. Let $\mathcal{L}(a)$ denote the sets of ridges that exist at scale a and contain a maximum at a scale $a' \leqslant a$. Then one defines finally the partition function

$$\mathcal{Z}(q, a) = \sum_{l \in \mathcal{L}(a)} \left(\sup_{(\vec{x}, a') \in l, a' \leqslant a} |\vec{T}[\mu](\vec{x}, a')| \right)^q. \tag{5.19}$$

In fact, introducing the "sup" amounts to adapting the size of the wavelet along the ridge so as to avoid divergences. Here again, the exponents $\tau(q)$ are defined from the power-law behavior of $\mathcal{Z}(q, a)$ as $a \to 0$:

$$\mathcal{Z}(q, a) \sim a^{\tau(q)}, \quad a \to 0^+.$$

(In the analogy with thermodynamics, q and $\tau(q)$ play the rôle of inverse temperature and free energy, respectively.)

The WTMM technique has been applied successfully to a wide variety of examples [43,44,47], that cover both artificial fractals (numerical snowflakes, diffusion limited aggregates, recursive fractal functions) and natural ones (electrodeposition clusters, various arborescent phenomena, fully developed turbulence data, clouds). The method permits the measurement of the fractal dimensions and the unraveling of universal laws (mean angle between branches, azimuthal Cantor structures, etc.). In 2-D, it should be remarked that the analysis uses exclusively an isotropic wavelet (usually a 2-D Mexican hat), and thus there is no θ dependence in (5.16). However, this may not be the end of the story. Indeed we shall exhibit in Section 4.5.2 below a fractal ("twisted snowflake") whose structure requires a *directional* wavelet for its complete determination.

In more recent work, the attention has focused on 2-D applications, around the theme of rough surfaces: fractional Brownian surfaces, anisotropic self-affine rough surfaces [Dec00,53], synthetic multifractal rough surfaces [129], cloud structure [52,54]. This last item opens a whole world of applications to physical processes, in particular to meteorology, since fractal objects abound there.

Another application, closely related to the previous analysis, concerns the analysis of real rough surfaces, that is, metal surfaces obtained after various kinds of machining processes. Here too, the 2-D wavelet transform yields a useful tool [235]. Actually the object to analyze is essentially the texture of the surface, which brings us to another field of application, namely, *texture analysis*, that we will discuss in Section 5.5.

A last point to notice in this context is that, considering the heavy computational cost of the 2-D CWT, the Bordeaux group has designed an ingenious hardware version, called the Optical WT [Arn95,46]. The technique, based on Fraunhofer diffraction, a familiar tool in optics, amounts to obtaining the WT with a binary approximation to the isotropic Mexican hat, that is, using as isotropic wavelet the Bessel filter described in (3.9)–(3.10). With this tool, they obtained beautiful pictures of CWT analyses of diffusion-limited aggregates [Arn95,43,46]. The technique could not, however, be pushed to full implementation for lack of a sufficiently fast CCD camera – or, equivalently, a sufficiently large budget!

Actually, several optical implementations of the WT have been proposed in the literature. However, they are often limited to 1-D data or, in the 2-D case, to Cartesian coordinates (translations and separate scaling factors in the x and y directions). We refer the interested reader, for instance, to a feature issue of *Applied Optics* dedicated to this topic [255]. An alternative optical approach, based on a special type of grating and able to reproduce many types of wavelets, has been presented in [269].

The isotropic Mexican hat has also found applications in optics proper, for instance, in the determination of the wave aberration coefficients of a rotationally invariant optical system, from the measured data of wave front deformations [340]. The result is that the wavelet method is more efficient and more robust to noise than standard least squares methods. Another application of the 2-D CWT in optics is to moiré interferometry, a well-established optical technique for measuring displacement and strain in materials, based on phase analysis of interference fringe patterns [238]. Thus a complex wavelet is needed, and the authors resort to a modified Morlet wavelet, obtained by replacing the Gaussian by a cubic spline, as advocated by Unser [357], and once again, the CWT method proves superior to the standard Fourier techniques.

5.4.2   Shape recognition and classification of patterns

The characterization of a 2-D shape from its outlines is an important problem in several applications of image analysis, such as character recognition, machine parts inspection for industrial applications, characterization of biological shapes such as chromosomes and neural cells, and so on. Furthermore, in the field of human vision and perception, 2-D shape analysis also plays a central role in psychophysics and neurophysiology.

There are two general approaches to shape characterization: *region based*, which deals with the region in the image corresponding to the analyzed object; and *boundary based*, where the shape is characterized in terms of its silhouette [Pav77]. The former is intrinsically 2-D, dealing directly with planar primitives and concepts, and thus 2-D wavelets may be used directly. The latter, however, mimics 2-D operations through 1-D representations, and is referred to henceforth as *contour characterization*. An alternative to standard techniques consists of representing the shape by the complex curve that describes its boundary, and applying the 1-D CWT to this complex signal [Ces97,21], as outlined in Figure 5.17. This leads to the so-called *W-representation*,

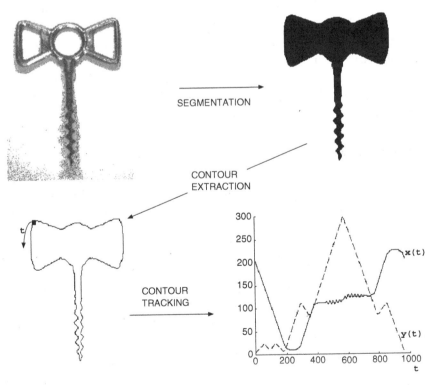

Fig. 5.17. Basic framework for 2-D shape characterization from its outline (from [21]).

which allows an easy way of performing a number of standard tasks (for instance, in machine vision), such as detection of dominant points, shape partitioning, natural scales analysis.

As compared to other techniques, the *W*-representation has the following useful characteristics, which are desirable for purposes of shape analysis [275] and follow directly from the basic properties of the CWT:

- uniqueness, because of the invertibility of the CWT;
- invariance under translation, scaling and rotation;
- robustness to local modifications of the shape;
- efficiency and ease of implementation.

Notice that an essential ingredient of the technique is the wavelet-based fractal analysis discussed above. In particular, one resorts systematically to the different types of local maxima lines defined in Chapter 2, Section 2.3.5. More precisely:

- The algorithm for the detection of *dominant points*, e.g. corners, is based on the vertical maxima lines, namely, their position and some relevance measure, for instance, their length.
- The detection of periodic patterns and the so-called *natural scales* is based on the horizontal maxima lines, since this amounts essentially to determining the instantaneous frequencies in the signal.
- Finally, the fractal behavior of certain contours is analyzed with the standard technique sketched in Section 5.4.1.

This analysis has numerous applications. An original one is the classification of neurons according to their complexity, that is, their fractal dimension [Ces97]. Of course, shape analysis is a whole different world. For an up-to-date review of it, we refer to the recent monograph of L. da F. Costa and R. M. Cesar, Jr. [Cos01].

5.5   Texture analysis

The determination and classification of textures in images is an old and difficult problem, with many potential applications, notably in computer vision. Numerous methods have been designed to that effect. Most of them are of a statistical nature, such as Markov random fields, but one has also used Gabor analysis [234] and various kinds of wavelet transforms. We shall concentrate on the latter, of course.

Some proposals have been made with the standard (Cartesian) discrete WT (see [166] for instance), but, since most textures possess directional features, it is more natural to use oriented wavelets for attacking the problem. An example is the steerable wavelet pyramid developed by Simoncelli and his collaborators (see Section 2.7). This technique was used in [317] as a basis for a parametric texture model based on joint statistics of wavelet coefficients. It has been considerably developed by Do and Vetterli [Do01,140,141], using appropriate generalizations of the ridgelets and curvelets (these will be described in Chapter 9 and Section 11.1), and combining them with wavelet-domain hidden Markov models.

Alternatively, one can use the plain 2-D CWT with an oriented wavelet. Actually a similar solution was proposed long ago by Rao and Schunck [323], using the first derivative of a 2-D Gaussian, (3.15), already considered by Canny [98] (see Section 3.3.1.1). This is indeed an oriented wavelet, but it is not directional in the technical sense. (In addition, one may find in [323] a survey of the early literature on texture determination.) Texture analysis with a genuine directional wavelet (a Morlet wavelet) was done first by Gonnet [195], using the characterization of the instantaneous frequency of the signal as a vector field [196] (Section 2.3.5). Further results were obtained by Murenzi *et al.* [290], using the same wavelet (truncated Morlet, also called Gabor filter) and the scale-angle representation. A more efficient technique yet is that of the directional dyadic wavelet packets [360] (Section 2.6.4). Substantial advances in the classification of textures have been obtained recently along this line by Menegaz *et al.* [270]. As usual in the framework of computer vision, a pattern is represented by a feature vector, as explained in Chapter 4, and the classification is made in defining similar images as those whose feature vectors are close to each other, in the sense of some notion of distance (often Euclidean or L^2). In order to use directional wavelet packets for this problem, one simply includes the directional details at each scale among the components of the feature vectors. The latter become longer, but the efficiency of the method for texture discrimination increases significantly.

A related topic is the so-called *shape from texture* problem, which can be formulated as follows. We are given a 2-D photograph of a 3-D surface, which displays a pattern or a texture, more or less regular. The image gives a distorted view of this texture, which depends of the geometry of the surface. The goal is to reconstruct the original surface from the distorted image.

The problem is usually split into two steps. First one measures the local distortion of the image, then one recovers the original surface. For the first step, some assumptions are usually made about the type of surface and texture. It turns out that the CWT is an efficient tool for the estimation of distortions, as first proposed by Super and Bovik [348] and Hwang *et al.* [229]. The planar surface is assumed to have a homogeneous texture, which is modeled as a linear superposition of plane waves. A further paper [231] considers a plane containing several textures with different orientations, a situation which requires first a segmentation step. In both cases, the authors rely on the properties of the ridges of the 2-D wavelet transform (Section 2.3.5). Further work along the same line was made in [179], exploiting the well-known covariance properties of the CWT under translation, rotation, and scaling. Here too, the authors consider a single homogeneous (sinusoidal) texture, and study its deformations.

An alternative to this deterministic model is to describe the texture by a stationary random process. This approach has been developed by Clerc and Mallat [Cle99,106,107] for the general case of a curved surface. In addition, they consider general distortions, for instance, anisotropic scalings. Since the conventional CWT is no longer covariant under such transformations, they introduce instead the so-called *warplet transform*, replacing the familiar global scaling by a distortion matrix which is not a multiple of the identity (the name refers to the fact that the image presents a warped view of the surface). This technique, while conceptually interesting, leads to high computational costs, in particular, the formalism of [179] cannot cope with it.

5.6   Applications of the DWT

For the sake of completeness, we conclude this chapter with some remarks on the applications of the DWT. As we said already, the latter is used in the majority of applications, but this is not the main subject of the present book. Thus we will give only a few indications.

As with other methods, wavelet bases may be applied to all the standard problems of image processing. The main problem of course is data compression, and for achieving useful rates one has to determine which information is really essential and which one may be discarded with acceptable loss of image quality. Significant results have been obtained in the following directions.

- Representation of images in terms of *wavelet maxima* [264], as a substitute for the familiar zero-crossing schemes [Mar82].
- In particular, application of this maxima representation to the detection of edges, and more generally detection of local singularities [262].
- Image compression and coding using vector quantization combined with the WT [41].
- Image compression, combining the previous wavelet maxima method for contours and biorthogonal wavelet bases for texture description [176].
- Image and signal denoising, by clever thresholding methods [144].

Some applications are less conventional. For instance, a technique based on the biorthogonal wavelet bases [108] has been adopted by the FBI for the identification of fingerprints. The advantages over more conventional tools are the ease of pattern identification and the superior compression rates, which allows one to store and transmit a much bigger amount of information in real time. The full story may be found in [87]. Another striking application is the deconvolution of noisy images from the Hubble Space Telescope, by a technique combining the DWT with a statistical analysis of the data [Bou93,85,328]. The results compare favorably in quality with those obtained by conventional methods, but the new method is much faster. One should also quote a large amount of work under development in the field of High Definition Television, where wavelet techniques are being actively exploited; here again the huge compression rates make them specially interesting.

As for applications of the multidimensional DWT more specifically oriented to physics, we like to mention two. The first one is in quantum field theory (although it was done before the wavelet techniques were born): various perturbation expansions (the so-called "cluster expansion") used in the analysis of Euclidean field theory models are in fact discrete wavelet expansions [65]. Actually the summation over scales, indexed by j, was originally motivated by renormalization group arguments. In the same domain, we may note that wavelet bases have been used also ([66] and references therein) for estimating the time evolution of solutions of some wave equations (Klein-Gordon, Dirac, Maxwell or the wave equation), or even to expand solutions of the equations in terms of dedicated "wavelets" (although the functions introduced in the last case seem rather far away from genuine wavelets [239]).

The other application resorts to solid state physics, namely the Quantum Hall Effect (quantization of the electric conductivity) that occurs when a 2-D electron gas is submitted to a strong transverse magnetic field. Here orthonormal wavelet bases may be used for generating localized orthonormal bases for the lowest Landau level, a necessary step towards the analysis of the Hall effect [14,36,57,58].

6 Matrix geometry of wavelet analysis. I

6.1 Group theory and matrix geometry of wavelets

In Chapters 1 and 2, we have studied systematically the continuous wavelet transform in one and two dimensions, respectively. As already emphasized there, the properties of the transforms in the two cases are remarkably similar. In 2-D we have formalized them in the three propositions 2.2.1, 2.2.2 and 2.2.3, and essentially the same statements may be made in 1-D. A moment's reflection shows that one could write out, without difficulty, an entirely parallel mathematical description in *any* dimension $n \geqslant 1$. Clearly there must be some unifying principle underlying the picture. The question is, of course, what is this principle? As so often in such situations, the answer is to be found in group representation theory, i.e., by looking at the underlying geometry of the space of signals. The various transformations (translation, rotation, zoom, etc.) that a signal may undergo, determine a set of mathematical symmetries, which, interestingly enough, can be expressed in simple matrix terms and, as will be made clear in the following, the signal space itself – as a mathematical object – emerges as a consequence of this geometry.

But we have been using group theory all along! Indeed, to draw on a literary analogy, like Molière's Monsieur Jourdain speaking in prose without knowing so, we have been using group-theoretical language throughout our analysis! It is the aim of the present chapter to demonstrate this fact. By so doing, we intend to achieve three goals. First, group theory will provide us with a unifying picture to view all the mathematical properties of signals that we have derived by hand, so to speak. Second, we hope to convince the reader that the group-theoretical approach is not only aesthetic, it is also *simpler*, in that it allows us to understand the deeper mathematical structures involved in a simple language. Finally, rewriting our results in this language allows us to easily extend the concept of a CWT to more general manifolds (i.e., more complex spaces of signal parameters), which we shall explore in later chapters: the n-dimensional space $\mathbb{R}^n$, space–time $\mathbb{R}^{1+1}$ or $\mathbb{R}^{1+3}$, the two-sphere S^2 or the n-sphere S^n, etc. Indeed, as recalled in the Prologue, this was the decisive factor in extending the CWT from one to two dimensions in R. Murenzi's thesis [Mur90], the key step being to identify the relevant group and its realization in the space of signals. To set the stage, let us analyze this mathematical structure in some detail. We shall do this first in 1-D, in this chapter, and then take up the 2-D problem in the next chapter.

We emphasize that the next three chapters, while interesting and intellectually satisfying, are *not* prerequisites for the last three, Chapters 9 to 11, which will treat various extensions of the wavelet transform studied so far. In addition, we have collected in the Appendix some pertinent definitions and results from group theory, that we hope would help the group theoretically uninitiated reader to understand the material and to appreciate better the breadth of its scope.

6.1.1   The 1-D CWT revisited

We start with the basic 1-D transformation (1.4):

$$\psi(x) \mapsto \psi_{b,a}(x) = |a|^{-1/2}\psi\left(\frac{x-b}{a}\right), \quad b \in \mathbb{R}, \ a \neq 0, \tag{6.1}$$

and rewrite it in the form

$$\psi_{b,a}(x) = |a|^{-1/2}\psi\left((b,a)^{-1}x\right), \tag{6.2}$$

where we have introduced the *affine* transformation of the line, consisting of a dilation (or scaling) by $a \neq 0$ and a (rigid) translation by $b \in \mathbb{R}$:

$$x = (b,a)y = ay + b, \tag{6.3}$$

and its inverse

$$y = (b,a)^{-1}x = \frac{x-b}{a}. \tag{6.4}$$

Writing $\phi = \psi_{b,a}$ and making a second transformation on ϕ we get

$$\begin{aligned}
\phi(x) \mapsto \phi_{b',a'}(x) &= |a'|^{-\frac{1}{2}}\phi((b',a')^{-1}x) \\
&= |aa'|^{-\frac{1}{2}}\psi((b',a')^{-1}(b,a)^{-1}x) \\
&= |aa'|^{-\frac{1}{2}}\psi\left(\frac{x-(b+ab')}{aa'}\right).
\end{aligned} \tag{6.5}$$

This shows us that the effect of two successive transformations is captured in the composition rule

$$(b,a)(b',a') = (b+ab', aa'), \tag{6.6}$$

which, if we represent these transformations by 2×2 matrices of the type

$$(b,a) \equiv \begin{pmatrix} a & b \\ 0 & 1 \end{pmatrix}, \quad a \neq 0, \quad b \in \mathbb{R}, \tag{6.7}$$

is reproduced by ordinary matrix multiplication. The point to be noted about these matrices is that the product of two of them is again a matrix of the same type and so also is the inverse,

$$(b, a)^{-1} = \begin{pmatrix} a & b \\ 0 & 1 \end{pmatrix}^{-1} = \begin{pmatrix} a^{-1} & -a^{-1}b \\ 0 & 1 \end{pmatrix}$$

of such a matrix. Furthermore, the 2×2 identity matrix is also in this class. In other words, the set of matrices (6.7) constitute a *group*, called the (full) *affine group* and denoted G_{aff}. Note also, that if we consider only those matrices in (6.7) for which $a > 0$, then this set is also stable under multiplication and inverse taking and hence, it constitutes a *subgroup* of G_{aff}, denoted G_{aff}^{+}. Coming back to the relation (6.2), we observe that G_{aff} or G_{aff}^{+} consists precisely of the transformations we apply to a signal: translation (time-shift) by an amount b and zooming in or out by the factor a. Hence, the group G_{aff} relates to the geometry of the signals.

Next let us study the effect of the transformation given by the group element (b, a) on the signal itself. Writing,

$$\psi \mapsto U(b, a)\psi \equiv \psi_{b,a},$$

we may interpret $U(b, a)$ as a linear operator on the space $L^2(\mathbb{R}, dx)$ of finite energy signals, with the explicit action,

$$(U(b, a)\psi)(x) = |a|^{-1/2} \psi \left(\frac{x - b}{a} \right). \tag{6.8}$$

Additionally, for each (b, a), the operator $U(b, a)$ is unitary, i.e., it preserves the Hilbert space norm of the signal:

$$\|\psi_{b,a}\|^2 = \|\psi\|^2 = \int_{-\infty}^{\infty} dx \; |\psi(x)|^2.$$

More interestingly, the association, $(b, a) \mapsto U(b, a)$ is a *group homomorphism*, preserving all the group properties. Indeed, the following relations are easily verified:

$$\begin{aligned} &U(b, a)U(b', a') = U(b + ab', aa') \\ &U((b, a)^{-1}) = U(b, a)^{-1} = U(b, a)^{\dagger} \\ &U(e) = I, \quad \text{with} \quad e = (0, 1), \quad \text{the unit element.} \end{aligned} \tag{6.9}$$

We say that the association $(b, a) \mapsto U(b, a)$ provides us with a *unitary representation* of G_{aff}. Note that we may also write,

$$U(b, a) = T_b \, D_a,$$

where T_b, D_a are the well-known shift and dilation operators, familiar from standard time-frequency analysis (see also (2.7)–(2.12)):

$$(T_b s)(x) = s(x - b), \qquad (D_a s)(x) = |a|^{-\frac{1}{2}} s(a^{-1}x).$$

We shall see later that the representation $U(b, a)$ is in a sense minimal or *irreducible*, in that the entire Hilbert space of finite energy signals $L^2(\mathbb{R}, dx)$ is needed to realize it completely. However, let us first attend to another question which is pertinent here, namely, why is it that G_{aff} is made to act as a transformation group on $\mathbb{R}$ (see (6.3) and (6.4)), even without manifestly identifying $\mathbb{R}$ with any set of signal parameters?

The answer to the above question lies in realizing that this space is intrinsic to the group itself. Indeed, let us factor an element $(b, a) \in G_{aff}$ in the manner

$$(b, a) = \begin{pmatrix} a & b \\ 0 & 1 \end{pmatrix} = \begin{pmatrix} 1 & b \\ 0 & 1 \end{pmatrix} \begin{pmatrix} a & 0 \\ 0 & 1 \end{pmatrix}, \tag{6.10}$$

and note that the first matrix on the right-hand side of this equation basically represents a point in $\mathbb{R}$. We also note that the set of matrices of the type appearing in the second term of the above product is a subgroup of G_{aff}. Dividing out by this matrix, we get $(b, a)(0, a)^{-1} = (b, 0)$, which enables us to identify the point $b \in \mathbb{R}$ with an element of the *quotient space* G_{aff}/H, (where H is the subgroup of matrices $(0, a)$, $a \neq 0$). Next we see that, since

$$\begin{pmatrix} a & b \\ 0 & 1 \end{pmatrix} \begin{pmatrix} 1 & x \\ 0 & 1 \end{pmatrix} = \begin{pmatrix} a & ax + b \\ 0 & 1 \end{pmatrix} = \begin{pmatrix} 1 & ax + b \\ 0 & 1 \end{pmatrix} \begin{pmatrix} a & 0 \\ 0 & 1 \end{pmatrix},$$

the action of the group G_{aff} on its quotient space G_{aff}/H is exactly the same as its action on $\mathbb{R}$ as given in (6.3). Thus, the parameter space $\mathbb{R}$ on which the signals $\psi(x)$ are defined is a quotient space of the group and hence intrinsic to the set of signal symmetries. We shall see below that the parameter space on which the wavelet transform of ψ is defined can also be identified with a quotient space of the group. In fact this space will turn out to be a *phase space*, in a sense to be specified later. Before moving on, let us re-emphasize that the philosophy which seems to be emerging here is that the group (of signal symmetries) is the determinative quantity and all aspects of the signal and its various transforms emanate from it.

We come back now to the point made earlier, that the representation $U(b, a)$ was irreducible. We shall see that it also enjoys a second crucial property, that of being *square integrable*. The group G_{aff} has a natural action on itself (by matrix multiplication from the left), according to which, for a given $(b_0, a_0) \in G_{aff}$, a general point $(b, a) \in G_{aff}$ is mapped to $(b', a') = (b_0, a_0)(b, a) = (b_0 + a_0b, a_0a)$. It is not hard to see that the measure

$$d\mu(b, a) = \frac{db\, da}{a^2}, \tag{6.11}$$

is invariant under this action:

$$\frac{db\, da}{a^2} = \frac{db'\, da'}{a'^2}. \tag{6.12}$$

We call the measure $d\mu$ the *left Haar measure* of G_{aff}. In a similar manner we could obtain a *right Haar measure* $d\mu_r$ (invariant under multiplication from the right), which would turn out to be $d\mu_r(b, a) = a^{-1}\, db\, da$. It is important to realize, that while these two measures are (measure theoretically) equivalent, they are not the same measure. The function $\triangle(b, a) = a^{-1}$, for which $d\mu(b, a) = \triangle(b, a)\, d\mu_r(b, a)$, is called the *modular function* of the group. The square-integrability of the representation $U(b, a)$ now means that there exist signals $\psi \in L^2(\mathbb{R}, dx)$ for which the matrix element $\langle U(b, a)\psi \mid \psi \rangle$ is square integrable as a function of the variables b, a, with respect to the left Haar measure $d\mu$, i.e.,

$$\iint_{G_{\text{aff}}} d\mu(b, a)\ |\langle U(b, a)\psi | \psi \rangle|^2 < \infty, \tag{6.13}$$

and a straightforward computation would then establish that the function is also square integrable with respect to the right Haar measure. Furthermore, it is a fact that the existence of one such (nonzero) vector implies the existence of an entire dense set of them. Indeed, the condition for a signal to be of this type is precisely the condition of *admissibility* required of mother wavelets.

To derive the admissibility condition, and also to verify our claim of irreducibility of the representation $U(a, b)$, it will be convenient to go over to the Fourier domain. It is not hard to see that, on the Fourier-transformed space, the unitary operator $U(b, a)$ transforms to $\widehat{U}(b, a)$, with explicit action,

$$\left(\widehat{U}(b, a)\widehat{\psi}\right)(\xi) = |a|^{1/2}\, \widehat{\psi}(a\xi)e^{-ib\xi} \quad (b \in \mathbb{R}, a \neq 0). \tag{6.14}$$

The Fourier transform is a linear isometry, and we denote by $L^2(\widehat{\mathbb{R}}, d\xi)$ the image of $L^2(\mathbb{R}, dx)$ under this map. It follows that the operators $\widehat{U}(b, a)$ are also unitary and that they again constitute a unitary representation of the group G_{aff}. Let $\widehat{\psi} \in L^2(\widehat{\mathbb{R}}, d\xi)$ be a fixed nonzero vector in the Fourier domain. We will now show that the set of all vectors $\widehat{U}(b, a)\widehat{\psi}$ as (b, a) runs through G_{aff} is dense in $L^2(\widehat{\mathbb{R}}, d\xi)$ and this is what will constitute the mathematically precise statement of the irreducibility of $\widehat{U}$. Indeed, let $\widehat{\chi} \in L^2(\widehat{\mathbb{R}}, d\xi)$ be a vector which is orthogonal to all the vectors $\widehat{U}(b, a)\widehat{\psi}$:

$$\langle \widehat{\chi} \mid \widehat{U}(b, a)\widehat{\psi} \rangle = 0, \forall b \in \mathbb{R}, a \neq 0.$$

Using (6.14) we get,

$$\langle \widehat{\chi} | \widehat{U}(b, a)\widehat{\psi} \rangle = |a|^{1/2} \int_{-\infty}^{\infty} d\xi\ \overline{\widehat{\chi}(\xi)}\, \widehat{\psi}(a\xi)\, e^{-ib\xi} = 0.$$

By the unitarity of the Fourier transform, this yields $\overline{\widehat{\chi}(\xi)}\, \widehat{\psi}(a\xi) = 0$, almost everywhere, for all $a \neq 0$. Since $\widehat{\psi} \neq 0$, this in turn implies $\widehat{\chi}(\xi) = 0$, almost everywhere. Thus, the only subspaces of $L^2(\widehat{\mathbb{R}}, d\xi)$ which are stable under the action of all the operators $\widehat{U}(b, a)$ are $L^2(\widehat{\mathbb{R}}, d\xi)$ itself and the trivial subspace containing just the zero vector. In other words, $L^2(\widehat{\mathbb{R}}, d\xi)$ is sort of a minimal space for the representation.

The unitarity of the Fourier transform also tells us that the representations $U(b, a)$ and $\widehat{U}(b, a)$ are equivalent and since $\widehat{U}(b, a)$ is irreducible, so also is $U(b, a)$. (Note, this is also clear from the fact that the linear isometry property of the Fourier transform implies that

$$\langle \chi \mid U(b, a)\psi \rangle = \langle \widehat{\chi} \mid \widehat{U}(b, a)\widehat{\psi} \rangle,$$

χ, ψ denoting the inverse Fourier transforms of $\widehat{\chi}, \widehat{\psi}$, respectively.)

Now we address the question of square integrability. We require that

$$\iint_{G_{\text{aff}}} \frac{da \, db}{a^2} \, |\langle \widehat{U}(b, a)\widehat{\psi} \mid \widehat{\psi} \rangle|^2 =$$

$$= \iiiint d\xi \, d\xi' \frac{da}{|a|} db \, e^{ib(\xi - \xi')} \overline{\widehat{\psi}(a\xi)} \, \widehat{\psi}(a\xi') \widehat{\psi}(\xi) \, \overline{\widehat{\psi}(\xi')}$$

$$= 2\pi \iint \frac{da}{|a|} d\xi \, |\widehat{\psi}(a\xi)|^2 \, |\widehat{\psi}(\xi)|^2$$

$$= 2\pi \|\psi\|^2 \int_{-\infty}^{\infty} \frac{d\xi}{|\xi|} |\widehat{\psi}(\xi)|^2 < \infty$$

(the integral over b yields a delta distribution, which can be used to perform the ξ' integration and the interchange of integrals can be justified using standard distribution theoretic arguments). This means that the vector ψ is admissible in the sense of (6.13) if and only if

$$c_\psi \equiv 2\pi \int_{-\infty}^{\infty} \frac{d\xi}{|\xi|} |\widehat{\psi}(\xi)|^2 \, < \infty. \tag{6.15}$$

From this discussion we draw two immediate conclusions. First, there is a dense set of vectors $\widehat{\psi}$ which satisfy the admissibility condition (6.15). Second, the admissibility condition (1.10) or (6.15), $c_\psi < \infty$, simply expresses the square integrability of the representation U. (Note that a vector $\psi \in L^2(\mathbb{R}, dx)$ is admissible if and only if its Fourier transform satisfies (6.15).) Defining an operator $\widehat{C}$ on $L^2(\widehat{\mathbb{R}}, d\xi)$,

$$(\widehat{C}\widehat{\psi})(\xi) = \left[\frac{2\pi}{|\xi|} \right]^{\frac{1}{2}} \widehat{\psi}(\xi), \tag{6.16}$$

and denoting by C its inverse Fourier transform, we see that the vector ψ is admissible if and only if

$$c_\psi = \|C\psi\|^2 < \infty. \tag{6.17}$$

This operator, known as the *Duflo–Moore operator*, is positive, self-adjoint and unbounded. It also has an inverse. A straightforward computation, using (6.14) and (6.16) now shows that if a vector ψ satisfies (6.17), then so also does the vector $U(b, a)\psi$, for any $(b, a) \in G_{\text{aff}}$.

A word now about the form of the representation $U(b, a)$. How does one arrive at it? In fact, given the way the group acts on $\mathbb{R}$, $x \mapsto ax + b$, the representation $U(b, a)$ is recognized as being the most natural, nontrivial way to realize a group homomorphism onto a set of unitary operators on the signal space $L^2(\mathbb{R}, dx)$. (Unitarity is required in order to ensure that the signal ψ and the transformed signal $U(b, a)\psi$ both have the same total energy.) Indeed, given any differentiable mapping $T : \mathbb{R}^n \to \mathbb{R}^n$, the operator $U(T)$, on the Hilbert space $L^2(\mathbb{R}^n, d^n\vec{x})$, defined as

$$(U(T)f)(\vec{x}) = |\det[J(T)]|^{-\frac{1}{2}} f(T^{-1}(\vec{x})),$$

where $J(T)$ is the Jacobian of the map T, is easily seen to be unitary. (Recall that

$$d(T(\vec{x})) = |\det[J(T)]| \, d\vec{x} \, .)$$

This provides the rationale for defining the representation $U(b, a)$ by (6.8). Of course, the interesting point here is that this representation turns out to be both irreducible and square integrable.

But then, why is square integrability of the representation a desirable criterion for wavelet analysis? In order to answer this question, let us take a vector ψ satisfying the admissibility condition (6.17) and use it to construct the wavelet transform of the signal s:

$$S(b, a) = \langle \psi_{b,a} \mid s \rangle.$$

As we already know, the total energy of the transformed signal is given by the integral

$$E(S) = \iint_{G_{\text{aff}}} d\mu(b, a) \mid S(b, a)\mid^2, \tag{6.18}$$

and we would like this to be finite, like that of the signal itself. An easy computation now shows that

$$E(S) = \|C\psi\|^2 \, \|s\|^2 = c_\psi \, \|s\|^2, \tag{6.19}$$

which means that the total energy of the wavelet transform will be finite if and only if the mother wavelet can be chosen from the domain of the operator C, i.e., if and only if it satisfies the square integrability condition (6.13). However, this is not the whole story, for let us rewrite the above equation in the expanded form

$$E(S) = \iint_{G_{\text{aff}}} d\mu(b, a) \, \langle s \mid \psi_{b,a}\rangle\langle\psi_{b,a} \mid s\rangle$$
$$= \langle s \mid \left[\iint_{G_{\text{aff}}} d\mu(b, a) \mid \psi_{b,a}\rangle\langle\psi_{b,a} \mid \right] s\rangle$$
$$= c_\psi \, \langle s \mid Is\rangle,$$

I being the identity operator on $L^2(\mathbb{R}, dx)$ and where, for any nonzero vector $\phi \in L^2(\mathbb{R}, dx)$, the quantity $\mid \phi\rangle\langle\phi \mid /\|\phi\|^2$ denotes the one-dimensional projection operator along this vector. (We have also interchanged two integrations with the taking of a scalar product, without justifying it, but the manipulation can easily be justified using standard absolute convergence techniques.) Using the well-known polarization identity for scalar products, we infer from this the operator relation

$$\frac{1}{c_\psi} \iint_{G_{\text{aff}}} d\mu(b,a) \ |\psi_{b,a}\rangle\langle\psi_{b,a}| = I, \tag{6.20}$$

also known as the *resolution of the identity* equation.

It is immediately clear that (6.20) is completely equivalent to the square integrability of the representation $U(b,a)$ as expressed in (6.13). The resolution of the identity also incorporates within it the possibility of reconstructing the signal $s(x)$ from its wavelet transform $S(b,a)$. To see this, let us act on the vector $s \in L^2(\mathbb{R}, dx)$ with both sides of (6.20). We get

$$\frac{1}{c_\psi} \iint_{G_{\text{aff}}} d\mu(b,a) \ \psi_{b,a}\langle\psi_{b,a}|s\rangle = Is = s,$$

implying

$$s(x) = \frac{1}{c_\psi} \iint_{G_{\text{aff}}} d\mu(b,a) \ S(b,a)\psi_{b,a}(x), \quad \text{almost everywhere,} \tag{6.21}$$

which is the celebrated *reconstruction formula* we encountered before. Summarizing, we conclude that square integrability (which is a group property) is precisely the condition which ensures, in this case, the very desirable consequences of (i) the finiteness of the energy of the wavelet transform, and (ii) the validity of the reconstruction formula (two properties also shared by the Fourier transform of a signal). The resolution of the identity condition (6.20) has independent mathematical interest. First of all, it implies that any vector in $L^2(\mathbb{R}, dx)$ which is orthogonal to all the wavelets $\psi_{b,a}$ is necessarily the zero vector, i.e., the linear span of the wavelets is dense in the Hilbert space of signals. This fact, which could also have been inferred from the irreducibility of the representation $U(b,a)$, is what enables us to use the wavelets as a basis set for expressing arbitrary signals. In fact we have here what is also known as an *overcomplete* basis. Second, this overcomplete basis is a continuously parametrized set, meaning that this is an example of a continuous basis and a continuous *frame*. There is a host of other useful mathematical properties of the wavelet transform and spaces of transforms, which emanate from square integrability. We proceed to examine a few.

6.1.2   The space of all wavelet transforms

A finite energy wavelet transform $S(b,a)$ is an element of the Hilbert space $L^2(\mathbb{H}, d\mu)$. Here we have written $\mathbb{H} = \mathbb{R} \times \mathbb{R}^*$ ($\mathbb{R}^* = $ real line with the origin removed). Although, $\mathbb{H}$ and G_{aff} are homeomorphic as topological spaces, we prefer to denote them by different symbols, for presently we shall identify $\mathbb{H}$ with a *phase space* of G_{aff}, arising from its matrix geometry. Using (6.18) and (6.19) to compare the L^2-norm of the wavelet transform S (as an element in $L^2(\mathbb{H}, d\mu)$) to the L^2-norm of the signal s (as an element in $L^2(\mathbb{R}, dx)$), we get

$$\|S\|^2 = c_\psi \|s\|^2, \tag{6.22}$$

which just means that, up to a constant, the wavelet transform preserves norms (i.e., energies). We define a map $W_\psi : L^2(\mathbb{R}, dx) \to L^2(\mathbb{H}, d\mu)$, by the relation

$$(W_\psi s)(a, b) = \left[c_\psi\right]^{-\frac{1}{2}} \langle \psi_{b,a} \mid s \rangle_{L^2(\mathbb{R}, dx)} = \left[c_\psi\right]^{-\frac{1}{2}} S(b, a). \tag{6.23}$$

This map is linear and, in view of (6.22), an isometry, so that its range, which is the set of all wavelet transforms corresponding to the mother wavelet ψ, is a closed subspace of $L^2(\mathbb{H}, d\mu)$. We denote the range by $\mathfrak{H}_\psi$:

$$\mathfrak{H}_\psi = W_\psi \left[L^2(\mathbb{R}, dx)\right] \subset L^2(\mathbb{H}, d\mu). \tag{6.24}$$

From the defining equation (6.23) we infer that $\mathfrak{H}_\psi$ consists of continuous functions over $\mathbb{H}$ and hence is a proper subspace of $L^2(\mathbb{H}, d\mu)$. It is worthwhile reiterating here the fact that the condition of W_ψ being an isometry implies, not only that the wavelet transform (with respect to the mother wavelet ψ) of any signal $s \in L^2(\mathbb{R}, dx)$ is an element in $\mathfrak{H}_\psi$, but also that every element in $\mathfrak{H}_\psi$ is the wavelet transform of some signal $s \in L^2(\mathbb{R}, dx)$.

6.1.2.1   An intrinsic characterization of the space of wavelet transforms

Is there some convenient, intrinsic way to characterize the subspace $\mathfrak{H}_\psi$? To answer this question we appeal to the resolution of the identity and a bit of group theory. The final characterization will be spelled out in Theorem 6.1.1. Multiplying each side of equation (6.20) by itself we find,

$$\frac{1}{c_\psi} \iint_{\mathbb{H} \times \mathbb{H}} d\mu(b', a') \, d\mu(b, a) \mid \psi_{b,a} \rangle \, K_\psi(b, a \, ; \, b', a') \, \langle \psi_{b',a'} \mid = I, \tag{6.25}$$

where we have written

$$K_\psi(b, a \, ; \, b', a') = \frac{1}{c_\psi} \langle \psi_{b,a} \mid \psi_{b',a'} \rangle. \tag{6.26}$$

Acting on the signal vector s with both sides of (6.25) and using (6.23), we obtain

$$\frac{1}{c_\psi} \int_{\mathbb{H}} d\mu(b, a) \left[\int_{\mathbb{H}} d\mu(b', a') \, K_\psi(b, a \, ; \, b', a') S(b', a') \right] \psi_{b,a}(x) = s(x),$$

almost everywhere (the change in the order of integrations being easily justified by Fubini's theorem). Comparing the above equation with the reconstruction formula (6.21) we obtain the interesting identity

$$\int_{\mathbb{H}} d\mu(b', a') \, K_\psi(b, a \, ; \, b', a') S(b', a') = S(b, a), \tag{6.27}$$

for almost all (b, a) in $\mathbb{H}$ (with respect to the measure $d\mu$). This then is the condition which characterizes wavelet transforms coming from the mother wavelet ψ. It is also known as the *reproducing property* of the integral kernel K_ψ.

As we know, the kernel $K_\psi : \mathbb{H} \times \mathbb{H} \to \mathbb{C}$ is called a reproducing kernel. It has the easily verifiable properties:

$$K_\psi(b, a\,;\, b, a) > 0, \ \text{for all } (b, a) \in \mathbb{H}, \tag{6.28}$$

$$K_\psi(b, a\,;\, b', a') = \overline{K_\psi(b', a'\,;\, b, a)}, \tag{6.29}$$

$$\iint_{\mathbb{H} \times \mathbb{H}} d\mu(b'', a'')\, K_\psi(b, a\,;\, b'', a'')\, K_\psi(b'', a''\,;\, b', a') = K_\psi(b, a\,;\, b', a'), \tag{6.30}$$

the last relation being again the reproducing property (6.27) in a different guise (and in fact following directly from it). The last two equations hold pointwise, for all (b, a), $(b', a') \in \mathbb{H}$.

Next let us compute the wavelet transforms of the wavelets $\psi_{b,a}$ themselves. Denoting the transforms by $S_{b,a}$ we find

$$S_{b,a}(b', a') = \langle \psi_{b',a'} \,|\, \psi_{b,a} \rangle = c_\psi\, K_\psi(b', a'\,;\, b, a). \tag{6.31}$$

Since the vectors $\psi_{b,a}$, $(b, a) \in \mathbb{H}$, are overcomplete in $L^2(\mathbb{R}, dx)$, the wavelet transforms $S_{b,a}$ must also be overcomplete in $\mathfrak{H}_\psi$. One also has the easily verifiable resolution of the identity,

$$\left[c_\psi \right]^{-2} \int_{\mathbb{H}} d\mu(b, a)\, |\, S_{b,a} \rangle \langle S_{b,a}\,| = I_\psi, \tag{6.32}$$

where we have written I_ψ for the identity operator on $\mathfrak{H}_\psi$. Thus, any vector $F \in L^2(\mathbb{H}, d\mu)$ which lies in the orthogonal complement of $\mathfrak{H}_\psi$ must satisfy

$$\int_{\mathbb{H}} d\mu(b', a')\, K_\psi(b, a\,;\, b', a') F(b', a') = 0.$$

All this goes to say that the reproducing kernel K_ψ defines the projection operator $\mathbb{P}_\psi$ from $L^2(\mathbb{H}, d\mu)$ to $\mathfrak{H}_\psi$:

$$(\mathbb{P}_\psi F)(b, a) = \int_{\mathbb{H}} d\mu(b', a')\, K_\psi(b, a\,;\, b', a') F(b', a'), \qquad F \in L^2(\mathbb{H}, d\mu), \tag{6.33}$$

equations (6.29) and (6.30) mirroring the conditions $\mathbb{P}_\psi = \mathbb{P}_\psi^* = \mathbb{P}_\psi^2$ (star denotes the adjoint). Stated differently, an arbitrary vector $F \in L^2(\mathbb{H}, d\mu)$ can be uniquely written as the sum

$$F = F_\psi + F_\psi^\perp,$$

of a part $F_\psi \in \mathfrak{H}_\psi$ and a part $F_\psi^\perp$ orthogonal to it. The operator $\mathbb{P}_\psi$, acting on F, projects out the part F_ψ (which is a wavelet transform). It is natural to ask at this point if $F_\psi^\perp$ could also be written as the wavelet transform with respect to some other mother wavelet. As will be seen below, generally $F_\psi^\perp$ can be written as an infinite sum of orthogonal wavelet transforms, corresponding to different mother wavelets.

To proceed further, we go back to the affine group, G_{aff}, and note that there is a natural unitary representation of it on the Hilbert space $L^2(\mathbb{H}, d\mu)$, given by its natural action on $\mathbb{H}$. This representation $\mathbf{U}_\ell(b, a)$, called the *left regular representation*, acts in the manner

$$(\mathbf{U}_\ell(b, a)F)(b', a') = F((b, a)^{-1}(b', a'))$$
$$= F\left(\frac{b' - b}{a}, \frac{a'}{a}\right), \qquad F \in L^2(\mathbb{H}, d\mu). \tag{6.34}$$

The unitarity of this representation,

$$\|\mathbf{U}_\ell(b, a)F\|^2_{L^2(\mathbb{H}, d\mu)} = \|F\|^2_{L^2(\mathbb{H}, d\mu)},$$

is guaranteed by the invariance of the measure $d\mu$ (see (6.12)). However, the left regular representation is by no means irreducible, since as we shall see below, the subspace $\mathfrak{H}_\psi$ carries a subrepresentation of it.

The isometry W_ψ (see (6.23)) maps the unitary operators $U(b, a)$ onto unitary operators $\mathbf{U}_\psi(b, a) = W_\psi U(b, a) W_\psi^{-1}$ on $L^2(\mathbb{H}, d\mu)$. Computing the action of these operators, using (6.8) and (6.23), we find

$$(\mathbf{U}_\psi(b, a)F)(b', a') = F\left(\frac{b' - b}{a}, \frac{a'}{a}\right), \qquad F \in \mathfrak{H}_\psi. \tag{6.35}$$

This is the same action as that of the operators $\mathbf{U}_\ell(b, a)$ of the left regular representation, except that now it is expressed exclusively in terms of vectors in $\mathfrak{H}_\psi$. This means, first of all, that the subspace $\mathfrak{H}_\psi$ is stable under the action of the operators $\mathbf{U}_\ell(b, a)$ and, secondly, that restricted to this subspace, it gives an irreducible unitary representation of G_{aff}.

6.1.2.2   Decomposition of the space of all finite energy wavelet transforms

Let ψ and ψ' be two different mother wavelets. This means that they are both vectors in the domain of the operator C (with their Fourier transforms satisfying (6.15)). How are wavelet transforms, taken with respect to these mother wavelets, related? In particular, denoting by S_ψ the wavelet transform of the signal s, taken with respect to the mother wavelet ψ, and by $S'_{\psi'}$ the wavelet transform of the signal s', taken with respect to the mother wavelet ψ', we would like to evaluate the overlap

$$I(\psi, \psi'; s', s) = \iint_\mathbb{H} d\mu(b, a) \, \overline{S'_{\psi'}(b, a)} \, S_\psi(b, a)$$
$$= \iint_\mathbb{H} d\mu(b, a) \, \langle s' \mid \psi'_{b,a}\rangle \langle \psi_{b,a} \mid s\rangle. \tag{6.36}$$

Let us begin by assuming that s, s' are taken from a class of smooth functions (e.g., the Schwartz class, $\mathcal{S}(\mathbb{R})$), which is dense in $L^2(\mathbb{R}, dx)$. Then

$$
\begin{aligned}
I(\psi, \psi'; s', s) &= \int_{\mathbb{R}} \int_{\mathbb{R}^*} \frac{db\, da}{a^2} \, \langle s' \mid \widehat{\psi}'_{b,a} \rangle \langle \widehat{\psi}_{b,a} \mid \widehat{s} \rangle \\
&= \int_{\mathbb{R}} \int_{\mathbb{R}^*} \frac{db\, da}{a^2} \left[\int_{\mathbb{R}} d\xi \, |a|^{\frac{1}{2}} \, e^{ib\xi} \, \overline{\widehat{s'}(\xi)} \, \widehat{\psi}'(a\xi) \right] \\
&\quad \times \left[\int_{\mathbb{R}} d\xi' \, |a|^{\frac{1}{2}} \, e^{-ib\xi'} \, \widehat{s}(\xi') \, \overline{\widehat{\psi}(a\xi')} \right].
\end{aligned}
$$

We exploit the smoothness of the functions s, s' to use the identity

$$
\frac{1}{2\pi} \int_{\mathbb{R}} db\, e^{ib(\xi - \xi')} = \delta(\xi - \xi'),
$$

which holds in the sense of distributions, and then perform the ξ'-integration to obtain

$$
I(\psi, \psi'; s', s) = 2\pi \int_{\mathbb{R}} \int_{\mathbb{R}} \frac{da}{a} \, d\xi \, \overline{\widehat{\psi}(a\xi)} \, \widehat{\psi}'(a\xi) \, \widehat{s}(\xi) \, \overline{\widehat{s'}(\xi)}.
$$

Changing variables, we get

$$
\begin{aligned}
I(\psi, \psi'; s', s) &= 2\pi \left[\int_{\mathbb{R}} dy \, \frac{\overline{\widehat{\psi}(y)} \, \widehat{\psi}'(y)}{|y|} \right] \cdot \left[\int_{\mathbb{R}} d\xi \, \widehat{s}(\xi) \, \overline{\widehat{s'}(\xi)} \right] \\
&= \langle \widehat{C\psi} \mid \widehat{C\psi'} \rangle \, \langle s' \mid s \rangle.
\end{aligned}
$$

Thus,

$$
\iint_{\mathbb{H}} d\mu(b, a) \, \overline{S'_{\psi'}(b, a)} S_{\psi}(b, a) = \iint_{\mathbb{H}} d\mu(b, a) \, \langle s' \mid \psi'_{b,a} \rangle \langle \psi_{b,a} \mid s \rangle
$$
$$
= \langle C\psi \mid C\psi' \rangle \, \langle s' \mid s \rangle. \tag{6.37}
$$

Using the continuity of the scalar product $\langle s' \mid s \rangle$ in s and s', we may now extend the above expression to all signals s, $s' \in L^2(\mathbb{R}, dx)$.

Equation (6.37) is a general *orthogonality relation* for wavelet transforms. In particular, if $C\psi$ and $C\psi'$ are orthogonal vectors, then the corresponding wavelet transforms are also orthogonal in $L^2(\mathbb{H}, d\mu)$. We may also write this equation in the form of an operator identity on $L^2(\mathbb{R}, dx)$:

$$
\iint_{\mathbb{H}} d\mu(b, a) \, |\psi'_{b,a} \rangle \langle \psi_{b,a} | = \langle C\psi \mid C\psi' \rangle \, I, \tag{6.38}
$$

which clearly is a generalization of the resolution of the identity (6.20). At the risk of being pedantic, we would still like to emphasize that the above orthogonality relation implies:

- If s and s' are signals which are orthogonal vectors in $L^2(\mathbb{R}, dx)$, then their wavelet transforms S and S', whether with respect to the same or different mother wavelets, are orthogonal as vectors in $L^2(\mathbb{H}, d\mu)$.

- Spaces of wavelet transforms, $\mathfrak{H}_\psi$, $\mathfrak{H}_{\psi'}$, corresponding to mother wavelets ψ, ψ' which satisfy the orthogonality condition $C\psi \perp C\psi'$, are orthogonal subspaces of $L^2(\mathbb{H}, d\mu)$.

Equation (6.38) is a remarkable result. Acting on a signal $s \in L^2(\mathbb{R}, dx)$ with both sides of this equation, and assuming that $\langle C\psi \mid C\psi' \rangle \neq 0$, we get

$$s = \frac{1}{\langle C\psi \mid C\psi' \rangle} \iint_{\mathbb{H}} d\mu(b, a)\, S_\psi(b, a) \psi'_{b,a}, \qquad \psi'_{b,a} = U(b, a)\psi',$$

where $S_\psi(b, a) = \langle \psi_{b,a} \mid s \rangle$ is the wavelet transform of s computed with respect to the mother wavelet ψ. Thus, although the wavelet transform is computed with respect to the mother wavelet ψ, it can be reconstructed using the wavelets of any other mother wavelet ψ', so long as $\langle C\psi \mid C\psi' \rangle \neq 0$. Moreover, up to a multiplicative constant, the reconstruction formula is exactly the same as that in which the same wavelet ψ is used both for analyzing and reconstructing (see (6.21)). This indicates, that in some sense, analysis and reconstruction are independent of the mother wavelet chosen.

Let us choose a set of mother wavelets $\{\psi_n\}_{n=1}^\infty$ such that the vectors $\phi_n = C\psi_n$ form an orthonormal basis of $L^2(\mathbb{R}, dx)$,

$$\langle \phi_n \mid \phi_m \rangle = \langle C\psi_n \mid C\psi_m \rangle = \delta_{nm}, \qquad n, m = 0, 1, 2, \dots, \infty. \tag{6.39}$$

Such a basis is easy to find and we shall construct one below. If $\mathfrak{H}_{\psi_n}$, $n = 0, 1, 2, \dots$, are the corresponding spaces of wavelet transforms and K_{ψ_n} the associated reproducing kernels, then $\mathfrak{H}_{\psi_n} \perp \mathfrak{H}_{\psi_m}$, for $n \neq m$, and

$$\iint_{\mathbb{H}} d\mu(b'', a'')\, K_{\psi_n}(b, a\,;\, b'', a'')\, K_{\psi_m}(b'', a''\,;\, b', a') = \delta_{nm}\, K_{\psi_m}(b, a\,;\, b', a'). \tag{6.40}$$

More interestingly, it is possible to show that the complete decomposition,

$$L^2(\mathbb{H}, d\mu) \simeq \bigoplus_{n=1}^\infty \mathfrak{H}_{\psi_n}, \tag{6.41}$$

of the space of all finite energy signals (on the parameter space $\mathbb{H}$) into an orthogonal direct sum of spaces of wavelet transforms, holds. Thus, in an L^2-sense, any element $S \in L^2(\mathbb{H}, d\mu)$ has the orthogonal decomposition,

$$S(b, a) = \bigoplus_{n=1}^\infty S_n(b, a), \qquad \text{almost everywhere,}$$

into orthogonal wavelet transforms S_n, with respect to a basis of mother wavelets. This result, which can be proved by direct computation in the present case, is actually a particular example of a much more general result on the decomposition of the left regular representation of a group into irreducibles. For a more detailed mathematical discussion of this point, we refer the reader to [Ali00]. The components $S_n(b, a)$ have the form

$$S_n(b, a) = \langle U(b, a)\psi_n \mid s_n \rangle, \qquad n = 0, 1, 2, \ldots \tag{6.42}$$

for some signal vectors $s_n \in L^2(\mathbb{R}, dx)$, which, in general, are different for different n. We also have the relations

$$\iint_{\mathbb{H}} d\mu(b', a') \, K_{\psi_n}(b, a\,;\, b', a') S_m(b', a') = \delta_{nm} \, S_m(b, a). \tag{6.43}$$

Finally, we construct an explicit example of a basis set of mother wavelets satisfying (6.39). Let $H_n(\xi)$, $n = 0, 1, 2, \ldots, \infty$, be the Hermite polynomials, normalized in the manner

$$\int_{\mathbb{R}} d\xi \, e^{-\xi^2} \, H_m(\xi) \, H_n(\xi) = \begin{cases} 0, & \text{if } m \neq n, \\ 2^n n! \sqrt{\pi}, & \text{if } m = n. \end{cases} \tag{6.44}$$

The first few are:

$$H_0(\xi) = 1, \qquad H_1(\xi) = 2\xi,$$
$$H_2(\xi) = 4\xi^2 - 2, \qquad H_3(\xi) = 8\xi^3 - 12\xi, \quad \text{etc.}$$

In the Fourier domain, define

$$\widehat{\psi}_n(\xi) = \frac{1}{\pi^{\frac{3}{4}} 2^{\frac{n+1}{2}} \sqrt{n!}} \, |\xi|^{\frac{1}{2}} \, e^{-\frac{\xi^2}{2}} \, H_n(\xi). \tag{6.45}$$

Then, it is easily verified that

$$\|\widehat{\psi}_n\|^2_{L^2(\widehat{\mathbb{R}}, d\xi)} < \infty,$$

and

$$\langle \widehat{C\psi}_m \mid \widehat{C\psi}_n \rangle = 2\pi \int_{\mathbb{R}} \frac{d\xi}{|\xi|} \, \overline{\widehat{\psi}_m(\xi)} \, \widehat{\psi}_n(\xi) = \delta_{mn}.$$

Thus, in the inverse Fourier domain, the vectors ψ_n are in the domain of the operator C, satisfying the condition for being mother wavelets, while from the well-known properties of Hermite polynomials, the vectors ϕ_n constitute an orthonormal basis of $L^2(\mathbb{R}, dx)$. More generally, since the range of C is dense in $L^2(\mathbb{R}, dx)$, we can take any orthonormal basis, $\{\phi_n\}_{n=1}^{\infty}$ of $L^2(\mathbb{R}, dx)$, chosen from vectors in this range and then $\{\psi_n = C^{-1}\phi_n\}_{n=1}^{\infty}$ will be the desired wavelet basis.

We collect the above results into a theorem:

Theorem 6.1.1 *The wavelet transform of the space of signals $L^2(\mathbb{R}, dx)$, with respect to the mother wavelet ψ, is a closed subspace of $L^2(\mathbb{H}, d\mu)$. This subspace has a reproducing kernel K_ψ, which is the integral kernel of the projection operator, $\mathbb{P}_\psi$: $L^2(\mathbb{H}, d\mu) \to \mathfrak{H}_\psi$. The Hilbert space $L^2(\mathbb{H}, d\mu)$ can be completely decomposed into an orthogonal direct sum of an infinite number of subspaces $\mathfrak{H}_{\psi_n}$, each a space of wavelet transforms with respect to a mother wavelet ψ_n. The vectors ψ_n are constructed by* *taking an orthonormal basis $\{\phi_n\}_{n=1}^{\infty}$ of $L^2(\mathbb{R}, dx)$, chosen from the range of the Duflo–Moore operator C, and writing $\psi_n = C^{-1}\phi_n$.*

6.1.2.3 Decomposition into orthogonal channels

The above theorem can be used to analyze a given wavelet transform into orthogonal channels along the lines of [105]. Referring back to (6.39), since the mother wavelets $\{\psi_n\}_{n=1}^{\infty}$ form a complete, linearly independent set, any mother wavelet ψ can be written as a linear combination,

$$\psi = \sum_{n=1}^{\infty} a_n \psi_n, \qquad a_n = \langle \phi_n \mid C\psi \rangle = \langle \psi_n \mid C^2\psi \rangle. \tag{6.46}$$

Hence if $s \in L^2(\mathbb{R}, dx)$ is any signal vector and $S_\psi(b, a)$ its wavelet transform with respect to ψ, then clearly

$$S_\psi(b, a) = \sum_{n=1}^{\infty} S_n(b, a), \quad \text{where} \quad S_n(b, a) = a_n \langle U(b, a)\psi_n \mid s \rangle. \tag{6.47}$$

In this way, the wavelet transform of the signal $S_\psi(b, a)$ has been decomposed into a set of mutually orthogonal wavelet transforms $S_n(b, a)$ (of this same signal). We call this a *decomposition into orthogonal channels*. Note that although the wavelet transforms are orthogonal in $L^2(\mathbb{H}, d\mu)$, the mother wavelets ψ_n are not orthogonal in $L^2(\mathbb{R}, dx)$ (see (6.39)). We shall see in the next chapter (at the end of Section 7.2.3), that in the case of a 2-D wavelet transform, it will actually be possible to obtain a decomposition into orthogonal (angular) channels using a family of mother wavelets which are themselves mutually orthogonal.

6.1.3 Localization operators

Let ψ be a mother wavelet and $\mathfrak{H}_\psi$ the corresponding space of wavelet transforms. Let $\Delta \subset \mathbb{H}$ be a measurable set (with respect to the measure $d\mu$). We associate to this set an integral kernel $a_\psi^\Delta : \mathbb{H} \times \mathbb{H} \to \mathbb{C}$:

$$a_\psi^\Delta(b, a; b', a') = \iint_\Delta d\mu(b'', a'') \, K_\psi(b, a; b'', a'') K_\psi(b'', a''; b', a'), \tag{6.48}$$

and an operator $a_\psi(\Delta)$ on $\mathfrak{H}_\psi$ acting via this kernel:

$$(a_\psi(\Delta)S)(b, a) = \iint_{\mathbb{H}} d\mu(b', a') \, a_\psi^\Delta(b, a; b', a')S(b', a'), \qquad S \in \mathfrak{H}_\psi. \tag{6.49}$$

If we compute the matrix element of this operator for the wavelet transform S, using the properties of the reproducing kernel (see (6.29)–(6.30)), we easily obtain,

$$\langle S \mid a_\psi(\Delta)S \rangle = \iint_\Delta d\mu(b, a) \mid S(b, a) \mid^2. \tag{6.50}$$

which shows that this operator is bounded, positive and self-adjoint. We call $a_\psi(\Delta)$ a *localization operator*, since in view of the above relation, the quantity

$$p_S(\Delta) = \frac{\langle S \mid a_\psi(\Delta)S \rangle}{\|S\|^2} \tag{6.51}$$

is the fraction of the wavelet transform which is localized in the region Δ of phase space. If $S(b, a) = \langle \psi_{b,a} \mid s \rangle$, then we shall also write $p_s(\Delta)$ for the above probability, for indeed, it measures the concentration of the phase space content (e.g., time–frequency content) of the signal s in the region Δ. As a set function, $p_S(\Delta)$ has the properties of a probability measure:

$$p_S(\mathbb{H}) = 1, \qquad p_S(\emptyset) = 0, \quad \emptyset = \text{ empty set,}$$

$$p_S(\bigcup_{i \in J} \Delta_i) = \sum_{i \in J} p_S(\Delta_i), \qquad \text{if } \Delta_i \bigcap \Delta_j = \emptyset, \text{ whenever } i \neq j, \tag{6.52}$$

J being some index set. Since this holds for all $S \in \mathfrak{H}_\psi$, we say that the operators $a_\psi(\Delta)$ themselves constitute a *positive operator-valued measure*, or *POV-measure*, satisfying the properties:

$$a_\psi(\mathbb{H}) = I_\psi, \qquad a_\psi(\emptyset) = 0,$$

$$a_\psi(\bigcup_{i \in J} \Delta_i) = \sum_{i \in J} a_\psi(\Delta_i), \qquad \text{if } \Delta_i \bigcap \Delta_j = \emptyset, \text{ whenever } i \neq j, \tag{6.53}$$

where the sum on the right-hand side of (6.53) has to be understood in the weak sense, i.e., in the sense of (6.52).

It is instructive to see how $p_S(\Delta)$ changes if the set Δ gets transformed under the action of the group G_{aff}. Since, for any $(b_0, a_0) \in G_{\text{aff}}$,

$$\langle U(b_0, a_0)\psi_{b,a} \mid \psi_{b',a'} \rangle_{L^2(\mathbb{R},dx)} = \langle \psi_{b,a} \mid U(b_0, a_0)^* \psi_{b',a'} \rangle_{L^2(\mathbb{R},dx)}, \tag{6.54}$$

using the group properties (6.9) and the definition of the reproducing kernel in (6.26), we find that it satisfies the following *covariance property*:

$$K_\psi(b_0 + a_0 b, a_0 a \; ; b', a') = K_\psi(b, a \; ; \frac{b' - b_0}{a_0}, \frac{a}{a_0}), \tag{6.55}$$

i.e.,

$$K_\psi((b_0, a_0)(b, a) \; ; b', a') = K_\psi(b, a \; ; (b_0, a_0)^{-1}(b', a')). \tag{6.56}$$

Let $(b_0, a_0)\Delta$ denote the translate of the set Δ by (b_0, a_0):

$$(b_0, a_0)\Delta = \{(b_0 + a_0 b, a_0 a) \in \mathbb{H} \mid (b, a) \in \Delta\}.$$

Then, taking note of the action (6.35) of the left regular representation of G_{aff} on wavelet transforms and exploiting the invariance of the measure $d\mu$, we easily find that

$$\langle \mathbf{U}_\psi(b, a)S \mid a_\psi(\Delta)\mathbf{U}_\psi(b, a)S' \rangle = \langle S \mid a_\psi((b, a)^{-1}\Delta)S' \rangle, \qquad S, S' \in \mathfrak{H}_\psi,$$

i.e., we have the operator identity

$$\mathbf{U}_\psi(b, a)a_\psi(\Delta)\mathbf{U}_\psi(b, a)^* = a_\psi((b, a)\Delta). \tag{6.57}$$

This is a group covariance condition satisfied by the localization operators $a_\psi(\Delta)$, and is generally known as an *imprimitivity relation* (see, e.g., [Ali00], for a more detailed discussion). For the probability measure $p_S(\Delta)$, this condition implies the transformation property

$$p_S((b, a)\Delta) = p_{\mathbf{U}_\psi(b,a)^{-1}S}(\Delta), \quad \text{or,} \quad p_s((b, a)\Delta) = p_{U(b,a)^{-1}s}(\Delta). \tag{6.58}$$

Physically, this relation means that the fraction of the signal s, localized in the transformed set $(b, a)\Delta$, is the same as the fraction of the transformed signal $U(b, a)^{-1}s$ localized in the original set Δ.

6.2   Phase space analysis

We turn our attention to a different way of understanding the wavelet transform, namely, as a function on a phase space (in a sense to be made clear in the sequel). First let us restrict the representation $\widehat{U}(b, a)$ of the full affine group (see (6.14)) to the connected affine group G^+_{aff} characterized by $a > 0$. We immediately see that this representation is no longer irreducible for this smaller group. Indeed, consider the two subspaces

$$\widehat{\mathfrak{H}}_+(\widehat{\mathbb{R}}) = \{\widehat{f} \in L^2(\widehat{\mathbb{R}}, d\xi) \mid \widehat{f}(\xi) = 0, \ \forall \xi < 0\},$$
$$\widehat{\mathfrak{H}}_-(\widehat{\mathbb{R}}) = \{\widehat{f} \in L^2(\widehat{\mathbb{R}}, d\xi) \mid \widehat{f}(\xi) = 0, \ \forall \xi > 0\}, \tag{6.59}$$

of the carrier space $L^2(\widehat{\mathbb{R}}, d\xi)$ of the representation $\widehat{U}(b, a)$. From (6.14) it is evident that vectors in any one of these subspaces are mapped to vectors in the same subspace under the action of the operators $\widehat{U}(b, a)$, when we only consider elements $(b, a) \in G^+_{\text{aff}}$. This means that each one of these subspaces carries a unitary representation of this smaller group and, as before, we can show that both these representations, which we denote by $\widehat{U}_+(b, a)$ and $\widehat{U}_-(b, a)$, respectively, are irreducible but unitarily inequivalent. In fact, these are the only two nontrivial, unitary irreducible representations of G^+_{aff}. Moreover, the Hilbert space $L^2(\widehat{\mathbb{R}}, d\xi)$ is the orthogonal direct sum of these two subspaces:

$$L^2(\widehat{\mathbb{R}}, d\xi) = \widehat{\mathfrak{H}}_+(\widehat{\mathbb{R}}) \oplus \widehat{\mathfrak{H}}_-(\widehat{\mathbb{R}}). \tag{6.60}$$

In other words, we have here a complete decomposition of the representation $\widehat{U}(b, a)$ of the connected affine group G^+_{aff} and we write

$$\widehat{U}(b, a) = \widehat{U}_+(b, a) \oplus \widehat{U}_-(b, a). \tag{6.61}$$

In the inverse Fourier domain, the representation $U(b, a)$ similarly breaks up (because of the unitarity of the Fourier transform) into two irreducible representations, $U_+(b, a)$ and $U_-(b, a)$, on the two subspaces

$$\mathfrak{H}_+(\mathbb{R}) = \{f \in L^2(\mathbb{R}, dx) \mid \widehat{f}(\xi) = 0, \ \forall \xi < 0\},$$
$$\mathfrak{H}_-(\mathbb{R}) = \{f \in L^2(\mathbb{R}, dx) \mid \widehat{f}(\xi) = 0, \ \forall \xi > 0\}, \tag{6.62}$$

of $L^2(\mathbb{R}, dx)$, respectively. These spaces are known as *Hardy spaces* [55,207,208]. Elements of $\mathfrak{H}_+(\mathbb{R})$ (respectively, $\mathfrak{H}_-(\mathbb{R})$) extend to functions analytic in the upper (respectively, lower) complex half-plane, and accordingly they are called upper (respectively, lower) analytic signals [Lyn82,Pap77].

6.2.1 Holomorphic wavelet transforms

It is an interesting fact that, for appropriate choices of a mother wavelets, the wavelet transforms of signals in the spaces $\mathfrak{H}_\pm(\mathbb{R})$ can (up to a factor) become holomorphic functions. We study this property in some detail in this section. For any $v \geqslant 0$, consider the mother wavelet $\widehat{\psi} \in \widehat{\mathfrak{H}}_+(\widehat{\mathbb{R}})$,

$$\widehat{\psi}(\xi) = \begin{cases} \left[\frac{2^v}{\pi \ \Gamma(v+1)}\right]^{\frac{1}{2}} \xi^{\frac{v+1}{2}} \ e^{-\xi}, & \text{for } 0 < \xi < \infty, \\ 0, & \text{otherwise}, \end{cases}$$
$$\|\widehat{C\psi}\|^2 = 1, \tag{6.63}$$

$\Gamma(v + 1)$ being the usual Gamma function. The wavelets for this vector have the form

$$\widehat{\psi}_{b,a}(\xi) = \left[\frac{2^v}{\pi \ \Gamma(v+1)}\right]^{\frac{1}{2}} a^{1+\frac{v}{2}} \ \xi^{\frac{v+1}{2}} \ e^{-i\xi\overline{z}}, \qquad \text{where } z = b + ia, \tag{6.64}$$

and the reproducing kernel is

$$K_\psi(b, a \, ; b', a') = \langle \widehat{\psi}_{b,a} \mid \widehat{\psi}_{b',a'} \rangle = \frac{2^v \ (v+1)}{i^{2+v} \ \pi} \frac{(aa')^{1+\frac{v}{2}}}{(\overline{z}' - z)^{2+v}}. \tag{6.65}$$

Computing the wavelet transform of a signal $\widehat{s} \in \mathfrak{H}_+(\mathbb{R})$,

$$S(b, a) = \langle \widehat{\psi}_{b,a} \mid \widehat{s} \rangle = \left[\frac{2^v}{\pi \ \Gamma(v+1)}\right]^{\frac{1}{2}} a^{1+\frac{v}{2}} \int_0^\infty d\xi \ \xi^{\frac{1+v}{2}} \ e^{i\xi z} \ \widehat{s}(\xi). \tag{6.66}$$

This, apart from the factor $a^{1+\frac{v}{2}}$, is a holomorphic function of $z = b + ia$. Indeed, writing this function as

$$F(z) = \left[\frac{2^v}{\pi \ \Gamma(v+1)}\right]^{-\frac{1}{2}} a^{-(1+\frac{v}{2})} S(b, a)$$

$$= \int_0^\infty d\xi \ e^{i\xi z} \ \xi^{\frac{1+v}{2}} \ \widehat{s}(\xi), \qquad z = b + ia, \tag{6.67}$$

we have,

$$| F(z) |^2 = \left| \int_0^\infty d\xi \; \xi^{\frac{1+\nu}{2}} \, e^{i\xi z} \, \widehat{s}(\xi) \right|^2 = \left| \int_0^\infty d\xi \; \xi^{\frac{1+\nu}{2}} \, e^{-\xi a} e^{i\xi b} \, \widehat{s}(\xi) \right|^2$$

$$\leqslant \|\widehat{s}\|^2 \int_0^\infty d\xi \; \xi^{1+\nu} \, e^{-2\xi a}, \qquad \text{by the Cauchy–Schwarz inequality}$$

$$= \frac{\|\widehat{s}\|^2 \, \Gamma(\nu + 2)}{(2a)^{2+\nu}}.$$

Thus, the convergence of the integral representing $F(z)$ is uniform in any bounded open set containing z and differentiation with respect to it, under the integral sign, is permissible, implying that F is holomorphic in z, on the complex, upper half plane which we identify with $\mathbb{H}_+ = \mathbb{R} \times \mathbb{R}_+^*$ (where, $\mathbb{R}_+^* = (0, \infty)$). We shall call $F(z)$ a *holomorphic wavelet transform*. Furthermore, since

$$\iint_{\mathbb{H}_+} \frac{db \, da}{a^2} \; | S(b, a) |^2 = \iint_{\mathbb{H}_+} d\mu_\nu(z, \overline{z}) \; | F(z) |^2,$$

where we have introduced the measure,

$$d\mu_\nu(z, \overline{z}) = \frac{(2a)^\nu}{\pi \, \Gamma(\nu + 1)} \, db \, da, \tag{6.68}$$

the set of all holomorphic wavelet transforms constitutes a closed subspace of the Hilbert space $L^2(\mathbb{H}_+ d\mu_\nu)$ of functions supported on the upper half plane. We denote this subspace by $\mathfrak{H}_{\text{hol}}^\nu$ and note that it is also a reproducing kernel Hilbert space, with reproducing kernel

$$K_{\text{hol}}^\nu(z, \; \overline{z}') = \frac{\Gamma(\nu + 2)}{i^{2+\nu}} \, \frac{1}{(z - \overline{z}')^{2+\nu}}. \tag{6.69}$$

One has indeed

$$\iint_{\mathbb{H}_+} d\mu_\nu(z, \overline{z}) \, K_{\text{hol}}^\nu(z, \; \overline{z}') \, F(z')$$

$$= \frac{2^\nu \, (\nu + 1)}{i^{2+\nu}} \iint_{\mathbb{H}_+} a^\nu \, da \, db \; \frac{F(z')}{(z - \overline{z}')^{2+\nu}}$$

$$= F(z).$$

The vectors $\eta_{\overline{z}}, \; z \in \mathbb{H}_+$, with

$$\eta_{\overline{z}}(z') = K_{\text{hol}}^\nu(z', \; \overline{z}), \tag{6.70}$$

which are the *holomorphic wavelets*, are again overcomplete in $\mathfrak{H}_{\text{hol}}^\nu$ and satisfy the resolution of the identity:

$$\iint_{\mathbb{H}_+} d\mu_\nu(z, \overline{z}) \; | \eta_{\overline{z}} \rangle \langle \eta_{\overline{z}} | = I_{\text{hol}}^\nu \quad (= \text{ identity operator of } \mathfrak{H}_{\text{hol}}^\nu). \tag{6.71}$$

There is also the holomorphic representation of G_{aff}^+ on $\mathfrak{H}_{\text{hol}}^\nu$, unitarily equivalent to $U_+(b, a)$. Denoting this by $U_{\text{hol}}^\nu(b, a)$, we easily compute its action:

$$(U_{\text{hol}}^\nu(b, a)F)(z) = a^{-(1+\frac{\nu}{2})} \, F\left(\frac{z - b}{a}\right). \tag{6.72}$$

The appearance of the holomorphic Hilbert spaces of wavelet transforms is remarkable in many ways. First of all, their existence is related to a geometrical property of the half plane $\mathbb{H}_+$, which is a differential manifold with a *complex Kähler structure*. This means, from a physical point of view, that it has all the properties of being a phase space of a classical mechanical system and, additionally, that this phase space can be given a complex structure (consistent with its geometry). In particular, it has a metric and a preferred differential two-form, which gives rise to the invariant measure $d\mu$ and using which classical mechanical quantities, such as Poisson brackets, may be defined. We will not go into the details of this here, but only point out the existence of a potential function in this context. Consider the function,

$$\Phi(z, \bar{z}') = -\log[-(z - \bar{z}')^2]. \tag{6.73}$$

This function is called a *Kähler potential* for the space $\mathbb{H}_+$ and it generates all the interesting quantities characterizing its geometry, such as the invariant two-form and the invariant measure. Indeed, we immediately verify that

$$e^{(1+\frac{\nu}{2}) \, \Phi(z, \bar{z}')} = \frac{i^{2-\nu}}{\Gamma(\nu + 2)} \, K_{\text{hol}}^\nu(z, \bar{z}'). \tag{6.74}$$

Next we define

$$\Omega(z, \bar{z}) = \frac{1}{i} \, \frac{\partial^2 \Phi(z, \bar{z})}{\partial z \, \partial \bar{z}} \, dz \wedge d\bar{z} = \frac{db \wedge da}{a^2}, \tag{6.75}$$

which is the invariant two-form (under the action of G_{aff}^+). This gives the invariant measure $d\mu$ of the group and furthermore,

$$e^{-(1+\frac{\nu}{2}) \, \Phi(z, \bar{z})} \, \Omega = 4(-1)^\nu \, (2a)^\nu \, db \wedge da, \tag{6.76}$$

from which follows the measure with respect to which the holomorphic functions $F(z)$ are square integrable and form a Hilbert space. (Recall that if u and v are two vectors in a vector space V, then $u \wedge v$ is the antisymmetric tensor product, $u \wedge v = u \otimes v - v \otimes u$. The differentials, dz, da, etc., are considered as being elements in the dual of the tangent space of the manifold – in this case G_{aff}^+ – at each point.) It ought to be emphasized here that $\mathfrak{H}_{\text{hol}}^\nu$ contains all holomorphic functions in $L^2(\mathbb{H}_+, \, d\mu_\nu)$. Note also that, in view of (6.67), any such function can be obtained by computing the Fourier transform of a function $f(\xi) = \xi^{\frac{1+\nu}{2}} \, \widehat{s}(\xi)$ and then analytically continuing it to the upper half plane, where $\widehat{s}$ is a signal in the Fourier domain, with support in $(0, \infty)$. Additionally, it ought to be noted that, for each $\nu > 0$, we get a family of holomorphic wavelet transforms, so that depending on the value of v, the same signal s can be represented by different holomorphic functions on phase space.

6.2.2   Matrix analysis of phase space

We have said earlier that the variables (b, a) parametrizing the space $\mathbb{H}_+$, and in terms of which the wavelet transform is written, should be identified as *phase space variables*. In this section, we proceed to elaborate on this. To begin, let us determine the *Lie algebra* of the group G^+_{aff}. This group has two subgroups, formed by matrices of the type

$$(e^t, 0) = \begin{pmatrix} e^t & 0 \\ 0 & 1 \end{pmatrix} \quad \text{and} \quad (1, t) = \begin{pmatrix} 1 & t \\ 0 & 1 \end{pmatrix}, \qquad t \in \mathbb{R},$$

and a general element of the group can be obtained by multiplying two such matrices. Consider now the following two matrices X_1, X_2, which generate the Lie algebra $\mathfrak{g}_{\text{aff}}$ of this group:

$$X_1 = \frac{d}{dt} (e^t, 0)\Big|_{t=0} = \begin{pmatrix} 1 & 0 \\ 0 & 0 \end{pmatrix}, \qquad X_2 = \frac{d}{dt} (1, t)\Big|_{t=0} = \begin{pmatrix} 0 & 1 \\ 0 & 0 \end{pmatrix}, \tag{6.77}$$

and satisfy the commutation relation,

$$[X_1, X_2] = X_1 X_2 - X_2 X_1 = X_2. \tag{6.78}$$

Exponentiating these matrices we get,

$$e^{(\log a) X_1} = \begin{pmatrix} a & 0 \\ 0 & 1 \end{pmatrix}, \qquad e^{b X_2} = \begin{pmatrix} 1 & b \\ 0 & 1 \end{pmatrix}. \tag{6.79}$$

The Lie algebra of the group is the two-dimensional vector space spanned by X_1 and X_2 and equipped with the commutation relation (6.78). A general element in the Lie algebra can be written as,

$$X = x^1 X_1 + x^2 X_2 = \begin{pmatrix} x^1 & x^2 \\ 0 & 0 \end{pmatrix}, \qquad x^1, x^2 \in \mathbb{R}. \tag{6.80}$$

Any group element can be obtained by exponentiating a suitable element of this Lie algebra. This is made clear if we write

$$(b, a) = e^X = \begin{pmatrix} e^{x^1} & \frac{x^2}{x^1}(e^{x^1} - 1) \\ 0 & 1 \end{pmatrix} = \begin{pmatrix} a & b \\ 0 & 1 \end{pmatrix}. \tag{6.81}$$

and note the inverse map from the group to the algebra:

$$X = \log(b, a) = x^1 X_1 + x^2 X_2, \qquad x^1 = \log a, \qquad x^2 = \frac{b \log a}{a - 1}. \tag{6.82}$$

Since every $X \in \mathfrak{g}_{\text{aff}}$ is mapped to an element $(b, a) \in G_{\text{aff}}^+$ by the exponential map (6.81), we identify the domain of this map with the full real plane and use $\vec{x} = (x^1, x^2) \in \mathbb{R}^2$ as the coordinates for the elements of the Lie algebra.

A group has a natural action on its Lie algebra, called the *adjoint action*. For $(b, a) \in G_{\text{aff}}^+$ this action, which we denote by $\text{Ad}_{(b,a)}$, is defined by

$$\text{Ad}_{(b,a)} X = (b, a) X (b, a)^{-1} = \begin{pmatrix} x^1 & -bx^1 + ax^2 \\ 0 & 0 \end{pmatrix}. \tag{6.83}$$

The matrix of this transformation, computed in the basis $\{X_1, X_2\}$, and acting on the vectors $\vec{x} = \begin{pmatrix} x^1 \\ x^2 \end{pmatrix} \in \mathbb{R}^2$ is

$$M(b, a) = \begin{pmatrix} 1 & 0 \\ -b & a \end{pmatrix}. \tag{6.84}$$

As a vector space, the Lie algebra has a dual space, which we denote by $\mathfrak{g}_{\text{aff}}^*$. On it the adjoint action of G_{aff}^+ induces, by duality, the *coadjoint action*, denoted $\text{Ad}_{(b,a)}^\sharp$. To compute this action we take the dual basis $\{X^{*1}, X^{*2}\}$ in $\mathfrak{g}_{\text{aff}}^*$ and write a general element in it as $X^* = \gamma_1 X^{*1} + \gamma_2 X^{*2}$. We identify X^* with the vector $\vec{\gamma} = \begin{pmatrix} \gamma_1 \\ \gamma_2 \end{pmatrix} \in \mathbb{R}^2$. On such vectors, the coadjoint action of the group is given (by definition) by the transposed inverse of the matrix $M(b, a)$. We write

$$M^\sharp(b, a) = \left[M(b, a)^{-1} \right]^T = \begin{pmatrix} 1 & ba^{-1} \\ 0 & a^{-1} \end{pmatrix}. \tag{6.85}$$

The determinants of these transformations are related as

$$\det[\text{Ad}_{(b,a)}] = a = \det[\text{Ad}_{(b,a)}^\sharp]^{-1}. \tag{6.86}$$

Explicitly, a point $\vec{\gamma} \in \mathbb{R}^2$ transforms under the coadjoint action as,

$$\vec{\gamma} \mapsto \vec{\gamma}' = M^\sharp(b, a)\vec{\gamma} = \begin{pmatrix} \gamma_1 + ba^{-1}\gamma_2 \\ a^{-1}\gamma_2 \end{pmatrix}, \tag{6.87}$$

and for fixed $\vec{\gamma}_0 \in \mathbb{R}^2$, its orbit under G_{aff}^+ is the set:

$$\mathcal{O}_{\vec{\gamma}_0}^* = \{ \vec{\gamma} = M^\sharp(b, a)\vec{\gamma}_0 \mid (b, a) \in G_{\text{aff}}^+ \} \subset \mathbb{R}^2. \tag{6.88}$$

Such an orbit is called a *coadjoint orbit* of the group G_{aff}^+. Orbits of different points either coincide entirely or are disjoint. In this way, the entire dual space $\mathfrak{g}_{\text{aff}}^*$ becomes the union over disjoint coadjoint orbits. Indeed, we easily establish the following orbit structure:

(i) The orbit of the vector $\begin{pmatrix} 0 \\ 1 \end{pmatrix}$,

$$\mathcal{O}_+^* = \{ \vec{\gamma} = \begin{pmatrix} \gamma_1 \\ \gamma_2 \end{pmatrix} \in \mathbb{R}^2 \,|\, \gamma_2 > 0 \} = \mathbb{R} \times \mathbb{R}_+^*. \tag{6.89}$$

(ii) The orbit of the vector $\begin{pmatrix} 0 \\ -1 \end{pmatrix}$ with the same matrices,

$$\mathcal{O}_-^* = \{ \vec{\gamma} = \begin{pmatrix} \gamma_1 \\ \gamma_2 \end{pmatrix} \in \mathbb{R}^2 \,|\, \gamma_2 < 0 \} = \mathbb{R} \times \mathbb{R}_-^*. \tag{6.90}$$

(iii) The orbits of vectors $\begin{pmatrix} \alpha \\ 0 \end{pmatrix}$, one for each $\alpha \in \mathbb{R}$. Such orbits are singletons, $\begin{pmatrix} \alpha \\ 0 \end{pmatrix}$, and we denote them by $\mathcal{O}_\alpha^*$.

(iv) It is obvious that

$$\mathcal{O}_+^* \cup \mathcal{O}_-^* \cup_{\alpha \in \mathbb{R}} \mathcal{O}_\alpha^* = \mathbb{R}^2. \tag{6.91}$$

We note also that each of the two orbits (i) and (ii) is homeomorphic to the group itself. Consider the first one of these, $\mathcal{O}_+^* = \mathbb{H}_+ = \mathbb{R} \times \mathbb{R}_+^*$. As a manifold, this space has a symplectic structure, i.e., there exists a preferred nondegenerate, antisymmetric, closed differential two-form on it, which is invariant under the coadjoint action. Indeed, it is trivially verified that the two-form

$$\Omega(\vec{\gamma}) = \frac{d\gamma_1 \wedge d\gamma_2}{\gamma_2} \tag{6.92}$$

satisfies this condition. This also gives the invariant measure on this phase space, in these coordinates. We look upon $\mathcal{O}_+^*$ as an abstract differential manifold, with (γ_1, γ_2) representing a particular choice of coordinates. In this context, we study two other possible choices of coordinates and see how the two-form (6.92) appears in these new coordinates.

As the first of these coordinate transformations, we write

$$\vec{\gamma} \mapsto \vec{\eta} = \begin{pmatrix} q \\ p \end{pmatrix} = \begin{pmatrix} \gamma_2^{-1}\gamma_1 \\ \gamma_2 \end{pmatrix}, \tag{6.93}$$

which maps $\mathbb{R} \times \mathbb{R}_+^*$ onto itself. Under the coadjoint action of the group element (b, a), these coordinates transform as $(q, p) \mapsto (q', p')$ with,

$$q' = aq + b, \qquad p' = a^{-1}p.$$

The invariant two-form is simply

$$\Omega(\vec{\eta}) = dq \wedge dp, \tag{6.94}$$

giving also the invariant Liouville measure in these coordinates. We call these coordinates the *canonical* or *Darboux* coordinates of the phase space $\mathcal{O}_+^*$, in view of the form, familiar from classical Hamiltonian mechanics, assumed by the two-form Ω. This also allows us to identify the variables q and p as position and momentum, respectively. We also immediately recognize the coadjoint action as inducing canonical transformations on phase space.

As the second coordinate transformation, we choose

$$\vec{\gamma} \mapsto \vec{\xi} = \begin{pmatrix} \xi_1 \\ \xi_2 \end{pmatrix} = \begin{pmatrix} \gamma_2^{-1}\gamma_1 \\ \gamma_2^{-1} \end{pmatrix}, \tag{6.95}$$

which again maps $\mathbb{R} \times \mathbb{R}_+^*$ onto itself. Under the coadjoint action, these coordinates change in the manner:

$$\xi_1' = a\xi_1 + b, \qquad \xi_2' = a\xi_2, \tag{6.96}$$

with the invariant two-form now being

$$\Omega(\vec{\xi}) = \frac{d\xi_1 \wedge d\xi_2}{\xi_2^2}. \tag{6.97}$$

The invariant measure arising from this should be compared to the left invariant measure, $d\mu$, of the group G_{aff}^+ (see (6.11)). Indeed, if we identify ξ_1, ξ_2 with group parameters, then the transformation (6.96) is just the left multiplication in the group: $(\xi_1', \xi_2') = (b, a)(\xi_1, \xi_2)$. It is because of the possibility of coordinatizing the orbit $\mathcal{O}_+^*$ in this particular way, that we may legitimately look upon the group elements themselves as phase space variables and the wavelet transform as a transform on phase space.

Similar considerations apply to the other nontrivial orbit $\mathcal{O}_-^*$. We shall see in the next chapter that the two-dimensional wavelet transform can also be analyzed in a completely analogous manner. Before concluding this section, we should point out one crucial fact about the orbit structure of G_{aff}^+. The composition rule (6.6) equips this group with the structure of a *semidirect product*, $\mathbb{R} \rtimes \mathbb{R}_+^*$, in which the subgroup $\mathbb{R}_+^*$ has an action on the subgroup $\mathbb{R}$, in this case by simple multiplication. This action induces an action (again by multiplication, $x \mapsto \alpha x$, for $x \in \widehat{\mathbb{R}}$ and $\alpha \in \mathbb{R}_+^*$), on the dual space $\widehat{\mathbb{R}}$ (which we naturally identify with $\mathbb{R}$). Under this action the dual space splits up into three orbits, the open half-spaces $\mathbb{R}_+^*$, $\mathbb{R}_-^*$ and the singleton $\{0\}$. The first two orbits are *open* and *free*, in the sense that they are both homeomorphic to the subgroup $\mathbb{R}_+^*$ itself and the map identifying the subgroup with the orbit is continuous and open. Furthermore, the union of the two open free orbits is dense in $\widehat{\mathbb{R}}$. The fact that the orbits $\mathcal{O}_\pm^*$ are open and free, is what leads to the representations $U_\pm(b, a)$ being square integrable (see, for example, [Ali00] for a detailed discussion of this point). Geometrically, the coadjoint orbits $\mathcal{O}_\pm^*$ (of the whole group G_{aff}^+) are the *cotangent bundles* of the orbits $\mathbb{R}_\pm^*$ (of the subgroup $\mathbb{R}_+^*$).

6.3   The case of Gabor wavelets

For the sake of completeness and comparison, we briefly look at the case of Gabor wavelets, or short-time Fourier transforms, for they too can be understood in group theoretical and phase space terms which are similar to those of the standard wavelets considered above. The one difference here is that the phase space of signals is not the entire group in question (the Weyl–Heisenberg group) and square integrability of the representation, giving rise to the Gabor wavelets, has also to be understood in the light of this fact. Another remarkable difference here is that for building Gabor type wavelets, any vector in the Hilbert space of signals can be used as a mother wavelet – the admissibility condition is trivial.

The Gabor transform was introduced in (1.3). It was obtained by taking a window function $\psi \in L^2(\mathbb{R}, dx)$, translating by b (in time) and modulating it in frequency $1/a$, to obtain the Gabor wavelets or *gaborettes*. It had the form:

$$\psi_{b,a}(x) = e^{i(x-b)/a}\,\psi(x-b).$$

The Gabor transform of a signal s is then defined in the same way as the wavelet transform:

$$S(b, a) = \langle \psi_{b,a} \mid s \rangle = \int_{\mathbb{R}} dx\, e^{-i(x-b)/a}\, \overline{\psi(x-b)}\, \phi(x). \tag{6.98}$$

It is a remarkable fact that the Gabor transform, and the 1-D continuous wavelet transform are built on exactly the same pattern (and this similarity persists in higher dimensions). Both are based on transformation groups of signals in $L^2(\mathbb{R}, dx)$, the action of the group being implemented by unitary operators, $s \mapsto U(b, a)s$:

- for the Gabor transform, the group is the Weyl–Heisenberg group G_{WH}, acting as in (1.3);
- for the continuous wavelet transform it is the $ax + b$ or affine group, acting as in (1.4),

and in fact the translation parts are identical in the two cases.

6.3.1   Group theoretical analysis

In order to understand the Gabor transform group theoretically, let us first rewrite the gaborettes by adopting a slightly different notation and also introducing an additional phase variable. We write q for b and p for $1/a$ and denote the phase variable by θ:

$$\psi_{\theta,q,p}(x) = e^{i\theta}\, e^{ip(x-\frac{q}{2})}\, \psi(x-q), \qquad \theta \in \mathbb{R}. \tag{6.99}$$

The gaborettes would then correspond to setting $\theta = -pq/2$. Writing

$$\psi_{\theta,q,p} = U(\theta, q, p)\psi, \tag{6.100}$$

an easy computation shows that $U(\theta, q, p)$ defines a unitary operator on the space of signals. Computing the effect of two such operators in succession, on a vector ψ yields the composition rule,

$$U(\theta_1, q_1, p_1)\, U(\theta_2, q_2, p_2) = U(\theta_1 + \theta_2 + \xi((q_1, p_1); (q_2, p_2)), \; q_1 + q_2, \; p_1 + p_2), \tag{6.101}$$

where,

$$\xi((q_1, p_1); (q_2, p_2)) = \frac{1}{2}(p_1 q_2 - p_2 q_1), \tag{6.102}$$

which in fact defines the multiplication rule between elements of the Weyl–Heisenberg group G_{WH}. The function ξ is called a *multiplier*. The Weyl–Heisenberg group is a three-parameter group, homeomorphic to $\mathbb{R}^3$. An arbitrary element g of G_{WH} is of the form

$$g = (\theta, q, p), \quad \theta \in \mathbb{R}, \quad (q, p) \in \mathbb{R}^2,$$

and the multiplication law in the group is

$$g_1 g_2 = (\theta_1 + \theta_2 + \xi((q_1, p_1); (q_2, p_2)), \; q_1 + q_2, \; p_1 + p_2). \tag{6.103}$$

The multiplier ξ equips the group with the structure of a *central extension*, of the group of translations of $\mathbb{R}^2$, corresponding to phase space translations (i.e., translations q and modulations p). This just means that the subgroup consisting of the elements $\Theta = \{g = (\theta, 0, 0) \mid \theta \in \mathbb{R}\}$, is the *center* of the group G_{WH}, i.e., these elements commute with every element g in the group, $g(\theta, 0, 0) = (\theta, 0, 0)g$, and it is the introduction of ξ which extends the commutative group $\mathbb{R}^2$, with elements (q, p), into the noncommutative Weyl–Heisenberg group. The Weyl–Heisenberg group is *unimodular*, the measure $d\mu = d\theta\, dq\, dp$ being invariant under both the left action, $(\theta, q, p) \mapsto (\theta_0, q_0, p_0)(\theta, q, p)$, and the right action, $(\theta, q, p) \mapsto (\theta, q, p)(\theta_0, q_0, p_0)$.

As with the affine group, the Weyl–Heisenberg group also has a matrix realization. It is given by the 4×4 matrices

$$(\theta, q, p) = \begin{pmatrix} 1 & \frac{1}{2}\vec{\zeta}^{\,T}\omega & \theta \\ \vec{0} & \mathbb{I}_2 & \vec{\zeta} \\ 0 & \vec{0}^{\,T} & 1 \end{pmatrix}, \qquad \omega = \begin{pmatrix} 0 & -1 \\ 1 & 0 \end{pmatrix}, \tag{6.104}$$

with

$$\theta \in \mathbb{R}, \quad \vec{\zeta} = \begin{pmatrix} q \\ p \end{pmatrix} \in \mathbb{R}^2, \quad \vec{0} = \begin{pmatrix} 0 \\ 0 \end{pmatrix}, \quad \mathbb{I}_2 = \begin{pmatrix} 1 & 0 \\ 0 & 1 \end{pmatrix}.$$

Also, as for the affine group, we can compute the generators of the Lie algebra of G_{WH}, by considering the one-parameter subgroups of elements of the type, $g_1(t) = (t, 0, 0)$, $g_2(t) = (0, t, 0)$ and $g_3(t) = (0, 0, t)$, with t ranging through $\mathbb{R}$. Computing

$$X_i \equiv \frac{d}{dt} g_i(t)\Big|_{t=0},$$

we obtain the three elements which generate the Lie algebra, $\mathfrak{g}_{\text{WH}}$:

$$X_0 = \begin{pmatrix} \vec{0}^T & 1 \\ \mathbb{O} & \vec{0} \end{pmatrix}, \quad X_1 = \begin{pmatrix} -\frac{1}{2}e_3^T & 0 \\ \mathbb{O} & e_1 \end{pmatrix}, \quad X_2 = \begin{pmatrix} \frac{1}{2}e_2^T & 0 \\ \mathbb{O} & e_2 \end{pmatrix}, \tag{6.105}$$

where

$$\vec{0} = \begin{pmatrix} 0 \\ 0 \\ 0 \end{pmatrix}, \quad e_1 = \begin{pmatrix} 1 \\ 0 \\ 0 \end{pmatrix}, \quad e_2 = \begin{pmatrix} 0 \\ 1 \\ 0 \end{pmatrix}, \quad e_3 = \begin{pmatrix} 0 \\ 0 \\ 1 \end{pmatrix},$$

and $\mathbb{O}$ is the 3×3 zero matrix. The matrices X_i satisfy the commutation relations

$$[X_0, X_1] = [X_0, X_2] = 0, \quad [X_1, X_2] = X_0. \tag{6.106}$$

A general element of $\mathfrak{g}_{\text{WH}}$ can be written as

$$X = x^0 X_0 + x^1 X_1 + x^2 X_2, \quad x^1, x^2, x^3 \in \mathbb{R};$$

the commutation relations (6.106) then define a *Lie bracket*, $[X, Y] = XY - YX$, between any two elements $X, Y \in \mathfrak{g}_{\text{WH}}$. Next, noting that for any $X \in \mathfrak{g}_{\text{WH}}$, $(X)^2$ is the null matrix (which, in group theoretical terms, is stated by saying that the group G_{WH} is *nilpotent*), we see that

$$e^X = (x^0, x^1, x^2) = \mathbb{I}_4 + X \in G_{\text{WH}}. \tag{6.107}$$

Thus, the group and the Lie algebra can be given the same parametrization.

In order to make the connection with gaborettes, we need to find unitary irreducible representations of G_{WH} on the Hilbert space of signals, $L^2(\mathbb{R}, dx)$. That is, we need to find a set of unitary operators, $U(\theta, q, p)$, for all $(\theta, q, p) \in G_{\text{WH}}$, which realize a group homomorphism, are stable under inverse taking and map the identity element $(0, 0, 0)$ of the group to the identity operator on $L^2(\mathbb{R}, dx)$. But this is already done by the operators defined in (6.100), so that $U(\theta, q, p)$ realize a unitary representation of the group. The fact that it is also irreducible can be proved in much the same way in which we proved irreducibility for the representation of the affine group in Section 6.1.1. More generally, it can be shown that any (nondegenerate) unitary irreducible representation of G_{WH} is characterized by a real number $\lambda \neq 0$ and may be realized on the Hilbert space $L^2(\mathbb{R}, dx)$ by the operators $U^\lambda(\theta, q, p)$:

$$(U^\lambda(\theta, q, p)s)(x) = e^{i\lambda\theta} e^{i\lambda p(x-\frac{q}{2})} s(x - q), \quad \phi \in L^2(\mathbb{R}, dx). \tag{6.108}$$

Two representations $U^\lambda(\theta, q, p)$ and $U^{\lambda'}(\theta, q, p)$ are unitarily inequivalent if $\lambda \neq \lambda'$. For the construction of gaborettes we shall mostly work with the representation for which $\lambda = 1$ and denote it simply by U instead of U^1. (The other representations will be used in our discussion of holomorphic Gabor transforms below.)

A general element (θ, q, p) in G_{WH} can be factorized as

$$(\theta, q, p) = (0, q, p)(\theta, 0, 0).$$

Thus, the quotient space G_{WH}/Θ (where the phase is factored out) is identifiable with $\mathbb{R}^2$. It is this space which plays the rôle of a phase space for the Weyl–Heisenberg group. We parametrize an element in G_{WH}/Θ by $(q, p) \in \mathbb{R}^2$ and since for a fixed element $(\theta_0, q_0, p_0) \in G_{\text{WH}}$ we again have the factorization

$$(\theta_0, q_0, p_0)(0, q, p) = (0_0, q + q_0, p + p_0)(\theta_0 + \frac{1}{2}(p_0 q - p q_0), 0, 0),$$

the action of the group on the quotient space is simply given by $(q, p) \mapsto (q + q_0, p + p_0)$. The invariant measure under this action is just the Lebesgue measure $dq\, dp$.

For any $\psi \in L^2(\mathbb{R}, dx)$, let us define the vectors

$$\psi_{q,p} = U(-\frac{qp}{2}, q, p)\psi, \qquad (q, p) \in G_{\text{WH}}/\Theta. \tag{6.109}$$

These are just gaborettes, expressed in the parameters q, p. The corresponding Gabor transform of a signal s is

$$S(q, p) = \langle \psi_{q,p} \mid s \rangle = \int_{-\infty}^{\infty} dx\, e^{-ip(x-q)}\, \overline{\psi(x - q)}\, s(x). \tag{6.110}$$

Once again, computing the total energy of the transform,

$$E(S) = \iint_{\mathbb{R}^2} dq\, dp\, |S(q, p)|^2 = 2\pi\, \|\psi\|^2\, \|s\|^2, \tag{6.111}$$

which should be compared to (6.19). From this there follows the resolution of the identity (exactly as in the derivation of (6.20))

$$\frac{1}{2\pi \|\psi\|^2} \iint_{\mathbb{R}^2} dq\, dp\, |\psi_{q,p}\rangle\langle\psi_{q,p}| = I, \tag{6.112}$$

and the reconstruction formula (see (6.21)),

$$s(x) = \frac{1}{2\pi \|\psi\|^2} \iint_{\mathbb{R}^2} dq\, dp\, S(q, p)\psi_{q,p}(x), \quad \text{for almost all } x. \tag{6.113}$$

In the physical literature, gaborettes, and indeed also wavelets, are referred to as *coherent states*, of their respective groups [Kla85, Ali00, 6]. Note, however, that unlike the wavelets $\psi_{b,a}$, which were defined for all elements (b, a) of the affine group, the gaborettes $\psi_{q,p}$ are only defined for points in the quotient space G_{WH}/Θ. Moreover, the two (equivalent) conditions (6.111) and (6.112) imply that the representation $U(\theta, q, p)$ is square integrable with respect to this space:

$$\iint_{G_{\text{WH}}/\Theta} dq\, dp\, |\langle U(-\frac{qp}{2}, q, p)\psi \mid \psi \rangle|^2 < \infty, \qquad \forall \psi \in \mathfrak{H}. \tag{6.114}$$

Again, one ought to emphasize here that the above admissibility condition is satisfied by all vectors ψ in the signal space, but that it is defined with respect to a quotient space. Indeed, it also follows from the above that, for any vector ψ in the Hilbert space of signals,

$$\iiint_{G_{WH}} dq\, dp\, d\theta \; |\langle U(\theta, q, p)\psi | \psi \rangle|^2 = \infty,$$

so that there is no vector which is admissible with respect to the entire group. We shall return to the subject of building wavelet transforms on general quotient spaces in the next chapter (see Section 7.1.5).

To complete this cycle of properties, we could again verify that the space $\mathfrak{H}_\psi$ of all Gabor transforms, corresponding to a particular window function ψ, is a closed subspace of $L^2(G_{WH}, dq\, dp)$, which is also a Hilbert space with a reproducing kernel:

$$K_\psi(q, p; q', p') = \frac{1}{2\pi \|\psi\|^2} \, \langle \psi_{q,p} \mid \psi_{q',p'} \rangle. \tag{6.115}$$

This relation is the exact analog of (6.26). The complete decomposition of the space $L^2(G_{WH}/\Theta, dq\, dp)$, of all Gabor transforms, could be worked out in the manner of (6.41), with very little change in the derivation.

6.3.1.1 Holomorphic Gabor wavelets

It is possible to construct spaces of holomorphic Gabor transforms and associated holomorphic gaborettes, in much the same way as we constructed holomorphic wavelet transforms in Section 6.2.1. We show below the existence of one such space and later we shall indicate how others may be obtained. However, in order to do so, it is first necessary to modify somewhat the definition of the Gabor transform. We proceed by first choosing the window function

$$\psi(x) = (\pi)^{-\frac{1}{4}}\, e^{-\frac{x^2}{2}}, \qquad \|\psi\|^2 = 1. \tag{6.116}$$

Next, we define the modified gaborettes,

$$\psi_{(0,q,p)}(x) = (U(0, q, p)\psi)(x) = (\pi)^{-\frac{1}{4}}\, e^{i(x-\frac{q}{2})p}\, e^{-\frac{(x-q)^2}{2}} = e^{i\frac{qp}{2}} \psi_{q,p} \tag{6.117}$$

and use them to define the modified Gabor transform of a signal s:

$$\widetilde{S}(q, p) = \langle \psi_{(0,q,p)} \mid s \rangle = e^{-i\frac{qp}{2}}\, S(q, p). \tag{6.118}$$

Then $\widetilde{S}$ obviously satisfies the finiteness of energy condition and moreover, the modified gaborettes also satisfy the resolution of the identity. Thus, signal reconstruction is possible using these modified gaborettes as well. Introducing the complex variable $z = q - ip$, it is easy to verify that

$$\widetilde{S}(q, p) = \frac{e^{-\frac{|z|^2}{4}}}{\pi^{\frac{1}{4}}} \int_{-\infty}^{\infty} dx\, e^{-\frac{1}{2}(x-z)^2}\, e^{\frac{z^2}{4}}\, s(x) \equiv \widetilde{S}(z). \tag{6.119}$$

Writing

$$\widetilde{S}(z) = \frac{e^{-\frac{|z|^2}{4}}}{\sqrt{2\pi}} \, F(z), \tag{6.120}$$

we see that $F(z)$ is an entire analytic function of z. These functions constitute the holomorphic Hilbert space $\mathfrak{H}_{\mathrm{hol}} = L^2(\mathbb{C}, d\nu(z, \overline{z}))$, where $d\nu$ is the measure

$$d\nu(z, \overline{z}) = e^{-\frac{|z|^2}{2}} \, \frac{dq \, dp}{2\pi}.$$

This Hilbert space is also a reproducing kernel Hilbert space, with the kernel given by

$$K_{\mathrm{hol}}(z, \overline{z}') = e^{\frac{z\overline{z}'}{2}}. \tag{6.121}$$

We call the $\psi_{(0,q,p)}$ holomorphic gaborettes and the functions $F(z)$ holomorphic Gabor transforms. Other spaces of holomorphic Gabor transforms can now be similarly constructed by replacing the window function (6.116) by a Gaussian with standard deviation λ^{-1}, $\lambda > 0$ and using the representation (6.108) of the Weyl–Heisenberg group.

Consider the window function

$$\psi(x) = \left[\frac{\lambda^2}{\pi} \right]^{\frac{1}{4}} e^{-\frac{\lambda^2 x^2}{2}}, \qquad \|\psi\|^2 = 1, \quad \lambda > 0, \tag{6.122}$$

and define the generalized gaborettes

$$\psi_{(0,q,p)}(x) = (U^\lambda(0, q, p)\psi)(x) = \left[\frac{\lambda^2}{\pi} \right]^{\frac{1}{4}} e^{i\lambda(x - \frac{q}{2})p} \, e^{-\frac{\lambda^2(x-q)^2}{2}}, \tag{6.123}$$

where U^λ is the unitary representation of the Weyl–Heisenberg group defined in (6.108). It can then be verified that the resolution of the identity

$$\frac{\lambda}{2\pi} \iint_{\mathbb{R}^2} dq \, dp \, |\psi_{(0,q,p)}\rangle \langle \psi_{(0,q,p)}| = I, \tag{6.124}$$

holds. The generalized Gabor transform is now

$$\widetilde{S}(q, p) = \langle \psi_{(0,q,p)} | s \rangle, \qquad s \in L^2(G_{\mathrm{WH}}/\Theta, dq \, dp), \tag{6.125}$$

which, by virtue of (6.124), enjoys the finiteness of energy condition

$$\|\widetilde{S}\|^2 = \frac{2\pi}{\lambda} \, \|s\|^2.$$

Introducing the complex variable

$$z = \sqrt{\lambda} \, q - i \frac{p}{\sqrt{\lambda}},$$

(6.123) may be rewritten as:

$$\psi_{(0,q,p)}(x) = \psi_{(0,\overline{z})}(x) = \left[\frac{\lambda^2}{\pi} \right]^{\frac{1}{4}} e^{-\frac{\lambda|z|^2}{4}} \, e^{\frac{\lambda \overline{z}^2}{4} - \frac{\lambda}{2}(\sqrt{\lambda}x - \overline{z})^2}. \tag{6.126}$$

The generalized Gabor transform now becomes

$$\widetilde{S}(q, p) \equiv \widetilde{S}(z) = \left[\frac{\lambda^2}{\pi}\right]^{\frac{1}{4}} e^{-\frac{\lambda|z|^2}{4}} e^{\frac{\lambda z^2}{4}} \int_{-\infty}^{\infty} dx \, e^{-\frac{\lambda}{2}(\sqrt{\lambda}x - z)^2} s(x), \tag{6.127}$$

so that

$$F(z) = \left[\frac{2\pi}{\lambda}\right]^{\frac{1}{2}} e^{\frac{\lambda|z|^2}{4}} \widetilde{S}(z) = e^{\frac{\lambda z^2}{4}} \int_{-\infty}^{\infty} dx \, e^{-\frac{\lambda}{2}(\sqrt{\lambda}x - z)^2} s(x)$$

is the analytic Gabor transform and it then follows that

$$\iint_{G_{WH}/\Theta} dq \, dp \, |\widetilde{S}(q, p)|^2 = \int_{\mathbb{C}} d\nu_\lambda(z, \bar{z}) \, |F(z)|^2,$$

where we have introduced the measure

$$d\nu_\lambda(z, \bar{z}) = \frac{\lambda}{2\pi} e^{-\frac{\lambda|z|^2}{2}} dq \, dp.$$

Thus, F is a vector in a Hilbert space of entire analytic functions, which we denote by $\mathfrak{H}_{hol}^\lambda$ and note that this space of holomorphic Gabor transforms is a subspace of $L^2(\mathbb{C}, d\nu_\lambda)$. The Hilbert space $\mathfrak{H}_{hol}^\lambda$ has the reproducing kernel

$$K_{hol}^\lambda(z, \bar{z}') = e^{\frac{\lambda z \bar{z}'}{2}}, \tag{6.128}$$

so that, for any $F \in \mathfrak{H}_{hol}^\lambda$,

$$\int_{\mathbb{C}} d\nu_\lambda(z, \bar{z}) \, K_{hol}^\lambda(z, \bar{z}') F(z') = F(z).$$

Finally, the complex plane $\mathbb{C}$ is also a Kähler manifold, with potential function

$$\Phi(z, \bar{z}) = \frac{z\bar{z}'}{2},$$

and invariant two-form

$$\Omega(z, \bar{z}) = \frac{1}{i} \frac{\partial^2 \Phi(z, \bar{z})}{\partial z \, \partial \bar{z}} dz \wedge d\bar{z} = dq \wedge dp,$$

so that, once again, we verify the relations (see (6.74)–(6.76))

$$K_{hol}^\lambda(z, \bar{z}') = e^{\lambda \Phi(z, \bar{z}')},$$

and

$$e^{-\lambda \Phi(z, \bar{z})} \Omega = e^{-\frac{\lambda|z|^2}{2}} dq \wedge dp,$$

from which follows the measure $d\nu_\lambda$.

   Thus, group theoretically, the analysis of Gabor wavelets using the Weyl–Heisenberg group runs entirely parallel to the analysis of the 1-D CWT using the affine group, except for the following two differences. First, here every vector in $L^2(\mathbb{R}, dx)$ is admissible (the group theoretical reason being that G_{WH} is a unimodular group). Second, the square integrability of the representation $U(\theta, q, p)$ is not on the entire Weyl–Heisenberg group G_{WH} itself, but on the quotient G_{WH}/Θ of the group by its center Θ. However here, as with the affine group, it is the square integrability of the representation in question which leads to the finiteness of the energy of the Gabor transform and enables us to reconstruct the signal from its transform.

6.3.2 Phase space considerations

It is instructive to carry out a phase space analysis of the Weyl–Heisenberg group, for it sheds light on both the differences and the similarities between it and the affine group. In particular, it will clearly emerge why the phase space over which the Gabor transform is built is two-dimensional, although the group itself is a three-dimensional manifold. Let $(\theta, q, p) \in G_{\text{WH}}$ and $X \in \mathfrak{g}_{\text{WH}}$. We compute the adjoint action of the group on the Lie algebra, using (6.107) to get

$$x^{0'} = x^0 + px^1 - qx^2,$$
$$x^{1'} = x^1,$$
$$x^{2'} = x^2,$$

where

$$X' = \text{Ad}_{(\theta,q,p)}X = x^{0'}X_0 + x^{1'}X_1 + x^{2'}X_2$$
$$= (\theta, q, p)\,[x^0 X_0 + x^1 X_1 + x^2 X_2]\,(\theta, q, p)^{-1}.$$

Representing X by the vector in $\mathbb{R}^3$ with components x^0, x^1, x^2, the adjoint action is given by the matrix

$$M(\theta, q, p) = \begin{pmatrix} 1 & p & -q \\ 0 & 1 & 0 \\ 0 & 0 & 1 \end{pmatrix}. \tag{6.129}$$

Note that there is no dependence of M on the phase θ. The adjoint action on the dual vectors $\vec{\gamma}$ is effected by the matrices $M^{\sharp}(\theta, q, p)$, which are the transposed inverses of the matrices $M(\theta, q, p)$. Writing $\vec{\gamma}' = M^{\sharp}(\theta, q, p)\vec{\gamma}$, we get the dual transformation equations

$$\gamma_0' = \gamma_0,$$
$$\gamma_1' = -p\gamma_0 + \gamma_1,$$
$$\gamma_2' = q\gamma_0 + \gamma_2.$$

Thus, the orbits of vectors under the coadjoint action fall into two categories:

(i) The orbits of vectors of $\begin{pmatrix} \lambda \\ \vec{0} \end{pmatrix} \in \mathbb{R}^3$, $\vec{0} = \begin{pmatrix} 0 \\ 0 \end{pmatrix}$, one for each $\lambda \neq 0$:

$$\mathcal{O}_\lambda^* = \{\vec{\gamma} = \begin{pmatrix} \lambda \\ \vec{x} \end{pmatrix} \mid \vec{x} \in \mathbb{R}^2\}, \tag{6.130}$$

which are planes orthogonal to the γ_0-axis.

(ii) The orbits of vectors $\begin{pmatrix} 0 \\ \vec{\lambda} \end{pmatrix}$, $\vec{\lambda} \in \mathbb{R}^2$. These orbits are singletons consisting of the

vector $\begin{pmatrix} 0 \\ \vec{\lambda} \end{pmatrix}$ itself. We denote them by $\mathcal{O}_{\vec{\lambda}}^*$ and note that together they form a set

of Lebesgue measure zero in $\mathbb{R}^3$.

Clearly, the union of all the orbits is the entire dual space, $\mathfrak{g}_{WH}^*$, of the Lie algebra, now identified with $\mathbb{R}^3$. The nontrivial orbits $\mathcal{O}_\lambda^*$, $\lambda \neq 0$, can each be identified with the quotient space G_{WH}/Θ, since under the coadjoint action the phase subgroup Θ (consisting of elements of the type $(\theta, 0, 0)$) is stable. These are the (two-dimensional) phase spaces of the problem and from the general theory of group representations it is known that each unitary irreducible representation of G_{WH} is associated to one such orbit. As stated earlier, these representations $U^\lambda(\theta, q, p)$ can be realized as in (6.108). The invariant two form under the coadjoint action on the orbits $\mathcal{O}_\lambda^*$ is just $d\gamma_1 \wedge d\gamma_2$ and for $\lambda = 1$, this action is simple: $\gamma_1 \mapsto \gamma_1 - p$, $\gamma_2 \mapsto \gamma_2 + q$. The preceding discussion makes clear the mathematical sense in which the space of parameters of the Gabor transform $S(q, p)$ is to be thought of as a phase space. The space of parameters is a coadjoint orbit of the group, which has the structure of a classical mechanical phase space.

7 Matrix geometry of wavelet analysis. II

The last chapter has already familiarized us with the use of group theoretical methods for the construction and analysis of wavelets and gaborettes. We aim in this chapter to first indicate the general applicability of these techniques and then to look at the case of the two-dimensional continuous transform, using the SIM(2) group. Later, we look at general matrix groups of the type that can be used for constructing other types of wavelet transforms in two dimensions. We shall be led, in this manner, to studying a class of semidirect product type groups, certain coadjoint orbits of which are isomorphic to the group itself. In all these cases, the common features of such a matrix-group analysis will be: (a) the group will refer to a set of possible symmetry transformations which the signal may undergo; (b) the space over which the signals are defined (as L^2-functions) is intrinsic to the group; (c) the parameters in terms of which the wavelet transform is expressed are the parameters of the group itself, i.e., symmetry parameters of the signal, and (d) these parameter spaces, which arise as coadjoint orbits of the group, are also identifiable with phase spaces of signals.

Referring back to the 2-D wavelet transform introduced in Chapter 2, we shall see that this transform is again related to a square integrable representation of a matrix group. The coadjoint orbit of this group (there is only one nontrivial orbit in this case) will again allow us to carry out a phase space analysis. As in the 1-D case, square integrability will enable us to obtain a resolution of the identity, lead to finiteness of the total energy of the wavelet transform and yield a reconstruction formula for the signal. In order to put the discussion in the context of a more general framework, we begin with a word about the choice of an appropriate group for building wavelet transforms and the general rationale for appealing to group theory in the first place.

7.1 A group-adapted wavelet analysis

As mentioned earlier, a group-theoretical approach enables one to exploit mathematically the symmetries underlying the particular geometry which the signal space may have. If symmetries exist, it is natural to try to build these into the wavelet transform itself. This generally implies finding a continuous wavelet transform by exploiting a representation of the group on some Hilbert space, the key ingredient required for such a construction being square integrability of the representation. Assume that the class of finite energy signals under consideration can be realized as functions on a manifold Y, i.e., $s \in L^2(Y, d\mu) \equiv \mathfrak{H}$. This manifold could in fact be the space of some parameters (e.g., frequency, time, position, etc.) of the signal. As examples, Y could be space $\mathbb{R}^n$, the 2-sphere S^2, space–time $\mathbb{R} \times \mathbb{R}$ or $\mathbb{R} \times \mathbb{R}^2$, and so on. Assume that we measure the signals with the help of a *probe* $\eta : s \mapsto \eta[s]$. Usually such probes are taken to be linear functionals, representing the action of a measuring apparatus, a reference frame, etc. Mathematically, the measurement of the signal is given usually by some sort of an overlap integral that is, in the present case, an inner product $s \mapsto \langle \psi \mid s \rangle$, with ψ representing the probe. Note that, if we were to restrict the signals to smooth functions on Y, measurements could also be represented by distributions of some type.

7.1.1 Some generalities

Suppose there is a group G of symmetries of the signal, which acts as a set of transformations of the manifold Y. This means that any element $g \in G$, representing a specific symmetry operation on the parameter space (e.g., rotation of the signal through some angle, translation by some amount, etc.), induces a transformation of $Y : y \mapsto g[y]$. Successive transformations of this kind can be composed: $g[g'[y]] = gg'[y]$ and the composite transformation gg' is again an element of G. To each transformation, $y' = g[y]$, there exists an inverse transformation, $y = g^{-1}[y']$, $g^{-1} \in G$ and in addition the identity transformation e for which $e[y] = y$, for all $y \in Y$ is also a member of the group G. We assume further that this action is *transitive*, i.e., for any pair $y, y' \in Y$, there is at least one $g \in G$ such that $g[y] = y'$. It should be noticed, however, that the transformation group G acting on Y is in general not unique, its choice may depend on the problem at hand. In the case of the 1-D CWT, for instance, we could choose between the full affine group G_{aff} ($a \neq 0$) and its connected subgroup G_{aff}^+ ($a > 0$). However, when talking about *wavelets*, the groups that we consider will always contain a *dilation* as a symmetry of the signal. Furthermore, in all the cases that will concern us here, the group action will be realized through matrices acting on a vector space into which Y can be embedded. Recall, this means that the set of matrices constituting this group is closed under multiplication, inverse taking and that the identity matrix is also a member of this set. The action of the group on Y induces an action on the signal space and there are two possible ways in which this could happen.

(1) Action on the signals themselves. This is the *active* point of view: $s \mapsto s_g$, in which one measures the transformed signal, which we denote by s_g, with the fixed probe η.

(2) Action on the probes. This is the *passive* point of view: $\eta \mapsto \eta_g$, in which one measures the fixed signal s with the transformed probe η_g.

Now, if the manifold Y is globally G-invariant, consistency requires that the two points of view be equivalent,

$$\eta_g[s_g] = \eta[s], \ \forall \, g \in G. \tag{7.1}$$

If now signals are taken to be vectors s in the Hilbert space $\mathfrak{H} \equiv L^2(Y, dv)$, where dv is some convenient measure, one naturally identifies probes with linear functionals on $\mathfrak{H}$, that is, $\eta[s] \equiv \langle \eta \,|\, s \rangle$, for some fixed vector $\eta \in \mathfrak{H}$. Next, if one imposes that the action of G be *linear*, one ends up with a linear representation of G in $\mathfrak{H}$, $s_g = U(g)s$, $\eta_g = U(g)\eta$. This means that to each $g \in G$ one associates an operator $U(g)$ on $\mathfrak{H}$, and the mapping $g \mapsto U(g)$ is a group homomorphism (see (6.9)). The consistency condition (7.1) now requires that the $U(g)$ be unitary operators, i.e., one has a unitary representation of G on the space $\mathfrak{H}$ of signals:

$$\langle U(g)s \,|\, U(g)s' \rangle = \langle s \,|\, s' \rangle, \ \forall \, g \in G, \ s, s' \in \mathfrak{H} \quad \Rightarrow \quad U(g)^{-1} = U(g)^*. \tag{7.2}$$

Being a representation of G, the operators $U(g)$ must also satisfy

$$U(g_1)U(g_2) = U(g_1 g_2), \qquad U(g^{-1}) = U(g)^{-1},$$
$$U(e) = I \ (= \text{identity operator of } L^2(Y, dv)). \tag{7.3}$$

for all g_1, g_2 in G and where e denotes the unit element of G (i.e., $ge = eg = g$, $\forall \, g$). As an additional requirement, the representation $U(g)$ will be assumed to be *irreducible*. Technically this means that for any nonzero vector $\psi \in \mathfrak{H}$, the set of vectors $U(g)\psi$, $g \in G$, span the Hilbert space. (At times a weaker condition, e.g., the existence of just one such nonzero vector ψ may be sufficient.) Irreducibility also means that $\mathfrak{H}$ is a minimal space for realizing the symmetries unitarily.

As a measure space, the group G has generally two invariant measures defined on it, a *left Haar measure*, $d\mu$, invariant under $g \mapsto g_0 g$, for fixed $g_0 \in G$ and a *right Haar measure*, $d\mu_r$, invariant under $g \mapsto g g_0$.

$$d\mu(g_0 g) = d\mu(g), \qquad d\mu_r(g g_0) = d\mu_r(g), \qquad g \in G.$$

The existence of these measures is a general property of all topological groups; for the affine group these were explicitly written down as $d\mu = db\, db/a^2$ and $d\mu_r = db\, da/a$ (see (6.11) and the discussion following). The two invariant measures on G are equivalent, but generally not equal. In particular, for all wavelet related groups, which include dilations, the two Haar measures are different. We shall usually work with the left Haar measure, although everything we do could just as well be done using the right Haar measure. The space $L^2(G, d\mu)$ is usually taken to be the Hilbert space of all wavelet transforms and it will turn out that finding wavelet transforms of signals implies mapping the signal space $L^2(Y, dv)$ isometrically into a (closed) subspace of $L^2(G, d\mu)$.

Given this setting, one may derive a wavelet analysis on Y, adapted to the symmetry group G, in which the wavelet transform of a signal s would be an associated function S, defined on the group (the group being identified with a phase space for the signal) or some other phase space related to the group. In case the group itself can be identified with a phase space (as was the situation with the affine group or as it will be for the 2-D wavelet transform discussed below), the obvious choice of the function S is to write it as $S(g) = \langle U(g)\eta \,|\, s \rangle$. Two questions immediately arise: first, does the transform $S(g)$ have finite energy and second, is it possible to reconstruct the signal s uniquely from its transform S?

7.1.2   Square integrability of representations

The first question posed above can be reformulated as follows. If the total energy of the transform S is identified with the square of its L^2-norm:

$$E(S) = \|S\|^2 = \int_G d\mu(g)\,|S(g)|^2,$$

then of course, we want $\|S\| < \infty$. In fact we require that the mapping $s \mapsto S$ be (up to a constant) an isometry. Thus, the original question about the finiteness of energy becomes: is it possible to find $\psi \in \mathfrak{H}$ such that $\|S\|$ is finite for all s or that the map $s \mapsto S$ be a multiple of the isometry? It will turn out that a positive answer to this question will also guarantee the possibility of reconstructing the signal from its transform, i.e., a positive answer to the second question.

Interestingly enough, from a mathematical point of view, in order to obtain finiteness of energy for all transforms S, it is enough to require this of the transform of the probe ψ only. Thus, we require the existence of a nonzero vector $\psi \in \mathfrak{H}$ such that

$$I(\psi) = \int_G d\mu(g)\,|\langle U(g)\psi \,|\, \psi \rangle|^2 < \infty. \tag{7.4}$$

In case such a vector exists, we call it an *admissible vector* or, in signal analytic language, a *generalized mother wavelet*. The representation $U(g)$ is then said to be *square integrable*. Square integrability is a property of both the representation $U(g)$ and of the group G itself. Not all groups have square integrable representations and the same group may have representations which are square integrable as well as other ones which are not. From the general theory of square integrable group representations (see, e.g., [Ali00] for a detailed account) one knows the following:

- The existence of one admissible vector guarantees the existence of a *dense set* of such vectors. In particular, if ψ is admissible, then so also are all the vectors $U(g)\psi$, $g \in G$. Let us denote the set of all admissible vectors by $\mathcal{A}$. Then there exists an operator C, in general unbounded, on the Hilbert space $\mathfrak{H}$ such that it is self-adjoint, has positive spectrum and such that its domain is precisely the set of all admissible vectors:

$$c_\psi \equiv \|C\psi\|^2 < \infty \quad \Leftrightarrow \quad \psi \in \mathcal{A}. \tag{7.5}$$

Moreover, C^{-1} also exists as a densely defined positive (spectrum) operator. As in the 1-D wavelet case (6.17), we call C the *Duflo–Moore operator* of the representation.

- If the group is nonunimodular (i.e., the left and right Haar measures are different) then C is an unbounded operator and $\mathcal{A}$ is a proper subset of $\mathfrak{H}$. If G is unimodular, then $C = \lambda I$, $\lambda > 0$, and $\mathcal{A}$ coincides with the entire Hilbert space $\mathfrak{H}$ (every vector is an admissible vector).

- For any two admissible vectors ψ, ψ' and arbitrary vectors s, $s' \in \mathfrak{H}$, the following *orthogonality relation* holds:

$$\int_G d\mu(g)\, \overline{\langle U(g)\psi' \,|\, s'\rangle}\, \langle U(g)\psi \,|\, s\rangle = \langle C\psi \,|\, C\psi'\rangle\, \langle s' \,|\, s\rangle. \tag{7.6}$$

Note that (6.37) is just a special case of this relation, when G is the affine group. Proceeding in the same way by which we arrived at (6.38) from (6.37), we derive from (7.6) the *resolution of the identity*,

$$\frac{1}{\langle C\psi \,|\, C\psi'\rangle} \int_G d\mu(g)\, |U(g)\psi'\rangle\langle U(g)\psi| = I, \tag{7.7}$$

provided $\langle C\psi \,|\, C\psi'\rangle \neq 0$. Taking $\psi = \psi'$ in the above gives,

$$\frac{1}{c_\psi} \int_G d\mu(g)\, |\psi_g\rangle\langle\psi_g| = I. \tag{7.8}$$

7.1.3 Construction of generalized wavelet transforms

Given the existence of a square integrable representation $U(g)$ of the group G, on the signal space $\mathfrak{H} = L^2(Y, dv)$, (generalized) wavelets and wavelet transforms can be constructed by exploiting the orthogonality relation (7.6), in much the same way as was done for the affine group in the previous chapter. We start by taking a mother wavelet $\psi \in \mathfrak{H}$ and defining *generalized wavelets* as the vectors $\psi_g = U(g)\psi \in \mathfrak{H}$, $g \in G$. In the physical literature, the vectors ψ_g are called *coherent states* of the representation $U(g)$. Our convention will be to call these vectors generalized wavelets only when the group contains some sort of a dilation transformation on the space Y and the group space itself can be identified with a phase space. Otherwise we shall use the term coherent state. The question of when the group also has the structure of a phase space will be analyzed later. Using the generalized wavelets ψ_g, the (generalized) wavelet transform of the signal $s \in \mathfrak{H}$ is defined to be the function $S(g) = \langle \psi_g \,|\, s\rangle$ on G. Since the group G is derived by analyzing symmetry transformations of signals, its elements g are defined in terms of these very symmetry parameters (e.g., rotation angle, translation distance, zoom factor, etc.) and hence the wavelet transform also becomes a function of these parameters.

Computing now the expectation value of both sides of (7.8) with respect to $s \in \mathfrak{H}$ yields

$$\|S\|^2_{L^2(G, d\mu)} = \int_G d\mu\, |S(g)|^2 = c_\psi \|s\|^2_\mathfrak{H}, \tag{7.9}$$

implying that the map $W_\psi : \mathfrak{H} \to L^2(G, d\mu)$, given by

$$(W_\psi s)(g) = c_\psi^{-1/2}\, \langle \psi_g \,|\, s\rangle, \tag{7.10}$$

is a linear isometry. Similarly, acting on a signal s with the operators appearing on both sides of (7.7), we derive the general reconstruction formula

$$s = \frac{1}{\langle C\psi \mid C\psi' \rangle} \int_G d\mu(g)\, S(g)\psi'_g. \tag{7.11}$$

Again, it ought to be emphasized here that the wavelet transform $S(g)$, appearing in this expression, is computed using the wavelets ψ_g, while the reconstruction is done using the different set ψ'_g. Of course, the formula is valid also if $\psi = \psi'$. The above discussion again illustrates how in the general situation, as in the case of the affine group discussed in the last chapter, it is the square integrability of the representation which leads to the finiteness of the energy of the transform, on the one hand, and to the reconstruction formula, on the other.

The orthogonality relation (7.6) acquires a more transparent physical meaning if we express it in terms of wavelet transforms. If $S_\psi(g)$ is the wavelet transform of the signal s, with respect to the mother wavelet ψ and $S'_{\psi'}(g)$ the transform of s' with respect to ψ', then (7.6) can be rewritten as,

$$\int_G d\mu(g)\, \overline{S'_{\psi'}(g)}\, S_\psi(g) = \langle C\psi \mid C\psi' \rangle\, \langle s' \mid s \rangle. \tag{7.12}$$

In other words, wavelet transforms of orthogonal signals are always orthogonal, independent of the mother wavelets chosen to represent them, while wavelet transforms of arbitrary signals, when computed with respect to a mother wavelet ψ, are all orthogonal to their transforms computed with respect to ψ', if $C\psi$ is orthogonal to $C\psi'$.

7.1.4   Reproducing kernels, partial isometries and localization operators

Let ψ be an admissible vector and let $W_\psi[L^2(Y, d\nu)] \equiv \mathfrak{H}_\psi$ be the range of the isometry W_ψ. This means that $\mathfrak{H}_\psi$ is a closed Hilbert subspace of $L^2(G, d\mu)$, consisting of all wavelet transforms $S(g)$ associated to the mother wavelet ψ. Let $\mathbb{P}_\psi$ be the projection operator onto $\mathfrak{H}_\psi$, i.e., $\mathbb{P}_\psi L^2(G, d\mu) = \mathfrak{H}_\psi$. If ψ and ψ' are two admissible vectors, chosen so that $\langle C\psi \mid C\psi' \rangle = 0$, then from (7.6) we infer that the corresponding spaces of wavelet transforms, $\mathfrak{H}_\psi$ and $\mathfrak{H}_{\psi'}$, are orthogonal. However, unlike in the case of the affine group (see Theorem 6.1.1), this fact cannot in general be used to obtain a complete decomposition of $L^2(G, d\mu)$ into orthogonal subspaces of wavelet transforms – generally the space $L^2(G, d\mu)$ contains more than just wavelet transforms. An interesting feature, connecting the spaces of wavelet transforms $\mathfrak{H}_\psi$ corresponding to different mother wavelets ψ, now emerges. All these spaces of transforms sit inside $L^2(G, d\mu)$ and the same signal s can be mapped into different transform spaces, by choosing different mother wavelets. It is natural to ask whether it is possible to interpolate between these spaces of transforms, for that would allow us to take the transform of a signal with respect to one mother wavelet and re-express it as a transform with respect to a different mother wavelet. The following discussion, in particular Theorem 7.1.1, will show exactly how this can be done.

7.1.4.1 Partial isometries

For two arbitrary mother wavelets ψ, ψ', define the function

$$K_{\psi,\psi'} : G \times G \to \mathbb{C}, \qquad K_{\psi,\psi'}(g, g') = (c_\psi \, c_{\psi'})^{-1/2} \, \langle \psi_g \mid \psi'_{g'} \rangle. \tag{7.13}$$

This function has the easily verifiable properties,

$$K_{\psi,\psi'}(g, g') = \overline{K_{\psi',\psi}(g', g)}\,,$$
$$\int_G d\mu(g'') \, K_{\psi,\psi''}(g, g'') \, K_{\psi'',\psi'}(g'', g') = K_{\psi,\psi'}(g, g'), \tag{7.14}$$

the first following from the definition of $K_{\psi,\psi'}$ in (7.13) and the second from the resolution of the identity in (7.8). Next let us define an integral operator $V_{\psi,\psi'}$, with $K_{\psi,\psi'}(g, g')$ as its kernel:

$$(V_{\psi,\psi'} F)(g) = \int_G d\mu(g') \, K_{\psi,\psi'}(g, g') \, F(g'), \qquad F \in L^2(G, d\mu). \tag{7.15}$$

We show below that $K_\psi \equiv K_{\psi,\psi}$ is the integral kernel of the projection operator $\mathbb{P}_\psi$ and hence defines a reproducing kernel of the type seen in the last chapter (see (6.28)–(6.30)), while the operator $V_{\psi,\psi'}$, for different ψ, ψ', is a *partial isometry* on $L^2(G, d\mu)$. This means that the range of $V_{\psi,\psi'}$ is the space $\mathfrak{H}_{\psi'}$, of wavelet transforms corresponding to the mother wavelet ψ', while its kernel is the orthogonal complement, $\mathfrak{H}_\psi^\perp$, of the space of transforms $\mathfrak{H}_\psi$. Between $\mathfrak{H}_\psi$ and $\mathfrak{H}_{\psi'}$ the operator interpolates as a linear isometry.

Theorem 7.1.1. *The operator $V_{\psi,\psi'}$ is a partial isometry on $L^2(G, d\mu)$, which maps the subspace $\mathfrak{H}_\psi$ isometrically onto $\mathfrak{H}_{\psi'}$, and has the properties*

$$V_{\psi,\psi'} V_{\psi',\psi''} = V_{\psi,\psi''}, \qquad V_{\psi,\psi'}^* = V_{\psi',\psi}, \tag{7.16}$$
$$V_{\psi,\psi} = \mathbb{P}_\psi, \qquad V_{\psi,\psi'} V_{\psi,\psi'}^* = \mathbb{P}_\psi, \qquad V_{\psi,\psi'}^* V_{\psi,\psi'} = \mathbb{P}_{\psi'}. \tag{7.17}$$

Proof. The two relations in (7.16) follow from (7.14). For any $F \in L^2(G, d\mu)$, it can be shown that

$$(\mathbb{P}_\psi F)(g) = \int_G d\mu(g') \, K_\psi(g, g') \, F(g'), \tag{7.18}$$

in exactly the same way as (6.33) was proved for the affine group. This establishes the first of the relations in (7.17). Also,

$$\int_G d\mu(g') \, K_{\psi,\psi'}(g, g') \, F(g') = (c_\psi \, c_{\psi'})^{-1/2} \, \langle \psi_g \mid \int_G d\mu(g') \, \psi'_{g'} \, F(g') \rangle$$
$$= \langle \psi_g \mid s \rangle,$$

where

$$s = (c_\psi \, c_{\psi'})^{-1/2} \int_G d\mu(g) \, \psi'_g \, F(g) \in L^2(Y, d\nu).$$

Thus, $V_{\psi,\psi'}F$ is the wavelet transform of s with respect to the mother wavelet ψ, implying that the range of $V_{\psi,\psi'}$ is contained in the subspace $\mathfrak{H}_\psi \subset L^2(G, d\mu)$. To see that it actually coincides with $\mathfrak{H}_\psi$, let F be an arbitrary element of $\mathfrak{H}_\psi$. Then there exists a signal $s \in L^2(Y, d\nu)$ for which $F(g) = \langle \psi_g \mid s \rangle$. But, by (7.8),

$$\langle \psi_g \mid s \rangle = \frac{1}{c_{\psi'}} \int_G d\mu(g') \, \langle \psi_g \mid \psi'_{g'} \rangle \langle \psi'_{g'} \mid s \rangle$$

$$= \int_G d\mu(g') \, K_{\psi,\psi'}(g, g') F'(g') = (V_{\psi,\psi'} F')(g),$$

where we have written

$$F'(g) = \left(\frac{c_\psi}{c_{\psi'}} \right)^{1/2} \langle \psi'_g \mid s \rangle .$$

Thus every vector in $\mathfrak{H}_\psi$ is also in the range of $V_{\psi,\psi'}$.

On the other hand, using (7.14) and (7.18),

$$\int_G d\mu(g') \, K_{\psi,\psi'}(g, g') \, F(g')$$

$$= \int_G\int_G d\mu(g') \, d\mu(g'') \, K_{\psi,\psi'}(g, g'') \, K_{\psi'}(g'', g') \, F(g')$$

$$= \int_G d\mu(g'') \, K_{\psi,\psi'}(g, g'') \, (\mathbb{P}_{\psi'} F)(g''),$$

implying that, for any vector $F \in \mathfrak{H}_{\psi'}^\perp$,

$$V_{\psi,\psi'} F = 0.$$

Thus $\mathfrak{H}_{\psi'}^\perp$ is contained in the kernel of $V_{\psi,\psi'}$. Finally, if $F' \in \mathfrak{H}_{\psi'}$, so that $F'(g) = \langle \psi'_g \mid s \rangle$, for some signal vector $s \in L^2(Y, d\nu)$, and $\|F'\|^2_{L^2(G,d\mu)} = c_{\psi'} \|s\|^2_{L^2(Y,d\nu)}$, then

$$(V_{\psi,\psi'} F')(g) = (c_\psi \, c_{\psi'})^{-1/2} \int_G d\mu(g') \, \langle \psi_g \mid \psi'_{g'} \rangle \langle \psi'_{g'} \mid s \rangle$$

$$= \left(\frac{c_{\psi'}}{c_\psi} \right)^{1/2} \langle \psi_g \mid s \rangle, \qquad \text{by (7.8).}$$

But the function $F(g) = \langle \psi_g \mid s \rangle$ is the wavelet transform of s with respect to the mother wavelet ψ. Hence,

$$\| V_{\psi,\psi'} F' \|^2_{L^2(G,d\mu)} = c_\psi \frac{c_{\psi'}}{c_\psi} \|s\|^2_{L^2(Y,d\nu)} = c_{\psi'} \|s\|^2_{L^2(Y,d\nu)} = \|F'\|^2_{L^2(G,d\mu)}.$$

Thus, $V_{\psi,\psi'}$ maps $\mathfrak{H}_{\psi'}$ isometrically onto $\mathfrak{H}_\psi$ and its kernel coincides with $\mathfrak{H}_{\psi'}^\perp$. $\qquad \square$

This theorem is useful for it shows, first of all, that the various transforms that can be obtained for the same signal but using different mother wavelets, are in a sense equivalent. Second, all these transforms have the same energy content as well as the same information content. From a theoretical point of view, therefore, it does not matter which mother wavelet is used. However, for practical or computational reasons, it may be preferable to use mother wavelets having specific properties.

7.1.4.2 Left regular representation and localization operators

We had started out by choosing an irreducible representation $U(g)$ of the group G, on the space of signals $L^2(Y, dv)$. Assuming U to be square integrable and taking different admissible vectors (or mother wavelets) ψ, we defined generalized wavelets ψ_g, using which we were able to map the signal space isometrically into subspaces $\mathfrak{H}_\psi$ of signal transforms. The isometric mapping W_ψ was defined (see (7.10)) by $(W_\psi s)(g) = c_\psi^{-1/2} \langle \psi_g \mid s \rangle$. As a consequence of this isometry, the unitary operators $U(g)$ are mapped to certain unitary operators on $\mathfrak{H}_\psi$ and it is interesting to obtain these image operators. Since, for arbitrary $s \in L^2(Y, dv)$ and fixed $g_0 \in G$,

$$(W_\psi U(g_0)s)(g) = c_\psi^{-1/2} \langle \psi_g \mid U(g_0)s \rangle = c_\psi^{-1/2} \langle U(g_0)^* U(g)\psi \mid s \rangle$$
$$= c_\psi^{-1/2} \langle U(g_0^{-1}g)\psi \mid s \rangle = c_\psi^{-1/2} \langle \psi_{g_0^{-1}g} \mid s \rangle,$$

we obtain, $W_\psi U(g_0) W_\psi^{-1} = \mathbf{U}_\psi(g_0)$ (the inverse of W_ψ being computed on its range), where, for any $S \in \mathfrak{H}_\psi$,

$$(\mathbf{U}_\psi(g_0)S)(g) = S(g_0^{-1}g), \qquad g \in G, \tag{7.19}$$

which ought to be compared with (6.35). Being the isometric image of a unitary irreducible representation, the representation given by these operators, on $\mathfrak{H}_\psi$, is also unitary and irreducible. In other words, $U(g)$ and $\mathbf{U}_\psi(g)$ are equivalent representations of the group G, however, expressed on different Hilbert spaces. What is important to note here is that the right-hand side of (7.19) is independent of ψ. In other words, one obtains the same form for the transformed operators $\mathbf{U}_\psi(g)$ on all the subspaces $\mathfrak{H}_\psi \subset L^2(G, d\mu)$, irrespective of the mother wavelet chosen. Using this fact let us define the operators $\mathbf{U}_\ell(g)$, $g \in G$, on the whole of $L^2(G, d\mu)$, adopting this same form:

$$(\mathbf{U}_\ell(g)F)(g') = F(g^{-1}g'), \qquad g, g' \in G, \quad F \in L^2(G, d\mu). \tag{7.20}$$

Since

$$\|\mathbf{U}_\ell(g)F\|^2 = \int_G d\mu(g')|(\mathbf{U}_\ell(g)F)(g')|^2$$
$$= \int_G d\mu(g') \, |F(g^{-1}g')|^2 = \|F\|^2,$$

the last step following from the invariance of the measure $d\mu$, the operators $\mathbf{U}_\ell(g)$ are unitary on $L^2(G, d\mu)$. Also, clearly, the map $g \mapsto \mathbf{U}_\ell(g)$ is a group homomorphism. Thus, we have obtained a unitary representation of G in terms of these operators, called the *left regular representation*. The situation is exactly what we also had in the case of the affine group (see (6.34) and the discussion leading up to and following it). The left regular representation is not in general irreducible, and indeed, (7.19) shows that, restricted to each one of the subspaces $\mathfrak{H}_\psi \subset L^2(G, d\mu)$, it admits the irreducible subrepresentation $\mathbf{U}_\psi(g)$. Another way to express this fact is to write

$$W_\psi U(g) = \mathbf{U}_\ell(g) W_\psi = \mathbf{U}_\psi(g). \tag{7.21}$$

Summarizing, the rôle of the representation $U(g)$ on the signal space is played by the left regular representation $\mathbf{U}_\ell(g)$ on the spaces of wavelet transforms.

The construction of localization operators on the group space G follows the same pattern as laid out for the affine group in Section 6.1.3. Let $\psi \in L^2(Y, dv)$ be an admissible vector, $K_\psi(g, g')$ the corresponding reproducing kernel and $\Delta \subset G$ a measurable set (with respect to the Haar measure $d\mu$). Associated to Δ, we define the integral kernel,

$$a_\psi^\Delta(g, g') = \int_\Delta d\mu(g'') \, K_\psi(g, g'') \, K_\psi(g'', g'), \tag{7.22}$$

and the resulting operator $a_\psi(\Delta)$ on $L^2(G, d\mu)$,

$$(a_\psi(\Delta)F)(g) = \int_G d\mu(g') \, a_\psi^\Delta(g, g') \, F(g'), \qquad F \in L^2(G, d\mu). \tag{7.23}$$

From (7.18) and the definition of $a_\psi(\Delta)$, it follows that

$$\langle F \mid a_\psi(\Delta)F \rangle = \int_\Delta d\mu(g) \, |(\mathbb{P}_\psi F)(g)|^2, \tag{7.24}$$

which shows that the operator is bounded, self-adjoint and has positive spectrum. In particular, for $S \in \mathfrak{H}_\psi$,

$$\langle S \mid a_\psi(\Delta)S \rangle = \int_\Delta d\mu(g) \, |S(g)|^2.$$

Thus, the quantity

$$p_S(\Delta) = \frac{\langle S \mid a_\psi(\Delta)S \rangle}{\|S\|^2}, \tag{7.25}$$

represents the fraction of the wavelet transform of s (where $S(g) = \langle \psi_g \mid s \rangle$) which is localized in the region Δ. This also motivates the term *localization operator* for $a_\psi(\Delta)$. Additionally, these operators have the measure theoretical properties as those obtained for the localization operators of the affine group (see (6.53)):

$$a_\psi(G) = \mathbb{P}_\psi, \qquad a_\psi(\emptyset) = 0,$$

$$a_\psi\left(\bigcup_{i \in J} \Delta_i\right) = \sum_{i \in J} a_\psi(\Delta_i), \qquad \text{if } \Delta_i \bigcap \Delta_j = \emptyset, \text{ whenever } i \neq j, \tag{7.26}$$

where, again, the sum in (7.26) is to be understood in the sense of scalar products. These relations also imply that $p_S(\Delta)$ is a probability measure.

Finally, the localization operators satisfy the *imprimitivity* or *covariance* condition,

$$\mathbf{U}_\ell(g) \, a_\psi(\Delta) \, \mathbf{U}_\ell(g)^* = a_\psi(g\Delta), \tag{7.27}$$

where $g\Delta$ is the shifted set,

$$g\Delta = \{gg' \in G \mid g' \in \Delta\}.$$

The above covariance relation can be derived in the same way as (6.57) and again leads to the corresponding relation,

$$p_S(g\Delta) = p_{U_\ell(g)^{-1}S}(\Delta), \tag{7.28}$$

for the probability measure $p_S(\Delta)$. We interpret this relation in the same way as in the case of the affine group namely, that the probability of localization in the transformed set is the same as the probability of localization of the transformed signal in the original set.

7.1.5   Wavelet transforms on general quotient spaces

The group theoretical analysis outlined above was intended to underscore the fact that, using purely symmetry arguments, one can arrive at a general wavelet transform, which then displays all the basic properties of the standard 1-D wavelet transform. However, the power of this general group theoretical analysis lies in its applicability to a vast number of other symmetry groups, thus opening up the possibility of constructing extremely general classes of wavelet transforms. Many of these turn out to be of enormous practical value, as well. It is in this light that we will undertake in Section 7.2 a general analysis of the 2-D wavelet transform.

Building the generalized wavelet transform, defined in (7.10), depended on the assumption that the underlying group representation $U(g)$ was square integrable – in the sense that there existed a vector ψ satisfying the admissibility condition (7.4). Furthermore, since we generally wish to identify the space of signal variables with the structure of a physical phase space, the group G itself would have to possess such a structure, if wavelet transforms are to be defined as functions over it. However, already in the case of the Gabor transform, we saw that these conditions were not fulfilled in the strict sense. Indeed in that case admissibility was only defined with respect to a quotient space of the group (see (6.114) and the discussion following it); it was this quotient space which had the structure of a phase space and on which the Gabor transform was defined. Let us briefly indicate here how this sort of a construction can be put against a more general setting.

As a first case, consider the situation where the analyzing wavelet ψ has a nontrivial isotropy subgroup $H_\psi \subset G$, up to a phase. This means that ψ satisfies the condition

$$U(h)\psi = e^{i\alpha(h)}\psi, \qquad h \in H_\psi, \tag{7.29}$$

where $\alpha(h)$ is a (real) phase factor, generally depending on h. (In the physical literature, this is the setting for the construction of Gilmore–Perelomov type of coherent states [191,192,Per86,312].) Clearly now, the integrand in (7.4) does not depend on g, but only on the coset $gH_\psi \in G/H_\psi$, so that the finiteness of the integral would force the subgroup H_ψ to be compact. Failing that, we assume that the quotient space $X = G/H_\psi$

carries an invariant measure ν. (Recall that elements of the quotient space X are the cosets gH_ψ, $g \in G$, and they transform under the action of an element g_0 of the group in the manner, $gH_\psi \mapsto g_0gH_\psi$.) We impose the weaker admissibility condition [191,192,312],

$$\int_X d\nu(x) \, |\langle U(g)\psi | \phi \rangle|^2 < \infty, \quad \forall \phi \in \mathfrak{H} \quad (x \equiv gH_\psi), \tag{7.30}$$

on ψ. The integrand in (7.30) manifestly does not depend on individual elements $g \in G$, only on their cosets modulo H_ψ, $x \equiv gH_\psi \in G/H_\psi$, and the condition (7.30) makes sense. This condition means that the representation U is square integrable on the coset space $X = G/H_\psi$ or, as it is called, *square integrable modulo the subgroup* H_ψ. Notice that the latter need no longer be compact. In order to define wavelets, it is necessary to go back to the group. We do this using the notion of a *section*. This is a map $\sigma : X \to G$, chosen so that if $\sigma(x) = g$ then $x = gH_\psi$. In these terms, the admissibility condition (7.30) may be rewritten in the slightly different, but completely equivalent form:

$$c_X(\psi, \phi) = \int_X d\nu(x) \, |\langle U(\sigma(x))\psi | \phi \rangle|^2 < \infty, \quad \forall \phi \in \mathfrak{H}, \tag{7.31}$$

where σ is an *arbitrary* section $\sigma : X \to G$. Indeed, since two different sections σ and σ' are related as $\sigma'(x) = \sigma(x)h(x)$, where $h(x) \in H_\psi$, it is obvious that the integrand does not depend on the choice of the section. Correspondingly, the wavelet vectors are written as $\psi_{\sigma(x)} = U(\sigma(x))\psi$, $x \in X$, which emphasizes that the proper index set is $X = G/H_\psi$ and not G.

Under the condition (7.30) or (7.31), the whole construction may be performed exactly as before [Ali00,6]. In particular, the map $W_\psi : \mathfrak{H} \to L^2(X, d\nu)$ given by $(W_\psi s)(x) \equiv c_X^{-1/2} \langle \psi_{\sigma(x)} | s \rangle$ is an isometry, where $c_X \equiv c_X(\psi, \psi)$; in other words, one has a resolution of the identity

$$c_X^{-1} \int_X d\nu(x) \, |\psi_{\sigma(x)}\rangle\langle\psi_{\sigma(x)}| = I. \tag{7.32}$$

From this follows, as before, that the range of W_ψ is a closed subspace $\mathfrak{H}_\psi$ of $L^2(X, d\nu)$, the corresponding projection $P_\psi = W_\psi W_\psi^*$ is an integral operator with (reproducing) kernel $K(x', x) = c_X^{-1} \langle \psi_{\sigma(x')} | \psi_{\sigma(x)} \rangle$, the familiar reconstruction formula holds, etc.

Coming back to the subject of this book, it is true that the continuous wavelet transform, both in one and two dimensions, are examples of wavelet transforms living directly on the associated group, $G_{\text{aff}}^{(+)}$ and SIM(2), respectively (see below). However, we have also seen in Section 6.3 that the Gabor transform is an example of a construction modulo a subgroup, in this case the phase subgroup Θ of the Weyl–Heisenberg group G_{WH}. In the same way, in dimensions higher than 2, the CWT with respect to an axisymmetric wavelet leads to wavelet transforms defined on a quotient of the above type (see Section 9.1). Physically, this means that while the total set of signal symmetries may be large, because of the needs of the problem at hand, the wavelet transform is defined over a smaller set of parameters. Moreover, it is this quotient space which turns out to be the relevant phase space of the problem.

Actually, one can go a step further, and extend the whole construction to the case of an arbitrary coset space $X = G/H$, where H is *not* the stability subspace of any vector ψ in the sense of (7.29). The main difference is that (i) the validity of the admissibility condition (7.31) may depend on the choice of the section σ; and (ii) when the condition holds, it reads

$$0 < \int_X d\nu(x) \, |\langle U(\sigma(x))\psi | \phi \rangle|^2 = \langle \phi | A_\sigma \phi \rangle, \ \forall \phi \in \mathfrak{H}, \tag{7.33}$$

where A_σ is a bounded positive invertible operator, sometimes called the *resolution operator*. Equivalently, the resolution of the identity (7.32) becomes

$$c_X^{-1} \int_X d\nu(x) \, |\psi_{\sigma(x)}\rangle \langle \psi_{\sigma(x)}| = A_\sigma. \tag{7.34}$$

Note that A_σ^{-1} may be unbounded in general. In the case where it is bounded, the system of wavelets $\{\psi_{\sigma(x)}, \ x \in X\}$ is called a (continuous) *frame*. (The unbounded case yields a far reaching generalization of the notion of a frame discussed in Section 2.4.1.) However, this extension of the theory of wavelets will not concern us in this book, with the sole exception of wavelets on the 2-sphere, discussed at length in Section 9.2.

7.2   The 2-D continuous wavelet transform

Group theoretically, we expect the 2-D continuous wavelet transform to arise from a group which should be an appropriate generalization of the affine group. This indeed is the case, and we are led to it by a relatively straightforward analysis of the symmetries which might be attributed to a two-dimensional signal. It turns out that the group in question, the 2-D *similitude group*, is a generalization of the affine group and in fact contains it as a subgroup.

7.2.1   The similitude group and 2-D wavelets

We begin with a model of a two-dimensional image. For our purposes, a 2-D image will be a finite energy signal $s \in L^2(\mathbb{R}^2, d^2\vec{x})$, as discussed in Section 2.1.1. The operations we want to apply to s are translations in the image plane ($\vec{b} \in \mathbb{R}^2$), global dilations (zooming in and out by $a > 0$) and rotations around the origin ($\theta \in [0, 2\pi)$). Together these transformations constitute a four-parameter group, called the *similitude group* of the plane and denoted by SIM(2). The action on the plane is

$$\vec{x} = (\vec{b}, a, \theta)\vec{y} = a r_\theta \vec{y} + \vec{b}, \tag{7.35}$$

where r_θ is the 2×2 rotation matrix,

$$r_\theta = \begin{pmatrix} \cos\theta & -\sin\theta \\ \sin\theta & \cos\theta \end{pmatrix}. \tag{7.36}$$

Our convention is to take a quantity such as $\vec{x}$ to be a column vector, the corresponding row vector being $\vec{x}^T$. A convenient representation of the joint transformation $(\vec{b}, a, \theta)$ is in the form of 3×3 matrices

$$(\vec{b}, a, \theta) \equiv \begin{pmatrix} ar_\theta & \vec{b} \\ \vec{0}^T & 1 \end{pmatrix}, \qquad \vec{0}^T = (0, 0). \tag{7.37}$$

Matrix multiplication then replicates the composition of successive transformations and thus we derive the group law,

$$(\vec{b}, a, \theta)(\vec{b}', a', \theta') = (\vec{b} + ar_\theta \vec{b}', aa', \theta + \theta')$$

$$e = (\vec{0}, 1, 0), \quad \text{(unit element)}$$

$$(\vec{b}, a, \theta)^{-1} = (-a^{-1}r_{-\theta}\vec{b}, a^{-1}, -\theta).$$

From this, we deduce the following.
- The set of rotations $(0, 1, \theta)$, $\theta \in [0, 2\pi)$, is a subgroup of SIM(2) and so also is the set of dilations $(0, a, 0)$, $a > 0$. Moreover, these two subgroups commute, i.e., $(0, 1, \theta)(0, a, 0) = (0, a, 0)(0, 1, \theta) = (0, a, \theta)$.
- The set of all translations $(\vec{b}, 1, 0)$, $\vec{b} \in \mathbb{R}^2$, is also a subgroup. Moreover, it has the structure of an invariant subgroup in the following sense: if $(\vec{b}, a, \theta) \in \mathrm{SIM}(2)$ is arbitrary and $(\vec{b_0}, 1, 0)$ any element of the translation subgroup, then

$$(\vec{b}, a, \theta)(\vec{b_0}, 1, 0)(\vec{b}, a, \theta)^{-1} = (ar_\theta \vec{b_0}, 1, 0),$$

which again is an element of the same subgroup.

Thus, the similitude group SIM(2) has the structure of a semidirect product:

$$\mathrm{SIM}(2) = \mathbb{R}^2 \rtimes (\mathbb{R}_*^+ \times \mathrm{SO}(2))$$

where $\mathbb{R}^2$ is the subgroup of translations, $\mathbb{R}_*^+$ that of dilations, and SO(2) of rotations. Topologically, we can identify $\mathbb{R}^2$ with $\mathbb{C}$, the complex plane and $\mathbb{R}_*^+ \times \mathrm{SO}(2)$ with $\mathbb{C}^*$, the complex plane with the origin removed. Thus we may write

$$\mathrm{SIM}(2) = \mathbb{C} \rtimes \mathbb{C}^*,$$

and denoting a group element by (z, w), where $z \in \mathbb{C}$ and $w \in \mathbb{C}^*$, the group composition law is, very simply,

$$(z_1, w_1)(z_2, w_2) = (z_1 + w_1 z_2, w_1 w_2).$$

In particular, if we only consider elements (z, w) for which $z = b + ic$, with $c = 0$, and $w = ae^{i\theta}$, with $\theta = 0$, then these elements clearly constitute a subgroup, which is just the affine group of the line. Thus, $G_{\mathrm{aff}} \subset \mathrm{SIM}(2)$, meaning that the similitude group is a generalization of the affine group, as indicated earlier. In fact, we may consider SIM(2) as being a *complexification* of G_{aff}.

Let us compute next the left (invariant) Haar measure on SIM(2). If $(\vec{b}_0, a_0, \theta_0)$ is a fixed element of the group and $(\vec{b}, a, \theta)$ arbitrary, then writing

$$(\vec{b}', a', \theta') = (\vec{b}_0, a_0, \theta_0)(\vec{b}, a, \theta) = (\vec{b}_0 + a_0 r_{\theta_0} \vec{b}, a_0 a, \theta_0 + \theta),$$

and noting that det $[r_{\theta_0}] = 1$, we get

$$d^2 \vec{b}' = a_0^2 \, d^2 \vec{b}, \qquad da' = a_0 \, da, \qquad d\theta' = d\theta.$$

Thus, the measure

$$d\mu(\vec{b}, a, \theta) = \frac{1}{a^3} \, d^2 \vec{b} \, da \, d\theta, \tag{7.38}$$

is invariant under left transformations. Similarly, the right Haar measure can be computed to be

$$d\mu_r(\vec{b}, a, \theta) = \frac{1}{a} \, d^2 \vec{b} \, da \, d\theta,$$

and thus like the affine group, the SIM(2) group is also nonunimodular.

Since $\vec{b}, a, \theta$ represent parameters in terms of which we want to analyze the signals $s \in L^2(\mathbb{R}^2, d^2\vec{x})$, we shall identify the Hilbert space $L^2(\text{SIM}(2), d\mu)$ with the space of all finite energy 2-D wavelet transforms. It will turn out that $\text{SIM}(2) \simeq \mathbb{R}^2 \times \mathbb{R}_*^+ \times S^1$ again has the structure of a phase space (S^1 being the unit circle). We shall later analyze the orbits of SIM(2) under the coadjoint action and we shall see that there is only one nontrivial orbit, which topologically is isomorphic to the group itself. Correspondingly, there is only one nontrivial unitary irreducible representation of SIM(2). This representation, which is a straightforward realization of the action (7.35) on the space of signals $L^2(\mathbb{R}^2, d^2\vec{x})$, is given by the operators $U(\vec{b}, a, \theta)$ (see (2.13)):

$$\left[U(\vec{b}, a, \theta)s \right](\vec{x}) = a^{-1} s(a^{-1} r_{-\theta}(\vec{x} - \vec{b})), \quad \vec{b} \in \mathbb{R}^2, a > 0, 0 \leqslant \theta < 2\pi. \tag{7.39}$$

The fact that these operators define a unitary representation is straightforward to verify. In the space $L^2(\widehat{\mathbb{R}}^2, d^2\vec{k})$ of Fourier-transformed signals $\widehat{s}$, this representation acquires the form (see (2.14))

$$\left[\widehat{U}(\vec{b}, a, \theta]\widehat{s} \right](\vec{k}) = a \, e^{-i\vec{b}\cdot\vec{k}} \, \widehat{s}(ar_{-\theta}\vec{k}). \tag{7.40}$$

This representation is also square integrable, with an admissibility condition on mother wavelets, see (2.16), which is analogous to (6.15). Indeed we have the result:

Theorem 7.2.1. *The family of operators $U(\vec{b}, a, \theta)$ defines a unitary irreducible representation of* SIM(2) *in the Hilbert space $L^2(\mathbb{R}^2, d^2\vec{x})$, which is unique up to unitary equivalence. This representation is square integrable, and a vector $\psi \in L^2(\mathbb{R}^2, d^2\vec{x})$ is admissible if, and only if, it verifies the condition:*

$$c_\psi \equiv (2\pi)^2 \int_{\mathbb{R}^2} \frac{d^2\vec{k}}{|\vec{k}|^2} \, |\widehat{\psi}(\vec{k})|^2 < \infty. \tag{7.41}$$

Proof. That U is a representation of SIM(2) results from explicit computation; its unitarity is obvious and its irreducibility follows from Proposition 2.1.2. As for the square integrability, it is proved by a direct calculation, using the Fourier-transformed realization $\widehat{U}$ (see above),

$$I(\psi) \equiv \int_{\text{SIM}(2)} \left| \langle \widehat{U}(\vec{b}, a, \theta) \widehat{\psi} \mid \widehat{\psi} \rangle \right|^2 d^2\vec{b} \, \frac{da}{a^3} \, d\theta$$

$$= \int_{\mathbb{R}^2} d^2 b \int_0^\infty \frac{da}{a} \int_0^{2\pi} d\theta \int_{\mathbb{R}^2} d^2\vec{k} \, e^{i\vec{b}\cdot\vec{k}} \, \overline{\widehat{\psi}(a\, r_{-\theta}(\vec{k}))} \, \widehat{\psi}(\vec{k})$$

$$\times \int_{\mathbb{R}^2} d^2\vec{k}' \, e^{-i\vec{b}\vec{k}'} \, \widehat{\psi}(a r_{-\theta}(\vec{k}')) \, \overline{\widehat{\psi}(\vec{k}')}.$$

Integrating first over $\vec{b}$ (the permutation of integrals is allowed by Fubini's theorem) yields a factor $(2\pi)^2\delta(\vec{k} - \vec{k}')$ and, therefore,

$$I(\psi) = (2\pi)^2 \int_0^\infty \frac{da}{a} \int_0^{2\pi} d\theta \int d^2\vec{k} \left| \widehat{\psi}(a\, r_{-\theta}\vec{k}) \right|^2 \left| \widehat{\psi}(\vec{k}) \right|^2.$$

From this, we get, exactly as in the proof of Proposition 2.2.1,

$$I(\psi) = c_\psi \, \|\psi\|^2,$$

with c_ψ given by (7.41), which proves the statement.   □

Introducing the Duflo–Moore operator $\widehat{C}$, in the Fourier transformed space:

$$c_\psi = \|\widehat{C}\widehat{\psi}\|^2,$$

we obtain,

$$(\widehat{C}\widehat{\psi})(\vec{k}) = \frac{2\pi}{|\vec{k}|} \, \widehat{\psi}(\vec{k}), \tag{7.42}$$

which should be compared to (6.16). Thus every function $\psi \in L^2(\mathbb{R}^2, d^2\vec{x})$, such that its Fourier transform lies in the domain of $\widehat{C}$ (i.e., satisfies (7.41)), is an admissible vector and can be used to build wavelets. Following our established practice, we shall call such vectors mother wavelets.

Choosing a mother wavelet ψ, we define the *2-D wavelets* as:

$$\psi_{\vec{b},a,\theta}(\vec{x}) = \left[U(\vec{b}, a, \theta)\psi \right](\vec{x}) = \frac{1}{a} \, \psi \left(\frac{1}{a} r_{-\theta}(\vec{x} - \vec{b}) \right), \tag{7.43}$$

and the *2-D continuous wavelet transform* as the inner product of the signal s with the wavelet $\psi_{\vec{b},a,\theta}$:

$$S(\vec{b},\, a,\, \theta) = \langle \psi_{\vec{b},a,\theta} \,|\, s \rangle, \tag{7.44}$$

which is a function on SIM(2) (see (2.18)–(2.20)). All the general properties of wavelets, as outlined in Section 7.1 and in particular the relations (7.6)–(7.11) and (7.21)–(7.27), follow in a straightforward manner. Some of these were worked out in detail in Section 2.2. Specific examples of 2-D wavelets with special symmetry properties have also been worked out in Chapter 2. For the sake of illustration, we display here the general resolution of the identity and reconstruction formula for signals. Following (7.7) we may write,

$$\frac{1}{\langle C\psi \,|\, C\psi' \rangle} \int_{-\infty}^{\infty}\int_{-\infty}^{\infty}\int_{0}^{\infty}\int_{0}^{2\pi} db_1\, db_2\, \frac{da}{a^3}\, d\theta\; |\psi'_{\vec{b},a,\theta}\rangle\langle\psi_{\vec{b},a,\theta}| = I, \tag{7.45}$$

(b_1, b_2 being the components of the vector $\vec{b}$), provided,

$$\langle C\psi \,|\, C\psi' \rangle = (2\pi)^2 \int_{\mathbb{R}^2} d^2\vec{k}\, \frac{\overline{\hat{\psi}(\vec{k})}\hat{\psi}'(\vec{k})}{|\vec{k}|^2} \neq 0.$$

From this we obtain the reconstruction formula for a signal,

$$s = \frac{1}{\langle C\psi \,|\, C\psi' \rangle} \int_{-\infty}^{\infty}\int_{-\infty}^{\infty}\int_{0}^{\infty}\int_{0}^{2\pi} db_1\, db_2\, \frac{da}{a^3}\, d\theta\; S_\psi(\vec{b}, a, \theta)\psi'_{\vec{b},a,\theta}, \tag{7.46}$$

where $S_\psi(\vec{b}, a, \theta) = \langle \psi_{\vec{b},a,\theta} \,|\, s \rangle$ is the wavelet transform of s in terms of the mother wavelet ψ.

One ought to comment here on the freedom that one has in designing the 2-D wavelet transform. On the one hand, one may ignore the rotation variable θ, for instance, if directions are irrelevant. This is achieved by choosing an isotropic or rotation invariant wavelet, $\psi(r_\theta(\vec{x})) = \psi(\vec{x})$. Equivalently, one may consider as transformations of the plane only translations and dilations, with the corresponding group $\mathbb{R}^2 \rtimes \mathbb{R}_*^+$. In this case, however, the representation structure is much more complicated, since every subspace of the form $L^2(C, d^2k)$ is invariant under the action of the operators $U(\vec{b}, a, \theta)$, where C is a cone with apex at the origin in the $\vec{k}$-plane. More interesting is the opposite move. If, besides the similitude operations, one considers also certain types of deformations, one gets a larger group, namely, the group obtained by replacing in (7.37) the matrix $ar_\theta \in \mathbb{R}_*^+ \times SO(2)$ by an arbitrary nonsingular 2×2 real matrix. This group is much more complicated and so also is its representation structure. In fact, it is unlikely that any of its representations would be of much use for our purposes. Thus, the similitude group SIM(2) seems to occupy a privileged position in the construction, although later, in Section 7.4, we shall study a second group in which the spatial rotations r_θ are replaced by hyperbolic rotations.

7.2.2   The group as the primary object

We started out by defining a two-dimensional image as a function on $\mathbb{R}^2$ and then obtained the group SIM(2) by considering a set of transformations on $\mathbb{R}^2$ (see (7.35)), which would lead to the physically desirable transformations (2.7)–(2.9) on the signal space. It is possible to reverse the argument, i.e., to start with the group SIM(2) as the primary object and then to obtain the space $\mathbb{R}^2$, over which the images are to be defined, as intrinsic to the group and on which it has the natural action given by (7.35). To see this, note first that the set of matrices in SIM(2),

$$\begin{pmatrix} ar_\theta & \vec{0} \\ \vec{0}^{\,T} & 1 \end{pmatrix}, \qquad a > 0, \quad \theta \in [0, 2\pi),$$

constitute the subgroup H of rotations and dilations and since,

$$\begin{pmatrix} ar_\theta & \vec{b} \\ \vec{0}^{\,T} & 1 \end{pmatrix} = \begin{pmatrix} \mathbb{I}_2 & \vec{b} \\ \vec{0}^{\,T} & 1 \end{pmatrix} \begin{pmatrix} ar_\theta & \vec{0} \\ \vec{0}^{\,T} & 1 \end{pmatrix}, \tag{7.47}$$

the set of matrices

$$\begin{pmatrix} \mathbb{I}_2 & \vec{y} \\ \vec{0}^{\,T} & 1 \end{pmatrix}, \qquad \vec{y} \in \mathbb{R}^2,$$

is identifiable with the quotient space $\mathrm{SIM}(2)/H$. Acting on such a matrix from the left, by an element of SIM(2), we see that,

$$\begin{pmatrix} ar_\theta & \vec{b} \\ \vec{0}^{\,T} & 1 \end{pmatrix} \begin{pmatrix} \mathbb{I}_2 & \vec{y} \\ \vec{0}^{\,T} & 1 \end{pmatrix} = \begin{pmatrix} ar_\theta & ar_\theta \vec{y} + \vec{b} \\ \vec{0}^{\,T} & 1 \end{pmatrix}$$

$$= \begin{pmatrix} \mathbb{I}_2 & ar_\theta \vec{y} + \vec{b} \\ \vec{0}^{\,T} & 1 \end{pmatrix} \begin{pmatrix} ar_\theta & \vec{0} \\ \vec{0}^{\,T} & 1 \end{pmatrix},$$

implying the transformation $\vec{y} \mapsto ar_\theta \vec{y} + \vec{b}$ on $\mathbb{R}^2$. Thus, the action of SIM(2) on the quotient space $\mathrm{SIM}(2)/H$ is the same as its action (7.35) on $\mathbb{R}^2$. The situation here is the same as that encountered in the case of the affine group (see (6.10) and the discussion following). One can just as well adopt the point of view that the group is the basic geometrical object, from which signals, their transformation properties and their representations in various spaces of functions, all follow as mathematical consequences.

7.2.3   Decomposition theory of 2-D wavelet transforms

As in the case of the affine group, we would like to identify the space $L^2(\mathrm{SIM}(2), d\mu)$ with the set of all finite energy 2-D wavelet transforms. In order to do this, we have to be able to decompose any vector $S \in L^2(\mathrm{SIM}(2), d\mu)$ into a sum (possibly infinite) of wavelet transforms of appropriate signals with respect to appropriate mother wavelets. For the affine group this was achieved in (6.41) and we would like to do the same in the 2-D case.

Since the Duflo–Moore operator C has an inverse, we can choose an orthonormal basis, $\{\phi_n\}_{n=1}^{\infty}$, in the signal space $L^2(\mathbb{R}^2, d^2\vec{x})$, such that each ϕ_n is in the domain of C^{-1}. Thus, the vectors $\psi_n = C^{-1}\phi_n$ are admissible. We recall next that the orthogonality condition (7.12) implies that all wavelet transforms are elements of $L^2(\mathrm{SIM}(2), d\mu)$. Specifically, if $S_n(\vec{b}, a, \theta)$ denotes the wavelet transform of a signal s with respect to the mother wavelet ψ_n and $S'_m(\vec{b}, a, \theta)$ that of a signal s' with respect to the mother wavelet ψ_m, then

$$\int_{\mathrm{SIM}(2)} d\mu(\vec{b}, a, \theta)\, \overline{S_n(\vec{b}, a, \theta)}\, S'_m(\vec{b}, a, \theta) = \delta_{nm}\, \langle s \,|\, s' \rangle. \tag{7.48}$$

For each $n = 1, 2, \ldots, \infty$, denoting by $\mathfrak{H}_{\psi_n}$ the space of all wavelet transforms with respect to the mother wavelet ψ_n, we infer from the above equation that these spaces are mutually orthogonal. Furthermore, $\oplus_{n=1}^{\infty} \mathfrak{H}_{\psi_n} \subseteq L^2(\mathrm{SIM}(2), d\mu)$, and since the group has only one irreducible unitary representation, one can in fact show that

$$\bigoplus_{n=1}^{\infty} \mathfrak{H}_{\psi_n} \simeq L^2(\mathrm{SIM}(2), d\mu), \tag{7.49}$$

where $\simeq$ denotes (unitary) equivalence. (A general discussion of such decompositions, and related results, may be found in [Ali00].) Thus, we have justified the expansion

$$S = \bigoplus_{n=1}^{\infty} S_n, \qquad S \in L^2(\mathrm{SIM}(2), d\mu), \tag{7.50}$$

of an arbitrary element of $L^2(\mathrm{SIM}(2), d\mu)$ in terms of wavelet transforms. Note that, in the above, $S_n(\vec{b}, a, \theta) = \langle U(\vec{b}, a, \theta)\psi_n \,|\, s \rangle$, for some signal vector $s \in L^2(\mathbb{R}^2, d^2\vec{x})$, where in general, s is different for different n. Note also that the sum in (7.50) holds in the sense of the L^2-norm, so that, although the functions $S_n(\vec{b}, a, \theta)$ are continuous in all variables, we only have $S(\vec{b}, a, \theta) = \sum_{i=1}^{\infty} S_n(\vec{b}, a, \theta)$ almost everywhere (with respect to $d\mu$).

We know, from the general theory outlined in Section 7.1.4, that each one of the subspaces $\mathfrak{H}_{\psi_n}$ is a reproducing kernel Hilbert space. Let

$$K_{\psi_n}(\vec{b}, a, \theta; \vec{b}', a', \theta') = \langle U(\vec{b}, a, \theta)\psi_n \,|\, U(\vec{b}', a', \theta')\psi_n \rangle$$

be the reproducing kernel for $\mathfrak{H}_{\psi_n}$. Then, given $S \in L^2(\mathrm{SIM}(2), d\mu)$, the component S_n appearing in (7.50) can be computed using (7.18):

$$S_n(\vec{b}, a, \theta) = \int_{\mathrm{SIM}(2)} d\mu(\vec{b}', a', \theta')\, K_{\psi_n}(\vec{b}, a, \theta; \vec{b}', a', \theta')\, S(\vec{b}', a', \theta'), \tag{7.51}$$

and, writing $S_n(\vec{b}, a, \theta) = \langle U(\vec{b}, a, \theta)\psi_n \,|\, s \rangle$, the signal vector s may be computed using (7.46):

$$s = \int_{\text{SIM}(2)} d\mu(\vec{b}, a, \theta) \, S_n(\vec{b}, a, \theta) \, \psi_{\vec{b}, a, \theta}. \tag{7.52}$$

If we introduce the basic wavelet transforms,

$$S_{nm}(\vec{b}, a, \theta) = \langle U(\vec{b}, a, \theta)\psi_n \mid \phi_m \rangle, \quad \phi_m = C\psi_m, \quad n, m = 1, 2, \ldots, \infty, \tag{7.53}$$

then, by (7.48) and the orthonormality of the ϕ_n, these functions are seen to satisfy

$$\int_{\text{SIM}(2)} d\mu(\vec{b}, a, \theta) \, \overline{S_{nm}(\vec{b}, a, \theta)} \, S_{k\ell}(\vec{b}, a, \theta) = \delta_{nk} \, \delta_{m\ell}. \tag{7.54}$$

Hence, any $S \in L^2(\text{SIM}(2), d\mu)$ has the orthogonal decomposition

$$S(\vec{b}, a, \theta) = \sum_{n,m=1}^{\infty} c_{nm} \, S_{nm}(\vec{b}, a, \theta), \tag{7.55}$$

with

$$c_{nm} = \int_{\text{SIM}(2)} d\mu(\vec{b}, a, \theta) \, \overline{S_{nm}(\vec{b}, a, \theta)} \, S(\vec{b}, a, \theta), \tag{7.56}$$

and

$$\|S\|^2 = \int_{\text{SIM}(2)} d\mu(\vec{b}, a, \theta) \, |S(\vec{b}, a, \theta)|^2 = \sum_{n=0}^{\infty} \sum_{m=-\infty}^{\infty} |c_{nm}|^2.$$

7.2.3.1   A concrete example

Finally, we give a concrete example of the decomposition (7.50), in terms of mother wavelets built out of the well-known trigonometric functions and Laguerre polynomials. In the Fourier-transformed signal space $L^2(\widehat{\mathbb{R}}^2, d^2\vec{k})$, we choose the basis vectors

$$\widehat{\phi}_{nm}(\vec{k}) = \frac{1}{(2\pi)^{1/2}} \, \varrho^{-1/2} \, e^{-\frac{\varrho}{2}} \, L_n(\varrho) \, e^{im\vartheta},$$
$$n = 0, 1, 2, \ldots, \infty, \quad m = 0, \pm 1, \pm 2, \ldots, \pm\infty, \tag{7.57}$$

where ϱ, ϑ are the polar coordinates of $\vec{k}$ and the $L_n(\varrho)$ are the Laguerre polynomials:

$$L_n(\varrho) = \sum_{k=1}^{n} \binom{n}{k} \frac{(-\varrho)^k}{k!}.$$

These satisfy the orthogonality relations,

$$\int_0^{\infty} L_m(\varrho) \, L_n(\varrho) \, e^{-\varrho} \, d\varrho = \delta_{mn},$$

implying that the $\widehat{\phi}_{nm}$ are orthonormal,

$$\langle \widehat{\phi}_{nm} \mid \widehat{\phi}_{k\ell} \rangle = \delta_{nk} \, \delta_{m\ell}. \tag{7.58}$$

The fact that they form a basis of $L^2(\widehat{\mathbb{R}}^2, d^2\vec{k})$, follows from well-known properties of Laguerre polynomials and trigonometric functions. Moreover, it is clear that the vectors $\widehat{\psi}_{nm} = \widehat{C}^{-1}\widehat{\phi}_{nm}$, where

$$\widehat{\psi}_{nm}(\vec{k}) = \frac{\varrho}{2\pi} \, \widehat{\phi}_{nm}(\vec{k})$$
$$= \frac{1}{(2\pi)^{\frac{3}{2}}} \varrho^{1/2} \, e^{-\frac{\varrho}{2}} \, L_n(\varrho) \, e^{im\vartheta},$$
$$n = 0, 1, 2, \dots, \infty, \quad m = 0, \pm 1, \pm 2, \dots, \pm\infty, \tag{7.59}$$

are also elements of the Hilbert space $L^2(\widehat{\mathbb{R}}^2, d^2\vec{k})$ and hence legitimate mother wavelets.

Using the mother wavelets $\widehat{\psi}_{nm}$, we can construct the spaces $\mathfrak{H}_{\psi_{nm}}$ of wavelet transforms $S_{nm}(\vec{b}, a, \theta) = \langle \widehat{U}(\vec{b}, a, \theta)\widehat{\psi}_{nm} \mid \widehat{s} \rangle$, $\widehat{s} \in L^2(\widehat{\mathbb{R}}^2, d^2\vec{k})$. The total space of all transforms would then decompose as:

$$L^2(\mathrm{SIM}(2), d\mu) \simeq \bigoplus_{n=0}^{\infty} \bigoplus_{m=-\infty}^{\infty} \mathfrak{H}_{\psi_{nm}}.$$

7.2.3.2   Decomposition into orthogonal angular channels

We saw, at the end of Section 6.1.2, how a 1-D wavelet transform can be analyzed into wavelet transforms in orthogonal channels. Here we carry out a similar decomposition of a 2-D wavelet transform into orthogonal angular channels, again following a suggestion in [105]. We use the mother wavelets $\widehat{\psi}_{nm}$ defined above. Let $\widehat{\psi}$ be an arbitrary mother wavelet in the Fourier domain. We may then write,

$$\widehat{\psi}(\vec{k}) = \sum_{n=1}^{\infty} \sum_{m=1}^{\infty} a_{mn} \widehat{\psi}_{mn}(\vec{k}) = \sum_{m=1}^{\infty} a_m(\varrho) \, e^{im\vartheta}, \tag{7.60}$$

where

$$a_m(\varrho) = \frac{1}{(2\pi)^{\frac{3}{2}}} \sum_{n=1}^{\infty} a_{nm}\varrho^{1/2} \, e^{-\frac{\varrho}{2}} \, L_n(\varrho), \qquad a_{mn} = \langle \widehat{\phi}_{mn} \mid C\widehat{\psi} \rangle. \tag{7.61}$$

The sum in (7.60) may be looked upon as a decomposition of the mother wavelet into angular channels. Next, writing

$$\widehat{\psi}_m(\vec{k}) = a_m(\varrho) \, e^{im\vartheta}, \tag{7.62}$$

we easily see that $\widehat{\psi}_m$ is a vector which is in the domain of the Duflo–Moore operator $\widehat{C}$ and hence it can be used as a mother wavelet. If $S_\psi(\vec{b}, a, \theta)$ is the 2-D wavelet transform of a signal $\widehat{s} \in L^2(\widehat{\mathbb{R}}^2, d^2\vec{k})$, then it is straightforward to verify that

$$S_\psi(\vec{b}, a, \theta) = \sum_{m=1}^{\infty} S_m(\vec{b}, a, \theta), \qquad S_m(\vec{b}, a, \theta) = \langle \widehat{U}(\vec{b}, a, \theta)\widehat{\psi}_m \mid \widehat{s} \rangle. \tag{7.63}$$

It is clear that, for $m \neq n$, the wavelet transforms S_m and S_n are orthogonal functions in $L^2(\mathrm{SIM}(2), d\mu)$, while the mother wavelets $\widehat{\psi}_m$ and $\widehat{\psi}_n$ are themselves orthogonal vectors in $L^2(\widehat{\mathbb{R}}^2, d^2\vec{k})$. Thus we call (7.63) a decomposition of the wavelet transform $S_\psi(\vec{b}, a, \theta)$ into orthogonal angular channels, the transform $S_m(\vec{b}, a, \theta)$ being the component along the m-th channel.

7.3   2-D wavelets on phase space

In Section 2.3.2 we pointed out how the 2-D wavelet transform could also be looked upon as a function on a physical phase space. Here we take up this point again and give a more exhaustive mathematical treatment of it. The SIM(2) group has only one nontrivial coadjoint orbit and hence only one phase space. Moreover, this phase space is topologically homeomorphic to the group itself, meaning that wavelet transforms may also be viewed upon as transforms built on this phase space. In order to analyze these features, it will first be necessary to study the matrix structure of the generators of the various transformations constituting the group.

7.3.1   Lie algebra and orbits

The four basic sets of operations of dilation, rotation and the two translations, each constitute one-parameter subgroups of SIM(2). More precisely, these subgroups are generated by group elements of the type

$$(\vec{b}, a, \theta) = (\vec{0}, e^t, 0), \qquad t \in \mathbb{R},$$
$$\text{or} \quad (\vec{b}, a, \theta) = (\vec{0}, 1, t), \qquad t \in [0, 2\pi),$$
$$\text{or} \quad (\vec{b}, a, \theta) = (e_i t, 1, 0), \qquad t \in \mathbb{R}, \quad i = 1, 2,$$

where

$$\vec{0} = \begin{pmatrix} 0 \\ 0 \end{pmatrix}, \quad e_1 = \begin{pmatrix} 1 \\ 0 \end{pmatrix}, \quad e_2 = \begin{pmatrix} 0 \\ 1 \end{pmatrix},$$

constitute one-parameter subgroups. A general element of SIM(2) can be written as a product of elements of these subgroups:

$$(\vec{b}, a, \theta) = (e_1 b_1, 1, 0) \, (e_2 b_2, 1, 0) \, (\vec{0}, 1, \theta) \, (\vec{0}, a, 0) \,, \quad \text{where} \quad \vec{b} = \begin{pmatrix} b_1 \\ b_2 \end{pmatrix}.$$

Generically, writing elements in any one of these subgroups as $g(t)$ and computing the derivative at the identity:

$$\frac{d}{dt} g(t) \Big|_{t=0},$$

as was done for the affine group in (6.77), we obtain the four 3×3 matrices

$$D = \begin{pmatrix} 1 & 0 & 0 \\ 0 & 1 & 0 \\ 0 & 0 & 0 \end{pmatrix}, \qquad J = \begin{pmatrix} 0 & -1 & 0 \\ 1 & 0 & 0 \\ 0 & 0 & 0 \end{pmatrix},$$

$$P_1 = \begin{pmatrix} 0 & 0 & 1 \\ 0 & 0 & 0 \\ 0 & 0 & 0 \end{pmatrix}, \qquad P_2 = \begin{pmatrix} 0 & 0 & 0 \\ 0 & 0 & 1 \\ 0 & 0 & 0 \end{pmatrix}. \tag{7.64}$$

Here D is the generator of dilations, $e^{tD} = (\vec{0}, e^t, 0)$, J that of rotations, $e^{tJ} = (\vec{0}, 1, t)$, and P_1, P_2 those of translations, $e^{tP_i} = (te_i, 1, 0)$, $i = 1, 2$. The four generators satisfy the commutation relations

$$[D, J] = 0, \qquad [D, P_i] = P_i, \quad i = 1, 2,$$
$$[J, P_1] = P_2, \qquad [J, P_2] = -P_1,$$
$$[P_1, P_2] = 0,$$

and together they generate the Lie algebra of SIM(2) which, in this case, is a four-dimensional real vector space. We denote this Lie algebra by $\mathfrak{sim}(2)$ and its dual space by $\mathfrak{sim}(2)^*$.

A general element of the Lie algebra $\mathfrak{sim}(2)$ has the form

$$X(\vec{\alpha}, \vec{\beta}) = \lambda D + \theta J + \beta_1 P_1 + \beta_2 P_2 = \begin{pmatrix} \lambda & -\theta & \beta_1 \\ \theta & \lambda & \beta_2 \\ 0 & 0 & 0 \end{pmatrix}, \tag{7.65}$$

with $\theta, \lambda, \beta_1, \beta_2 \in \mathbb{R}$ and

$$\vec{\alpha} = \begin{pmatrix} \lambda \\ \theta \end{pmatrix}, \qquad \vec{\beta} = \begin{pmatrix} \beta_1 \\ \beta_2 \end{pmatrix}. \tag{7.66}$$

Corresponding to a 2-vector $\vec{v}$, consider again, as in (2.52), the 2×2 matrix $\mathfrak{s}(\vec{v})$,

$$\mathfrak{s}(\vec{v}) = \begin{pmatrix} v_1 & -v_2 \\ v_2 & v_1 \end{pmatrix}, \qquad \vec{v} = \begin{pmatrix} v_1 \\ v_2 \end{pmatrix}. \tag{7.67}$$

Then, for any two vectors $\vec{v}$ and $\vec{w}$,

$$\mathfrak{s}(\vec{v})\mathfrak{s}(\vec{w}) = \mathfrak{s}(\vec{w})\mathfrak{s}(\vec{v}), \qquad \mathfrak{s}(\vec{v})\vec{w} = \mathfrak{s}(\vec{w})\vec{v}. \tag{7.68}$$

Using $\mathfrak{s}$, the general Lie algebra element (7.65) may be rewritten as

$$X(\vec{\alpha}, \vec{\beta}) = \begin{pmatrix} \mathfrak{s}(\vec{\alpha}) & \vec{\beta} \\ \vec{0} & 0 \end{pmatrix}. \tag{7.69}$$

Exponentiating the matrix $X(\vec{\alpha}, \vec{\beta})$ yields the group element

$$g = e^{X(\vec{\alpha},\vec{\beta})} = \begin{pmatrix} e^\lambda \, r_\theta & F(\mathfrak{s}(\vec{\alpha}))\vec{\beta} \\ \vec{0} & 1 \end{pmatrix}, \tag{7.70}$$

where now $0 \leqslant \theta < 2\pi$ and the 2×2 matrix $F(\mathfrak{s}(\vec{\alpha}))$ is defined as the sum of an infinite series:

$$F(\mathfrak{s}(\vec{\alpha})) = \mathbb{I}_2 + \frac{\mathfrak{s}(\vec{\alpha})}{2!} + \frac{[\mathfrak{s}(\vec{\alpha})]^2}{3!} + \frac{[\mathfrak{s}(\vec{\alpha})]^3}{4!} + \cdots . \tag{7.71}$$

We shall see below that every group element can be so obtained, by exponentiating an appropriate Lie algebra element.

It will be useful for the sequel to express the operator $F(\mathfrak{s}(\vec{\alpha}))$ in a somewhat different form. Let us define a function, sinch, of a real variable u as

$$\text{sinch } u = \frac{\sinh u}{u}, \qquad \text{sinch } 0 = 1. \tag{7.72}$$

This is a positive, infinitely differentiable function and so also is the related function

$$F(u) = e^{\frac{u}{2}} \text{ sinch } (\frac{u}{2}). \tag{7.73}$$

This latter function has the Taylor expansion

$$F(u) = e^{\frac{u}{2}} \text{ sinch } (\frac{u}{2}) = 1 + \frac{u}{2!} + \frac{u^2}{3!} + \frac{u^3}{4!} + \cdots . \tag{7.74}$$

For $u \neq 0$, we may also write

$$F(u) = e^{\frac{u}{2}} \text{ sinch } (\frac{u}{2}) = u^{-1}(e^u - 1). \tag{7.75}$$

Using the function F we now define the 2×2 matrix valued function

$$e^{\frac{A}{2}} \text{ sinch } \left(\frac{A}{2}\right) = F(A) = \mathbb{I}_2 + \frac{A}{2!} + \frac{A^2}{3!} + \frac{A^3}{4!} + \cdots , \quad F(\mathbb{O}_2) = \mathbb{I}_2, \tag{7.76}$$

for any 2×2 real matrix A and where $\mathbb{O}_2$ and $\mathbb{I}_2$ are, respectively, the 2×2 null and identity matrices. If $\det A \neq 0$, then

$$F(A) = e^{\frac{A}{2}} \text{ sinch } \left(\frac{A}{2}\right) = A^{-1}[e^A - \mathbb{I}_2], \tag{7.77}$$

and if, $\det[e^A - \mathbb{I}_2] \neq 0$, we shall also write

$$F(A)^{-1} = \frac{e^{-\frac{A}{2}}}{\text{sinch } \left(\frac{A}{2}\right)} = [e^A - \mathbb{I}_2]^{-1} \, A. \tag{7.78}$$

Hence, for $|\vec{\alpha}| \neq 0$,

$$F(\mathfrak{s}(\vec{\alpha})) = \frac{1}{\lambda^2 + \theta^2} \begin{pmatrix} \lambda & \theta \\ -\theta & \lambda \end{pmatrix} \begin{pmatrix} e^\lambda \cos\theta - 1 & -e^\lambda \sin\theta \\ e^\lambda \sin\theta & e^\lambda \cos\theta - 1 \end{pmatrix}$$

$$= e^{\frac{\mathfrak{s}(\vec{\alpha})}{2}} \operatorname{sinch}\left(\frac{\mathfrak{s}(\vec{\alpha})}{2}\right), \tag{7.79}$$

and

$$[F(\mathfrak{s}(\vec{\alpha}))]^{-1} = \frac{1}{2(\cosh\lambda - \cos\theta)} \begin{pmatrix} \cos\theta - e^{-\lambda} & \sin\theta \\ -\sin\theta & \cos\theta - e^{-\lambda} \end{pmatrix} \begin{pmatrix} \lambda & -\theta \\ \theta & \lambda \end{pmatrix}$$

$$= \frac{e^{-\frac{\mathfrak{s}(\vec{\alpha})}{2}}}{\operatorname{sinch}\left(\frac{\mathfrak{s}(\vec{\alpha})}{2}\right)}. \tag{7.80}$$

Going back to (7.70), we rewrite it as

$$g = e^{X(\vec{\alpha}, \vec{\beta})} = \begin{pmatrix} e^\lambda \, r_\theta & e^{\frac{\mathfrak{s}(\vec{\alpha})}{2}} \operatorname{sinch}\left(\frac{\mathfrak{s}(\vec{\alpha})}{2}\right) \vec{\beta} \\ \vec{0}^{\,T} & 1 \end{pmatrix}, \tag{7.81}$$

and writing $(\vec{b}, \theta, a)$ in the form given in (7.37), we find the relations

$$\lambda = \log a, \qquad \vec{\beta} = [F(\mathfrak{s}(\vec{\alpha}))]^{-1}\vec{b}, \tag{7.82}$$

between the group parameters and those of the Lie algebra. This also shows that any group element $(\vec{b}, \theta, a)$ can be written as the exponential of some element $X(\vec{\alpha}, \vec{\beta})$ in the Lie algebra.

The group SIM(2) acts on its Lie algebra $\mathfrak{sim}(2)$ via the adjoint action:

$$\operatorname{Ad}_{(\vec{b},a,\theta)}X(\vec{\alpha}, \vec{\beta}) = (\vec{b}, a, \theta)\, X(\vec{\alpha}, \vec{\beta})\, (\vec{b}, a, \theta)^{-1} = X'(\vec{\alpha}, \vec{\beta})$$

$$= \begin{pmatrix} \mathfrak{s}(\vec{\alpha}) & -\mathfrak{s}(\vec{b})\vec{\alpha} + a\, r_\theta\vec{\beta} \\ \vec{0}^{\,T} & 0 \end{pmatrix}, \tag{7.83}$$

(in computing the above, we have made use of the fact that $r_\theta \mathfrak{s}(\vec{\alpha})r_{-\theta} = \mathfrak{s}(\vec{\alpha})$.) Writing $X'(\vec{\alpha}, \vec{\beta}) = X(\vec{\alpha}', \vec{\beta}')$, we get the transformation rules for the components:

$$\vec{\alpha}' = \vec{\alpha}$$
$$\vec{\beta}' = -\mathfrak{s}(\vec{b})\vec{\alpha} + a\, r_\theta\vec{\beta}, \tag{7.84}$$

so that the matrix $M(\vec{b}, \theta, a)$ of the adjoint action in the $\{D, J, P_1, P_2\}$ basis becomes

$$M(\vec{b}, a, \theta) = \begin{pmatrix} \mathbb{I}_2 & \mathbb{O}_2 \\ -\mathfrak{s}(\vec{b}) & a\, r_\theta \end{pmatrix}, \qquad \begin{pmatrix} \vec{\alpha}' \\ \vec{\beta}' \end{pmatrix} = M(\vec{b}, a, \theta)\begin{pmatrix} \vec{\alpha} \\ \vec{\beta} \end{pmatrix}. \tag{7.85}$$

The adjoint action induces an action on the dual $\mathfrak{sim}(2)^*$ of the Lie algebra, which we now proceed to determine. Let $\{D^*, J^*, P_1^*, P_2^*\}$ denote the elements of the basis in $\mathfrak{sim}(2)^*$ which is dual to the basis $\{D, J, P_1, P_2\}$ of $\mathfrak{sim}(2)$. We write a general element $X^* \in \mathfrak{sim}(2)^*$ as

$$X^* = \alpha_1^* D^* + \alpha_2^* J^* + \beta_1^* P_1^* + \beta_2^* P_2^*, \quad \alpha_1^*, \alpha_2^*, \beta_1^*, \beta_2^* \in \mathbb{R}. \tag{7.86}$$

We also set

$$\vec{\alpha}^* = \begin{pmatrix} \alpha_1^* \\ \alpha_2^* \end{pmatrix}; \qquad \vec{\beta}^* = \begin{pmatrix} \beta_1^* \\ \beta_2^* \end{pmatrix} \quad \text{and} \quad \vec{\gamma} = \begin{pmatrix} \vec{\alpha}^* \\ \vec{\beta}^* \end{pmatrix}, \tag{7.87}$$

so that that the dual pairing between X and X^* is given by

$$\langle X^* ; X \rangle = \vec{\alpha}^* \cdot \vec{\alpha} + \vec{\beta}^* \cdot \vec{\beta}. \tag{7.88}$$

The matrix of the coadjoint action of a group element $(\vec{b}, a, \theta)$ on $\mathfrak{sim}(2)^*$, in the above basis, denoted $M^\sharp(\vec{b}, a, \theta)$, is the transposed inverse of the matrix (7.85) of the adjoint action:

$$M^\sharp(\vec{b}, a, \theta) = [M((\vec{b}, a, \theta)^{-1})]^T = \begin{pmatrix} \mathbb{I}_2 & a^{-1}\mathfrak{s}(\vec{b})^T r_\theta \\ \mathbb{O}_2 & a^{-1} r_\theta \end{pmatrix}. \tag{7.89}$$

Writing

$$\vec{\gamma}' = \begin{pmatrix} \vec{\alpha}^{*\prime} \\ \vec{\beta}^{*\prime} \end{pmatrix} = M^\sharp(\vec{b}, a, \theta)\vec{\gamma} = M^\sharp(\vec{b}, a, \theta) \begin{pmatrix} \vec{\alpha}^* \\ \vec{\beta}^* \end{pmatrix}, \tag{7.90}$$

we obtain the transformation rules for the components of the dual vectors X^*:

$$\vec{\alpha}^{*\prime} = \vec{\alpha}^* + a^{-1} r_\theta \mathfrak{s}(\vec{b})^T \vec{\beta}^*,$$
$$\vec{\beta}^{*\prime} = a^{-1} r_\theta \vec{\beta}^*, \qquad |\vec{\beta}^{*\prime}| = a^{-1}|\vec{\beta}^*|. \tag{7.91}$$

Since orbits under the coadjoint action are the sets

$$\mathcal{O}^* = \{M^\sharp(\vec{b}, a, \theta)\vec{\gamma}_0 \mid (\vec{b}, a, \theta) \in \text{SIM}(2)\}, \tag{7.92}$$

for fixed vectors $\vec{\gamma}_0 \in \mathbb{R}^4$, it is easy to see that there are exactly two types of orbits.

(i) *Trivial orbits:* these are degenerate orbits, which are single point sets $\{\vec{\gamma}_0\}$, obtained by choosing

$$\vec{\gamma}_0 = \begin{pmatrix} \alpha_1^* \\ \alpha_2^* \\ 0 \\ 0 \end{pmatrix}, \qquad \alpha_1^*, \alpha_2^* \in \mathbb{R}.$$

The isotropy subgroup of any such point $\vec{\gamma}_0$, i.e., the subgroup which leaves it invariant, is of course the entire group SIM(2).

(ii) *The open free orbit:* this is the only nontrivial orbit of SIM(2) and is obtained by choosing

$$\vec{\gamma_0} = \begin{pmatrix} 0 \\ 0 \\ 1 \\ 0 \end{pmatrix},$$

(7.93)

or by taking for $\vec{\gamma_0}$ any other vector such that at least one of its last two components is nonvanishing. This is the only orbit which concerns us here and we denote it by $\mathcal{O}^*$. Note that in this case the isotropy subgroup of $\vec{\gamma_0}$ (i.e., the subgroup of SIM(2) elements for which this vector is a fixed point) is just the trivial subgroup consisting of the identity element of SIM(2). This also means that topologically, the orbit is homeomorphic to the group space itself (the map, $(\vec{b}, a, \theta) \mapsto \vec{\gamma} = M^{\sharp}(\vec{b}, a, \theta)\vec{\gamma_0}$, from the group to the orbit, is open and free).

7.3.2   The coadjoint orbit $\mathcal{O}^*$ as a phase space

Since the orbit $\mathcal{O}^*$ is homeomorphic to the group SIM(2) itself, wavelet transforms $S(\vec{b}, a, \theta)$ may be considered as being transforms defined on this space. In other words, the parameters $\vec{b}, a, \theta$ in terms of which the signal is being analyzed, can be looked upon as phase space parameters. This is completely in line with the situation encountered earlier, for 1-D wavelets and Gabor transforms.

In order to understand better, the structure of the coadjoint orbit $\mathcal{O}^*$ as a physical phase space, let us first note that points on the orbit are obtained from (7.91) upon setting $\vec{\alpha}^* = \vec{0}$ and taking for $\vec{\beta}^*$ the two dimensional unit vector $e_1 = \begin{pmatrix} 1 \\ 0 \end{pmatrix}$ (see (7.93)). Thus a generic point $\vec{\gamma} \in \mathcal{O}^*$ is given as

$$\vec{\gamma} = \begin{pmatrix} \vec{\alpha}^* \\ \vec{\beta}^* \end{pmatrix} = \frac{1}{a}\begin{pmatrix} \sigma_3 \, r_\theta \, \vec{b} \\ r_\theta \, e_1 \end{pmatrix} = M^{\sharp}(\vec{b}, a, \theta)\begin{pmatrix} \vec{0} \\ e_1 \end{pmatrix},$$

$$\sigma_3 = \begin{pmatrix} 1 & 0 \\ 0 & -1 \end{pmatrix}, \qquad \sigma_3 \, r_{-\theta} = r_\theta \, \sigma_3.$$

(7.94)

Explicitly, with the group element $(\vec{b}, a, \theta)$ written as in (7.37), we find for the general phase space point $\vec{\gamma} \in \mathcal{O}^*$,

$$\vec{\gamma} = \begin{pmatrix} \alpha_1^* \\ \alpha_2^* \\ \beta_1^* \\ \beta_2^* \end{pmatrix} = \frac{1}{a}\begin{pmatrix} b_1 \cos\theta + b_2 \sin\theta \\ -b_1 \sin\theta + b_2 \cos\theta \\ \cos\theta \\ \sin\theta \end{pmatrix}.$$

(7.95)

The above relations can be solved to express the group parameters $(\vec{b}, a, \theta)$ in terms of the phase space variables $\vec{\alpha}^*$, $\vec{\beta}^*$:

$$\vec{b} = a \, r_\theta \, \sigma_3 \, \vec{\alpha}^*, \qquad a = \frac{1}{|\vec{\beta}^*|}, \qquad \theta = \tan^{-1}\left(\frac{\beta_2^*}{\beta_1^*}\right). \qquad (7.96)$$

This also reflects the fact that the orbit $\mathcal{O}^*$ is topologically homeomorphic to the group space, and moreover, from the form of (7.95) one infers that, geometrically, $\mathcal{O}^*$ is also the cotangent bundle of $\mathbb{R}_*^2$ (2-D plane with the origin removed), i.e.,

$$\mathcal{O}^* \simeq \mathrm{SIM}(2) \simeq \mathbb{R}^2 \times \mathbb{R}_*^2 = T^*\mathbb{R}_*^2. \qquad (7.97)$$

Indeed, consider the point

$$\vec{\beta}^* = \frac{1}{a}\begin{pmatrix} \cos\theta \\ \sin\theta \end{pmatrix} \in \mathbb{R}_*^2.$$

Differentiating with respect to a and θ, we get the two tangent vectors at $\vec{\beta}^*$:

$$\vec{t}_a = -\frac{1}{a^2}\begin{pmatrix} \cos\theta \\ \sin\theta \end{pmatrix}, \qquad \vec{t}_\theta = \frac{1}{a}\begin{pmatrix} -\sin\theta \\ \cos\theta \end{pmatrix}.$$

We may thus take, as basis for the tangent space at $\vec{\beta}^*$, the columns of the matrix,

$$\mathbf{T} = \frac{1}{a}\begin{pmatrix} \cos\theta & -\sin\theta \\ \sin\theta & \cos\theta \end{pmatrix}.$$

The columns of the transposed matrix:

$$\vec{t}_a^* = \frac{1}{a}\begin{pmatrix} \cos\theta \\ -\sin\theta \end{pmatrix}, \qquad \vec{t}_\theta^* = \frac{1}{a}\begin{pmatrix} \sin\theta \\ \cos\theta \end{pmatrix},$$

then form a basis for the cotangent (i.e., dual of the tangent) space at $\vec{\beta}^*$. An arbitrary element of this dual space has, therefore, the form

$$\vec{\alpha}^* = b_1\vec{t}_a^* + b_2\vec{t}_\theta^* = \frac{1}{a}\begin{pmatrix} b_1\cos\theta + b_2\sin\theta \\ -b_1\sin\theta + b_2\cos\theta \end{pmatrix}, \qquad b_1, b_2 \in \mathbb{R}.$$

Thus, we shall interpret $\vec{\beta}^*$ in (7.95) as representing a point in the manifold $\mathbb{R}_*^2$ and $\vec{\alpha}^*$ as a vector in its cotangent space at this point. Physically, one calls the $\vec{\alpha}^*$ *configuration space vectors*, while the $\vec{\beta}^*$ are *momentum vectors*.

Coadjoint orbits carry natural invariant measures under the group action (see, for example, [Kir76]). Using the transformation rules (7.91) under the coadjoint action, it is straightforward to compute the invariant measure on $\mathcal{O}^*$. Expressed in terms of the $\vec{\alpha}^*$, $\vec{\beta}^*$, it is

$$d\Omega(\vec{\alpha}^*, \vec{\beta}^*) = \frac{d^2\vec{\alpha}^* \, d^2\vec{\beta}^*}{|\vec{\beta}^*|^2} = \frac{d\alpha_1^* \, d\alpha_2^* \, d\beta_1^* \, d\beta_2^*}{\beta_1^{*2} + \beta_2^{*2}}. \tag{7.98}$$

(In the physical literature, this would be called the Liouville measure for this phase space.)

If we express the position vector $\vec{\beta}^*$ in polar coordinates ($\rho = |\vec{\beta}^*|$, θ), the coadjoint-invariant measure (7.98) transforms to

$$d\Omega(\vec{\alpha}^*, \rho, \theta) = d^2\vec{\alpha}^* \, \frac{d\rho}{\rho} \, d\theta. \tag{7.99}$$

In these coordinates it is easy to verify that the differential 2-form

$$\omega(\vec{\alpha}^*, \rho, \theta) = \frac{1}{\rho} \, d\alpha_1^* \wedge d\rho + d\alpha_2^* \wedge d\theta, \tag{7.100}$$

is invariant under the coadjoint action. Moreover, the coordinate transformations on phase space,

$$\vec{\alpha}^* \mapsto -\vec{\alpha}^*, \qquad \rho \mapsto \frac{1}{\rho}, \qquad \theta \mapsto -\theta, \tag{7.101}$$

leave this 2-form invariant, meaning that they constitute a canonical transformation of the phase space. (We might point out that the 2-form (7.100) is just the well-known Kirillov–Kostant–Souriau symplectic structure [Kir76], carried by coadjoint orbits.)

On the other hand, if we parametrize $\mathcal{O}^*$ by means of the group parameters $(\vec{b}, a, \theta)$ using (7.96), the coadjoint action (7.91) transforms to group multiplication from the left. In other words, the coadjoint action $\mathrm{Ad}_{g_0}^\sharp$, corresponding to the group element $g_0 = (\vec{b}_0, a_0, \theta_0)$ transforms the point in $\mathcal{O}^*$ represented by $(\vec{b}, a, \theta)$ as follows:

$$(\vec{b}, a, \theta) \mapsto (\vec{b}', a', \theta') = (\vec{b}_0, a_0, \theta_0)(\vec{b}, a, \theta) = (\vec{b}_0 + a_0 r_\theta \vec{b}, \; a_0 a, \; \theta_0 + \theta), \tag{7.102}$$

and thus the invariant measure (7.98) on $\mathcal{O}^*$, changes to precisely the left Haar measure $d\mu$ in (7.38) under this transformation [compare (2.51)]:

$$d\Omega(\vec{b}, a, \theta) = d\mu(\vec{b}, a, \theta) = \frac{1}{a^3} \, d^2\vec{b} \, da \, d\theta. \tag{7.103}$$

There is yet another parametrization of the points of the orbit $\mathcal{O}^*$, which in a sense is more natural than either the $(\vec{\alpha}^*, \vec{\beta}^*)$ or the $(\vec{b}, a, \theta)$ parametrizations. This other parametrization is given in terms of the so-called *Darboux coordinates*, which we denote by $(\vec{q}, \vec{p})$ and which are related to the other two sets of coordinates as

$$\vec{q} = \begin{pmatrix} q_1 \\ q_2 \end{pmatrix} = \frac{1}{\rho} r_\theta \, \sigma_3 \, \vec{\alpha}^* = \vec{b}$$

$$\vec{p} = \begin{pmatrix} p_1 \\ p_2 \end{pmatrix} = \vec{\beta}^* = \rho \begin{pmatrix} \cos\theta \\ \sin\theta \end{pmatrix} = \frac{1}{a} \begin{pmatrix} \cos\theta \\ \sin\theta \end{pmatrix}. \tag{7.104}$$

The transformation properties of these coordinates under the coadjoint action are also easily obtained. Once again, if $(\vec{q}\,', \vec{p}\,')$ is the transform of $(\vec{q}, \vec{p})$ under $\mathrm{Ad}^{\sharp}_{g_0}$, $g_0 = (\vec{b}_0, a_0, \theta_0)$, then

$$\vec{q}\,' = \vec{b}_0 + a_0\, r_{\theta_0}\, \vec{q}, \qquad \vec{p}\,' = a_0^{-1}\, r_{\theta_0}\, \vec{p}. \tag{7.105}$$

One can verify that the differential 2-form,

$$\omega(\vec{q}, \vec{p}) = dq_1 \wedge dp_1 + dq_2 \wedge dp_2 \tag{7.106}$$

is invariant under the above coadjoint action and hence the corresponding Liouville measure on $\mathcal{O}^*$

$$d\Omega(\vec{q}, \vec{p}) = d^2\vec{q}\, d^2\vec{p} = dq_1\, dq_2\, dp_1\, dp_2, \tag{7.107}$$

is also invariant under this action. It is not hard to see that (7.106) and (7.107) are precisely the transforms of the 2-form (7.100) and the measure (7.103), respectively, under the coordinate change (7.104).

This last choice of coordinates, and in particular the differential form (7.106) makes evident the phase space structure of the coadjoint orbit $\mathcal{O}^*$. The components of $\vec{q}$ refer to the position of the system on the configuration space $\mathbb{R}^2$, while at each such point the vector $\vec{p} \in \mathbb{R}^2_*$ denotes its canonical momentum. On the other hand, the fact that the group parameters $(\vec{b}, a, \theta)$ can also be used as coordinates for the orbit $\mathcal{O}^*$, shows that the group itself can be identified with the phase space as well. In this case, $\vec{b}$ denotes a point in configuration space and (a^{-1}, θ) are the polar coordinates of a momentum vector.

7.4   The affine Poincaré group

While the SIM(2) group is the most natural generalization of the affine group for building 2-D wavelets, it is by no means the only group which could be used. As a matter of fact, any group of the type $G = \mathbb{R}^2 \rtimes H$, where H is a group consisting of 2×2 matrices, which acting on $\mathbb{R}^2$ generates an *open free orbit*, can be used to build wavelets. In other words, if the group H is such that for some fixed 2-vector $\vec{x}$, the set,

$$\mathcal{O}_{\vec{x}} = \{\vec{y} = \mathsf{h}^T \vec{x} \mid \mathsf{h} \in H\},$$

is an open set in $\mathbb{R}^2$ and for all $\vec{x} \neq \vec{0}$, $\mathsf{h}\vec{x} = \vec{x}$ if and only if h is the identity matrix, then such a group G has square integrable representations and hence can be used to build wavelets. As an example, we briefly look at the *affine Poincaré group*. This group, which we denote by $\mathcal{P}_{\mathrm{aff}}$, is a semidirect product of the above type. It differs from the SIM(2) group in that the spatial rotations r_θ are replaced by *hyperbolic rotations* Λ_ϑ:

$$\Lambda_\vartheta = \begin{pmatrix} \cosh\vartheta & \sinh\vartheta \\ \sinh\vartheta & \cosh\vartheta \end{pmatrix}. \tag{7.108}$$

The set of matrices $\{\Lambda_\vartheta \mid \vartheta \in \mathbb{R}\}$ (note that $\det\Lambda_\vartheta = 1$ and $\Lambda_\vartheta^{-1} = \Lambda_{-\vartheta}$) constitutes a group, denoted $SO_0(1,1)$. (In physics, this is the group of relativistic transformations of a space–time having only one spatial dimension.) In dealing with this group, we shall use the physicists' convention of writing the components of a vector $\vec{x}$ as

$$\vec{x} = \begin{pmatrix} x_0 \\ \mathbf{x} \end{pmatrix}, \qquad x_0, \mathbf{x} \in \mathbb{R},$$

and use the *Minkowski inner product* between two such vectors:

$$\langle \vec{x} \; ; \; \vec{x}' \rangle = x_0 x_0' - \mathbf{x}\mathbf{x}' = \vec{x}^{\,T} g \vec{x}', \qquad g = \begin{pmatrix} 1 & 0 \\ 0 & -1 \end{pmatrix}, \qquad g^2 = \mathbb{I}_2. \tag{7.109}$$

Duality between $\mathbb{R}^2$ and $\widehat{\mathbb{R}}^2$ will also be defined using this inner product. If $\vec{x}' = \Lambda_\vartheta \vec{x}$, then $\langle \vec{x} \; ; \; \vec{x} \rangle = \langle \vec{x}' \; ; \; \vec{x}' \rangle$, so that hyperbolas $x_0^2 - \mathbf{x}^2 = const$ are mapped into themselves by $SO_0(1,1)$.

7.4.1 Group structure and representations

A general element of $\mathcal{P}_{\text{aff}}$ has the matrix representation,

$$(\vec{b}, a, \vartheta) = \begin{pmatrix} a\Lambda_\vartheta & \vec{b} \\ \vec{0}^{\,T} & 1 \end{pmatrix}, \qquad a > 0, \quad \vartheta \in \mathbb{R}, \quad \vec{b} = \begin{pmatrix} b_0 \\ \mathbf{b} \end{pmatrix} \in \mathbb{R}^2, \tag{7.110}$$

giving the group the structure of the semidirect product,

$$\mathcal{P}_{\text{aff}} = \mathbb{R}^2 \rtimes (\mathbb{R}_*^+ \times SO_0(1,1)).$$

Topologically, $\mathcal{P}_{\text{aff}} \simeq \mathbb{R}^2 \times \mathcal{C}$, where $\mathcal{C}$ is any one of the four open cones:

$$\mathcal{C}_\pm^\uparrow = \{\vec{x} \in \mathbb{R}^2 \mid x_0^2 > \mathbf{x}^2, \; \pm x_0 > 0\},$$
$$\mathcal{C}_\pm^\downarrow = \{\vec{x} \in \mathbb{R}^2 \mid x_0^2 < \mathbf{x}^2, \; \pm x_0 > 0\}. \tag{7.111}$$

The set of elements of the type $(\vec{b}, 1, 0)$, $\vec{b} \in \mathbb{R}^2$, is a commutative subgroup of $\mathcal{P}_{\text{aff}}$ and so also is the set of elements, $(\vec{0}, a, \vartheta)$, $a > 0$, $\vartheta \in \mathbb{R}$.

The affine Poincaré group is nonunimodular; the left and right Haar measures can be computed in exactly the same way that we computed them for the similitude group in Section 7.2.1. This time the two measures are

$$d\mu(\vec{b}, a, \vartheta) = \frac{1}{a^3}\, d^2\vec{b}\, da\, d\vartheta, \quad \text{and} \quad d\mu_r(\vec{b}, a, \vartheta) = \frac{1}{a}\, d^2\vec{b}\, da\, d\vartheta, \tag{7.112}$$

which look exactly the same as those for the SIM(2) group (to be expected, since the group $SO_0(1, 1)$ is unimodular). The group $\mathcal{P}_{\text{aff}}$ acts on the plane in the manner, $\vec{y} \mapsto a\Lambda_\vartheta \vec{y} + \vec{b}$ and therefore, we can again look for its representations in the signal space $L^2(\mathbb{R}^2, d^2\vec{x})$. As already noted, the situation here is largely similar to that of the similitude group. The signal symmetries again include translations and dilations; however, we have hyperbolic rotations now and not rigid rotations of space. Such signal symmetries could be expected in problems involving the detection of extremely fast moving objects (such as occurs, for example, in high energy physical experiments). Unlike the rotations r_θ, the action of Λ_ϑ actually has the effect of deforming the shapes of objects: the disc, $x_0^2 + \mathbf{x}^2 \leqslant r^2$, is transformed into the interior of the rotated ellipse, $x_0^2 + \mathbf{x}^2 + \tanh(2\vartheta)\, x_0\mathbf{x}_0 \leqslant \operatorname{sech}(2\vartheta)\, r^2$. Thus, if images are scanned using instruments which distort them in this manner, a group such as this could be more appropriate for their analysis than the similitude group.

The natural unitary representation of $\mathcal{P}_{\text{aff}}$ on the signal space $L^2(\mathbb{R}^2, d^2\vec{x})$, which reflects its action $\vec{y} \mapsto a\Lambda_\vartheta \vec{y} + \vec{b}$ on $\mathbb{R}^2$, is carried by the unitary operators $U(\vec{b}, a, \vartheta)$:

$$[U(\vec{b}, a, \vartheta)s](\vec{x}) = a^{-1}\, s(a^{-1}\Lambda_\vartheta(\vec{x} - \vec{b})), \tag{7.113}$$

an expression which should be compared to that for the similitude group in (2.13). The unitarity of these operators is straightforward to prove; however, in contrast with the operators (2.13), the operators $U(\vec{b}, a, \vartheta)$ do not carry an irreducible representation of $\mathcal{P}_{\text{aff}}$. For isolating the irreducible sectors, it is best to work in the Fourier domain. In order to do this, it will be convenient to adopt the physicists' convention for defining the Fourier transform, using the Minkowski inner product. Accordingly, we define $\mathcal{F} : L^2(\mathbb{R}^2, d^2\vec{x}) \to L^2(\widehat{\mathbb{R}}^2, d^2\vec{k})$,

$$(\mathcal{F}s)(\vec{k}) = \widehat{s}(\vec{k}) = \frac{1}{2\pi} \int_{\mathbb{R}^2} d^2\vec{x}\; e^{i\langle \vec{k}\,;\,\vec{x}\rangle}\, s(\vec{x}), \qquad s \in L^2(\mathbb{R}^2, d^2\vec{x}),$$

$$(\mathcal{F}^{-1}\widehat{s})(\vec{k}) = s(\vec{x}) = \frac{1}{2\pi} \int_{\widehat{\mathbb{R}}^2} d^2\vec{k}\; e^{-i\langle \vec{k}\,;\,\vec{x}\rangle}\, \widehat{s}(\vec{k}), \qquad \widehat{s} \in L^2(\widehat{\mathbb{R}}^2, d^2\vec{k}). \tag{7.114}$$

Using this Fourier transform, and the matrix identity, $g\Lambda_\vartheta g = \Lambda_{-\vartheta}$, the $U(\vec{b}, a, \vartheta)$ are seen to transform into the operators $\widehat{U}(\vec{b}, a, \vartheta)$ on $L^2(\widehat{\mathbb{R}}^2, d^2\vec{k})$:

$$\left[\widehat{U}(\vec{b}, a, \vartheta)\widehat{s}\right](\vec{k}) = a\, e^{i\langle \vec{k}\,;\,\vec{b}\rangle}\, \widehat{s}(a\Lambda_{-\vartheta}\vec{k}). \tag{7.115}$$

Let $\mathcal{C}$ be any one of the four open cones defined in (7.111). A quick computation shows that if $\vec{k} \in \mathcal{C}$, then also $a\Lambda_{-\vartheta}\vec{k} \in \mathcal{C}$. From the nature of the operator $\widehat{U}(\vec{b}, a, \vartheta)$ in (7.115) we see that if $\widehat{s}$ has support inside this cone, then so also does the transformed function $\widehat{U}(\vec{b}, a, \vartheta)\widehat{s}$. Thus, the Hilbert space $L^2(\mathcal{C}, d^2\vec{k})$, which is a subspace of $L^2(\widehat{\mathbb{R}}^2, d^2\vec{k})$, carries a subrepresentation of $\widehat{U}(\vec{b}, a, \vartheta)$, i.e., restricted to this subspace the operators $\widehat{U}(\vec{b}, a, \vartheta)$ again define a unitary representation of $\mathcal{P}_{\text{aff}}$. This representation can also be shown to be irreducible (see, for example, [Ali00] for a detailed proof of such results). Also, we have the obvious Hilbert space decomposition,

$$L^2(\widehat{\mathbb{R}}^2, d^2\vec{k}) = L^2(\mathcal{C}_+^\uparrow, d^2\vec{k}) \oplus L^2(\mathcal{C}_-^\uparrow, d^2\vec{k}) \oplus L^2(\mathcal{C}_+^\downarrow, d^2\vec{k}) \oplus L^2(\mathcal{C}_-^\downarrow, d^2\vec{k}),$$

and using a self-evident notation for the restrictions of $\widehat{U}(\vec{b}, a, \vartheta)$ to these four subspaces, we may write

$$\widehat{U}(\vec{b}, a, \vartheta) = \widehat{U}_+^\uparrow(\vec{b}, a, \vartheta) \oplus \widehat{U}_-^\uparrow(\vec{b}, a, \vartheta) \oplus \widehat{U}_+^\downarrow(\vec{b}, a, \vartheta) \oplus \widehat{U}_-^\downarrow(\vec{b}, a, \vartheta). \tag{7.116}$$

This shows that the representation $\widehat{U}(\vec{b}, a, \vartheta)$ is a direct sum of four irreducible representations, carried by four orthogonal subspaces of $L^2(\widehat{\mathbb{R}}^2, d^2\vec{k})$. Each one of these subspaces consists of signals whose supports are contained in a cone. Returning to the inverse Fourier domain, the signal space decomposes as:

$$L^2(\mathbb{R}^2, d^2\vec{x}) = \mathfrak{H}_+^\uparrow \oplus \mathfrak{H}_-^\uparrow \oplus \mathfrak{H}_+^\downarrow \oplus \mathfrak{H}_-^\downarrow,$$

where, for example, $\mathfrak{H}_+^\downarrow$ consists of all functions in $L^2(\mathbb{R}^2, d^2\vec{x})$ whose Fourier transforms have supports contained in $\mathcal{C}_+^\downarrow$. Correspondingly, the representation decomposes as

$$U(\vec{b}, a, \vartheta) = U_+^\uparrow(\vec{b}, a, \vartheta) \oplus U_-^\uparrow(\vec{b}, a, \vartheta) \oplus U_+^\downarrow(\vec{b}, a, \vartheta) \oplus U_-^\downarrow(\vec{b}, a, \vartheta). \tag{7.117}$$

Let us generically represent any one of these subrepresentations by $U_\mathcal{C}(\vec{b}, a, \vartheta)$ and in the Fourier domain by $\widehat{U}_\mathcal{C}(\vec{b}, a, \vartheta)$. This latter representation acts on the Hilbert space $L^2(\mathcal{C}, d^2\vec{k})$ of all square integrable functions supported in the cone $\mathcal{C}$.

7.4.2   Affine Poincaré wavelets

The representation $\widehat{U}_\mathcal{C}(\vec{b}, a, \vartheta)$ is known to be square integrable [Ali00]. Indeed, if we compute the integral

$$I(\widehat{\psi}) = \int_{\mathcal{P}_{\text{aff}}} d\mu(\vec{b}, a, \vartheta) \, |\langle \widehat{U}_\mathcal{C}(\vec{b}, a, \vartheta)\widehat{\psi} \mid \widehat{\psi} \rangle|^2,$$

for some $\widehat{\psi} \in L^2(\mathcal{C}, d^2\vec{k})$, we easily find

$$I(\widehat{\psi}) = (2\pi)^2 \, \|\widehat{\psi}\|^2 \int_\mathcal{C} \frac{d^2\vec{k}}{|k_0^2 - \boldsymbol{k}^2|} \, |\widehat{\psi}(\vec{k})|^2. \tag{7.118}$$

From this follows the admissibility condition for an *affine Poincaré wavelet*: a vector $\widehat{\psi} \in L^2(\mathcal{C}, d^2\vec{k})$ is admissible if and only if it satisfies the integrability condition $I(\widehat{\psi}) < \infty$, i.e., if and only if it is in the domain of the unbounded operator $\widehat{C}$ (the Duflo–Moore operator), where

$$(\widehat{C}\widehat{\psi})(\vec{k}) = \frac{2\pi}{|k_0^2 - \boldsymbol{k}^2|^{1/2}} \, \widehat{\psi})(\vec{k}). \tag{7.119}$$

Setting $c_{\widehat{\psi}} = \|C\widehat{\psi}\|^2$, we may also write down the resolution of the identity

$$\frac{1}{c_{\widehat{\psi}}} \int_{\mathcal{P}_{\text{aff}}} d\mu(\vec{b}, a, \vartheta) \, |\widehat{\psi}_{\vec{b}, a, \vartheta}\rangle\langle\widehat{\psi}_{\vec{b}, a, \vartheta}| = I, \qquad \widehat{\psi}_{\vec{b}, a, \vartheta} = \widehat{U}_\mathcal{C}(\vec{b}, a, \vartheta)\widehat{\psi}, \tag{7.120}$$

or, more generally,

$$\frac{1}{\langle \widehat{C\psi'} \mid \widehat{C\psi} \rangle} \int_{\mathcal{P}_{\text{aff}}} d\mu(\vec{b}, a, \vartheta) \, |\widehat{\psi}_{\vec{b}, a, \vartheta}\rangle \langle \widehat{\psi'}_{\vec{b}, a, \vartheta}| = I, \tag{7.121}$$

for two admissible vectors $\widehat{\psi}$, $\widehat{\psi'}$ such that $\langle \widehat{C\psi'} \mid \widehat{C\psi} \rangle \neq 0$. These equations should be compared to (7.41), (7.42), and (7.45).

We can now go ahead and define the *affine Poincaré wavelet transform* of an arbitrary signal $\widehat{s} \in L^2(\mathcal{C}, d^2\vec{k})$ by the quantity

$$S(\vec{b}, a, \vartheta) = \langle \widehat{\psi}_{\vec{b}, a, \vartheta} \mid \widehat{s} \rangle = a \int_{\mathcal{C}} d^2\vec{k} \, e^{-i\langle \vec{k}\,;\,\vec{b}\rangle} \, \overline{\widehat{\psi}(a\Lambda_{-\vartheta}\vec{k})} \, \widehat{s}(\vec{k}). \tag{7.122}$$

All the analysis carried out for the 2-D wavelet transform (obtained using the similitude group), including the phase space considerations, can again be repeated in the present case. In particular, any two-dimensional signal can be decomposed using affine Poincaré wavelets. If the signal $\widehat{s}$ is, for example, the quantum mechanical wave function of a fast-moving elementary particle (in a space–time of one time and one spatial dimension), the analyzing parameters, $\vec{b}$, a, and ϑ could represent its position (in space–time), its mass and its *rapidity*.

8 Minimal uncertainty and Wigner transforms

This chapter is devoted to a brief examination of two topics. The first concerns a certain minimality property of gaborettes and how it generalizes to wavelets in one and two dimensions. The second is an analysis of the Wigner transform, as an alternative to the wavelet transform. This latter transform is extensively used in certain physical computations and in the analysis of radar signals. Notice that neither of these topics is a prerequisite for the study of the more general wavelets described in Chapters 9 and 10.

8.1   Phase space distributions and minimal uncertainty gaborettes

The generalized gaborettes defined in (6.123), which give rise to holomorphic Gabor transforms, have a well-known *minimal uncertainty* property, related to localization in phase space. In (7.23) we had introduced the localization operators $a_\psi(\Delta)$. As discussed in Chapter 7, these operators can be used to measure the proportion of the signal transform S which is concentrated in the (phase space) region Δ. Consider the case of Gabor wavelets and let $\psi_{q,p} \in L^2(\mathbb{R}, \, dx)$ be the family of gaborettes defined in (6.109), using the window function ψ. These vectors satisfy the resolution of the identity (6.112). Assuming the normalization $\|\psi\|^2 = 1/2\pi$, we see that the operators

$$a_\psi(\Delta) = \int_\Delta dq \, dp \, |\psi_{q,p}\rangle\langle\psi_{q,p}|,$$

give rise to the probability measure

$$p_S(\Delta) = \frac{\langle s|a_\psi(\Delta)s\rangle}{\|s\|^2} = \frac{1}{2\pi \|s\|^2} \int_\Delta dq \, dp \, |S(q,p)|^2, \tag{8.1}$$

for any signal $s \in L^2(\mathbb{R}, \, dx)$ with Gabor transform S. In Section 6.3 it was noted that the Gabor transform $S(q,p)$ is a time–frequency transform, with q being the time and p the frequency parameter. On the other hand, if the signal $s = s(q)$ is given as a function over time, then its Fourier transform, $\widehat{s} = \widehat{s}(p)$ is a function over frequency. However, as is well known, the density distribution in time, $|s(q)|^2/\|s\|^2$, gives no information on the frequency content of the signal, while the frequency distribution, $|\widehat{s}(p)|^2/\|s\|^2$, gives no information on the variation of the signal with time. The phase space density, $|S(q,p)|^2$, does however, carry information on both time and frequency, but this information is not expected to be sharp, i.e., we do not expect $|s(q)|^2$ to be the marginal density of $|S(q,p)|^2$ in time or $|\widehat{s}(p)|^2$ to be its marginal in frequency. Indeed, computing these marginal densities, we find

$$S_1(q) = \int_{-\infty}^{\infty} dp \, |S(q,p)|^2 = \int_{-\infty}^{\infty} dx \, \chi_q(x) \, |s(x)|^2,$$

$$S_2(p) = \int_{-\infty}^{\infty} dq \, |S(q,p)|^2 = \int_{-\infty}^{\infty} dx \, \widetilde{\chi}_q(\xi) \, |\widehat{s}(\xi)|^2, \tag{8.2}$$

where

$$\chi_q(x) = 2\pi \, |\psi(x-q)|^2 \quad \text{and} \quad \widetilde{\chi}_p(\xi) = 2\pi \, |\widehat{\psi}(\xi - p)|^2, \tag{8.3}$$

are the shifted weight functions in time and frequency, respectively, generated by the window ψ. Thus, the marginal density $S_1(q)$ appears as a weighted average, at each point, over the *sharp* density $|s(x)|^2$ in time, while $S_2(p)$ appears as a similarly weighted average over the *sharp* density $|\widehat{s}(\xi)|^2$ in frequency. It is in this sense that we should think of $|S(q,p)|^2$ as being an averaged phase space density, the variables q and p representing averages of their "sharp" values, computed with respect to the probability densities $2\pi|\psi(x)|^2$ and $2\pi|\widehat{\psi}(\xi)|^2$, respectively. It then becomes pertinent to ask for what choice of window function, this averaging would be optimal, i.e., entail a minimum of unsharpness, for it is clear from (8.2) and (8.3) that there is no L^2-function ψ for which $S_1(q) = |s(q)|^2$ and $S_2(p) = |\widehat{s}(p)|^2$. Consider the standard deviations of the (sharp) time and frequency variables, measured with respect to the probability distributions $2\pi|\psi(x)|^2$ and $2\pi|\widehat{\psi}(\xi)|^2$, respectively:

$$\sigma_\psi = \left[2\pi \int_{-\infty}^{\infty} dx \, x^2 |\psi(x)|^2 - \left(2\pi \int_{-\infty}^{\infty} dx \, x |\psi(x)|^2 \right)^2 \right]^{\frac{1}{2}},$$

$$\sigma_{\widehat{\psi}} = \left[2\pi \int_{-\infty}^{\infty} d\xi \, \xi^2 |\widehat{\psi}(\xi)|^2 - \left(2\pi \int_{-\infty}^{\infty} d\xi \, \xi |\widehat{\psi}(\xi)|^2 \right)^2 \right]^{\frac{1}{2}}. \tag{8.4}$$

It is well known, from the theory of Fourier transforms, that their product satisfies the inequality $\sigma_\psi . \sigma_{\widehat{\psi}} \geqslant \frac{1}{2}$ and that equality is attained when $\psi(x)$ is a Gaussian as in (6.116) (or one of the other modified gaborettes $\psi_{(0,q,p)}$, constructed using such a Gaussian). Hence the choice of a Gaussian for the window function ψ would, in the light of the present analysis, lead to a Gabor transform $S(q,p)$ which measures the variables q and p with optimal accuracy.

In order to put the above discussion in operator terms and to make the connection with group theory again, let us assume that the chosen window function ψ satisfies the symmetry property $|\psi(x)|^2 = |\psi(-x)|^2$, almost everywhere. A fairly straightforward computation (see, for example, [5] for details) then leads to the following interesting average values:

$$\overline{q} = \int_{\mathbb{R}} dp_S(q, p) \, q = \frac{\langle s | Qs \rangle}{2\pi \|s\|^2}, \qquad \overline{p} = \int_{\mathbb{R}} dp_S(q, p) \, p = \frac{\langle s | Ps \rangle}{2\pi \|s\|^2}. \qquad (8.5)$$

where Q and P are (unbounded) self-adjoint operators, defined by the integral relations

$$Q = \int_{\mathbb{R}^2} dq \, dp \, q | \psi_{q,p} \rangle \langle \psi_{q,p} |, \qquad P = \int_{\mathbb{R}^2} dq \, dp \, p | \psi_{q,p} \rangle \langle \psi_{q,p} |. \qquad (8.6)$$

By (8.1), the phase space probability distribution p_S has the density

$$\rho(q, p) = \frac{1}{2\pi \|s\|^2} \, |S(q, p)|^2$$

and hence it follows from (8.5) that the operator Q (respectively, P) gives the mean value of the phase space position parameter q (respectively, momentum parameter p), computed using the probability distribution determined by the Gabor transform $S(q, p)$ of the signal s. Indeed, for an arbitrary signal vector s, having Gabor transform S corresponding to the window ψ, we get

$$\langle s | Qs \rangle = \int_{\mathbb{R}^2} dq \, dp \, q | S(q, p) |^2, \qquad \langle s | Ps \rangle = \int_{\mathbb{R}^2} dq \, dp \, p | S(q, p) |^2. \qquad (8.7)$$

It is remarkable that the values of the two integrals in (8.7) depend only on the signal vector and not on the window ψ. To see this, we note that the actions of these operators on the Hilbert space $L^2(\mathbb{R}, dx)$ are easily calculated. One obtains,

$$(Qs)(x) = x s(x), \qquad (Ps)(x) = -\frac{i}{\lambda} \frac{d}{dx} s(x), \qquad (8.8)$$

on vectors s taken from appropriate domains. Thus, for $\lambda^{-1} = \hbar$, Q and P are the well-known position and momentum operators of quantum mechanics, satisfying the commutation relations,

$$[Q, P] = QP - PQ = \frac{i}{\lambda} I, \qquad (8.9)$$

but now appearing explicitly as measures of average phase space position and momentum localization. This shows that Q and P are independent of the window function ψ, although their expressions in (8.6) appear in terms of it. Once again, this brings out a point we noted earlier: although the Gabor transform (like the wavelet transform) depends on the window, intrinsic quantities, such as mean values of the parameters of the transform, turn out to be independent of its choice. It is also possible to rewrite the unitary operators U^λ in (6.108), realizing an irreducible representation of the Weyl–Heisenberg group, in terms of the operators Q and P [Ali00]. We get

$$U^\lambda(\theta, q, p) = e^{i\lambda(\theta + pQ - qP)}. \qquad (8.10)$$

This means that Q, P (together with the identity operator I) also constitute a Hilbert space representation of the generators (6.106) of the Lie algebra of the Weyl–Heisenberg group.

Given any self-adjoint operator A on a Hilbert space $\mathfrak{H}$, we define its mean value $\langle A \rangle$ in the state (i.e., normalized vector) $\phi \in \mathfrak{H}$ by $\langle A \rangle \equiv \langle A \rangle_\phi = \langle \phi | A\phi \rangle$ and its standard deviation ΔA (in the state ϕ) by $\Delta A \equiv \Delta_\phi A = \sqrt{\langle A^2 \rangle - \langle A \rangle^2}$. In standard quantum mechanical lore, it well-known [Coh89,Got66], that given two self-adjoint operators A and B, the product of their standard deviations obeys the *uncertainty relation*

$$\Delta A . \Delta B \geqslant \frac{1}{2} |\langle [A, B] \rangle|, \qquad [A, B] = AB - BA. \tag{8.11}$$

The state ϕ is said to have *minimum uncertainty* if equality holds in (8.11), which happens if and only if

$$(A - \langle A \rangle)\phi = -i\lambda_o(B - \langle B \rangle)\phi, \tag{8.12}$$

for some $\lambda_o > 0$. By (8.9), for the operators Q and P, the uncertainty relation (8.11) assumes the form:

$$\Delta Q . \Delta P \geqslant \frac{1}{2\lambda}, \tag{8.13}$$

which in this case is exactly the same relation as (8.4). Thus, minimal uncertainty is attained for vectors of the type:

$$\phi(x) = \left[\frac{\lambda^2}{\pi} \right]^{\frac{1}{4}} e^{i\lambda(x - \frac{q}{2})p} \, e^{-\frac{\lambda^2(x-q)^2}{2}}, \tag{8.14}$$

for fixed $\lambda \neq 0$ and $q, p \in \mathbb{R}$. Of course, these vectors are precisely the generalized gaborettes defined in (6.123), which minimize the product of the standard deviations (8.4) and which give rise to holomorphic Gabor transforms. Thus, referring to our previous discussion, for a signal vector s, the absolute square of its Gabor transform $|S(q, p)|^2$ can be interpreted as giving a sort of "unsharp" joint probability distribution of the position (which now appears as the spectrum of the operator Q) and momentum (the spectrum of P) variables, the uncertainty relation (8.13) forbidding the existence of a sharp joint distribution. However, since the uncertainty is the smallest when the window is a vector of the type (8.13), the corresponding Gabor transform is in a sense optimal. In the following section, we shall use the ideas developed here to construct minimal uncertainty wavelets. We shall proceed group theoretically and isolate two generators of the corresponding representation and use their commutation relation to compute and minimize the uncertainty.

8.2   Minimal uncertainty wavelets

We turn our attention now to determining minimal uncertainty wavelets in one and two dimensions. As we know, the relevant groups are $G_{\text{aff}}^{(+)}$ and SIM(2) and we have to work with the representations of their Lie algebras on the Hilbert spaces of irreducible representations [24,116]. Consider first the one-dimensional case. We know from Section 6.2.2 that the Lie algebra of $G_{\mathrm{aff}}^{(+)}$ is two-dimensional. The two matrix generators of this algebra, X_1 and X_2, computed in (6.77) are represented on the Hilbert space $\widehat{\mathfrak{H}}_+(\mathbb{R})$ of the irreducible representation $\widehat{U}_+(b, a)$ (see (6.59)–(6.61)) by the generators $\widehat{D}$ and $\widehat{P}$, of dilation and translation, respectively. They act on Hilbert space vectors in the manner

$$(\widehat{D}\,\hat{s})(\xi) = -i\,([\tfrac{1}{2} + \xi\tfrac{\xi}{d\xi}]\,\hat{s})(\xi), \qquad (\widehat{P}\,\hat{s})(\xi) = \xi\,\hat{s}(\xi), \tag{8.15}$$

where $\hat{s} \in \widehat{\mathfrak{H}}_+(\mathbb{R})$ are chosen from the appropriate domains of the unbounded operators. (Note, this is the "momentum space representation" of these operators.) The operators $\widehat{D}$ and $\widehat{P}$ satisfy the commutation relations (see (6.78))

$$[\widehat{D}, \widehat{P}] = i\,\widehat{P}. \tag{8.16}$$

The minimal uncertainty vectors in this case are found to be [246,Pau85,305] the 1-D Cauchy wavelets, namely, $\widehat{\psi}_m(\xi) = \xi^m\,e^{-\xi}$, for $\xi \geqslant 0$ $(m > 0)$ and 0, otherwise. Note that these are also the wavelets which lead to holomorphic wavelet transforms.

A similar analysis applies for 2-D wavelets. The Lie algebra is now four-dimensional. The four generators, denoted by P_1, P_2 for translations, D for dilations and J for rotations, may be derived explicitly from the transformation (7.35), or its action (2.13) on signals or its equivalent (2.14) in $\vec{k}$-space. (Note that we are using the same notation as for the corresponding matrix generators of the Lie algebra (see (7.64)).) Among these four operators, there are four nonzero commutators, namely

$$[D, P_1] = i\,P_1; \quad [J, P_2] = -i\,P_1; \quad [D, P_2] = i\,P_2; \quad [J, P_1] = i\,P_2, \tag{8.17}$$

but the first two transform into the last two under a rotation by $\pi/2$. More generally, defining $P_\gamma = P_1 \cos\gamma + P_2 \sin\gamma$, we replace both pairs of commutators by the relations:

$$[D, P_\gamma] = i\,P_\gamma; \quad [J, P_{\gamma+\pi/2}] = -i\,P_\gamma, \tag{8.18}$$

and we look for wavelets which are minimal with respect to this pair. Thus, minimality has to be defined with respect to a fixed direction $\vec{e}_\gamma$, and it is impossible to do it for two directions at the same time, for instance, for all four relations (8.17) simultaneously [116]. For fixing ideas, we consider the uncertainty relations for the first pair in (8.17), corresponding to $\gamma = 0$:

$$\Delta D.\Delta P_1 \geqslant \tfrac{1}{2}|\langle P_1\rangle|; \qquad \Delta J.\Delta P_2 \geqslant \tfrac{1}{2}|\langle P_1\rangle|. \tag{8.19}$$

Then, according to (8.12), a vector $\widehat{\psi}$ saturates these inequalities iff it satisfies the following system of equations

$$
\begin{aligned}
(D + i\lambda_1 P_1)\widehat{\psi}(\vec{k}) &= (\langle D \rangle + i\lambda_1 \langle P_1 \rangle)\widehat{\psi}(\vec{k}) \\
(J + i\lambda_2 P_2)\widehat{\psi}(\vec{k}) &= (\langle J \rangle + i\lambda_2 \langle P_1 \rangle)\widehat{\psi}(\vec{k})
\end{aligned}
\qquad (\lambda_1,\ \lambda_2 > 0).
\tag{8.20}
$$

This system of partial differential equations may be solved, in polar coordinates $\vec{k} = (\rho, \phi)$, imposing successively five conditions: (i) Integrability of the system requires $\lambda_1 = \lambda_2 = \lambda > 0$. (ii) 2π-periodicity in ϕ implies that $\langle P_2 \rangle = 0$ and $\langle J \rangle = m \in \mathbb{Z}$. (iii) Square integrability, $\widehat{\psi} \in L^2$, implies that the support of $\widehat{\psi}$ is restricted to a convex cone in the right half-plane. (iv) Admissibility of $\widehat{\psi}$ implies $\lambda \langle P_1 \rangle > 1$. (v) Finally, imposing the condition that $\widehat{\psi}(\vec{k})$ be real implies that $\langle J \rangle = \langle D \rangle = 0$.

The result is that a real wavelet $\widehat{\psi}$ is minimal with respect to the first pair of the commutation relations (8.17) iff it vanishes outside some convex cone $\mathcal{C}$ in the half-plane $k_x > 0$ and is exponentially decreasing inside:

$$
\widehat{\psi}(\vec{k}) =
\begin{cases}
c\,|\vec{k}|^\kappa\, e^{-\lambda k_x} & (\kappa > 0,\ \lambda > 0), \quad \vec{k} \in \mathcal{C}, \\
0, & \text{otherwise.}
\end{cases}
\tag{8.21}
$$

$$
\widehat{\psi}(\vec{k}) = c\, \chi_c(\vec{k})\, |\vec{k}|^\kappa\, e^{-\lambda k_x} \qquad (\kappa > 0,\ \lambda > 0),
\tag{8.22}
$$

where χ_c is the characteristic function of $\mathcal{C}$, or a smoothened version thereof.

More generally, if one chooses the commutation relations in (8.18), one obtains a similar result, rotated by γ, that is, a wavelet supported in a convex cone $\mathcal{C}_\gamma$ with axis in the direction $\vec{e}_\gamma$, and exponentially decreasing inside.

We may now impose some degree of regularity (vanishing moments) at the boundary of the cone, by taking an appropriate linear superposition of such minimal wavelets $\widehat{\psi}$. Thus we obtain finally:

$$
\widehat{\psi}^C(\vec{k}) = c\, \chi_c(\vec{k})\, F(\vec{k})\, e^{-\lambda k_x}, \qquad (\lambda > 0)
\tag{8.23}
$$

where $F(\vec{k})$ is a polynomial in k_x, k_y, vanishing at the boundaries of the cone $\mathcal{C}$, including the origin. Clearly a Cauchy wavelet is of this type but, of course, one uses in practice a narrow support cone $\mathcal{C}$, in order to obtain good directional selectivity, as discussed in Section 3.3.4.

Other minimal wavelets may be obtained if one includes commutators with elements of the enveloping algebra, i.e., polynomials in the generators. For instance, if one requires the wavelet to be rotation invariant, one may start from the commutator between D and the Laplacian $-\Delta = P_1^2 + P_2^2$. Then one finds a whole family of minimal isotropic wavelets, among them all powers of the Laplacian, Δ^n, acting on a Gaussian, i.e., the wavelets (3.7) [22]. For $n = 2$, this gives the 2-D isotropic Mexican hat (3.6) [116]. There exist more general solutions of the minimizing equations, but most of them are not square integrable.

We ought to emphasize at this point, that the property of minimality for wavelets is a mathematical one, and it is not clear whether it implies an operational meaning in the same way as was discussed for gaborettes. Cauchy wavelets are linear combinations of minimal wavelets, but they are not the most efficient conical wavelets for directional analysis. This is not new: in 1-D too, the Cauchy–Paul wavelet ([Pau85]) is minimal, but many others are as least as useful in practice, for instance the derivatives of the Gaussian or the Morlet wavelet.

As a last remark, it may be interesting to note that a concept closely related to minimality has been developed by Simoncelli *et al.* [341] under the name of *jointly shiftable filters*. First, shiftable filters are the natural generalization of steerable filters to other variables than rotations, such as translation or scaling. Then a filter is jointly shiftable in two variables simultaneously if and only if the corresponding operations commute (i.e., "are independent"). Thus strict joint shiftability is impossible for position and spatial frequency; only approximately shiftable filters exist and the optimal ones, that is, those that minimize the "joint aliasing", are the same as our minimal wavelets.

8.3   Wigner functions

In Chapter 1, we mentioned the Wigner–Ville transform as an example of a signal transform which is quadratic (or generally, sesquilinear) in the signal vector(s). We now take a closer look at this transform. Wigner functions have long been used in signal analysis as phase space transforms of, generally, a pair of signal vectors. The original transform, due to Wigner [373], was introduced as a phase space *quasi-probability* density for computations in atomic physics. We begin by introducing the *cross-Wigner function*, $W^\lambda(\psi, \phi | q, p)$, of two signal vectors $\psi, \phi \in L^2(\mathbb{R}, dx)$:

$$W^\lambda(\psi, \phi | q, p) = \frac{\lambda}{2\pi} \int_{\mathbb{R}} dx \; \overline{\psi(q - \frac{x}{2})} e^{-i\lambda \, xp} \phi(q + \frac{x}{2}), \qquad \lambda > 0, \qquad (8.24)$$

adopting a somewhat different notational convention than used in (1.5). The phase space here is $\mathbb{R}^2$, with the variables q and p identified either as time and frequency or position and momentum. The superscript λ will eventually be identified with the parameter labeling the representation of the Weyl–Heisenberg group (see (6.108)). The form of the expression for the cross-Wigner function makes it plausible to look upon it as a mapping of the rank-one operator $\rho = |\phi\rangle\langle\psi|$ to a function of the phase space variables q, p. (Recall that the operator ρ acts on an arbitrary vector $\chi \in L^2(\mathbb{R}, dx)$ by $\rho\chi = \langle\psi|\chi\rangle \, \phi$.) Indeed, the integral on the right-hand side of (8.24) can be manipulated to be brought into the form

$$W^\lambda(\psi, \phi | q, p) \equiv W^\lambda(\rho | q, p)$$
$$= \frac{\lambda}{2\pi} \int_{\mathbb{R}^2} dq' \, dp' \, e^{i\lambda(qp' - pq')} \, \mathrm{Tr}[e^{-i\lambda(Qp' - Pq')} \rho], \qquad (8.25)$$

where Q and P are the operators defined in (8.8), and "Tr" denotes the trace of an operator:

$$\text{Tr}\rho = \sum_{i=k}^{\infty} \langle \phi_k | \rho \phi_k \rangle,$$

$\{\phi_k\}_{k=1}^{\infty}$ being an orthonormal basis of $L^2(\mathbb{R}, dx)$.

A fairly straightforward computation also shows that the unitary operator appearing within the square brackets on the right-hand side of (8.25) is none other than the representation operator of the Weyl–Heisenberg group, appearing in (6.108). Indeed,

$$e^{i\lambda(Qp-Pq)} = U^\lambda(0, q, p),$$

and hence

$$W^\lambda(\rho|q, p) = \frac{\lambda}{2\pi} \iint_{\mathbb{R}^2} dq' \, dp' \, e^{i\lambda(qp'-pq')} \, \text{Tr}[U^\lambda(0, q', p')^* \rho], \qquad (8.26)$$

which clarifies the relationship of the superscript λ, in the definition of the Wigner function, to representations of the Weyl–Heisenberg group. Let us introduce the *symplectic Fourier transform*, $\widetilde{f}$ of a function $f \in L^2(\mathbb{R}^2, dq\, dp)$:

$$\widetilde{f}(q, p) = \frac{\lambda}{2\pi} \iint_{\mathbb{R}^2} dq' \, dp' \, e^{i\lambda(qp'-pq')} \, f(q', p'),$$

with inverse,

$$f(q, p) = \frac{\lambda}{2\pi} \iint_{\mathbb{R}^2} dq' \, dp' \, e^{i\lambda(qp'-pq')} \, \widetilde{f}(q', p').$$

Clearly, the symplectic Fourier transform is a Hilbert space isometry. Then (assuming for the moment that $W^\lambda(\rho|q, p)$ is an L^2-function), we may write

$$\text{Tr}[U^\lambda(0, q, p)^* \rho] = \widetilde{W}^\lambda(\rho|q, p). \qquad (8.27)$$

Comparing with (6.125), we see that the symplectic Fourier transform of the cross-Wigner function $W^\lambda(\psi, s|q, p)$ is just the generalized Gabor transform of the signal s, computed using the window ψ:

$$\widetilde{S}(q, p) = \widetilde{W}^\lambda(\psi, s|q, p). \qquad (8.28)$$

However, it ought to be emphasized that, while the Gabor transform $\widetilde{S}(q, p)$ is an L^2-transform, on phase space, of the signal s, the cross-Wigner function $W^\lambda(\psi, s|q, p)$ is to be looked upon as an L^2-transform, on the same phase space, but of the rank-one operator $\rho = |s\rangle\langle\psi|$.

The original Wigner function [373] was defined with $\psi = \phi$ and $\rho = |\psi\rangle\langle\psi|$. For this case, we shall use the simpler notation $W^\lambda(\psi|q, p)$, and call it the Wigner function for the wave function ψ (or more accurately, for the operator ρ). This function has a number of well-known properties, which make it resemble a probability distribution. However, as is clear from its definition, for a general ψ, its Wigner function $W^\lambda(\psi|q, p)$ is not positive for all q, p. In fact, it is only when ψ is a Gaussian of the type in (6.122), that $W^\lambda(\psi|q, p)$ is everywhere positive. It is for this reason that the Wigner function is also called a *quasi-probability distribution*. Nevertheless, it is still possible to think of it as a phase space transform, which in a certain sense, is the signature of the signal vector. A few properties of the Wigner function $W^\lambda(s|q, p)$ of a signal vector s are now in order.

(i) **Reality:** The Wigner function is real-valued,

$$W^\lambda(s|q, p) = \overline{W^\lambda(s|q, p)}. \tag{8.29}$$

This is not generally true of the cross-Wigner function, for which one has the *hermiticity condition*,

$$W^\lambda(\psi, \phi|q, p) = \overline{W^\lambda(\phi, \psi|q, p)}. \tag{8.30}$$

(ii) **Trace condition:**

$$\iint_{\mathbb{R}^2} dq\, dp\, W^\lambda(s|q, p) = \mathrm{Tr}\rho = \|s\|^2 \qquad \rho = |s\rangle\langle s|. \tag{8.31}$$

This condition is reminiscent of the fact that for a probability distribution, the total probability equals one (as would be the case when $\|s\|^2 = 1$).

(iii) **Marginality:**

$$\int_{\mathbb{R}} dp\, W^\lambda(s|q, p) = |s(q)|^2, \qquad \int_{\mathbb{R}} dq\, W^\lambda(s|q, p) = |\widehat{s}(p)|^2. \tag{8.32}$$

If we think of $|s(q)|^2$ as giving the distribution of the signal in time (or position) and $|\widehat{s}(q)|^2$ its distribution in frequency (or momentum), then the above relations make $W^\lambda(s|q, p)$ formally look like a joint time–frequency (or position–momentum) distribution. Again, the fact that this "joint distribution" is not everywhere positive is a reflection of the uncertainty relations (8.13).

(iv) **Covariance:** An important property, which the Wigner function inherits from the Weyl–Heisenberg group, is reflected in the covariance relation:

$$W^\lambda(U^\lambda(0, q_0, p_0)s|q, p) = W^\lambda(s|q - q_0, p - p_0). \tag{8.33}$$

Identifying the phase space variables with coordinates on a coadjoint orbit of the Weyl–Heisenberg group, the transformation $q \to q - q_0$, $p \to p - p_0$ is the coadjoint action discussed in Section 6.3.2. Thus, the above relation expresses symmetry under the natural phase space transformations.

We come now to the problem of reconstructibility. In general it is not possible to reconstruct the signal s itself from its Wigner function; however, the operator $|s\rangle\langle s|$ can be recovered from it. More generally, the operator $|\phi\rangle\langle\psi|$ can be reconstructed from the cross-Wigner function $W^\lambda(\psi, \phi|q, p)$. In order to obtain an inversion formula, we go back to the general orthogonality relation (7.12) and see that in the present case it leads to the relation

$$\frac{\lambda}{2\pi} \iint_{\mathbb{R}^2} dq\, dp\, \overline{\widetilde{S}_{\psi'}(q, p)}\, \widetilde{S}_\psi(q, p) = \langle \psi | \psi' \rangle \langle s' | s \rangle, \tag{8.34}$$

where $\widetilde{S}_{\psi'}(q, p)$ is the generalized Gabor transform of the signal s, computed using the window ψ' and $\widetilde{S}_\psi(q, p)$ that of s computed using ψ. We note, however, that in this case, the above relation holds for any four vectors ψ, ψ', s, and s'. Moreover, we immediately get from it the resolution of the identity

$$\frac{\lambda}{2\pi} \iint_{\mathbb{R}^2} dq\, dp\, |\psi'_{(0,q,p)}\rangle\langle \psi_{(0,q,p)}| = \langle \psi | \psi' \rangle\, I, \tag{8.35}$$

again for arbitrary $\psi, \psi' \in L^2(\mathbb{R}, dx)$, and where, of course,

$$\psi_{(0,q,p)} = U^\lambda(0, q, p)\psi, \qquad \psi'_{(0,q,p)} = U^\lambda(0, q, p)\psi'.$$

Let $\rho = |\psi\rangle\langle\phi|$ and consider $\widetilde{W}^\lambda(\rho|q, p)$, the symplectic Fourier transform of its cross-Wigner function. Then, for any $\psi' \in L^2(\mathbb{R}, dx)$ we have,

$$\left[\frac{\lambda}{2\pi} \iint_{\mathbb{R}^2} dq\, dp\, \widetilde{W}^\lambda(\rho|q, p) U^\lambda(0, q, p) \right] \psi'$$

$$= \frac{\lambda}{2\pi} \iint_{\mathbb{R}^2} dq\, dp\, |\psi'_{(0,q,p)}\rangle\langle\phi_{(0,q,p)}|\psi\rangle = \langle\phi|\psi'\rangle\, \psi$$

$$= \rho\psi'.$$

Since ψ' is arbitrary in the above expression, we obtain the reconstruction formula,

$$\rho = \frac{\lambda}{2\pi} \iint_{\mathbb{R}^2} dq\, dp\, \widetilde{W}^\lambda(\rho|q, p) U^\lambda(0, q, p), \tag{8.36}$$

with

$$\widetilde{W}^\lambda(\rho|q, p) = \frac{\lambda}{2\pi} \iint_{\mathbb{R}^2} dq'\, dp'\, e^{i\lambda(qp' - pq')} W^\lambda(\rho|q', p').$$

We can exploit the orthogonality relations (8.34) to extend the definition of the cross-Wigner function to arbitrary Hilbert–Schmidt operators on $L^2(\mathbb{R}, dx)$. Recall that these operators form a Hilbert space with respect to the scalar product $\langle \rho_1 | \rho_2 \rangle_2 = \text{Tr}[\rho_1^* \rho_2]$. Let us denote this Hilbert space by $\mathcal{B}_2(\mathbb{R})$. It is well-known that finite linear combinations of rank-one operators form a dense set in $\mathcal{B}_2(\mathbb{R})$. Next note that using (8.28) and the fact that the symplectic Fourier transform is an isometry, we may transform (8.34) into

$$\frac{\lambda}{2\pi} \iint_{\mathbb{R}^2} dq\, dp\, \overline{W^\lambda(\rho'|q, p)}\, W^\lambda(\rho|q, p) = \text{Tr}[\rho'^* \rho], \tag{8.37}$$

where

$$\rho = |s\rangle\langle\psi|, \qquad \rho' = |s'\rangle\langle\psi'|.$$

The relation (8.37) remains valid if we replace ρ, ρ' by finite linear combinations of rank-one operators, meaning that the map given by the integral on the right-hand side of (8.25), associating an operator $\rho \in \mathcal{B}_2(\mathbb{R})$, taken from this dense set, to the function $W^\lambda(\rho|q, p)$, is linear and an isometry (up to a factor of $\lambda/2\pi$). Using this fact we can extend the map to a unitary transformation between the Hilbert spaces $\mathcal{B}_2(\mathbb{R})$ and $L^2(\mathbb{R}, dq\, dp)$. Thus, to any Hilbert–Schmidt operator ρ, we can associate a general Wigner function $W^\lambda(\rho|q, p)$. However, the explicit expression for this function is given by the integral in (8.25) only for operators with a well-defined trace. Otherwise, it has to be obtained as an L^2-limit of such functions. Conversely, every function f in $L^2(\mathbb{R}, dq\, dp)$ is the general Wigner function of a unique Hilbert–Schmidt operator. If f is also L^1-integrable, then this operator is given by (see (8.36))

$$\rho = \frac{\lambda}{2\pi} \iint_{\mathbb{R}^2} dq\, dp\, \widetilde{f}(q, p) U^\lambda(0, q, p), \tag{8.38}$$

where again $\widetilde{f}$ is the symplectic Fourier transform of f. If f is not L^1-integrable, then the corresponding Hilbert–Schmidt operator is obtained as a Hilbert space limit (in $\mathcal{B}_2(\mathbb{R})$) of operators obtained using (8.38). The Wigner function of a general Hilbert–Schmidt operator probably does not have a natural meaning in signal analysis. It is only Wigner functions of rank-one operators that have been used directly, as signature functions of signals. In quantum mechanics, however, a Hilbert–Schmidt operator, which is trace-class and of unit trace, represents a mixed state and its Wigner function again has the interpretation of a phase space quasi-probability distribution for this state.

We end this section by repeating what we said already in Chapter 1, that the Wigner function $W^\lambda(s|q, p)$ of a signal is in a sense more intrinsic than its Gabor transform, since the former does not depend on an arbitrarily chosen window. On the other hand, the reconstruction formula (8.36) only gives back the operator $|s\rangle\langle s|$, and not the function s itself. Still, both are transforms carrying information about the signal in terms of phase space variables. A second point to be borne in mind is that while the cross-Wigner function is sesquilinear (see (8.30)), and the Wigner function is quadratic, when looked upon as a transform on signal vectors, it is in fact linear when looked upon as a transform on the space of Hilbert–Schmidt operators.

8.4   Wigner functions for the wavelet groups

In view of the fact that the Wigner function has proved itself to be an extremely useful tool, both in signal analysis and in atomic and quantum optical computations [63,64,376], it makes sense to look for similar signal transforms related to groups other than the Weyl–Heisenberg group. In particular, one would like to construct such transforms for the one- and two-dimensional wavelet groups. These could then provide one with alternatives to the wavelet transforms discussed in the previous two chapters. We now proceed to obtain such transforms, with the proviso that these *generalized* Wigner functions should also be phase space functions (i.e., functions defined on coadjoint orbits of the relevant groups). Also, we would like to preserve as many of the properties (8.29)–(8.33) as possible. A general procedure for constructing such functions has been proposed in [7–9,248]. (See also [72–74].) Again, the objective is to find a one-to-one linear correspondence between Hilbert–Schmidt operators on the carrier space of an irreducible representation of the group and L^2-functions on phase space. As with wavelet transforms, the square integrability of the group representations, used in the construction, will turn out to be of crucial importance and as before, the orthogonality relations (7.12) will guarantee a reconstructibility condition. We shall keep the discussion here mainly descriptive, without venturing into too many mathematical details.

Just as while constructing two- (or higher) dimensional wavelet transforms, we exploited the symmetry groups of signals, so also for constructing generalized Wigner transforms we look at semi-direct product groups of a particular type. Suppose that our signal vectors are elements of $L^2(\mathbb{R}^n, d^n\vec{x})$, where for the rest of this discussion $n = 1$ or 2. Following our discussion in Section 7.1.1, we assume that the allowable transformation symmetries of our signals are of the following types:

• Translations:

$$s(\vec{x}) \mapsto s(\vec{x} + \vec{b}), \qquad \vec{b} \in \mathbb{R}^n.$$

• Dilations:

$$s(\vec{x}) \mapsto s(a\vec{x}), \qquad a > 0.$$

• Matrix transformations of $\mathbb{R}^n$:

$$s(\vec{x}) \mapsto s(\mathsf{h}\vec{x}),$$

where h is an $n \times n$ nonsingular matrix. We shall assume that the set of all such admissible matrices form a group H and, furthermore, that the following technical condition is satisfied: there exists a vector $\vec{x} \in \mathbb{R}^n$ such that the set,

$$\mathcal{O}_{\vec{x}} = \{\vec{y} = a\mathsf{h}^T\vec{x} \mid a > 0, \ \mathsf{h} \in H\},$$

is open in $\mathbb{R}^n$ and $\vec{y} = \mathsf{h}\vec{y}$, for any $\vec{y} \neq \vec{0}$, is true only when h is the identity matrix. In this case $\mathcal{O}_{\vec{x}}$ is called an open free orbit. (This condition is satisfied by all the wavelet groups used in the current literature, including of course, the affine group, the 2-D wavelets group and the affine Poincaré group, studied in the previous two chapters.)

Thus, we are assuming that the full symmetry group of allowed transformations on our signals is $G = \mathbb{R}^n \rtimes (\mathbb{R}_*^+ \times H)$. Elements of this group can be conveniently represented by the matrices

$$g \equiv (\vec{b}, a, \mathsf{h}) = \begin{pmatrix} a\mathsf{h} & \vec{b} \\ \vec{0}^T & 1 \end{pmatrix}, \qquad a > 0, \ \mathsf{h} \in H, \ \vec{b} \in \mathbb{R}^n. \tag{8.39}$$

A point $\vec{x} \in \mathbb{R}^n$ undergoes the transformation $\vec{x} \mapsto a\mathsf{h}\vec{x} + \vec{b}$ under its action. This group is nonunimodular and its left invariant (Haar) measure is,

$$d\mu_G(\vec{b}, a, \mathsf{h}) = \frac{1}{a^{n+1}\det\mathsf{h}} \, d^n\vec{b} \, da \, d\mu_H(\mathsf{h}), \tag{8.40}$$

where $d^n\vec{b}$ is the Lebesgue measure on $\mathbb{R}^n$ and $d\mu_H(\mathsf{h})$ the left invariant measure of the group H. The Hilbert space of signals $L^2(\mathbb{R}^n, d^n\vec{x})$ then carries a natural unitary representation of the group G by unitary operators $U(\vec{b}, a, \mathsf{h})$, reflecting the transformation properties of the signals under the group action

$$\left[U(\vec{b}, a, \mathsf{h})s \right](\vec{x}) = \frac{1}{(a^n \det\mathsf{h})^{\frac{1}{2}}} \, s(a^{-1}\mathsf{h}^{-1}(\vec{x} - \vec{b})). \tag{8.41}$$

In the Fourier space $L^2(\widehat{\mathbb{R}}^n, d^n\vec{k})$, the corresponding operators act in the manner

$$\left[\widehat{U}(\vec{b}, a, \mathsf{h})\widehat{s} \right](\vec{k}) = (a^n \det\mathsf{h})^{\frac{1}{2}} \, e^{-i\vec{b}\cdot\vec{k}} \, \widehat{s}(a^{-1}\mathsf{h}^T\vec{k}). \tag{8.42}$$

The representation $\widehat{U}(\vec{b}, a, \mathsf{h})$ is in general not irreducible. However, if $\mathcal{O}$ is an open free orbit in $\widehat{\mathbb{R}}^n$, and $L^2(\mathcal{O}, d^n\vec{k})$ the subspace of $L^2(\widehat{\mathbb{R}}^n, d^n\vec{k})$, consisting of functions supported on this orbit, then the representation $\widehat{U}(\vec{b}, a, \mathsf{h})$, restricted to this subspace, is irreducible and square-integrable; hence it is appropriate for constructing generalized wavelet transforms. It can also be shown [69] that all irreducible subrepresentations of $\widehat{U}(\vec{b}, a, \mathsf{h})$ are of this type and that there is only a finite number of them (the connected affine group, G_{aff}^+, has two, while the SIM(2) group has only one, etc.). Generalized Wigner functions, which bear a strong resemblance to the ones defined in (8.24) and (8.26), can also be constructed now using these same irreducible representations. The phase space on which the Wigner functions are defined is $\mathcal{O}^* = \mathbb{R}^n \times \mathcal{O}$, which can be identified with a coadjoint orbit of the group G and is associated to an irreducible representation. The dimension of the phase space is $2n$. The exact construction relies rather heavily on the properties of the Lie algebra $\mathfrak{g}$ of G and its dual $\mathfrak{g}^*$, which we prefer to omit, displaying and discussing only the final expressions for the two wavelet groups G_{aff}^+ and SIM(2). For details, the reader may refer to [7,8,248].

Let $X_1, X_2, \ldots, X_{2n}$ be a (vector space) basis of the Lie algebra and X^{1*}, $X^{2*}, \ldots, X^{2n*}$ the dual basis for the vector space $\mathfrak{g}^*$. The basis elements are $(n + 1) \times (n + 1)$ matrices and we assume that the last n elements, $X_{n+1}, X_{n+2}, \ldots, X_{2n}$ correspond to translations of $\mathbb{R}^n$ (i.e., they are the generators of the one-parameter translation subgroups). A general element in the Lie algebra is a linear combination $X = \sum_{i=1}^{2n} x^i X_i = \vec{x} \cdot \vec{X}$, in an obvious notation. Similarly, a general element in the dual is a linear combination, $X^* = \sum_{i=1}^{2n} \gamma_i X^{i*} = \vec{\gamma} \cdot \vec{X}^*$. The identification of the coadjoint orbit $\mathcal{O}^*$ with $\mathbb{R}^n \times \mathcal{O} \subset \mathbb{R}^{2n}$ is then done with respect to the basis $\{X^{i*}\}_{i=1}^{2n}$ and we denote a point in the orbit by a column vector $\vec{\gamma} = (\gamma_1, \gamma_2, \ldots, \gamma_{2n})^T$. Let $U(g)$, $g \in G$, be an irreducible (square-integrable) representation of G on the Hilbert space $\mathfrak{H}$. Denote by C the Duflo–Moore operator defined in (7.5) (see also (6.16) and

(7.42)) and by $\widehat{X}_i$ the self-adjoint operator which represents X_i on the Hilbert space $\mathfrak{H}$ (via the representation $U(g)$). Also, $\vec{\widehat{X}}$ will denote a vector operator with components, $\widehat{X}_i$. The generalized Wigner function, corresponding to the irreducible representation $U(g)$ is then explicitly given, for any Hilbert–Schmidt operator ρ on $\mathfrak{H}$, such that the operator ρC^{-1} is of trace class, by the expression

$$W(\rho \mid \vec{\gamma}) = \frac{1}{(2\pi)^n} \int_{\widehat{N}_0} d^{2n}\vec{x} \; \mathrm{Tr}\left[e^{i(\vec{\widehat{X}}-\vec{\gamma}\,)\cdot\vec{x}}\rho C^{-1}\right] [\sigma(\vec{\gamma})\, m(\vec{x})]^{\frac{1}{2}}, \tag{8.43}$$

where $\widehat{N}_0$ is an appropriate subset of $\mathbb{R}^{2n}$, depending on certain properties of the Lie algebra. In the two examples given below, $\widehat{N}_0 = \mathbb{R}^{2n}$. The function $m(\vec{x})$ is needed to transform the Haar measure on the group to a measure on the Lie algebra parameters x_i, while the function $\sigma(\vec{\gamma})$ is a density which converts the invariant measure $d\Omega(\vec{\gamma})$ on the phase space $\mathcal{O}^*$ to the Lebesgue measure on $\mathbb{R}^{2n}$. The orthogonality relations (7.6) play a crucial role in the derivation of the above expression for the Wigner function, which can in fact be extended (by taking limits) to all Hilbert–Schmidt operators on $\mathfrak{H}$. In other words, the generalized Wigner function is again a transform of a Hilbert–Schmidt operator (on the Hilbert space of signals) to a function on phase space. If $\rho = |\phi\rangle\langle\psi|$, we obtain the cross-Wigner function,

$$W(\psi, \phi \mid \vec{\gamma}) = \frac{1}{(2\pi)^n} \int_{\widehat{N}_0} d^{2n}\vec{x} \; \langle C^{-1}\psi \mid e^{i(\vec{\widehat{X}}-\vec{\gamma}\,)\cdot\vec{x}}\phi\rangle [\sigma(\vec{\gamma})\, m(\vec{x})]^{\frac{1}{2}}, \tag{8.44}$$

provided the vector ψ is chosen from the domain of the operator C^{-1}, implying thereby an admissibility condition on the operator ρ. The inversion formula is then,

$$\rho = \frac{1}{(2\pi)^n} \int_{\mathcal{O}^*} d\Omega(\vec{\gamma}) \int_{\widehat{N}_0} d^{2n}\vec{x} \; e^{-i(\vec{\widehat{X}}-\vec{\gamma}\,)\cdot\vec{x}} \; C^{-1} \; W(\rho|\vec{\gamma})\, [\sigma(\vec{\gamma})\, m(\vec{x})]^{\frac{1}{2}}. \tag{8.45}$$

8.4.1   Wigner functions for the affine group

We take the irreducible representation $\widehat{U}_+(b, a)$ of the connected affine group G^+_{aff}, defined in Section 6.2 (see (6.61)–(6.62)). The Hilbert space of this representation is $\widehat{\mathfrak{H}}_+(\mathbb{R})$ of L^2-functions supported on the positive real axis (see (6.62)). The open free orbit is $\mathcal{O} = \mathbb{R}^*_+$ and the phase space is the coadjoint orbit $\mathcal{O}^*_+ = \mathbb{R} \times \mathbb{R}^*_+$, identified in (6.89). This orbit is equipped with the measure (6.92), which is invariant under the coadjoint action spelled out in (6.87). If $(\gamma_1, \gamma_2) \in \mathbb{R} \times \mathbb{R}^*_+$, and $\widehat{\psi}, \widehat{\phi} \in \widehat{\mathfrak{H}}_+(\mathbb{R})$, inserting into (8.44) the cross-Wigner function becomes,

$$W(\widehat{\psi}, \widehat{\phi}|\gamma_1, \gamma_2) = \frac{1}{(2\pi)^{\frac{1}{2}}} \int_{-\infty}^{\infty} dx \; \overline{\widehat{\psi}\left(\frac{\gamma_2\, e^{\frac{x}{2}}}{\operatorname{sinch} \frac{x}{2}}\right)} \frac{\gamma_2\, e^{-i\gamma_1 x}}{\operatorname{sinch} \frac{x}{2}} \widehat{\phi}\left(\frac{\gamma_2\, e^{-\frac{x}{2}}}{\operatorname{sinch} \frac{x}{2}}\right), \tag{8.46}$$

where (see (7.72)),

$$\operatorname{sinch} u = \frac{\sinh u}{u}.$$

For the purposes of signal analysis, the parameters γ_1 and γ_2 would be identified with time and frequency, respectively. The following facts should be noted about this function.

- W is sesquilinear in $\widehat{\phi}$, $\widehat{\psi}$ and indeed, it is again a transform of the rank-one operator $\rho = |\widehat{\phi}\rangle\langle\widehat{\psi}|$ to a phase space function. Moreover, the support of this function is contained entirely in the coadjoint orbit $\mathcal{O}^*$.
- Writing $W(\widehat{\psi}|\gamma_1, \gamma_2)$ for the case when $\widehat{\phi} = \widehat{\psi}$, this function is seen to be real, although in general not everywhere positive, and again, the trace condition is satisfied:

$$\frac{1}{(2\pi)^{\frac{1}{2}}} \int_{-\infty}^{\infty} d\gamma_1 \int_0^{\infty} \frac{d\gamma_2}{\gamma_2} \, W(\widehat{\psi}|\gamma_1, \gamma_2) = \text{Tr}\,[\rho], \qquad \rho = |\widehat{\psi}\rangle\langle\widehat{\psi}|. \tag{8.47}$$

- As anticipated, the covariance condition assumes the form:

$$W(\widehat{U}_+(b, a)\widehat{\psi}|\vec{\gamma}) = W(\widehat{\psi}|M^{\sharp}(b, a)^{-1}\vec{\gamma}) = W(\widehat{\psi}|\gamma_1 - b\gamma_2, a\gamma_2), \tag{8.48}$$

with $M^{\sharp}(b, a)$ being the matrix of the coadjoint action of the group on the phase space, obtained in (6.85)–(6.87).

- The definition of the Wigner function can be extended, using the trace condition, to all Hilbert–Schmidt operators ρ, on the Hilbert space of the representation $\widehat{U}_+(b, a)$ and then, the orthogonality relation

$$\frac{1}{(2\pi)^{\frac{1}{2}}} \int_{-\infty}^{\infty} d\gamma_1 \int_0^{\infty} \frac{d\gamma_2}{\gamma_2} \, \overline{W(\rho_1|\gamma_1, \gamma_2)} \, W(\rho_2|\gamma_1, \gamma_2) = \text{Tr}\,[\rho_1^* \rho_2], \tag{8.49}$$

holds. From this, or using (8.45) it is easy to write down a reconstruction formula for ρ, given its Wigner function. (For details, the reader is referred to [7].)

- Unlike the original Wigner function (8.24), only one of the two marginality conditions (8.32) is satisfied in this case. One gets,

$$\frac{1}{(2\pi)^{\frac{1}{2}}} \int_{-\infty}^{\infty} \frac{d\gamma_1}{\gamma_2} \, W(\widehat{\psi}|\gamma_1, \gamma_2) = |\widehat{\psi}(\gamma_2)|^2. \tag{8.50}$$

The nonexistence of a simple form for the second marginality condition is related to the fact that there does not seem to be a "natural" choice of coordinates on the phase space, in terms of which the condition could be stated.

8.4.2   Wigner functions for the similitude group

To construct Wigner functions for the SIM(2) group, we use its only irreducible representation $U(\vec{b}, a, \theta)$, carried by the Hilbert space $L^2(\mathbb{R}^2, d^2\vec{x})$ (see (2.13)) or equivalently, its Fourier-transformed version $\widehat{U}(\vec{b}, a, \theta)$, carried by the Hilbert space $L^2(\widehat{\mathbb{R}}^2, d^2\vec{k})$ (see (2.14)). The Duflo–Moore operator is the one given in (7.42) and we adhere to the notation and terminology introduced in Sections 7.3.1 and 7.3.2. The computation of the cross-Wigner function follows in several steps, which we now summarize.

(i) Taking a ψ in the domain of C^{-1} and an arbitrary vector ϕ in the Hilbert space we may write

$$\langle C^{-1}\psi | e^{i(\vec{\hat{X}} - \vec{\gamma})\cdot \vec{x}}\phi\rangle = e^{-i[\vec{\alpha}\cdot\vec{\alpha}^* + \vec{\beta}\cdot\vec{\beta}^*]}\,\langle U(e^X)C^{-1}\psi | \phi\rangle, \tag{8.51}$$

where we have set $\vec{x} = (\vec{\alpha}, \vec{\beta})^T$. (Recall that in the above equation, e^X indicates an element of the group SIM(2), while $e^{-i\vec{\hat{X}}\cdot\vec{x}}$ is a unitary operator giving its Hilbert space representation.)

(ii) Using the explicit form of the representation, the expression for the Duflo–Moore operator and equations (7.70) and (7.82), we obtain,

$$(U(e^X)C^{-1}\psi)(\vec{k}) = \frac{e^{2\lambda}}{2\pi}|\vec{k}|\,e^{-i[F(\mathfrak{s}(\vec{\alpha}))\vec{\beta}]\cdot\vec{k}}\,\psi(e^\lambda\,r_{-\theta}\,\vec{k}). \tag{8.52}$$

(iii) Next we note that, if V is a 2×2 real matrix of the type in (7.67) and $\vec{u}$ a 2-vector, then $|V\vec{u}| = [\det V]^{\frac{1}{2}}\,|\vec{u}|$. Thus,

$$|F(\mathfrak{s}(\vec{\alpha}))^{-1}\vec{k}| = \left[\frac{\lambda^2 + \theta^2}{2e^\lambda(\cosh\lambda - \cos\theta)}\right]^{\frac{1}{2}}|\vec{k}|. \tag{8.53}$$

Using this fact to effect a change of variables in the integration involved in the scalar product, we obtain

$$\langle U(e^X)C^{-1}\phi | \psi\rangle = \frac{e^{2\lambda}}{2\pi}\left[\frac{(\lambda^2 + \theta^2)}{2e^\lambda(\cosh\lambda - \cos\theta)}\right]^{\frac{3}{2}}\iint_{\mathbb{R}^2}d^2\vec{k}\left[e^{i\vec{\beta}\cdot\vec{k}}|\vec{k}|\right.$$
$$\left.\times\phi\left(\frac{e^{\mathfrak{s}(\vec{\alpha})^T/2}}{\mathrm{sinch}(\mathfrak{s}(\vec{\alpha})^T/2)}\vec{k}\right)\psi\left(\frac{e^{-\mathfrak{s}(\vec{\alpha})^T/2}}{\mathrm{sinch}(\mathfrak{s}(\vec{\alpha})^T/2)}\vec{k}\right)\right]. \tag{8.54}$$

(iv) Inserting the above expression into the formula for the Wigner function in (8.44) and using the expressions (see [8] for details of derivation),

$$m(\vec{x}) = m(\vec{\alpha}, \vec{\beta}) = 2e^{-\lambda}\frac{\cosh\lambda - \cos\theta}{\lambda^2 + \theta^2},$$
$$\sigma(\vec{\gamma}) = \sigma(\vec{\alpha}^*, \vec{\beta}^*) = |\vec{\beta}^*|^2 = (\beta_1^*)^2 + (\beta_2^*)^2, \tag{8.55}$$

leads to

$$W(\psi, \phi | \vec{\alpha}^*, \vec{\beta}^*) = \frac{1}{2(2\pi)^3}\iint_{\mathbb{R}^2}d^2\vec{k}\int_{-\infty}^{\infty}d\lambda\int_0^{2\pi}d\vartheta\iint_{\mathbb{R}^2}d^2\vec{\beta}$$
$$\left[e^{-i\vec{\alpha}^*\cdot\vec{\alpha}}e^{i(\vec{k}-\vec{\beta}^*)\cdot\vec{\beta}}|\vec{k}|\,|\vec{\beta}^*|\,\frac{\lambda^2 + \vartheta^2}{\cosh\lambda - \cos\vartheta}\right.$$
$$\left.\times\psi\left(\frac{e^{\mathfrak{s}(\vec{\alpha})^T/2}}{\mathrm{sinch}(\mathfrak{s}(\vec{\alpha})^T/2)}\vec{k}\right)\phi\left(\frac{e^{-\mathfrak{s}(\vec{\alpha})^T/2}}{\mathrm{sinch}(\mathfrak{s}(\vec{\alpha})^T/2)}\vec{k}\right)\right],$$

$$\mathfrak{s}(\vec{\alpha})^T = \begin{pmatrix} \lambda & \vartheta \\ -\vartheta & \lambda \end{pmatrix}, \qquad \vec{\alpha} = \begin{pmatrix} \lambda \\ \vartheta \end{pmatrix}. \tag{8.56}$$

The integration over $\vec{\beta}$ represents a delta measure in $\vec{k} - \vec{\beta}^*$. Using this fact to perform the $\vec{k}$-integration, we obtain finally

$$W(\psi, \phi | \vec{\alpha}^*, \vec{\beta}^*) = \frac{1}{4\pi} \int_{-\infty}^{\infty} d\lambda \int_0^{2\pi} d\vartheta \left[e^{-i(\alpha_1^* \lambda + \alpha_2^* \vartheta)} |\vec{\beta}^*|^2 \frac{\lambda^2 + \vartheta^2}{\cosh \lambda - \cos \vartheta} \right.$$
$$\left. \times \overline{\psi \left(\frac{e^{\mathfrak{s}(\vec{\alpha})^T/2}}{\mathrm{sinch}(\mathfrak{s}(\vec{\alpha})^T/2)} \vec{\beta}^* \right)} \phi \left(\frac{e^{-\mathfrak{s}(\vec{\alpha})^T/2}}{\mathrm{sinch}(\mathfrak{s}(\vec{\alpha})^T/2)} \vec{\beta}^* \right) \right].$$
$$(8.57)$$

This then is the final expression of the cross-Wigner function for SIM(2) at a phase space point, written in the $(\vec{\alpha}^*, \vec{\beta}^*)$ coordinates. More compactly, using the matrices defined in (7.79)–(7.80), we may write

$$W(\psi, \phi | \vec{\alpha}^*, \vec{\beta}^*) = \frac{|\vec{\beta}^*|^2}{2\pi} \iint_{\mathbb{R}^2} d^2 \vec{\alpha} \; e^{-i\vec{\alpha}^* \cdot \vec{\alpha}} \; \overline{\psi(e^\lambda \; r_{-\vartheta} \; F(\mathfrak{s}(\vec{\alpha})^T)^{-1} \vec{\beta}^*)}$$
$$\times e^\lambda \; [\det F(\mathfrak{s}(\vec{\alpha})^T)]^{-1} \; \phi(F(\mathfrak{s}(\vec{\alpha})^T)^{-1} \vec{\beta}^*). \qquad (8.58)$$

Also, it is straightforward to express this function in terms of the other coordinates discussed at the end of Section 7.3.2. For example, in the Darboux (or canonical) coordinates, introduced in (7.104), the cross-Wigner function assumes the form:

$$W(\psi, \phi | \vec{q}, \vec{p}) = \frac{\rho^2}{2\pi} \iint_{\mathbb{R}^2} d^2 \vec{\alpha} \; e^{-i[\rho \, r_\varphi \, \sigma_3 \, \vec{q} \,] \cdot \vec{\alpha}} \; \overline{\psi(e^\lambda \; r_{-\vartheta} \; F(\mathfrak{s}(\vec{\alpha})^T)^{-1} \vec{p})}$$
$$\times e^\lambda \; [\det F(\mathfrak{s}(\vec{\alpha})^T)]^{-1} \; \phi(F(\mathfrak{s}(\vec{\alpha})^T)^{-1} \vec{p}), \qquad (8.59)$$

where we have written $\vec{p} = (\rho, \varphi)$ in polar coordinates while, as before, $\vec{\alpha}$ has the Cartesian coordinates λ, ϑ.

By its very construction, for each pair of wave functions ψ, ϕ, the cross-Wigner function W is a function on phase space. The exact variables of phase space may however be chosen to suit our convenience. Moreover, it is clear, that even when $\phi = \psi$, the Wigner function may not be an everywhere positive function. The covariance property, in the present context, can be expressed as

$$W(U(g_0)\psi, U(g_0)\phi | \vec{\gamma}) = W(\psi, \phi | M^\sharp(g_0)^{-1} \vec{\gamma}), \; g_0 = (a_0, \theta_0, \vec{b}_0). \qquad (8.60)$$

Using (7.85), this can be written more explicitly as

$$W(U(a_0, \theta_0, \vec{b}_0)\psi, \; U(a_0, \theta_0, \vec{b}_0)\phi \mid \vec{\alpha}^*, \vec{\beta}^*) = W(\psi, \phi | \vec{\alpha}^{*\prime}, \vec{\beta}^{*\prime})$$
$$\vec{\alpha}^{*\prime} = \vec{\alpha}^* - \mathfrak{s}(\vec{b}_0)^T \vec{\beta}^*, \qquad \vec{\beta}^{*\prime} = a_0 \, r_{-\theta_0} \, \vec{\beta}^*. \qquad (8.61)$$

The orthogonality condition is reflected in the cross-Wigner function through the following relation:

$$\int_{\mathcal{O}^*} \frac{d^2 \vec{\alpha}^* \, d^2 \vec{\beta}^*}{|\vec{\beta}^*|^2} \; \overline{W(\psi_1, \phi_1 | \vec{\alpha}^*, \vec{\beta}^*)} \; W(\psi_2, \phi_2 | \vec{\alpha}^*, \vec{\beta}^*) = \langle \psi_2 | \psi_1 \rangle \langle \phi_1 | \phi_2 \rangle. \quad (8.62)$$

Although this relation is guaranteed by the very way the Wigner function is constructed, it is still instructive to demonstrate it directly, as we now proceed to do. Denoting the left-hand member of (8.62) by $I(\psi_1, \phi_1, \psi_2, \phi_2)$ and substituting from (8.58) we get

$$
I(\psi_1, \phi_1, \psi_2, \phi_2) = \int_{\mathcal{O}^*} d^2\vec{\alpha}^* \, d^2\vec{\beta}^* \left[|\vec{\beta}^*|^2 \int_{\mathbb{R}^2 \times \mathbb{R}^2} d^2\vec{\alpha}' \, d^2\vec{\alpha} \, e^{i\vec{\alpha}^* \cdot (\vec{\alpha}' - \vec{\alpha})} \right.
$$
$$
\times \, \psi_1(e^{\lambda'} \, r_{-\vartheta'} \, F(\mathfrak{s}(\vec{\alpha}')^T)^{-1} \vec{\beta}^*) \, \overline{\phi_1(F(\mathfrak{s}(\vec{\alpha}')^T)^{-1} \vec{\beta}^*)}
$$
$$
\times \, e^{(\lambda' + \lambda)} \, [\det F(\mathfrak{s}(\vec{\alpha}'))^T]^{-1} \, [\det F(\mathfrak{s}(\vec{\alpha}))^T]^{-1}
$$
$$
\left. \times \, \overline{\psi_2(e^{\lambda} \, r_{-\vartheta} \, F(\mathfrak{s}(\vec{\alpha})^T)^{-1} \vec{\beta}^*)} \, \phi_2(F(\mathfrak{s}(\vec{\alpha})^T)^{-1} \vec{\beta}^*) \right]. \qquad (8.63)
$$

The integral in $\vec{\alpha}^*$ represents a delta measure in $\vec{\alpha}' - \vec{\alpha}$ which allows us to perform the $\vec{\alpha}'$-integration. Next changing variables, $\vec{\beta}^* \mapsto F(\mathfrak{s}(\vec{\alpha})^T)^{-1} \vec{\beta}^*$, we obtain

$$
I(\psi_1, \phi_1, \psi_2, \phi_2) = \int_{\mathcal{O}^*} d^2\vec{\alpha}^* \, d^2\vec{\beta} \left[\overline{\phi_1(\vec{\beta}^*)} \, \phi_2(\vec{\beta}^*) \, |\vec{\beta}^*|^2 \, e^{2\lambda} \right.
$$
$$
\left. \times \, \psi_1(e^{\lambda} \, r_{-\vartheta} \, \vec{\beta}^*) \, \overline{\psi_2(e^{\lambda} \, r_{-\vartheta} \, \vec{\beta}^*)} \right]. \qquad (8.64)
$$

Noting that λ, ϑ are the components of $\vec{\alpha}$, we change variables as $\vec{\alpha} \mapsto e^{\lambda} \, r_{-\vartheta} \, \vec{\beta}^*$ to rewrite the $\vec{\alpha}^*$-integration. The determinant of this transformation is precisely $|\vec{\beta}^*|^2 \, e^{2\lambda}$, implying that

$$
I(\psi_1, \phi_1, \psi_2, \phi_2) = \langle \psi_2 | \psi_1 \rangle \, \langle \phi_1 | \phi_2 \rangle, \qquad (8.65)
$$

as asserted in (8.62).

Finally, we again obtain in this case only one marginality condition for the Wigner function $W(\psi | \vec{\alpha}^*, \vec{\beta}^*)$ (obtained by taking $\phi = \psi$ in the cross-Wigner function):

$$
\frac{1}{2\pi} \iint_{\mathbb{R}^2} \frac{d^2\vec{\alpha}^*}{|\vec{\beta}^*|^2} \, W(\psi | \vec{\alpha}^*, \vec{\beta}^*) = |\psi(\vec{\beta}^*)|^2. \qquad (8.66)
$$

Note also the trace condition,

$$
\frac{1}{2\pi} \int_{\mathcal{O}^*} \frac{d^2\vec{\alpha}^* \, d^2\vec{\beta}^*}{\|\vec{\beta}^*\|^2} \, W(\psi | \vec{\alpha}^*, \vec{\beta}^*) = |\psi|^2. \qquad (8.67)
$$

8.4.3   The Wigner function and the wavelet transform

As already noted, there is a close connection between the Wigner function and the wavelet transform arising from a general Lie group. In this section we explicitly demonstrate this relationship for the SIM(2) group. The starting point is the resolution of the identity (7.45):

$$
\frac{1}{c_\psi} \int_{\text{SIM}(2)} d\mu(a, \theta, \vec{b}) \, U(a, \theta, \vec{b}) |\psi\rangle \langle \psi| U(a, \theta, \vec{b})^* = I, \qquad c_\psi = \|C\psi\|^2. \qquad (8.68)
$$

Let s be a signal in the Fourier domain, i.e., an element of the Hilbert space $L^2(\mathbb{R}^2_*, d^2\vec{k})$. Its wavelet transform, corresponding to the mother wavelet ψ, is

$$S_\psi(\vec{b}, a, \theta) = \langle \psi_{a,\theta,\vec{b}} \mid s \rangle = \langle U(a, \theta, \vec{b})\psi \mid s \rangle. \tag{8.69}$$

Using the group parameters as the phase space variables for the Wigner function (see (7.96)), we easily deduce

$$W(C\psi, s|\vec{b}, a, \theta) = \frac{\sqrt{2}}{(2\pi)^2 a} \int_\mathbb{R} d\lambda \int_0^{2\pi} d\vartheta \iint_{\mathbb{R}^2} d^2\vec{\beta} \left[e^{-\frac{i}{a}[\vec{\alpha}\cdot(r_\theta \sigma_3 \vec{b}) + \vec{\beta}\cdot\hat{e}_\theta]} \right.$$
$$\left. \times \frac{(\cosh\lambda - \cos\vartheta)^{\frac{1}{2}}}{|\vec{\alpha}|} S_\psi(F(\mathfrak{s}(\vec{\alpha}))\vec{\beta}, e^\lambda, \vartheta) \right], \tag{8.70}$$

with $\hat{e}_\theta = \left(\cos\theta, \sin\theta\right)^T$, $\vec{\alpha} = \left(\vartheta, \lambda\right)^T$ and $F(\mathfrak{s}(\vec{\alpha}))$ given, as before, by (7.71). Thus, the relationship between the wavelet transform and the Wigner function is an integral transform on the Hilbert space $L^2(\mathrm{SIM}(2), d\mu)$ of all square integrable (with respect to the Haar measure) functions of the group, and it is then straightforward to check that

$$\int_{\mathrm{SIM}(2)} d\mu(\vec{b}, a, \theta) \, |S_\psi(\vec{b}, a, \theta)|^2 = \int_{\mathrm{SIM}(2)} d\mu(\vec{b}, a, \theta) \, |W(C\psi, s \mid \vec{b}, a, \theta)|^2. \tag{8.71}$$

If the signal s is a Hilbert space vector which is in the domain of the inverse operator C^{-1}, then taking $s = C\psi$ we also obtain

$$W(s|\vec{b}, a, \theta) = \frac{\sqrt{2}}{(2\pi)^2 a} \int_\mathbb{R} d\lambda \int_0^{2\pi} d\vartheta \iint_{\mathbb{R}^2} d^2\vec{\beta} \left[e^{-\frac{i}{a}[\vec{\alpha}\cdot(r_\theta \sigma_3 \vec{b}) + \vec{\beta}\cdot\hat{e}_\theta]} \right.$$
$$\left. \times \frac{(\cosh\lambda - \cos\vartheta)^{\frac{1}{2}}}{|\vec{\alpha}|} S_{C^{-1}s}(F(\mathfrak{s}(\vec{\alpha}))\vec{\beta}, e^\lambda, \vartheta) \right]. \tag{8.72}$$

The above results show that any information about the signal which can be obtained using the wavelet transform, can also be derived using the Wigner function. The use of one rather than the other is therefore more a matter of practical convenience or computational ease, in any given situation.

9 Higher-dimensional wavelets

In the previous chapters, we have thoroughly discussed the 2-D CWT and some of its applications. Then we have made the connection with the group theoretical origins of the method, thus establishing a general framework, based on the coherent state formalism. In the present chapter, we will apply the same technique to a number of different situations involving higher dimensions: wavelets in 3-D space $\mathbb{R}^3$, wavelets in $\mathbb{R}^n$ ($n > 3$), and wavelets on the 2-sphere S^2. Then, in the next chapter, we will treat time-dependent wavelets, that is, wavelets on space–time, designed for motion analysis.

In all cases, the technique is the same. First one identifies the manifold on which the signals are defined and the appropriate group of transformations acting on the latter. Next one chooses a square integrable representation of that group, possibly modulo some subgroup. Then one constructs wavelets as admissible vectors and derives the corresponding wavelet transform.

9.1 Three-dimensional wavelets

Some physical phenomena are intrinsically multiscale and three-dimensional. Typical examples may be found in fluid dynamics, for instance the appearance of coherent structures in turbulent flows, or the disentangling of a wave train in (mostly underwater) acoustics, as discussed above. In such cases, a 3-D wavelet analysis is clearly more adequate and likely to yield a deeper understanding [56]. The same is true for many problems in astrophysics, such as galaxy/void counting or grouping, or cluster structure analysis. Hence we will also describe briefly the 3-D CWT, following the general pattern of the previous section.

9.1.1 Constructing 3-D wavelets

Given a 3-D signal $s \in L^2(\mathbb{R}^3, d^3\vec{x})$, with finite energy, one may act on it by translation, dilation and rotation:

$$s_{\vec{b},a,\varrho}(\vec{x}) \equiv \left[U(\vec{b}, a, \varrho)s \right](\vec{x}) = a^{-3/2} s(a^{-1}\varrho^{-1}(\vec{x} - \vec{b})), \tag{9.1}$$

where $\vec{b} \in \mathbb{R}^3$, $a > 0$ and $\varrho \in SO(3)$. The 3×3 rotation matrix $\varrho \in SO(3)$ may be parametrized, for instance, in terms of three Euler angles. These three operations generate the 3-D Euclidean group with dilations, i.e., the similitude group of $\mathbb{R}^3$, $SIM(3) = \mathbb{R}^3 \rtimes (\mathbb{R}_*^+ \times SO(3))$. Then (9.1) is a unitary representation of $SIM(3)$ in $L^2(\mathbb{R}^3, d^3\vec{x})$, which is irreducible and square integrable, hence it generates a CWT exactly as before.

Wavelets are taken in $L^2(\mathbb{R}^3, d^3\vec{x})$ and the admissibility condition is now

$$\int_{\mathbb{R}^3} \frac{d^3\vec{k}}{|\vec{k}|^3} \, |\widehat{\psi}(\vec{k})|^2 < \infty. \tag{9.2}$$

As in 2-D, a necessary, and almost sufficient, condition for admissibility is the familiar zero mean condition:

$$\int_{\mathbb{R}^3} d^3\vec{x} \, \psi(\vec{x}) = 0. \tag{9.3}$$

The two standard wavelets have a 3-D realization.

- *The 3-D Mexican hat* is given by

$$\psi_{\text{H}}(\vec{x}) = (3 - |A\vec{x}|^2) \, \exp(-\tfrac{1}{2}|A\vec{x}|^2). \tag{9.4}$$

where $A = \text{diag}[\epsilon_1^{-1/2}, \epsilon_2^{-1/2}, 1]$, $\epsilon_1 \geqslant 1, \epsilon_2 \geqslant 1$, is a 3×3 anisotropy matrix. We distinguish three cases: (1) If $\epsilon_1 \neq \epsilon_2 \neq 1$, one has the fully anisotropic 3-D Mexican hat (the stability subgroup H_ψ is trivial); (2) If $\epsilon_1 = \epsilon_2 = 1$, one has the isotropic, $SO(3)$-invariant, 3-D Mexican hat ($H_\psi = SO(3)$); (3) If $\epsilon_1 = \epsilon_2 \equiv \epsilon \neq 1$, the wavelet is *axisymmetric*, i.e., $SO(2)$-invariant, but not isotropic ($H_\psi = SO(2)$). Thus, in this case, wavelets are coherent states of Gilmore–Perelomov type, whose parameter space is indeed the quotient $SO(3)/SO(2) \simeq S^2$ (see the discussion in Section 7.1.5).

- *The 3-D Morlet wavelet* is given by

$$\psi_{\text{M}}(\vec{x}) = \exp(i\vec{k}_o \cdot \vec{x}) \, \exp(-\tfrac{1}{2}|A\vec{x}|^2) + \text{corr.}, \tag{9.5}$$

where A is the same 3×3 anisotropy matrix as in the first example. Here again, for $\epsilon_1 = \epsilon_2 \equiv \epsilon \neq 1$ and $\vec{k}_o$ along the z-axis, the wavelet ψ is invariant under $SO(2)$ and we obtain coherent states of Gilmore–Perelomov type.

- *A 3-D Cauchy wavelet* is defined by a straightforward generalization of the 2-D case, as follows. Consider the convex simplicial (or pyramidal) cone $\mathcal{C}(\alpha, \beta, \gamma)$ defined by the three unit vectors $\vec{e}_\alpha, \vec{e}_\beta, \vec{e}_\gamma$, the angle between any two of them being smaller than π. The dual cone is also simplicial, namely $\widetilde{\mathcal{C}} = \mathcal{C}(\tilde{\alpha}, \tilde{\beta}, \tilde{\gamma})$, where $\vec{e}_{\tilde{\alpha}} = \vec{e}_\beta \wedge \vec{e}_\gamma$ is orthogonal to the β-γ face, etc. With these notations, given a vector $\vec{\eta} \in \widetilde{\mathcal{C}}$ and $l, m, n \in \mathbb{N}^*$, we define a 3-D Cauchy wavelet in spatial frequency space as:

$$\widehat{\psi}_{lmn}^{(\mathcal{C}, \eta)}(\vec{k}) = \begin{cases} (\vec{k} \cdot \vec{e}_{\tilde{\alpha}})^l \, (\vec{k} \cdot \vec{e}_{\tilde{\beta}})^m \, (\vec{k} \cdot \vec{e}_{\tilde{\gamma}})^n \, e^{-\vec{k} \cdot \vec{\eta}}, & \vec{k} \in \mathcal{C}(\alpha, \beta, \gamma), \\ 0, & \text{otherwise.} \end{cases} \tag{9.6}$$

As in the 2-D case, the factors $(\vec{k} \cdot \vec{e}_{\tilde{\alpha}})^l$, etc. represent vanishing moments on the faces of the cone, and thus determine the regularity of the wavelet in $\vec{k}$-space.

Here too the expression for the 3-D wavelet in position space may be obtained explicitly:

$$\psi_{lmn}^{(C,\eta)}(\vec{x}) = \frac{i^{l+m+n+3}}{2\pi} \, l! \, m! \, n! \cdot \det A \cdot \frac{\left(\text{vol}\,[\vec{e}_\alpha, \vec{e}_\beta, \vec{e}_\gamma]\right)^{l+m+n}}{(\vec{z} \cdot \vec{e}_\alpha)^{l+1} \, (\vec{z} \cdot \vec{e}_\beta)^{m+1} \, (\vec{z} \cdot \vec{e}_\gamma)^{n+1}}, \tag{9.7}$$

where A is the matrix that transforms the unit vectors $\vec{e}_1, \vec{e}_2, \vec{e}_3$ into the triple $\vec{e}_\alpha, \vec{e}_\beta, \vec{e}_\gamma$, vol $[\cdot, \cdot, \cdot]$ denotes the volume of the parallelepiped generated by the three vectors, and we have written $\vec{z} = \vec{x} + i\vec{\eta}$. Note that the direct calculation, following the pattern of the 2-D case (Section 3.3.4) yields for the numerator in (9.7) the expression $(\vec{e}_{\tilde{\alpha}} \cdot \vec{e}_\alpha)^l$ $(\vec{e}_{\tilde{\beta}} \cdot \vec{e}_\beta)^m \, (\vec{e}_{\tilde{\gamma}} \cdot \vec{e}_\gamma)^n$, but then one has

$$\vec{e}_{\tilde{\alpha}} \cdot \vec{e}_\alpha = \vec{e}_\beta \wedge \vec{e}_\gamma \cdot \vec{e}_\alpha = \vec{e}_\beta \cdot \vec{e}_\gamma \wedge \vec{e}_\alpha = \vec{e}_\beta \cdot \vec{e}_{\tilde{\beta}} = \vec{e}_\gamma \cdot \vec{e}_{\tilde{\gamma}} = \text{vol}\,[\vec{e}_\alpha, \vec{e}_\beta, \vec{e}_\gamma],$$

which proves (9.7).

From the expressions (9.6) and (9.7), one may then obtain other 3-D Cauchy wavelets, for instance, one supported in a circular cone. Take a circular convex cone, aligned on the positive k_z-axis, with total opening angle $2\theta_o$ $(0 < \theta_o < \pi/2)$. In spherical polar coordinates $\vec{k} = (|k|, \theta, \phi)$, the interior of the cone is simply $\mathcal{C}(\theta_o) = \{\vec{k} \in \mathbb{R}^3 \,|\, \theta \leqslant \theta_o\}$. Then an axisymmetric 3-D Cauchy wavelet supported in this cone may be defined, for instance, by

$$\widehat{\psi}_m^{(\theta_o)}(\vec{k}) = \begin{cases} |k|^l (\tan^2 \theta_o - \tan^2 \theta)^m \, e^{-|k| \cos \theta}, & 0 \leqslant \theta \leqslant \theta_o; \\ 0, & \text{otherwise.} \end{cases} \tag{9.8}$$

Again $m \in \mathbb{N}$ defines the number of vanishing moments on the surface of the cone, that is, the regularity of the wavelet. For θ_o very small, this wavelet lives inside a narrow pencil: it clearly evokes the beam of a searchlight – a vivid illustration of the wavelet as a directional probe!

If we note that the expression on the right-hand side of (9.8) may be written as $|k|^{l'} (\tan^2 \theta_o \, k_z^2 - k_x^2 + k_y^2)^m \, e^{-k_z}$, we see that all these wavelets are built on the same model, namely $F(\vec{k})^m \, e^{-k_z}$, where $F(\vec{k}) = 0$ is the equation of the cone.

9.1.2   The 3-D continuous wavelet transform

Then, given a signal $s \in L^2(\mathbb{R}^3)$, its CWT with respect to the admissible wavelet ψ is given as

$$S(\vec{b}, a, \varrho) = a^{-3/2} \int_{\mathbb{R}^3} d^3 \vec{x} \, \overline{\psi(a^{-1} \varrho^{-1}(\vec{x} - \vec{b}))} \, s(\vec{x}). \tag{9.9}$$

As compared with (2.19), the only differences are in the normalization factors and the rotation matrices. Since the structure of the formulas is the same as before, so are the

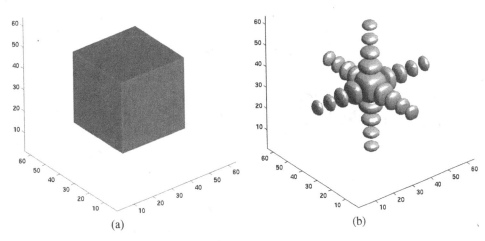

Fig. 9.1. The 3-D cube and its Fourier transform.

interpretation and the consequences (local filtering, reproducing kernel, reconstruction formula, etc.). Thus the CWT (9.9) may be interpreted as a mathematical *camera* with *magnification* $1/a$, *position* $\vec{b}$ and *directional selectivity* given, in the axisymmetric case, by the rotation parameters $\zeta \equiv (\theta, \varphi)$. As for the visualization, the full CWT $S(\vec{b}, a, \varrho)$ is a function of seven variables. However, if the wavelet ψ is chosen *axisymmetric*, i.e., SO(2)-invariant, S depends on six variables only, $\vec{b} \in \mathbb{R}^3$, $a > 0$, and $\zeta \in S^2 \simeq$ SO(3)/SO(2), the unit sphere in $\mathbb{R}^3$. In this case again, (a^{-1}, ζ) may be interpreted as polar coordinates in spatial frequency space. This is in fact true in any number of dimensions. It follows that, here too, there are two natural representations for the visualization of the WT: the position representation (a, ζ fixed) and the scale-orientation (or spatial frequency) representation ($\vec{b}$ fixed). Of course, there are many other possible representations that may be useful.

As an example, we present in Figure 9.1 (the characteristic function of) a 3-D cube and its Fourier transform (note the occurrence of sinc functions among all three co-ordinate axes), then in Figure 9.2 the wavelet transform of the cube signal, with a Morlet wavelet, at four successively finer scales, $a = 1, 0.5, 0.25, 0.125$. As in 2-D, the net result, for a sufficiently fine scale, is the contour of the cube. The latter has become totally transparent, only the edges survive! A similar, slightly more complicated example is that of a cube with a small box removed, see Figure 9.3. Here one faces the well-known visual ambiguity: depending of the angle of view, the small cube appears either as removed (concave) or added (convex). As shown in Figure 9.4, the ambiguity is resolved with the WT, the part that has been removed yields a negative WT, whereas an added, convex, portion would yield a positive contribution to the WT. This is true, of course, if one uses a real wavelet (here the 3-D Mexican hat) and plots the WT itself, not its modulus. We recover exactly the 2-D situation, discussed in Section 4.1.2.

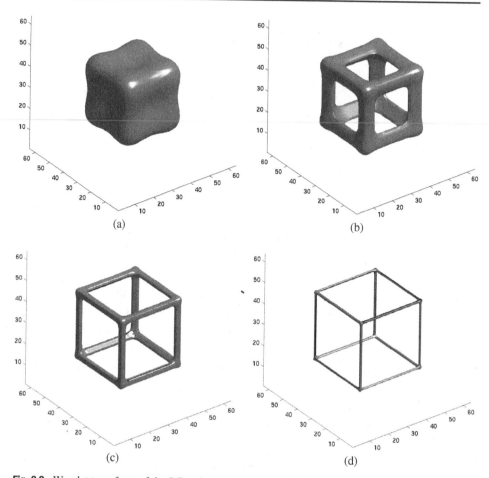

Fig. 9.2. Wavelet transform of the 3-D cube at four successively finer scales.

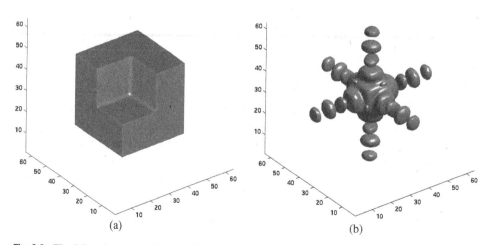

Fig. 9.3. The 3-D cube minus a box and its Fourier transform.

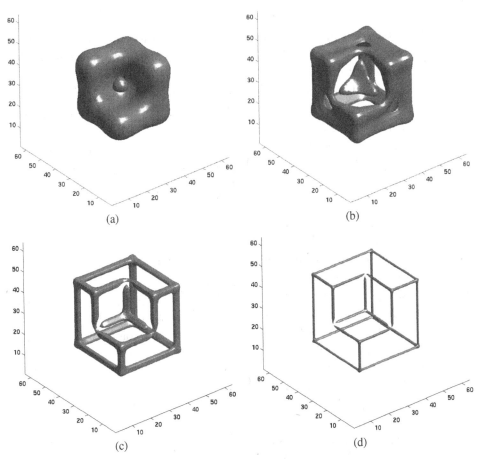

Fig. 9.4. Wavelet transform of the 3-D cube minus a box at four successively finer scales ($a = 1$, $0.5, 0.25, 0.125$).

A practical application of this 3-D CWT is the detection of 3-D objects in a cluttered medium. We consider a scene with 3-D objects (targets) immersed in a cluttered medium, modeled by the signal:

$$s(\vec{x}) = \sum_{l=1}^{L} s_l(\vec{x}) + n(\vec{x}), \tag{9.10}$$

where $s_l(\vec{x})$ denotes the density of the target l, and $n(\vec{x})$ the density of the medium. Since the density of the targets is very different from that of the medium, there will be a high density gradient at the boundary between the objects and the medium. In this situation, the wavelet transform $S(\vec{b}, a, \zeta)$ may be used for extracting the 3-D objects and determining their characteristics, position (range and orientation) and spatial frequency. This is, of course, nothing but a 3-D version of the ATR problem discussed already in Section 4.2 (see also [16]). In particular, 3-D directional wavelets (e.g. Morlet or Cauchy) will behave as their 2-D counterparts. If a 3-D image $s(\vec{x})$ contains elongated objects, a visualization of $S(\vec{b}, a, \zeta)$ shows "sausage" images for all the objects which have the same direction as the wavelet, and small spheres at the tips of all the objects which are misaligned. With an appropriate thresholding, we may filter out all the objects which are not in the desired direction.

Another domain of application is in computational fluid dynamics. For instance, the technique has been used in [56] for analyzing two types of simulated 3-D flows: (i) detection of elongated vortex structures in homogeneous 3-D turbulence, and (ii) interaction of a shock wave with a 3-D supersonic mixing layer between two parallel streams. In both situations, the use of scale as a fundamental variable offers distinct advantages over traditional methods based on sophisticated graphic tools. In the first case, the selection of structures is based on characteristic scales rather than on vorticity values. For the shear/shock interaction, the simultaneous detection of main steady flow features with unsteady multiscale mixing features is a practical way to the study of compressibility controlled mixing phenomena. Clearly this opens interesting perspectives for wavelet analysis in fluid dynamics.

As a final remark, we may note that the 3-D *discrete* WT has also been used successfully in several applications. One of them is the analysis of the substructure of galaxy clusters [300]. Another one is the lossless compression of 3-D images produced by computer tomography or radar scanning [81]. The lossless character of the transform is based, as in 1-D or 2-D, on the use of an integer WT (see Section 2.5.2.4).

9.1.3   Extension to higher dimensions

From the analysis of the previous section, it should be clear that the extension to higher dimensions is straightforward – although it has probably an academic interest only (except perhaps in quantum mechanics, since an N-particle wave function belongs to $L^2(\mathbb{R}^{3N}, d^{3N}\vec{x})$).

Given a finite energy signal $s \in L^2(\mathbb{R}^n, d^n\vec{x})$, the transformation group to consider is again the similitude group, now $\mathrm{SIM}(n) = \mathbb{R}^n \rtimes (\mathbb{R}_*^+ \times \mathrm{SO}(n))$. This group has a (unique) natural unitary irreducible representation in $L^2(\mathbb{R}^n, d^n\vec{x})$, given by

$$s_{\vec{b},a,\varrho}(\vec{x}) \equiv \left[U(\vec{b}, a, \varrho)s\right](\vec{x}) = a^{-n/2}s(a^{-1}\varrho^{-1}(\vec{x} - \vec{b})), \tag{9.11}$$

where $\vec{b} \in \mathbb{R}^n$ is a translation, $a > 0$ is a global dilation and $\varrho \in \mathrm{SO}(n)$ is a rotation. This representation is square integrable, with admissibility condition

$$\int_{\mathbb{R}^n} \frac{d^n\vec{k}}{|\vec{k}|^n} |\widehat{\psi}(\vec{k})|^2 < \infty, \tag{9.12}$$

reducing as usual to the necessary condition $\int_{\mathbb{R}^n} d^n\vec{x}\, \psi(\vec{x}) = 0$. The n-D wavelet transform is, as before,

$$S(\vec{b}, a, \varrho) = a^{-n/2} \int_{\mathbb{R}^n} d^n\vec{x}\, \overline{\psi(a^{-1}\varrho^{-1}(\vec{x} - \vec{b}))}\, s(\vec{x}). \tag{9.13}$$

Standard wavelets are the same, namely,

- *The n-D Mexican hat:*

$$\psi_{\text{H}}(\vec{x}) = (n - |A\vec{x}|^2) \exp(-\tfrac{1}{2}|A\vec{x}|^2).$$

- *The n-D Morlet wavelet:*

$$\psi_{\text{M}}(\vec{x}) = \exp(i\vec{k}_o \cdot \vec{x}) \exp(-\tfrac{1}{2}|A\vec{x}|^2) + \text{ corr.}$$

Once again, the anisotropy matrix $A = \text{diag}[\epsilon_1^{-1/2}, \epsilon_2^{-1/2}, \ldots, \epsilon_{n-1}^{-1/2}, 1]$, $\epsilon_1 \geqslant 1, \epsilon_2 \geqslant 1, \ldots, \epsilon_{n-1} \geqslant 1$, leads to various situations, depending on the number of different values taken by the parameters ϵ_j. The interesting case is, of course, the axisymmetric or $SO(n-1)$-invariant case, where all the ϵ_j coincide and $\epsilon_j = \epsilon \neq 1$. Then, the parameter space reduces to $\mathbb{R}^n \times \mathbb{R}_*^+ \times S^{n-1} \simeq \mathbb{R}^n \times \mathbb{R}_*^n$ and the WT reads $S(\vec{b}, a, \zeta)$, $\zeta \in S^{n-1}$. As before, (a^{-1}, ζ) may be interpreted as polar coordinates in spatial frequency space, and thus the n-D CWT yields a phase space representation of signals. The rest is as before.

Cauchy wavelets also extend to $\mathbb{R}^n$, as follows. We start with the n-simplex $\mathcal{C}(\vec{e}_1, \vec{e}_2, \ldots, \vec{e}_n)$, generated by the unit vectors $\vec{e}_1, \vec{e}_2, \ldots, \vec{e}_n$. Define successively

$$\vec{e}_{\bar{1}} = \vec{e}_2 \wedge \vec{e}_3 \wedge \ldots \wedge \vec{e}_{n-1} \wedge \vec{e}_n$$
$$\vec{e}_{\bar{2}} = \vec{e}_3 \wedge \vec{e}_4 \wedge \ldots \wedge \vec{e}_n \wedge \vec{e}_1$$

$$\ldots$$

$$\vec{e}_{\bar{n}} = \vec{e}_1 \wedge \vec{e}_2 \wedge \ldots \wedge \vec{e}_{n-2} \wedge \vec{e}_{n-1}.$$

Then the only nonzero inner products are

$$\vec{e}_{\bar{1}} \cdot \vec{e}_1 = \vec{e}_{\bar{2}} \cdot \vec{e}_2 = \ldots = \vec{e}_{\bar{n}} \cdot \vec{e}_n = \text{vol}\,[\vec{e}_1, \vec{e}_2, \ldots, \vec{e}_n].$$

Thus, for $\vec{\eta} \in \tilde{\mathcal{C}}$, we define the n-dimensional Cauchy wavelet as

$$\widehat{\psi}^{(C,\eta)}_{l_1 l_2 \ldots l_n}(\vec{k}) = \begin{cases} \prod_{j=1}^{n} (\vec{k} \cdot \vec{e}_{\bar{j}})^{l_j} e^{-\vec{k}\cdot\vec{\eta}}, & \vec{k} \in \mathcal{C}(\vec{e}_1, \vec{e}_2, \ldots, \vec{e}_n), \\ 0, & \text{otherwise,} \end{cases} \tag{9.14}$$

and, in position space,

$$\psi^{(C,\eta)}_{l_1 l_2 \ldots l_n}(\vec{x}) = \text{const}\,\frac{\left(\text{vol}\,[\vec{e}_1, \vec{e}_2, \ldots, \vec{e}_n]\right)^{l_1+l_2+\ldots+l_n}}{\prod_{j=1}^{n}(\vec{z} \cdot \vec{e}_j)^{l_j+1}}. \tag{9.15}$$

In these formulas, $l_1, l_2, \ldots, l_n \in \mathbb{N}^*$ and the factors $(\vec{k} \cdot \vec{e}_{\bar{j}})^{l_j}$ represent vanishing moments on the faces of the cone.

Exactly as in three dimensions, one may also construct Cauchy or conical wavelets adapted to a general cone of equation $F(\vec{k}) = 0$ and, in particular, n-D axisymmetric wavelets in the case of a circular cone.

9.2   Wavelets on the 2-sphere and other manifolds

9.2.1   The problem

In most cases of physical interest, experimental data are given on the line (signal processing), on the plane (image analysis), or occasionally in $\mathbb{R}^3$ (e.g., in fluid dynamics). However, there are situations where data are given on a *sphere*, for instance, in geophysics or astronomy, of course, but also in statistical problems [Fis87], computer vision or medical imaging (see [216] for precise references). If one is interested only in very local features, one may forget the curvature and work on the tangent plane with standard methods, that is, Fourier or time–frequency analysis, in particular the CWT. However, when global aspects become important (description of plate tectonics on the Earth, for instance, or structure of the Universe as a whole), curvature can no longer be ignored, so that one needs a genuine generalization of wavelet analysis to the sphere (or more general manifolds).

Let us first make that statement precise. We may speak of a genuine spherical CWT if (i) the signals and the wavelets live *on* the sphere; (ii) the transform involves (local) dilations of some kind; and (iii) for small scales, the spherical CWT reduces to the usual CWT on the (tangent) plane (Euclidean limit).

Several authors have studied this problem, with various techniques, mostly discrete. For instance:

- One may extend to S^2 the discrete wavelet scheme based on a multiresolution analysis, but this approach leads often to numerical difficulties around the poles [114,197,318].

- A different technique is to use the lifting scheme and second generation wavelets, as described in Chapter 2, Section 2.5.2. An efficient solution has been obtained in this way by Schröder and Sweldens [336], but this obviously misses the particular symmetry of the sphere.

- One may exploit the geometry of the sphere, as encoded in the system of spherical harmonics [Fre97,Fre99,169,294,319], but the resulting analyzing functions are poorly localized (in fact they do not really resemble wavelets). On the other hand, this approach leads to good approximation methods for spherical functions; we shall come back to this problem in Section 9.3.

However, to fully preserve the rotational invariance of the sphere, a continuous approach is clearly necessary. Here too, several authors have proposed solutions.

- Considering the fact that the sphere does not admit global dilations, since it is compact, one resorts to a wavelet transform on the tangent bundle of the sphere [116] or, instead, to a Gabor transform on the sphere itself [355,356].

- The most satisfactory approach is that of Holschneider [225], who produces a CWT on the sphere that satisfies the three criteria above. However, the rôle of dilation is played by an abstract parameter that satisfies a number of *ad hoc* assumptions. The correct Euclidean limit is obtained, but it is essentially put by hand.

As can be seen from this brief description, none of the proposed solutions fully qualifies for a genuine CWT on S^2. It turns out that the general formalism developed in [Ali00,6] and sketched in Chapter 7 yields an elegant solution to the problem, entirely derived from group theory, and in particular allows one to derive all the assumptions of [225]. Although the discussion is too technical to be given here in detail, it is interesting to outline the main ideas, because they lead to significant generalizations. A detailed treatment may be found in [23,28,29].

9.2.2   The continuous wavelet transform on S^2

9.2.2.1   Affine transformations on the sphere S^2

We consider the 2-sphere S^2, with polar spherical coordinates $\zeta = (\theta, \varphi)$. As usual, finite energy signals are taken as square integrable functions on the 2-sphere, $s \in \mathcal{H} = L^2(S^2, d\mu)$, where $d\mu(\zeta) \equiv d\mu(\theta, \varphi) = \sin\theta \, d\theta \, d\varphi$ is the usual (rotation invariant) measure on S^2. The first step for constructing a CWT on S^2 is to identify the natural operations on such signals. These are of two types:

(i) Motions or displacements, given by elements of the rotation group SO(3), which indeed acts transitively on S^2, and $S^2 \simeq$ SO(3)/SO(2).

(ii) Dilations, that may be derived in two steps. First, dilations around the North Pole are obtained by considering usual dilations in the tangent plane and lifting them to S^2 by inverse stereographic projection from the South Pole. This gives:

$$D_a^{(N)}(\theta, \varphi) = (\theta_a, \varphi), \quad \text{with} \quad \tan\frac{\theta_a}{2} = a \cdot \tan\frac{\theta}{2}. \tag{9.16}$$

Then a dilation around any other point $\zeta \in S^2$ is obtained by moving ζ to the North Pole by a rotation $\varrho \in$ SO(3), performing a dilation $D_a^{(N)}$ as before and going back by the inverse rotation: $D_a^{(\zeta)} = \varrho^{-1} D_a^{(N)} \varrho$. Clearly the dilations act also transitively on S^2.

Next we have to identify a group of affine transformations on S^2. First we note that motions $\varrho \in$ SO(3) and dilations by $a \in \mathbb{R}_*^+$ do *not* commute. Also it is impossible to build a semidirect product from SO(3) and $\mathbb{R}_*^+$, and therefore the only extension of SO(3) by $\mathbb{R}_*^+$ is their *direct product*. A way out is to embed the two factors into the Lorentz group $SO_o(3, 1)$, by the so-called *Iwasawa decomposition*:

$$SO_o(3, 1) = SO(3) \cdot A \cdot N, \tag{9.17}$$

where $A \sim SO_o(1, 1) \sim \mathbb{R} \sim \mathbb{R}_*^+$ is the subgroup of Lorentz boosts in the z-direction and $N \sim \mathbb{C}$ is two-dimensional and abelian (under the stereographic projections, N corresponds to translations in the tangent plane). The appearance of the Lorentz group

$SO_o(3, 1)$ in this context is not fortuitous, since it is the *conformal* group of the sphere S^2 (and of the plane $\mathbb{R}^2$ as well).

It turns out that the stability subgroup of the North Pole is the so-called minimal parabolic subgroup $P = M = SO(2) \cdot A \cdot N$, where $M = SO(2)$ is the subgroup of rotations around the z-axis. Thus we get

$$S^2 \simeq SO_o(3, 1)/P \simeq SO(3)/SO(2). \tag{9.18}$$

This shows that $SO_o(3,1)$ acts transitively on S^2 as well. This action may be computed explicitly using the Iwasawa decomposition (9.17). For a pure dilation by a, the result is precisely the usual dilation lifted on S^2 by inverse stereographic projection, given in (9.16).

9.2.2.2 Spherical wavelets

The next step towards constructing wavelets (affine coherent states) on S^2 is to find a suitable unitary irreducible representation (UIR) of the Lorentz group $SO_o(3, 1)$ acting in the Hilbert space $L^2(S^2, d\mu)$. Natural candidates are the representations of the continuous principal series, also called class I representations [Kna96,351]. The simplest one, that we shall use, is given by the operators:

$$[U(g)f](\zeta) = \lambda(g, \zeta)^{1/2} f\left(g^{-1}\zeta\right), \ g \in SO_o(3, 1), \ f \in L^2(S^2, d\mu), \tag{9.19}$$

where $g = \varrho an$ by the Iwasawa decomposition and the multiplier $\lambda(g, \zeta)$ is a Radon–Nikodym derivative (or a 1-cocycle), expressing the fact that the measure $d\mu$ is not invariant under the full group $SO_o(3, 1)$:

$$\lambda(g, \zeta) = \frac{d\mu\left(g^{-1}\zeta\right)}{d\mu(\zeta)}, \ g \in SO_o(3, 1). \tag{9.20}$$

This representation U of $SO_o(3, 1)$ is unitary and irreducible.

Since we are only interested in the action of dilations and motions, we quotient out the subgroup N. In other words, the parameter space of the spherical wavelets is $X = SO_o(3, 1)/N \simeq SO(3) \cdot \mathbb{R}_*^+$. Then, introducing a suitable section $\sigma : X = SO_o(3, 1)/N \to SO_o(3, 1)$, we concentrate on the reduced expression

$$[U(\sigma(x))f](\zeta) = \lambda(\sigma(x), \zeta)^{1/2} f\left(\sigma(x)^{-1}\zeta\right), \ x \equiv (\varrho, a). \tag{9.21}$$

We choose the natural (Iwasawa) section $\sigma(\varrho, a) = \varrho a$, $\varrho \in SO(3)$, $a \in A$. Using the action (9.16) of dilations, one gets easily

$$\lambda(\sigma(\varrho, a), \zeta) \equiv \lambda(a, \theta) = \frac{4a^2}{\left[(a^2 - 1)\cos\theta + (a^2 + 1)\right]^2}, \quad \zeta = (\theta, \varphi). \tag{9.22}$$

The function $\lambda(a, \theta)$ satisfies the so-called cocycle relation (which guarantees that U is indeed a representation):

$$\lambda(a^{-1}, \theta)\lambda(a, \theta_a) = \lambda(1, \theta) = 1. \tag{9.23}$$

In addition, from the choice of the section, we have $U(\sigma(\varrho, a)) = U(\varrho\, a) = U(\varrho)U(a)$, and therefore the representation (9.21) factorizes as

$$U(\sigma(\varrho, a)) = R_\varrho\, D_a. \tag{9.24}$$

In this relation, $R_\varrho \equiv U_{\mathrm{qr}}(\varrho)$, $\varrho \in SO(3)$, where U_{qr} is the quasi-regular representation of $SO(3)$ in $L^2(S^2, d\mu)$, and D_a, $a \in \mathbb{R}_*^+$, is an operator of pure dilation, that is, $(D_a f)(\zeta) = \lambda(a, \theta)^{1/2} f(\zeta_{1/a})$, with $\zeta_a \equiv (\theta_a, \varphi)$. The quasi-regular representation of $SO(3)$, $(U_{\mathrm{qr}}(\varrho)f)(\zeta) = f(\varrho^{-1}\zeta)$, is infinite dimensional and decomposes into the direct sum of all the familiar $(2l + 1)$-dimensional representations, $l = 0, 1, 2, \ldots$.

Following the general approach of [Ali00,6], we build now a system of spherical wavelets, realized as coherent states for the Lorentz group, indexed by points of $X = SO_o(3, 1)/N$. Since N is not the isotropy subgroup of a particular vector in the representation Hilbert space, the resulting coherent states are not of the Gilmore–Perelomov type [Per86] (see Section 7.1.5). First we show that the UIR (9.21) is indeed square integrable on X, that is, we check that there exists a nonzero vector $\psi \in L^2(S^2, d\mu)$ such that

$$\int_X d\nu(\varrho, a) \, |\langle U(\sigma(\varrho, a))\psi | \phi \rangle|^2 < \infty, \quad \forall \phi \in L^2(S^2, d\mu),$$

where $d\nu(\varrho, a) = a^{-3}\, d\varrho\, da$, and $d\varrho$ is the invariant (Haar) measure on $SO(3)$.

Proposition 9.2.1 *The representation U given in (9.21) is square integrable modulo the subgroup N and the section σ. A nonzero vector $\psi \in L^2(S^2, d\mu)$ is admissible $\mathrm{mod}(N, \sigma)$ iff there exists $c > 0$, independent of l, such that, for all $l \geqslant 0$:*

$$G_l \equiv \frac{8\pi^2}{2l + 1} \sum_{m=-l}^{l} \int_0^\infty \frac{da}{a^3} \, |\widehat{\psi_a}(l, m)|\psi_a)|^2 < c. \tag{9.25}$$

Here $\widehat{f}(l, m) \equiv \langle Y_l^m | f \rangle$ denotes a Fourier coefficient of $f \in L^2(S^2)$, with Y_l^m the usual spherical harmonic, and $\psi_a = D_a\psi = U(\sigma(e, a))\psi$ corresponds to a pure dilation.

The proof, as usual, consists in an explicit calculation, using the properties of Fourier analysis on the sphere.

Thus any admissible ψ generates a continuous family $\{\psi_{a,\varrho} \equiv U(\sigma(\varrho, a))\psi, (\varrho, a) \in X\}$ of spherical wavelets, but in fact we have more.

Proposition 9.2.2 *For any admissible vector ψ such that $\int_0^{2\pi} d\varphi \, \psi(\theta, \varphi) \neq 0$ (for instance, axisymmetric), the family $\{\psi_{a,\varrho}, (\varrho, a) \in X\}$ is a continuous frame, that is, there exist constants $A > 0$ and $B < \infty$ such that*

$$A \, \|\phi\|^2 \leqslant \int_X d\nu(\varrho, a) \, |\langle \psi_{a,\varrho} | \phi \rangle|^2 \leqslant B \, \|\phi\|^2, \quad \forall \phi \in L^2(S^2, d\mu). \tag{9.26}$$

Thus, for most admissible vectors ψ, we get a continuous frame, but not necessarily a tight frame. We conjecture that the resulting frame is never tight, that is, $A \neq B$.

9.2.2.3   The spherical wavelet transform

Proposition 9.2.1 yields the basic ingredient for writing the CWT on S^2. Given an admissible vector $\psi \in L^2(S^2, d\mu)$, our wavelets on the sphere are the functions $\psi_{a,\varrho} = U(\sigma(\varrho, a))\psi = R_\varrho D_a \psi = R_\varrho \psi_a.$[†] Then, the spherical CWT of a signal $s \in L^2(S^2)$ is defined as

$$
\begin{aligned}
S(\varrho, a) &= \langle \psi_{a,\varrho} | s \rangle \\
&= \int_{S^2} d\mu(\zeta)\, \overline{\psi_{a,\varrho}(\zeta)}\, s(\zeta) \\
&= \int_{S^2} d\mu(\zeta)\, \overline{\psi_a(\varrho^{-1}\zeta)}\, s(\zeta).
\end{aligned}
\tag{9.27}
$$

This relation gives the spherical CWT as a convolution on the sphere S^2: $\psi_a(\zeta)$ and $s(\zeta)$ are two functions on S^2, and their convolution $\psi_a \tilde{\star} s$ is a function on $SO(3)$. Such a formulation leads both to mathematical subtleties and to numerical difficulties. The former will be treated in Section 9.3 (see also Appendix A) and the latter in Section 9.2.5, where we will discuss the numerical implementation of the spherical CWT.

According to the general theory, the admissibility of the wavelet ψ is sufficient to guarantee the invertibility of the spherical CWT on its range, that is, we can reconstruct the signal $s(\zeta)$ from its wavelet transform $S(\varrho, a)$:

$$
s(\zeta) = \int_X d\nu(\varrho, a)\, S(\varrho, a)\, A_\sigma^{-1} \psi_{a,\varrho}(\zeta).
\tag{9.28}
$$

In this relation, A_σ denotes the resolution operator (7.34), whose action is a multiplication in Fourier space (as it is the case with most Duflo–Moore operators, for instance, (7.42)),

$$
\widehat{A_\sigma f}(l, m) = G_l \widehat{f}(l, m),
$$

with G_l defined in the admissibility condition (9.25). As usual, the integral in (9.28) is to be taken in the weak sense. Of course, as in the flat case, one may consider more general reconstruction formulas, with two different wavelets for the analysis and the synthesis (see (2.31) in Section 2.2).

These formulas gets simpler if the wavelet ψ is axisymmetric, i.e., $\psi(\zeta) \equiv \chi(\theta)$, for, then, we may exploit the fact that $S^2 \simeq SO(3)/SO(2)$. First, since the dilation is purely radial, ψ_a is also axisymmetric and $\psi_a(\zeta) = \chi_a(\theta)$. Then the action of R_ϱ on ψ_a has for sole effect to transport its center from the North Pole ζ_o to some point $\zeta' = \varrho\,\zeta_o$.

[†] As the notation $\psi_{a,\varrho}$ suggests, all the operations involved in the CWT consist in manipulating the function ψ_a at a fixed scale a. This is consistent with [Fre97,169] and [35], but not with [29], where the same wavelets were denoted $\psi_{\varrho,a}$.

We can thus characterize $R_\varrho \psi_a$ by its center point ζ', which is independent of a, and we may write $\psi_{a,\varrho} \equiv \psi_{a,\zeta'}$.

A more precise way of achieving this is to split $\varrho \in SO(3)$ into $\varrho = (\chi, [\zeta'])$ with $\chi \in SO(2)$ and $\zeta' \in S^2$. This is formally done through a projection $\varrho \mapsto \zeta'(\varrho)$ in the fiber bundle $SO(3) \to S^2 \simeq SO(3)/SO(2)$, followed by an arbitrary choice of section $\zeta' \mapsto [\zeta']$ in $SO(3)$. The splitting corresponds to decomposing the motion R_ϱ of the wavelet ψ_a into an initial rotation of angle χ around the North Pole ζ_0 followed by a transport to the point $\zeta' = \varrho \zeta_o$ on the sphere. In other words,

$$R_\varrho \psi_a(\zeta) = R_\chi \psi_a([\zeta']^{-1}\zeta)$$

where R_χ is a rotation around the North Pole. Accordingly, the spherical wavelet transform will also be denoted by $S(\chi, \zeta', a)$. This amounts simply to decomposing the parameter space $SO(3) \times \mathbb{R}_*^+$ into a more appropriate form, including "translations" ζ', dilations a, and "rotations" χ, exactly as in the Euclidean case. Of course, when the wavelet ψ is axisymmetric, the dependence on χ can be dropped and the spherical wavelet transform will be written simply as $S(\zeta', a)$:

$$S(\zeta', a) = \int_{S^2} d\mu(\zeta) \, \overline{\psi_{a,\zeta'}(\zeta)} \, s(\zeta). \tag{9.29}$$

(A related statement is given in Proposition 9.3.1, in Section 9.3.) In that case, the parameter space of the spherical CWT reduces to $S^2 \times \mathbb{R}_*^+$, with measure $a^{-3} d\mu(\zeta) \, da$. Hence, the integral over $SO(3)$ in the reconstruction formula is replaced by an integral over S^2:

$$s(\zeta) = \int_{S^2} \int_{\mathbb{R}_*^+} \frac{d\mu(\zeta') \, da}{a^3} \, S(\zeta', a) \, A_\sigma^{-1} \psi_{a,\zeta'}(\zeta). \tag{9.30}$$

According to the general coherent state formalism, the reconstruction formulas (9.28) and (9.30) are valid only in the weak sense. In the flat case, however, we have seen in Section 2.6 that the corresponding formula holds in the strong L^2 sense. This guarantees that it can be used for approximating functions on the plane through an approximate identity. That means, the approximating function is obtained by convolution with a smoothing kernel, which tends to the identity (δ function) as the parameter goes to 0. We will show in Section 9.3 that exactly the same situation prevails on the sphere. In order to prove these results, we will have to switch to an L^1 formalism (as already mentioned in [29]), by introducing a modified dilation operator D^a that preserves the L^1 norm of functions. As a consequence, we will have at our disposal *two* types of spherical CWT.

An important aspect of the flat space CWT is its covariance property (Proposition 2.2.3). In the present case, an explicit calculation [29] shows that the spherical CWT (9.27) is covariant under motions on S^2, but *not* covariant under dilations.

- It is covariant under motions on S^2, namely, for any $\varrho_o \in SO(3)$, the transform of the rotated signal $s(\varrho_o^{-1}\zeta)$ is the function $S(\varrho_o^{-1}\varrho; a)$.

- But it is *not* covariant under dilations. Indeed the wavelet transform of the dilated signal $\lambda(a_o, \zeta)^{1/2} s(a_o^{-1}\zeta)$ is $\langle U(g)\psi|s\rangle$, with $g = a_o^{-1}\varrho a$, and the latter, while a well-defined element of $SO_o(3, 1)$, is *not* of the form $\sigma(\varrho', a')$.

For applications, of course, it is the covariance under motions that is essential, since it reduces to translation covariance in the Euclidean limit, as we shall see in Section 9.2.3. As for dilations, the negative result reflects the fact that the parameter space $X \simeq SO(3) \times \mathbb{R}_*^+$ of the spherical CWT is not a group.

The condition (9.25), which was derived in [225] in a different way, is necessary and sufficient for the admissibility of ψ, but it is somewhat complicated to use in practice, since it requires the evaluation of nontrivial Fourier coefficients. Instead, there is a simpler, although only necessary, condition.

Proposition 9.2.3 *A function $\psi \in L^2(S^2, d\mu)$ is admissible only if it satisfies the condition*

$$\int_{S^2} d\mu(\zeta) \frac{\psi(\theta, \varphi)}{1 + \cos\theta} = 0, \quad \zeta \equiv (\theta, \varphi). \tag{9.31}$$

This necessary condition is the exact equivalent of the usual necessary condition for wavelets in the plane, $\int d^2\vec{x}\, \psi(\vec{x}) = 0$, and it reduces to the latter in the Euclidean limit (see Section 9.2.3). The interesting point is that (9.31) is a zero mean condition, as in the flat case. As such it ensures that the CWT on S^2 given in (9.27) acts as a local filter. This is crucial for applications and it is one of the main reasons of the efficiency of the CWT, and the same holds here.

Using Proposition 9.2.3, it is easy to build explicit wavelets on the sphere, namely "Difference wavelets", similar to the ones described for the flat case in Chapter 3, Section 3.2.2. For that purpose, we notice the following easy result [28].

Proposition 9.2.4 *Let $\phi \in L^2(S^2, d\mu)$. Then*

$$\frac{1}{a} \int_{S^2} d\mu(\zeta) \frac{D_a\phi(\theta, \varphi)}{1 + \cos\theta} = \int_{S^2} d\mu(\zeta) \frac{\phi(\theta, \varphi)}{1 + \cos\theta},$$

where $D_a = U(\sigma(e, a))$ is again a pure (covariant) dilation.

Given a square integrable (smoothing) function ϕ, we define

$$\psi_\phi^{(\alpha)}(\theta, \varphi) = \phi(\theta, \varphi) - \frac{1}{\alpha} D_\alpha \phi(\theta, \varphi) \quad (\alpha > 1). \tag{9.32}$$

By Proposition 9.2.4, $\psi_\phi^{(\alpha)}$ satisfies the admissibility condition (9.31), that is, it is a spherical wavelet, and it is fully admissible if ϕ is sufficiently regular at the poles. The simplest difference wavelet is obtained with the choice $\phi_G(\theta, \varphi) = \exp(-\tan^2 \frac{\theta}{2})$, which is essentially the inverse stereographic projection of a Gaussian in the tangent plane. The corresponding spherical wavelet, called the spherical DOG wavelet and

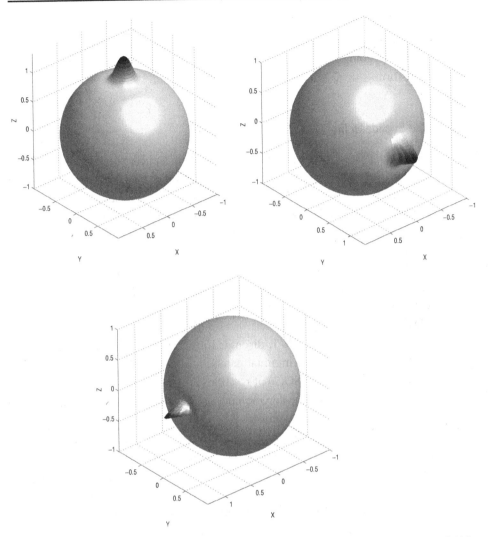

Fig. 9.5. The spherical DOG $\psi_G^{(\alpha)}$ wavelet, for $\alpha = 1.25$. (Top line) The wavelet at scale $a = 0.125$ and positioned at the North Pole $\theta = 0°$ (left) and on the equator $\theta = 90°$, $\varphi = 90°$ (right); (Bottom) The same at scale $a = 0.0625$ and position $\theta = 90°$, $\varphi = 0°$. As mentioned in the text, "at scale a" means the function $D_a \psi_\phi^{(\alpha)}$.

denoted $\psi_G^{(\alpha)}$ in the sequel, is an axisymmetric wavelet, which is shown in Figure 9.5, for different values of the scale a and in various positions (θ, φ) on the sphere. Note that here "ψ at scale a" means that the function being plotted is $D_a \psi$, i.e., one must always use the covariant dilation operator D_a. This wavelet yields an efficient detection of discontinuities on the sphere [28]. Explicit examples of spherical wavelet transforms based on it will be given in Section 9.2.5.3.

Now, the construction of the scaling function ϕ_G by inverse stereographic projection from the tangent plane suggests a general procedure for generating spherical wavelets.

We will implement it in Section 9.2.4, where we will construct *directional* wavelets on the sphere by the same method.

Before going into that, let us discuss the Euclidean limit of our spherical wavelet transform.

9.2.3   The Euclidean limit

As said above, a good wavelet transform on the sphere should be asymptotically Euclidean, that is, the spherical WT should match the usual CWT in the plane (in this case, the tangent plane at the North Pole) at small scales or, what amounts to the same, for large values of the radius of curvature. This statement may be given a precise mathematical meaning, using the technique of group contractions (or deformations). Without entering into technical details, we sketch the successive steps.

First, we reformulate the theory on a sphere of radius R and let $R \to \infty$. Then S_R^2 becomes the plane $\mathbb{R}^2$, the group SO(3) contracts into the Euclidean group of $\mathbb{R}^2$ and the Lorentz group $SO_o(3, 1)$ into the (semidirect) product $G_E = \mathbb{R}^2 \rtimes \text{SIM}(2)$, where $\text{SIM}(2) = \mathbb{R}^2 \rtimes (\mathbb{R}_*^+ \times SO(2))$ is the similitude group of $\mathbb{R}^2$, that is, the invariance group of the Euclidean CWT, discussed in Chapter 7, Section 7.1.2. Notice that the contraction preserves the minimal parabolic subgroup $P = MAN \sim \text{SIM}(2)$, and, in particular, the subgroup SO(2) of rotations around the z-axis.

The next step is to transfer the contraction process to the relevant homogeneous spaces. On one hand, the manifolds $S^2 = SO_o(3, 1)/MAN$ and $\mathbb{R}^2 = G_E/MAN$, that carry the respective CWT, are related through contraction. On the other hand, since the abelian subgroup N is preserved under the contraction, the parameter space $X = SO_o(3, 1)/N \simeq SO(3) \times A$ of the spherical CWT goes into that of the Euclidean CWT, namely $\text{SIM}(2) = G_E/N$. Notice that the former is not a group (and this forces us to use the general formalism of [Ali00,6], described in Chapter 7), whereas, after contraction, we get SIM(2), that is, the missing group structure is restored by the contraction!

In this geometrical context, the Euclidean limit itself can be formulated as a contraction at the level of group representations. Whereas contractions of Lie algebras and Lie groups are relatively ancient and well-known [232,331], the extension of the procedure to group representations is rather recent [272]. A rigorous version has been given by Dooley [150–152], that was followed in [23,29]. The additional difficulty here is that the representation space itself varies during the procedure.

Let $\mathcal{H}_R = L^2(S_R^2, d\mu_R)$ be the Hilbert space of square integrable functions on a sphere of radius R (where $d\mu_R = R^2 d\mu$) and $\mathcal{H} = L^2(\mathbb{R}^2, d^2\vec{x})$. The two spaces are related by the unitary map $I_R : \mathcal{H}_R \to \mathcal{H}$, given as

$$(I_R f)(r, \varphi) = \frac{4R^2}{4R^2 + r^2} f\left(2 \arctan \frac{r}{2R}, \varphi\right), \tag{9.33}$$

where (r, φ) are polar coordinates in the plane. Clearly, the map I_R just describes the stereographic projection of the sphere S_R^2 onto its tangent plane at the North Pole. The inverse map, that is, the inverse stereographic projection, reads as

$$(I_R^{-1} f)(\theta, \varphi) = \frac{2}{1 + \cos\theta} \, f(2R \tan\frac{\theta}{2}, \varphi), \tag{9.34}$$

(in the case $R = 1$, this map was given already in (5.1)). Now let U be the usual wavelet representation (2.13) of SIM(2) in $\mathcal{H}$ and U_R the representation (9.21) of $SO_o(3, 1)$ realized in $\mathcal{H}_R$. For each R, we choose the corresponding representation U_R. Then the precise statement is that the representation U of SIM(2) is a contraction, in the sense of Dooley [150–152] of the family of representations U_R of $SO_o(3, 1)$ as $R \to \infty$. This means that, for every $g \in$ SIM(2), the following strong limit holds in $\mathcal{H}$:

$$\lim_{R\to\infty} \| I_R U_R \left(\tilde{\Pi}_R(g) \right) I_R^{-1} \phi - U(g)\phi \|_{\mathcal{H}} = 0, \tag{9.35}$$

where $\tilde{\Pi}_R :$ SIM(2) $\to X$ is the so-called reduced contraction map (see [29] for details).

This theorem yields the expected result that local wavelet analysis on the sphere as defined here is equivalent to local wavelet analysis in flat space. Indeed the whole structure on the sphere S_R^2 goes into the corresponding one in $\mathbb{R}^2$ as $R \to \infty$. Since $U_R \to U$, the corresponding matrix elements converge to one another, and so the square integrability condition (9.25) converges into the corresponding one for the CWT in $\mathbb{R}^2$, namely

$$\int_{\mathbb{R}^2} d^2\vec{k} \, \frac{|\widehat{\psi}(\vec{k})|^2}{|\vec{k}|^2} < \infty.$$

Admissible wavelets on S^2 converge to admissible wavelets on $\mathbb{R}^2$ (Proposition 9.2.5), and the necessary condition (9.31) also goes into the usual one in the plane, namely $\int_{\mathbb{R}^2} d^2\vec{x} \, \psi(\vec{x}) = 0$.

9.2.4   Directional wavelets on the sphere

The general mathematical setting for designing *directional* wavelets on S^2 has been introduced in the previous sections, but the construction and the properties of these particular spherical wavelets have not been discussed. Yet these wavelets are quite important in practice, since directional features (roads, streams, geological faults, ...) abound on the spherical Earth! Thus one really needs the additional degree of freedom they offer for characterizing signals.

9.2.4.1   General remarks

Whenever the wavelet ψ is not axisymmetric, the continuous spherical wavelet transform depends on the additional parameter χ, the rotation angle around the North Pole, thus (9.27) may be rewritten as

$$S(\chi, \zeta', a) = \int_{S^2} d\mu(\zeta) \, \overline{R_\chi \psi_a([\zeta']^{-1}\zeta)} \, s(\zeta). \tag{9.36}$$

In this formula, there is an arbitrariness in the way the rotation $[\zeta']$ of $SO(3)$ is associated to the point ζ' on the sphere. The section $[\cdot] : S^2 \to SO(3)$ can be depicted as mapping the sphere to a tangent vector field of unit length defined on it. Intuitively, one would like this mapping to be smooth to correspond to the idea of *direction* defined on the sphere, and the arbitrariness in the choice of the section may be exploited to that effect, at least locally. When this is the case, we expect that the values of the wavelet transform correspond to filtering in a given direction χ and at a given scale a as in the case of the 2-D wavelet transform in the plane.

Some caution must be exercised, however, when dealing with directions on the sphere. Indeed, it is a classical result in topology that there exists no differentiable vector field of constant norm on S^2, which means there is no *global* way of defining directions. There will always be some singular point where the definition fails.[†] In other words, one cannot comb a perfectly spherical porcupine! Nevertheless, testing orientations using directional wavelets is a small-scale operation, that is, a *local* procedure [24]. Then, around any point on S^2, the support of the wavelet at small scale defines a neighborhood in which the following reasoning holds. This ability to perform local analysis is definitely one the most important properties of wavelet analysis. We will see more about this in the examples below.

From now on, we will make use of the classical parametrization of $SO(3)$ in terms of Euler angles, $\varrho \equiv (\chi, \theta', \varphi')$, which corresponds to the choice of section $(\theta', \varphi') \mapsto (0, \theta', \varphi')$, which in turn defines a direction on the sphere. The singular points are the North and South Poles: it makes no sense to define cardinal points at the poles!

For this choice of parametrization, we may write

$$R_\chi \psi_a([\zeta']^{-1}\zeta) = \psi_{a,\chi,\zeta'}(\zeta) \equiv \psi_{a,\chi,\theta',\varphi'}(\theta, \varphi), \tag{9.37}$$

which implies

$$\psi_{a,\chi,\theta',\varphi'}(\theta, \varphi) = \psi_{a,\chi,\theta',0}(\theta, \varphi - \varphi'). \tag{9.38}$$

Therefore, (9.36) becomes a convolution in φ which, by means of the convolution theorem, takes the form

$$S(\chi, \theta', \varphi', a) = \int_0^\pi \int_0^{2\pi} \overline{\psi_{a,\chi,\theta',0}(\theta, \varphi - \varphi')} s(\theta, \varphi) \, \sin\theta \, d\theta \, d\varphi \tag{9.39}$$

$$= 2\pi \sum_{k=-\infty}^{\infty} e^{ik\varphi'} \int_0^\pi \overline{\check{\psi}_{a,\chi,\theta',0}(\theta)[k]} \, \check{s}(\theta)[k] \, \sin\theta \, d\theta, \tag{9.40}$$

[†] This statement is valid for S^2, but not in the case of the circle S^1 and the higher dimensional spheres S^3 and S^7.

where, for any function h on the sphere, $\check{h}(\theta)[k]$ denotes its Fourier coefficient with respect to the longitudinal coordinate φ:

$$\check{h}(\theta)[k] = \int_0^{2\pi} d\varphi \, h(\theta, \varphi) \, e^{-i\,k\varphi}. \tag{9.41}$$

In the discretization method of Section 9.2.5, the relations (9.39)–(9.40) will give us a tool for reducing the computational time of the spherical CWT. Indeed, they will allow us to use the fast Fourier transform (FFT).

9.2.4.2   Estimating the angular selectivity of a wavelet

Given a wavelet ψ, it is very important in practice to know how well it will discriminate between two close directions. In other words, we would like to quantify the angular resolving power (ARP) of a spherical wavelet (see Section 3.4.1). A tempting definition is simply to look at the correlation between ψ and its rotated version:

$$
\begin{aligned}
K_\psi(\chi) &= \frac{\langle \psi | R_\chi \psi \rangle}{\langle \psi | \psi \rangle} = \|\psi\|^{-2} \sum_{l \geqslant 0} \sum_{|m| \leqslant l} \langle \psi | Y_l^m \rangle \langle Y_l^m | R_\chi \psi \rangle \\
&= \|\psi\|^{-2} \sum_{l \geqslant 0} \sum_{|m| \leqslant l} \langle \psi | Y_l^m \rangle \langle R_\chi Y_l^m | \psi \rangle \\
&= \|\psi\|^{-2} \sum_{l \geqslant 0} \sum_{|m| \leqslant l} \langle \psi | Y_l^m \rangle e^{im\chi} \langle Y_l^m | \psi \rangle \\
&= \sum_{m \in \mathbb{Z}} \left(\|\psi\|^{-2} \sum_{l \geqslant |m|} |\widehat{\psi}(l, m)|^2 \right) e^{im\chi} \\
&= \sum_{m \in \mathbb{Z}} a_m e^{im\chi}.
\end{aligned}
$$

As a function of χ, this expression reduces to a Fourier series where the coefficients are the same as in the spherical harmonic expansion of ψ. If the wavelet is highly sensitive to changes in the orientation, K_ψ should be peaked around $\chi = 0$. As we shall see later, the problem with this definition is that it does not depend on the scaling parameter a. This is not a problem for wavelets in $\mathbb{R}^2$, but is rather counterintuitive on the sphere, because we know that a direction cannot be defined at large scales. This motivates the study of a more general indicator. Let us then introduce the following operator-valued function:

$$R_{\psi,a}(\chi) = \int_{S^2} d\mu(\zeta) \, |\psi_{a,\chi,\zeta}\rangle \langle \psi_{a,\chi,\zeta}|, \tag{9.42}$$

where we choose again for $[\cdot]$ the Euler angles section. A good candidate for the ARP would then be the mean value:

$$\langle R_{\psi,a}(\chi) \rangle_{\psi_a} = \frac{\langle \psi_a | R_{\psi,a}(\chi) \psi_a \rangle}{\langle \psi_a | \psi_a \rangle}. \tag{9.43}$$

This time the inspection of $\langle R_{\psi,a}(\chi) \rangle_{\psi_a}$ for different values of a should reveal a lack of angular precision at large scales. This is precisely what is observed in the case of the spherical Morlet wavelet, as shown later in Figure 9.8.

9.2.4.3 Designing directional spherical wavelets

We have not yet addressed the problem of constructing good directional wavelets on S^2. We will show now that this job is very naturally handled in our framework. First of all, we recall that the very definition of a direction on S^2 forces us to work at small scales. As is well known, the geometry of S^2 at small scales, or for large radii of the sphere, is closer and closer to that of $\mathbb{R}^2$. As discussed in Section 9.2.3, the spherical wavelet transform respects one's intuition by nicely approximating the Euclidean wavelet transform at small scales (the Euclidean limit property). We may remark that the notation used in (9.37) is consistent with it: roughly speaking, as the radius of the sphere goes to infinity, $\psi_{a,\chi,\zeta'}(\zeta)$ goes to $\psi_{a,\chi,\vec{b}}(\vec{x})$, where $\vec{b} \in \mathbb{R}^2$ is the translation parameter. Moreover, it is a simple application of the Euclidean limit that small-scale Euclidean wavelets can be mapped to the sphere and yield small-scale admissible *spherical* wavelets. These can then be dilated to larger scales using the spherical dilation. This is neatly summarized by the following result, where we repeat for convenience the definition (5.1) of the inverse stereographic projection, namely, in polar coordinates,

$$(I^{-1}f)(\theta, \varphi) = \frac{2}{1 + \cos\theta} \, f(2\tan\frac{\theta}{2}, \varphi). \tag{9.44}$$

Proposition 9.2.5 *The inverse stereographic projection (9.44) is a unitary map* I^{-1} : $L^2(\mathbb{R}^2, d\vec{x}) \to L^2(S^2, \sin\theta\, d\theta\, d\varphi)$ *between the respective Hilbert spaces. Moreover, if* $\psi \in L^2(\mathbb{R}^2)$ *is an admissible 2-D Euclidean wavelet, then the function* $I^{-1}\psi$ *is an admissible spherical wavelet.*

This results tells us that we can construct a spherical wavelet starting from any Euclidean wavelet. Now what does this tell us about directional wavelets? Since directional sensitivity is a local or small-scale attribute, it should intuitively survive this process. Yet there is more than intuition in this result. The stereographic projection and both spherical and Euclidean dilations are conformal mappings. Thus Proposition 9.2.5 defines a conformal application that, by definition, preserves angles. The directional sensitivity of the Euclidean wavelet is thus transported to the spherical wavelet.

A natural candidate for building a directional spherical wavelet is to start with the (truncated) Euclidean Morlet or Gabor wavelet (3.19):

$$\psi_{\mathrm{G}}(\vec{x}) = e^{i\vec{k}_0 \cdot \vec{x}} e^{-|\vec{x}|^2}. \tag{9.45}$$

Using Proposition 9.2.5, we find the following spherical wavelet:

$$\psi_{\mathrm{G}}(\theta, \phi) = \frac{e^{ik_0 \tan\frac{\theta}{2} \cos(\phi_0 - \phi)} e^{-\frac{1}{2}\tan^2\frac{\theta}{2}}}{1 + \cos\theta}. \tag{9.46}$$

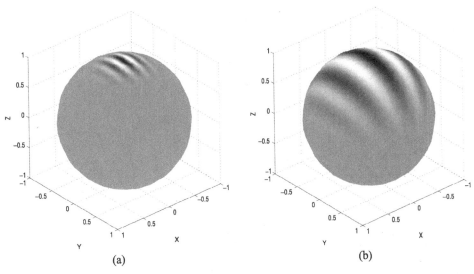

Fig. 9.6. Real part of the spherical Morlet wavelet (9.46) at scale (a) $a = 0.03$ and (b) $a = 0.3$ (from [35]).

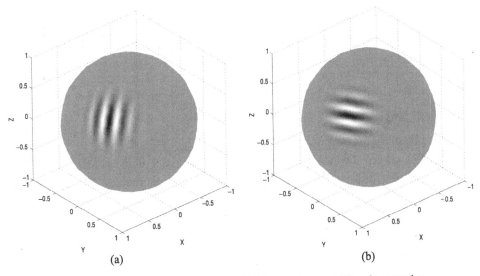

Fig. 9.7. Real part of the spherical Morlet wavelet (9.46) at scale $a = 0.03$ and centered at $(\pi/3, \pi/3)$: (a) $\theta' = 0$ and (b) $\theta' = \pi/2$ (from [35]).

This function is represented in Figures 9.6 and 9.7 for various values of the scale and rotation parameters. An example illustrating the directional sensitivity of this wavelet will be presented in Section 9.2.5.3 below (Figure 9.14), where we will compare it with that of the spherical DOG wavelet function.

Of course, a spherical wavelet analysis should always be performed at small scales (see Section 9.2.5.2). As a confirmation, we show in Figure 9.8 the ARP (9.43) of

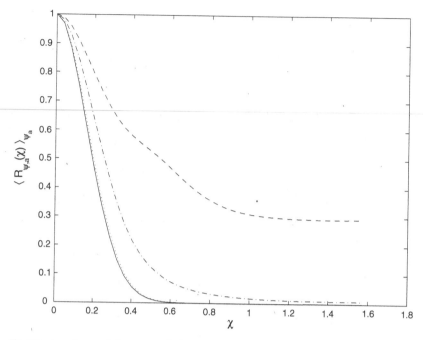

Fig. 9.8. Angular resolving power (9.43) of the spherical Morlet wavelet (9.46) for various values of the scale parameter: $a = 0.03$ (plain), $a = 0.1$ (dotted), $a = 0.3$ (dashed-dotted) and $a = 1$ (dashed).

the spherical Morlet wavelet for different values of the scale parameter. One clearly notices that, as the scale gets bigger, angles that are far apart become more and more correlated.

9.2.4.4 Representation of object surfaces

The spherical wavelets just described are the basis of a novel representation of object surfaces [89] – a longstanding problem in computer graphics. The idea is the following. The surface of a 3-D (star-shaped) object is treated as a function on the 2-sphere (a human head is a good example, with a lot of neurophysiological interest). In a first step, the *coarse structure* of the surface is described by a truncated expansion of that function into spherical harmonics. Thus we get a band-limited function:

$$f_c(\theta, \varphi) = \sum_{l=0}^{L} \sum_{|m| \leqslant l} \widehat{f}(l, m) \, Y_l^m(\theta, \varphi), \tag{9.47}$$

for some small value of L (the authors choose $L = 5$). The justification is that the small l terms in the Fourier expansion are low-frequency components, which are well represented by functions supported on the whole sphere, such as spherical harmonics. Then the remainder $f_{\text{res}} = f - f_c$, which represents the *fine structure* of the surface, is described by well-localized spherical wavelets, in particular, spherical Morlet wavelets, with the help of an appropriate optimization procedure. The crucial step here is to treat the low- and high-frequency components separately. We will see in Section 9.3 a mathematically precise version of this procedure. Namely, in the L^1 formalism, the reconstruction formula for the spherical CWT is based on precisely the same idea, with the scale variable a as the relevant parameter to distinguish between low and high frequencies (Theorems 9.3.8 and 9.3.9, which yield indeed approximation schemes for functions on the sphere).

The same technique may be used also [90] to define smoothing on S^2 by a spherical Gaussian kernel, obtained again by inverse stereographical projection from a plane Gaussian (essentially our function ϕ_G of Section 9.2.2.3). Actually this Gaussian is nothing but a heat kernel, which generates a diffusion process on S^2. This approach opens interesting new perspectives in object deformation ("morphing"), along the lines of the work of Sweldens *et al.* on digital geometry processing [242].

9.2.5   Implementation of the spherical wavelet transform

The spherical CWT (9.27) is given as a convolution over the sphere S^2. This creates numerical difficulties, for no really fast algorithm exists today. In particular, it is difficult to find an appropriate discretization of the sphere. Several methods have been proposed in the literature, mostly based on Fourier and spherical harmonics techniques [Moh97,154,169,216,274], but none of them is fully satisfactory. A possible exception is the method introduced by Wandelt and Górski [367] and based on the use of the FFT. Interestingly enough, this approach was motivated by the analysis of cosmic microwave background (CMB) data, that we have mentioned in Section 5.1.2 – and for which spherical wavelets have been proposed! The new algorithm presented here, however, seems to answer the question rather well [35].

For a practical implementation of the spherical CWT, the first step is that of discretization. This means finding a suitable grid in the parameter space, so as to allow a fast calculation *and* a good approximation of the continuous theory. As we shall see, the key to the algorithm presented below is to use an FFT in the (periodic) longitude angle φ. Actually we also need some sort of criterion on the grid density for controlling aliasing problems, as indicated already in [13]. More precisely, we have to specify the scale interval in which the spherical wavelet transform makes sense. A possible answer will be suggested in Section 9.2.5.2. Then several examples will be discussed, both academic and real-life.

9.2.5.1   Discretization and algorithm

Following an approach similar to that in [Win95], the first step is to discretize the integral (9.39) on a regular spherical grid $M \times N$,

$$\mathcal{G} = \{(\theta_t = \frac{\pi}{M}t, \varphi_p = \frac{2\pi}{N}p) \mid 0 \leqslant t \leqslant M - 1, \, 0 \leqslant p \leqslant N - 1\}, \tag{9.48}$$

by a weighted sum (χ and a are fixed throughout)

$$S(\chi, \theta_{t'}, \varphi_{p'}, a) \equiv S[\chi, t', p', a] \tag{9.49}$$

$$= \sum_{\substack{0 \leqslant t \leqslant M-1 \\ 0 \leqslant p \leqslant N-1}} \overline{\psi_{a,\chi,t'}[t, p - p']}\, s[t, p]\, w_t, \tag{9.50}$$

where:

- $s[t, p] \equiv s(\theta_t, \varphi_p)$;
- $\psi_{a,\chi,t'}[t, p - p'] \equiv \psi_{a,\chi,\theta_{t'},0}(\theta_t, \varphi_{p-p'})$;
- the index of φ is extended to $\mathbb{Z}$ by angular periodicity with the rule $\varphi_{p+N} = \varphi_p$;
- $w_t = (2\pi^2/MN) \sin \theta_t$ are the weights suggested in [Win95] for the discretization of the Lebesgue measure on the particular grid $\mathcal{G}$.

Evaluating the sums in (9.50) requires MN additions and multiplications for each (t', p'), that is, $M^2 N^2$ operations altogether.

However, an easy simplification can be obtained for the longitudinal coordinates by the use of a Fourier series and the Plancherel formula. Indeed, denoting by

$$\check{h}[t, k] = \sum_{0 \leqslant p \leqslant N-1} h[t, p] \exp(-i\, kp\, \frac{2\pi}{N}), \tag{9.51}$$

the longitudinal Fourier coefficients of a given discrete function h, we obtain

$$S[\chi, t', p', a] = 2\pi \sum_{0 \leqslant t \leqslant M-1} w_t\, \mathcal{F}[\chi, t', p', a, t] \tag{9.52}$$

with

$$\mathcal{F}[\chi, t', p', a, t] = \sum_{0 \leqslant k \leqslant N-1} \overline{\check{\psi}_{a,\chi,t'}[t, k]}\, \check{s}[t, k] \exp(i\, kp'\, \frac{2\pi}{N}). \tag{9.53}$$

The quantity $\mathcal{F}$ may be computed with the inverse fast Fourier transform (IFFT), which leads to a reduction of the computational time from $O(M^2 N^2)$ to $O(M^2 N \log N)$. On a grid $\mathcal{G}$ of 256×256, the gain is a factor of 46.

Notice that other discretization techniques than a plain Riemann sum, as used in (9.50), would be beneficial only if one imposes additional regularity conditions on the signal s. Furthermore, other weights w_t could be chosen to achieve a better approximation of (9.49). An example of a different choice, both for the weights and for the discretization technique, is that of a band-limited spherical function, as considered in [216] and [367], that is, a function $f(\theta, \varphi)$ whose expansion in spherical harmonics has only finitely many nonzero terms, $\widehat{f}(l, m) = 0$, $\forall l > L$.

One can also ask whether a better algorithm could be designed on another type of grid. For instance, what is the use in the continuous case of an icosahedral grid, as introduced in [336,337] (see Figure 9.9) for the starting point of a lifting procedure? The great advantage of such a grid (or one built on a fullerene type of grid) is its better isotropy. The point is that the usual spherical coordinates necessarily introduce a preferred direction, the polar axis, and this is at the origin of the various tentatives reported in the literature

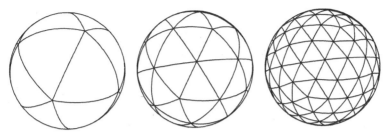

Fig. 9.9. The geodesic sphere construction, starting with the icosahedron on the left (subdivision level 0) and the next two subdivision levels (from [336]).

of designing an equidistribution of points on the sphere [113,169,330]. The last paper, in particular, contains many references to related works, from pure mathematics to physics (electrostatics).

9.2.5.2   Numerical criterion for the scale range

The discretization of the continuous spherical wavelet transform gives rise to a sampling problem. Since the grid $\mathcal{G}$ is fixed, if we contract or dilate our wavelet too much, we obtain a function which is very different from the original ψ. In other words, aliasing occurs and the wavelet is not numerically admissible. This problem has been noted also in the flat space CWT (see Section 4.1.1), but only an empirical criterion was given [13].

We have seen in Proposition 9.2.3 that a function $\psi \in L^2(S^2, d\mu)$ is admissible only if it satisfies the zero mean condition (9.31). Approximating the integral by its Riemann sum, we get the quantity

$$C[\psi] = \sum_{\substack{1 \leqslant t \leqslant M-1 \\ 1 \leqslant p \leqslant N-1}} \frac{\psi(\theta_t, \varphi_p)}{1 + \cos \theta_t} \, \mu(\theta_t, \varphi_p). \tag{9.54}$$

Because of the discretization, even if ψ verifies (9.31), it is not necessarily true that $C[\psi]$ vanishes. However, we may suppose that this quantity is very close to zero when ψ is sampled sufficiently, that is, if the grid $\mathcal{G}$ is fine enough.

Nevertheless, it is difficult to give a quantitative meaning to the value of $C[\psi]$. How small is "very close to zero"? Here is a possible solution to this problem. Since the spherical measure μ and the function $1 + \cos \theta$ are positive, it is clear that

$$C[\psi] \leqslant C[|\psi|] \tag{9.55}$$

for any $\psi \in L^2(S^2, d\mu)$. So we can define a numerical normalized admissibility by

$$\widetilde{C}[\psi] = \frac{C[\psi]}{C[|\psi|]}, \tag{9.56}$$

a quantity always contained in the interval [-1,1].

We can now give a precise definition of numerical admissibility.

Definition 9.2.6 *A spherical wavelet of $L^2(S^2, d\mu)$ is numerically admissible with threshold p% or p% - admissible, if the numerical normalized admissibility (9.56) is smaller than $(100 - p)/100$ in absolute value:*

$$|\widetilde{C}[\psi]| \leqslant \frac{100 - p}{100}. \tag{9.57}$$

As an example, we present in Figure 9.10 the behavior of the dilated spherical DOG wavelet, $D_a \psi_G^{(\alpha)}$ ($\alpha = 1.25$), as a function of $a > 0$, discretized on a 128×128 grid (notice that, in the flat case, $\alpha = 1.6$ is the value for which the DOG wavelet is almost indistinguishable from the Mexican hat).

According to this plot, the wavelet $D_a \psi_G^{(\alpha)}$ is 99 % - admissible on the scale interval $a \in [0.072, 24.71]$. The lower limit is due to the fact that, for small a, $D_a \psi$ is not sampled enough. The upper limit comes from the subsampling of the area far from the North Pole which, according to the spherical dilation, gets more and more contracted. Figure 9.11 presents three typical behaviors of $D_a \psi$ discretized on a 22 point θ sampling. For $a = 0.5$, the sampling is correct. For $a = 0.05$, that is, below the lower admissibility bound, subsampling occurs, so that negative parts of $D_a \psi$ are completely missed. Clearly, this discretized wavelet is no longer admissible. Exactly the same effect was observed long ago in the flat case [13]. The third case, with $a = 3.5$, thus

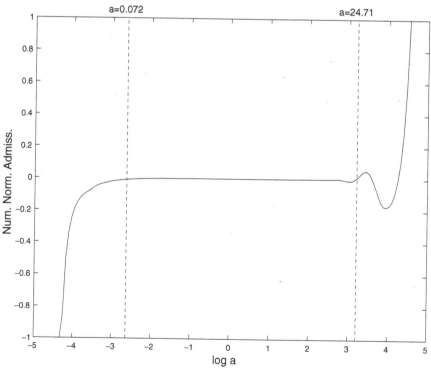

Fig. 9.10. $\widetilde{C}[D_a \psi_G^{(\alpha)}]$ as a function of $\log a$ for $\alpha = 1.25$ (from [35]).

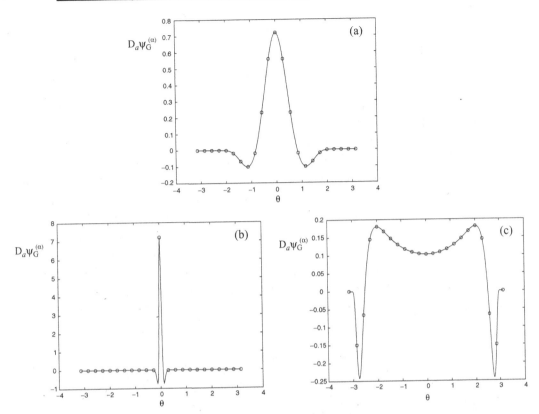

Fig. 9.11. Three typical behaviors of $D_a \psi_G^{(\alpha)}$ discretized on a 22×22 grid $\mathcal{G}$. (a) For $a = 0.5$, the sampling is correct; (b) for $a = 0.05$, subsampling occurs, negative parts of $D_a \psi_G^{(\alpha)}$ are completely missed; (c) subsampling on the negative parts of $D_a \psi_G^{(\alpha)}$ for $a = 3.5$. Notice the minimum at $\theta = 0$ (from [35]).

beyond the upper bound, is less intuitive. Here the subsampling takes place for *large* values of θ, that is, close to the South Pole, but the result is the same; the discretized wavelet does not have a zero mean, it is not admissible. In addition, the curve presents a *minimum* at $\theta = 0$. This somewhat unexpected effect is in fact due to the cocycle, as is the dependence of the height on a. Indeed, if one performs the same calculation *without* the cocycle, all curves show a maximum at $\theta = 0$, with the same height. Here again we see that curvature, which requires the presence of the cocycle, has a nontrivial effect.

As a further illustration of this behavior, it is instructive to consider the function ι identically equal to 1. In the flat case, this function has a vanishing WT, by the admissibility condition $\int_{\mathbb{R}^2} d^2\vec{x} \, \psi(\vec{x}) = 0$ on the wavelet, but it is not square integrable and thus cannot be reconstructed. In the present case, however, the situation is different. The function ι is square integrable, since the sphere S^2 is compact, but its WT does *not* vanish, because of the presence of the cocycle. Indeed, the function ι is invariant under rotation, but *not* under dilation:

$$(D_a \iota)(\theta, \phi) = \lambda(a, \theta)^{1/2} \not\equiv 1, \tag{9.58}$$

and, therefore,

$$I(\varrho, a) = \langle R_\varrho \, D_a \psi | \iota \rangle \; = \; \langle \psi | D_a \iota \rangle \; \equiv \; I(a)$$

$$= \int_{S^2} d\mu(\zeta) \, \overline{\psi(\zeta)} \, \lambda(a, \theta)^{1/2} \neq 0. \tag{9.59}$$

Thus, for fixed a, the WT $I(a)$ of the unit function is constant, and essentially negligible for $a \ll 1$. Significant values appear only for $a > 2$, and these scales are irrelevant for the analysis of signals such as contours. This behavior has been checked numerically [35], with the familiar DOG wavelet $\psi_G^{(\alpha)}$, discretized on a 128×128 grid. Because of the discretization, however, the function $I(\varrho, a) \equiv I(\theta, a)$ *does* depend on a, but very little, and its value remains extremely small. As a matter of fact, for $a < 0.1$, the WT of ι is numerically negligible over the whole sphere, and may be taken as zero to a very good approximation. As a consequence, the spherical CWT does have the familiar local filtering effect, provided small scales are considered. This will be confirmed by the examples below. Once again, we see that the CWT is useful only as a *local* analysis.

9.2.5.3   Examples of spherical wavelet transforms

As a first example, we analyze in Figure 9.12 an academic picture, namely, (the characteristic function of) a spherical triangle on S^2, with one of the corners sitting at the North Pole. The triangle is given by $0° \leqslant \theta \leqslant 70°$, $0° \leqslant \varphi \leqslant 90°$ and is discretized on a 128×128 grid in (θ, φ). We use again the spherical Gaussian wavelet $\psi_G^{(\alpha)}$, for $\alpha = 1.25$, discretized on the same grid. According to the admissibility analysis presented above (Figure 9.10), the wavelet is 95 % - admissible on the scale interval $a \in [0.033, 29.27]$. Thus we can evaluate the spherical CWT of this picture for various scales in the allowed range, and we have chosen four successive scales from $a = 0.5$ to $a = 0.035$. Figure 9.12 shows that the spherical WT behaves here exactly as, in the flat case, the WT of the characteristic function of a square, as shown in Figure 4.1, in Section 4.1.1. For large a, the WT sees only the object as a whole, thus allowing us to determine its position on the sphere. When a decreases, increasingly finer details appear; in this simple case, eventually only the contour remains, and it is perfectly seen at $a = 0.035$. The transform vanishes in the interior of the triangle, as it should, only the "walls" remain, with a negative value (black) just outside, a zero-crossing right on the boundary and a sharp positive maximum (white) just inside. In addition, each corner gives a neat peak, which is positive, since the corner is convex [13]. Notice that the three corners are alike, so that indeed the poles play no special rôle in our spherical WT, contrary to what occurs often in the classical spherical analysis based on spherical harmonics [Fre97,Fre99,169,294,319]. On the other hand, the scales are different in the four cases, since the amplitude of the transform diminishes for decreasing values of a.

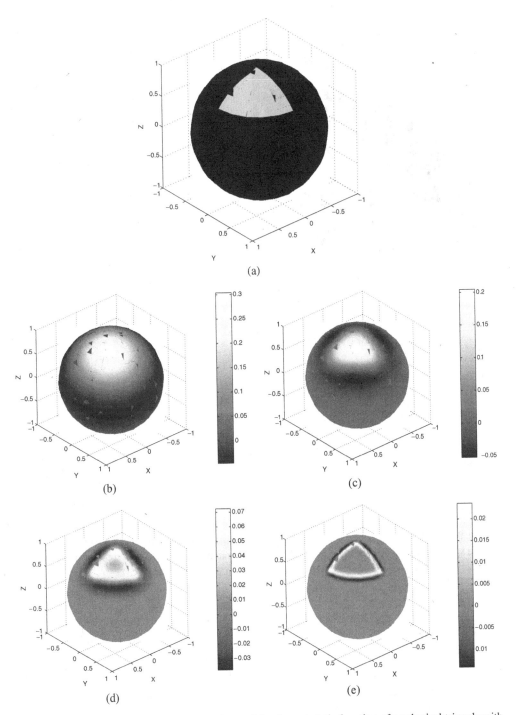

Fig. 9.12. Spherical wavelet transform of the characteristic function of a spherical triangle with apex at the North Pole, $0° \leqslant \theta \leqslant 70°$, $0° \leqslant \varphi \leqslant 90°$, obtained with the difference wavelet (9.32), at scale $a = 0.25$. (a) Original image. The transform is shown at four successive scales, (b) $a = 0.5$; (c) $a = 0.2$; (d) $a = 0.1$; and (e) $a = 0.035$. As expected, it vanishes inside the triangle, and presents a "wall" along the contour, with sharp peaks at each summit. Notice that the amplitudes are different in the four cases (from [35]).

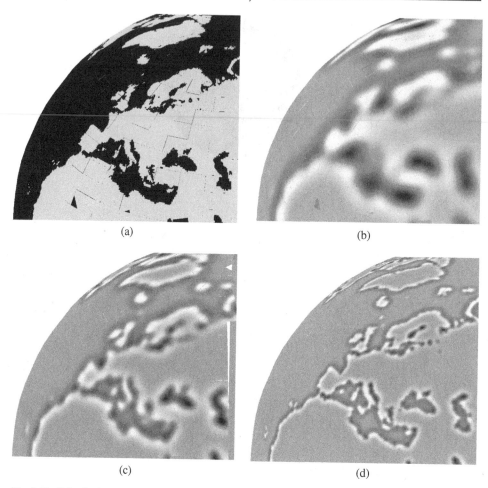

(a)

(b)

(c)

(d)

Fig. 9.13. Spherical wavelet transform of the spherical map of the European area, computed with the spherical DOG wavelet for $\alpha = 1.25$. (a) The original picture; (b) wavelet transform at $a = 0.032$; (c) the same at $a = 0.016$; (d) the same at $a = 0.0082$ (from [35]).

As a second, real life example, we present in Figure 9.13 the wavelet transform of a significant piece of the terrestrial globe, covering Europe, Greenland and North Africa. As before, we use the spherical DOG wavelet $\psi_G^{(\alpha)}$ for $\alpha = 1.25$. The transforms are shown again at three successive scales, $a = 0.032, 0.016, 0.0082$ (the grid used here is finer than the one used in the previous examples, so that smaller values of a are admissible). As expected, the resolution improves with diminishing a. However, at $a = 0.0082$, the discretization grid used for the computation of the transform coincides with that of the original picture, so that one sees exactly the same artifacts, such as a closed strait of Gibraltar, an unresolved complex Corsica–Sardinia, ragged coastlines, etc. Of course, we cannot hope to *improve* on the resolution of the original!

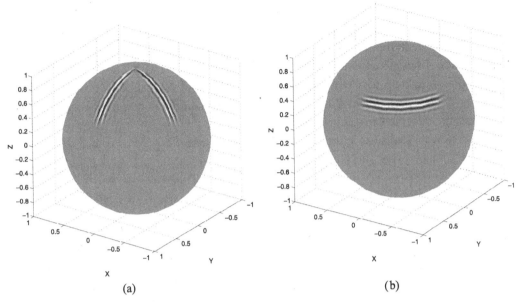

Fig. 9.14. Spherical CWT of the spherical triangle at scale $a = 0.03$, using spherical Morlet wavelet (9.46): (a) $\theta' = 0$ and (b) $\theta' = \pi/2$ (from [35]).

As a final example, we take again the spherical triangle of Figure 9.12(a) and analyze it with the spherical Morlet wavelet (9.46). The results confirm our analysis: the spherical CWT is able to detect the local orientation of the edges of the triangle (Figure 9.14).

9.2.6   Extension to other manifolds

The whole construction made so far extends almost verbatim to the $(n-1)$-dimensional sphere $S^{n-1} = SO(n)/SO(n-1)$, with help of a similar class I representation of the generalized Lorentz group $SO_o(n, 1)$ [23]. Although the spheres are the manifolds on which a CWT is most desirable for applications, the mathematical analysis made here invites to consider other manifolds with similar geometrical properties.

We take first $n = 3$. The sphere $S^2 = SO(3)/SO(2)$ is a compact Riemannian symmetric space of constant curvature $\kappa = 1$. It has a noncompact dual, $H^2 = SO_o(2, 1)/SO(2)$, of constant curvature $\kappa = -1$ [Hel78]. H^2 is a two-sheeted hyperboloid, symmetric around the x_3-axis. Duality corresponds to the fact that $SO(3)$ and $SO_o(2, 1)$ are the two real forms of the complex group $SO(3)^{\mathbb{C}} \sim SL(2, \mathbb{C})$.

Exactly as in the case of the sphere, we can perform a stereographic projection Φ from the South Pole onto the equatorial plane $x_3 = 0$ (or, equivalently, to the plane tangent at the other pole). Then Φ maps the upper sheet H^2_+ onto the interior $\mathcal{D}_+$ of the unit disk, and the lower sheet H^2_- onto the exterior. The domain $\mathcal{D}_+$, called the Lobachewskian disk, is conformally equivalent to H^2_+, and both manifolds have $SO_o(2, 1)$ as isometry group.

As we did for the sphere, dilations on H_+^2 may be obtained by lifting dilations in the equatorial plane by inverse stereographic projection. The resulting map has all the required properties for a dilation, but does not come directly from a linear group action. Thus it can only be used for constructing wavelets on H_+^2 if one puts it by hand. It remains to obtain a suitable representation of the resulting set $SO_o(2, 1) \cdot \mathbb{R}_*^+$ in $L^2(H_+^2, d\mu)$, where $d\mu$ is the $SO_o(2, 1)$-invariant measure, and to show that it is square integrable in a suitable sense.

Now this suggests a further generalization. In both cases, S^2 as well as H^2, the unit disk, image of one sheet or one hemisphere, is a classical domain. Also the stereographic projection has a group-theoretical origin [Per86]. This paves the way to the generalization of the CWT to a whole class of homogeneous spaces (Riemannian symmetric spaces). For instance,

$$S^{n-1} = SO(n)/SO(n-1) \quad \text{and} \quad H^{n-1} = SO_o(n-1, 1)/SO(n-1)$$

are dual Riemannian symmetric spaces, with constant curvature $\kappa = \pm 1$, respectively. Again $SO(n)$ and $SO_o(n-1, 1)$ are two real forms of the complexified group $SO(n)^{\mathbb{C}}$. Their isometry groups are $SO(n)$ and $SO_o(n-1, 1)$, respectively, so that suitable representations of the generalized Lorentz group $SO_o(n, 1)$ should provide the corresponding wavelets. While this has been obtained explicitly in the case of the $(n-1)$-sphere S^{n-1} in [23], the problem is still open in the noncompact case, i.e., the hyperboloid H^{n-1}.

9.3   Wavelet approximations on the sphere

The central theme of approximation theory is the representation of a function by a truncated series expansion into a family of basis functions, for instance, the elements of a frame. Thus, in the flat case, one- or two-dimensional wavelets are widely used for approximation in various function spaces [Mal99]. The crucial advantage is their multiresolution character, which is optimally adapted to local perturbations. A natural framework is given by the Lebesgue spaces $L^p(\mathbb{R}^n)$, $1 \leqslant p < \infty$. One of the reasons is that approximation is often formulated in terms of convolution with an *approximate identity*, and many useful convolution identities are available in L^p.

Therefore, in order to apply these considerations to the sphere S^2, it is necessary to have a good notion of convolution on S^2. For that purpose, it is useful to represent the sphere as the quotient $SO(3)/SO(2)$, since the convolution machinery extends almost verbatim to locally compact groups, and then partly to homogeneous spaces. For the convenience of the reader, we have collected in the Appendix the main definitions and essential properties of convolution on a locally compact group. In what follows, we will need two different cases. For simplicity, we write $L^2(SO(3)) \equiv L^2(SO(3), d\varrho)$, where $d\varrho$ is the Haar measure on $SO(3)$, and $L^p(S^2) \equiv L^p(S^2, d\mu)$.

- If $f \in L^2(SO(3))$ and $g \in L^1(S^2)$, then their *spherical convolution* is the function on S^2 defined as

$$(f \star g)(\zeta) = \int_{SO(3)} f(\varrho) \, g(\varrho^{-1}\zeta) \, d\varrho. \tag{9.60}$$

Then $f \star g \in L^2(S^2)$ and one has

$$\|f \star g\|_2 \leqslant \|f\|_2 \, \|g\|_1, \tag{9.61}$$

where the norms refer to the corresponding spaces.
- If $f \in L^2(S^2)$ and $g \in L^1(S^2)$, their *spherical convolution* is the function on $SO(3)$ defined as

$$(f \,\widetilde{\star}\, g)(\varrho) = \int_{S^2} f(\varrho^{-1}\zeta) \, g(\zeta) \, d\mu(\zeta). \tag{9.62}$$

Then $f \,\widetilde{\star}\, g \in L^2(SO(3))$ and

$$\|f \,\widetilde{\star}\, g\|_2 \leqslant \|f\|_2 \, \|g\|_1, \tag{9.63}$$

Here, however, we are only interested in functions on the sphere S^2, that is, functions on $SO(3)$ that are $SO(2)$-invariant. In particular, we will deal mostly with axisymmetric, or zonal, functions on S^2, that is, functions of θ alone. Thus, we will focus on elements of $L^2([-1, +1], dt)$, where $t = \cos\theta$, for which the Fourier series reduces to a Legendre expansion:

$$\psi(t) = \sum_{l=0}^{\infty} \frac{2l+1}{4\pi} \, \widehat{\psi}(l) \, P_l(t),$$

$$\widehat{\psi}(l) = 2\pi \int_{-1}^{+1} dt \, P_l(t) \, \psi(t) = \sqrt{\frac{4\pi}{2l+1}} \, \widehat{\psi}(l, 0),$$

where $\widehat{\psi}(l, m) \equiv \langle Y_l^m | \psi \rangle$ denotes the Fourier coefficient of ψ. If f is an axisymmetric function, the spherical convolution (9.62) takes a simpler form [Fre97]:

Proposition 9.3.1 *Let f and g be two measurable functions on S^2. If f is axisymmetric, the spherical convolution of f and g is a function on S^2, which can be written:*

$$(f \star g)(\zeta') = \int_{S^2} d\mu(\zeta) \, f(\widehat{\zeta'} \cdot \widehat{\zeta}) \, g(\zeta), \tag{9.64}$$

where $\widehat{\zeta'} \cdot \widehat{\zeta}$ is the $\mathbb{R}^3$ scalar product of unit vectors of directions ζ' and ζ.

The proof amounts to a straightforward application of harmonic analysis (Fourier series) on S^2 and of the addition theorem for spherical harmonics, to the effect that $f(\varrho^{-1}\zeta') = f(\widehat{\zeta'} \cdot \widehat{\zeta})$, where $\zeta' \equiv \dot{\varrho} \in S^2$ denotes the left coset of $\varrho \in SO(3)$ (see the geometrical discussion in Section 9.2.4.1).

Now we may turn to the approximation problem proper. As in the Euclidean case [Lie97,Ste71], a convenient technique is to perform a convolution with a smoothing kernel, that acts as an approximate identity, that is, a kernel which tends to the identity (δ function) as the parameter goes to 0. For the sake of simplicity, we will only deal with zonal kernels, following mainly [Fre97].

Definition 9.3.2 Let $\mathcal{K}_\tau$, $\tau \in (0, \tau_o]$, $\tau_o \in \mathbb{R}_*^+$, be a family of elements of $L^1([-1, +1], dt)$ satisfying $\widehat{\mathcal{K}_\tau}(0) = 1$. The functional $S_\tau[f]$ defined by

$$S_\tau[f] = \mathcal{K}_\tau \star f, \quad f \in L^p(S^2), \quad 1 \leqslant p < \infty,$$

is called a singular integral. It is called an approximate identity of $L^p(S^2)$ if

$$\lim_{\substack{\tau \to 0 \\ \tau > 0}} \|f - S_\tau[f]\|_p = 0, \quad \forall f \in L^p(S^2). \tag{9.65}$$

The following theorem characterizes those spherical kernels which are associated with an approximate identity.

Theorem 9.3.3 Let $\{\mathcal{K}_\tau\}$ be a uniformly bounded spherical kernel, that is, there exists a constant M, independent of τ, such that

$$\int_{-1}^{+1} dt \, |\mathcal{K}_\tau(t)| \leqslant M, \quad \forall \tau \in (0, \tau_o].$$

Then the associated singular integral is an approximate identity of $L^p(S^2)$ if and only if

$$\lim_{\substack{\tau \to 0 \\ \tau > 0}} \widehat{\mathcal{K}_\tau}(n) = 1, \quad \forall n \geqslant 0. \tag{9.66}$$

A proof may be found in [Fre97]. A particularly interesting case is given by positive definite kernels. In this case, since $|P_l(t)| \leqslant 1$, $\{\mathcal{K}_\tau\}$ is uniformly bounded, with bound $M = \sup_{\tau \in (0, \tau_o]} \widehat{\mathcal{K}_\varrho}(0)$. The following theorem gives a nice characterization of approximate identities associated with positive kernels.

Theorem 9.3.4 Let $\{\mathcal{K}_\tau\}$, $\tau \in (0, \tau_o]$, be a positive kernel associated with a singular integral of $L^p(S^2)$. Then each of the following conditions is equivalent to (9.65) and (9.66), which means that $\{\mathcal{K}_\tau\}$ is the kernel of an approximate identity:

(i) $\lim_{\substack{\tau \to 0 \\ \tau > 0}} \widehat{\mathcal{K}_\tau}(0) = 1$,

(ii) $\lim_{\substack{\tau \to 0 \\ \tau > 0}} \int_{-1}^{\delta} dt \, \mathcal{K}_\tau(t) = 0, \, \delta \in (-1, +1)$.

Condition (ii) is in fact a constraint on the localization of the kernel, as we shall see in an explicit example below (see Figure 9.15).

Approximate identities are a very useful tool for harmonic analysis on the sphere and many applications can be found in [Fre97]. Thus it is gratifying that the spherical wavelet transform naturally yields a systematic way of deriving approximate identities on the sphere. This is actually an interesting way of handling functions on the sphere, because it allows us to represent information by means of localized, and hierarchically organized, coefficients. With such a representation, a local modification of the function would only result in a slight local perturbation of the original coefficients, a definite advantage over Fourier series (exactly as in the case of flat space).

As a matter of fact, many examples of approximate identities are given in the textbook of Freeden *et al.* [Fre97], and they are applied extensively by these authors to geophysical data [Fre99]. Most of these examples are based on families of kernels indexed by a parameter which behaves like a dilation. However, since the latter is introduced directly as a parameter in those kernels, there is no unique way of generating approximate identities, as in $\mathbb{R}^n$ [Ste71]. But this problem disappears naturally if one uses the spherical dilation, since, as we shall see, the dilation operator generates an approximate identity in $L^2(S^2)$.

However, we have to modify it first and adapt it to the L^1 environment. Using the notation of Section 9.2.2, we define, instead of D_a, a new dilation operator:

$$(D^a f)(\zeta) \equiv f^a(\zeta) = \lambda(a, \theta) f(\zeta_{1/a}),\tag{9.67}$$

and this operator clearly conserves the L^1 norm. Notice that the situation is more complicated here than in the flat case. There, indeed, changing the dilation operator from L^2 to L^1 simply amounts to changing the power of a in front of the transform [29]. Here, one replaces the factor $\lambda(a, \theta)^{1/2}$ by its square $\lambda(a, \theta)$, but this modifies the CWT itself in a nontrivial way. In particular, the admissibility condition (9.25) becomes

$$\frac{8\pi^2}{2l+1} \sum_{|m|\leqslant l} \int_0^\infty \frac{da}{a} \, |\widehat{\psi^a}(l, m)|^2 < c\,.\tag{9.68}$$

First, we notice that our new dilation operator does not change the mean of a function, thus simplifying the statement of Proposition 9.2.4.

Proposition 9.3.5 *If $\psi \in L^1(S^2)$, then*

$$\int_{S^2} d\mu(\zeta) \, \psi^a(\zeta) = \int_{S^2} d\mu(\zeta) \, \psi(\zeta).\tag{9.69}$$

The proof reduces to a simple change of variables, taking into account the cocycle relation (9.23).

Acting with the new dilation D^a on a suitable function, one can now easily construct an approximate identity, as shown in the next proposition.

Proposition 9.3.6 *Let* $f \in \mathcal{C}([-1, +1])$, *with* $\widehat{f}(0) = 1$. *Then the family* $\{f^a \equiv D^a f, a > 0\}$ *is the kernel of an approximate identity.*

In view of Theorem 9.3.3, the proof consists in two steps. First, one shows that the family $\{f^a\}$, $a \in (0, 1]$, is uniformly bounded, which is obvious since $\int_{-1}^{+1} dt \ |f^a(t)| = \|f\|_1$. Next, it remains to verify that

$$\lim_{\substack{a \to 0 \\ a > 0}} \widehat{f^a}(l) = 1,$$

and this is done again by a change of variables and applying the cocycle relation (9.23) [35].

This technique is applied in Figure 9.15 to a zonal function of Gaussian shape, namely the mother wavelet of the spherical DOG wavelet, $\phi_G(\theta, \phi) = \exp(-\tan^2(\theta/2))$, $\theta \in [-\pi, \pi]$. One clearly sees how dilation localizes the kernel better and better as $a \to 0$.

In the L^1 formalism, we recall from [29] that the necessary condition for admissibility becomes a genuine zero mean condition, exactly as in the flat case:

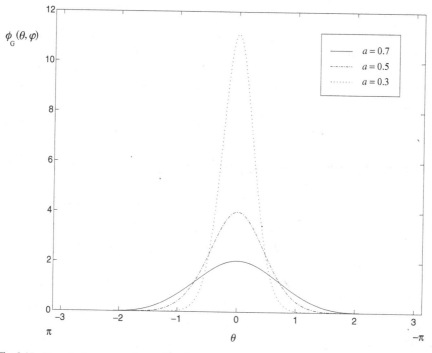

Fig. 9.15. Kernel of an approximate identity obtained by dilating a Gaussian mother function with scaling factor $a = 0.7, 0.5$ and 0.3 (from [35]).

$$\int_{S^2} d\mu(\theta, \varphi) \, \psi(\theta, \varphi) = 0, \tag{9.70}$$

Correspondingly, in view of Proposition 9.3.5, the difference wavelet $\psi_\phi^{(\alpha)}$ given in (9.32) is replaced by

$$\check{\psi}_\phi^{(\alpha)}(\theta, \varphi) = \phi(\theta, \varphi) - D^\alpha \phi(\theta, \varphi) \quad (\alpha > 1).$$

Now, combining the modified dilation operator D^a with the usual rotation operator R_ϱ, we define a new set of spherical wavelets, starting from an admissible ψ, namely, $\psi_\varrho^a \equiv R_\varrho D^a \psi = R_\varrho \psi^a$. Accordingly, we redefine as follows the spherical wavelet transform of a signal $s \in L^2(S^2)$:

$$\check{S}_\psi(\varrho, a) = \int_{S^2} d\mu(\zeta) \, \overline{\psi_\varrho^a(\zeta)} \, s(\zeta). \tag{9.71}$$

In particular, if the wavelet ψ is zonal, we get

$$\check{S}_\psi(\zeta, a) = \int_{S^2} d\mu(\zeta') \, \overline{\widehat{\psi^a(\zeta \cdot \zeta')}} \, s(\zeta'). \tag{9.72}$$

In addition, the correspondence between spherical wavelets and their stereographic projections on the tangent plane (Proposition 9.2.5) is modified, as follows.

Proposition 9.3.7 *If $\psi \in L^2(\mathbb{R}^2)$ is an admissible 2-D Euclidean wavelet, and I^{-1} denotes the inverse stereographic projection (9.44), then the function $(1 + \cos\theta)^{-1} I^{-1} \psi$ is an admissible spherical wavelet for the transform defined with the L^1 norm-preserving dilation operator D^a.*

After this preparation, we proceed to show that the spherical CWT admits a reconstruction formula, valid in the strong L^2 topology, exactly as the usual CWT in $\mathbb{R}^n$. As in the flat case, described in Section 2.6.1, we may distinguish between a bilinear and a linear formalism [Tor95]. But there is a crucial difference. In the flat case, it is advantageous, but not compulsory, to treat the large scales or low frequencies separately, in terms of a scaling function. Here, however, we are *forced* to do it. The reason is that, geometrically, only small scales are relevant and lead to the expected filtering behavior, as discussed in Section 9.2.5.2. We arbitrarily choose $a = a_o$ as reference scale and define the scales $a > a_o$ as large. Notice that we recover here, in precise mathematical terms, the argument behind the mixed spherical harmonics/spherical wavelets representation of object surfaces from [89], discussed in Section 9.2.4.4.

Let us begin with the bilinear analysis. Given a wavelet $\psi \in L^1(S^2)$, we define the corresponding scaling function $\Phi \equiv \Phi^{(a_o)}$ by its Fourier coefficients:

$$|\widehat{\Phi}(l,m)|^2 = \int_{a_o}^{\infty} \frac{da}{a} \, |\widehat{\psi^a}(l,m)|^2, \quad l \geqslant 1, \tag{9.73}$$

$$|\widehat{\Phi}(0,0)|^2 = \frac{1}{8\pi^2} \tag{9.74}$$

(the integral in (9.73) converges in virtue of the admissibility condition (9.68) satisfied by ψ). Of course, (9.73) does not define the function Φ uniquely. We can, for instance, assume in addition that $\widehat{\Phi}(l,m) \geqslant 0$, $\forall\, l,m$, as in [Fre97]. Corresponding to (9.71), we define the large-scale part of a signal s as

$$\check{\Sigma}_{\Phi}(\varrho, a_o) = \int_{S^2} d\mu(\zeta) \, \overline{\Phi_{\varrho}^{(a_o)}(\zeta)} \, s(\zeta), \tag{9.75}$$

where we have put $\Phi_{\varrho}^{(a_o)}(\zeta) \equiv \Phi^{(a_o)}(\varrho^{-1}\zeta)$.

Theorem 9.3.8 (Bilinear analysis) *Let $\psi \in L^1(S^2)$ be a spherical wavelet and let $\Phi \equiv \Phi^{(a_o)}$, $a_o > 0$, denote the associated scaling function. Assume the following two conditions are satisfied:*
*   *for all $l = 1, 2, \ldots$,*

$$\frac{8\pi^2}{2l+1} \sum_{|m| \leqslant l} \int_0^{\infty} \frac{da}{a} \, |\widehat{\psi^a}(l,m)|^2 = 1, \tag{9.76}$$

*   *for all $\epsilon \in (0, a_o)$, there is a constant $M > 0$, independent of ϵ, such that*

$$\int_{\epsilon}^{a_o} \frac{da}{a} \, \|\psi^a\|^2 \leqslant M. \tag{9.77}$$

Then, for all $s \in L^2(S^2)$, we have the equality

$$s = \int_0^{a_o} \frac{da}{a} \int_{SO(3)} d\varrho \, \check{S}_{\psi}(\varrho, a) \, \psi_{\varrho}^a + \int_{SO(3)} d\varrho \, \check{\Sigma}_{\Phi}(\varrho, a_o) \, \Phi_{\varrho}^{(a_o)}, \tag{9.78}$$

where $\check{S}_{\psi}$ is the spherical CWT of s with respect to the wavelet ψ, $\check{\Sigma}_{\Phi}$ is the large-scale part of s and the integral is understood in the strong sense in $L^2(S^2)$.

Proof. We consider the first term in (9.78). Since $\psi \in L^1(S^2)$ and $s \in L^2(S^2)$, Young's convolution inequality (9.62) shows that $\check{S}_{\psi} \in L^2(SO(3))$. As in the flat case, we define the infinitesimal detail at scale a:

$$d^a(\zeta) = \int_{SO(3)} d\varrho \, \check{S}_{\psi}(\varrho, a) \, \psi_{\varrho}^a(\zeta).$$

This is a convolution on $SO(3)$ and Young's inequality (9.61) shows that $d^a \in L^2(S^2)$. Explicitly, we have

$$d^a(\zeta) = \int_{S^2} d\mu(\zeta') \, s(\zeta') \int_{SO(3)} d\varrho \, \overline{\psi^a(\varrho^{-1}\zeta')} \, \psi^a(\varrho^{-1}\zeta). \tag{9.79}$$

Using the relation

$$\psi^a(\varrho^{-1}\zeta') = \sum_{l=0}^{\infty} \sum_{|m|\leqslant l} \sum_{|n|\leqslant l} \mathcal{D}^l_{mn}(\varrho)\,\widehat{\psi^a}(l,n)\,Y^m_l(\zeta'), \tag{9.80}$$

where $\mathcal{D}^l_{mn}(\varrho)$ denotes a Wigner function [Ros57,Tal68], we find

$$d^a(\zeta) = \int_{S^2} d\mu(\zeta')\,s(\zeta') \sum_{\substack{lmn \\ l'm'n'}} \overline{Y^m_l}(\zeta')\,Y^{m'}_{l'}(\zeta)\,\overline{\widehat{\psi^a}(l,n)}\,\widehat{\psi^a}(l',n')$$

$$\times \int_{SO(3)} d\varrho\,\overline{\mathcal{D}^l_{mn}(\varrho)}\,\mathcal{D}^{l'}_{m'n'}(\varrho).$$

Using the orthogonality of Wigner functions and the addition theorem for spherical harmonics (see Section A.4), this gives:

$$d^a(\zeta) = 2\pi \int_{S^2} d\mu(\zeta')\,s(\zeta') \sum_{l=0}^{\infty} \sum_{|m|\leqslant l} P_l\left(\widehat{\xi}\cdot\widehat{\zeta'}\right)\,|\widehat{\psi^a}(l,m)|^2.$$

Now consider the following expression:

$$s^{\varrho}_{\epsilon}(\zeta) = \int_{\epsilon}^{a_o} \frac{da}{a}\,d^a(\zeta)$$

$$= 2\pi \int_{S^2} d\mu(\zeta')\,s(\zeta') \int_{\epsilon}^{a_o} \frac{da}{a} \sum_{l=0}^{\infty} \sum_{|m|\leqslant l} P_l\left(\widehat{\xi}\cdot\widehat{\zeta'}\right)\,|\widehat{\psi^a}(l,m)|^2.$$

In virtue of condition (9.77), the double summation on the right-hand side of this equation is absolutely and uniformly convergent, since it is majorized by

$$\int_{\epsilon}^{a_o} \frac{da}{a} \sum_{l=0}^{\infty} \sum_{|m|\leqslant l} |\widehat{\psi^a}(l,m)|^2 = \int_{\epsilon}^{a_o} \frac{da}{a}\,\|\psi^a\|^2.$$

Now let us introduce the quantity:

$$\mathcal{K}^{(a_o)}_{\epsilon}(t) = 2\pi \sum_{l=0}^{\infty} \sum_{|m|\leqslant l} \left(\int_{\epsilon}^{a_o} \frac{da}{a}\,|\widehat{\psi^a}(l,m)|^2\right) P_l(t),$$

so that

$$s^{(a_o)}_{\epsilon} = \mathcal{K}^{(a_o)}_{\epsilon} \star s.$$

By (9.77), we see that $\mathcal{K}^{(a_o)}_{\epsilon} \in L^1([-1,+1])$, for all $0 < \epsilon \leqslant a_o$, and $\|\mathcal{K}^{(a_o)}_{\epsilon}\|_1 \leqslant 2\pi M$.
Next, we show in the same way that the second term in (9.78) equals $\mathcal{H}^{(a_o)} \star s$, where

$$\mathcal{H}^{(a_o)}(t) = 2\pi \sum_{l=0}^{\infty} \sum_{|m|\leqslant l} |\widehat{\Phi}(l,m)|^2\,P_l(t).$$

Again, $\mathcal{H}^{(a_o)} \in L^1([-1,+1])$. Finally, we define the kernel $\mathcal{K}_{\epsilon} = \mathcal{K}^{(a_o)}_{\epsilon} + \mathcal{H}^{(a_o)}$, which also belongs to $L^1([-1,+1])$. Condition (9.77) shows that $\mathcal{K}_{\epsilon}$ is a uniformly bounded kernel. In addition, from (9.76) and the definition (9.73)–(9.74) of $\widehat{\Phi}(l, m)$, we deduce the following constraint on its Legendre coefficients :

$$
\lim_{\epsilon \to 0} \widehat{\mathcal{K}_\epsilon}(l) = \frac{8\pi^2}{2l+1} \sum_{|m| \leqslant l} \left(\int_0^{a_o} \frac{da}{a} \, |\widehat{\psi^a}(l, m)|^2 + |\widehat{\Phi}(l, m)|^2 \right)
$$

$$
= \begin{cases} \dfrac{8\pi^2}{2l+1} \displaystyle\sum_{|m| \leqslant l} \int_0^{\infty} \dfrac{da}{a} \, |\widehat{\psi^a}(l, m)|^2 = 1, & l \geqslant 1, \\[2ex] 8\pi^2 \, |\widehat{\Phi}(0, 0)|^2 = 1, & l = 0. \end{cases}
$$

Then Theorem 9.3.3 shows that $\mathcal{K}_\epsilon$ is the kernel of an approximate identity, which proves the strong convergence in $L^2(S^2)$ of the approximation:

$$
\lim_{\epsilon \to 0} (\mathcal{K}_\epsilon \star s) = s.
$$

$\square$

As a check of the reconstruction formula (9.78), let us consider the unit function ι. Contrary to the case of the L^2 formalism, the L^1-normalized CWT of ι vanishes identically, as a consequence of Proposition 9.3.5:

$$
\check{\iota}_\psi(\varrho, a) = \int_{S^2} d\mu(\zeta) \, \overline{\psi^a(\zeta)} = \int_{S^2} d\mu(\zeta) \, \overline{\psi(\zeta)} = 0.
$$

Hence only the second term, the large-scale part, subsists in (9.78). Using again the expansion (9.80), we find successively:

$$
\check{\iota}_\Phi(\varrho, a_o) = \int_{S^2} d\mu(\zeta) \, \overline{\Phi(\varrho^{-1}\zeta)} = \overline{\widehat{\Phi}(0, 0)},
$$

and, for (9.78),

$$
\iota(\zeta) = \overline{\widehat{\Phi}(0, 0)} \int_{SO(3)} d\varrho \, \overline{\Phi(\varrho^{-1}\zeta)} = 8\pi^2 \, |\widehat{\Phi}(0, 0)|^2 = 1.
$$

This result shows that the large-scale part of a signal must be treated separately, because constant functions on the sphere are square integrable, and hence must be reconstructible, although their CWT vanishes identically. In practice, of course, large scales should be irrelevant, since wavelet analysis is local, and we expect the second term in (9.78) to be numerically negligible (that is, one must choose a_o large enough for this to be true).

Theorem 9.3.8 applies, in particular, to a zonal wavelet. The only change is the parameter space of the spherical CWT which takes the form of the product $S^2 \times \mathbb{R}_*^+$, with the measure $a^{-1} da \, d\mu(\zeta)$. A further simplification yet is to consider a singular reconstruction wavelet and build a framework similar to the Morlet linear analysis. As in the bilinear case, we begin by defining, through its Legendre coefficients, a scaling function $\phi \equiv \phi^{(a_o)}$ that takes care of the large scales :

$$\widehat{\phi}(l) = \int_{a_o}^{\infty} \frac{da}{a}\, \widehat{\psi^a}(l), \quad l \geqslant 1, \tag{9.81}$$

$$\widehat{\phi}(0) = 1. \tag{9.82}$$

The corresponding large part of a signal s is then

$$\check{\sigma}_\phi(\zeta, a_o) = \int_{S^2} d\mu(\zeta')\, \overline{\widehat{\phi(\widehat{\zeta} \cdot \widehat{\zeta'})}}\, s(\zeta'). \tag{9.83}$$

In these notations, the linear reconstruction formula is given by the following theorem.

Theorem 9.3.9 (Linear analysis) *Let $\psi \in L^1(S^2)$ be a zonal spherical wavelet satisfying the following two conditions:*
- *for all $l = 1, 2, \ldots,$*

$$\int_0^{\infty} \frac{da}{a}\, \widehat{\psi^a}(l) = 1, \tag{9.84}$$

- *for all $\epsilon \in (0, a_o)$,*

$$\sum_{l=0}^{\infty} \frac{2l+1}{4\pi} \int_{\epsilon}^{a_o} \frac{da}{a}\, \widehat{\psi^a}(l) < \infty. \tag{9.85}$$

Then, for all $s \in L^2(S^2)$, we have the equality

$$s(\zeta) = \int_0^{a_o} \frac{da}{a}\, \check{S}_\psi(\zeta, a) + \check{\sigma}_\phi(\zeta, a_o),$$

the integral being again understood in the strong sense in L^2.

Proof. The same arguments as in the proof of Theorem 9.3.8 show that the partial sum

$$s_\epsilon^{(a_o)}(\zeta) = \int_\epsilon^{a_o} \frac{da}{a}\, \check{S}_\psi(\zeta, a)$$

belongs to $L^2(S^2)$. Expanding this expression and adding the large-scale term, we find

$$s_\epsilon(\zeta) = \int_{S^2} d\mu(\zeta') \int_\epsilon^{a_o} \frac{da}{a}\, \overline{\widehat{\psi^a(\widehat{\zeta} \cdot \widehat{\zeta'})}}\, s(\zeta') + \check{\sigma}_\phi(\zeta, a_o)$$

$$= \int_{S^2} d\mu(\zeta')\, s(\zeta') \left(\int_\epsilon^{a_o} \frac{da}{a}\, \overline{\widehat{\psi^a(\widehat{\zeta} \cdot \widehat{\zeta'})}} + \overline{\widehat{\phi(\widehat{\zeta} \cdot \widehat{\zeta'})}} \right)$$

$$= \int_{S^2} d\mu(\zeta')\, s(\zeta') \sum_{l=0}^{\infty} \frac{2l+1}{4\pi} \left(\int_\epsilon^{a_o} \frac{da}{a}\, \overline{\widehat{\psi^a}(l)} + \overline{\widehat{\phi}(l)} \right) P_l(\widehat{\zeta} \cdot \widehat{\zeta'})$$

$$= (\kappa_\epsilon \star s)(\zeta),$$

where we have used (9.85) and set

$$\kappa_\epsilon(t) = \sum_{l=0}^{\infty} \frac{2l+1}{4\pi} \left(\int_\epsilon^{a_o} \frac{da}{a} \overline{\widehat{\psi^a(l)}} + \overline{\widehat{\phi(l)}} \right) P_l(t).$$

The Legendre coefficients of this kernel are

$$\widehat{\kappa_\epsilon}(l) = \int_\epsilon^{a_o} \frac{da}{a} \overline{\widehat{\psi^a(l)}} + \overline{\widehat{\phi(l)}}.$$

As in the proof of Theorem 9.3.8, we deduce from condition (9.84) that $\lim_{\epsilon \to 0} \widehat{\kappa_\epsilon}(l) = 1$, $\forall l = 0, 1, \ldots$. Thus we have again an approximate identity, which allows us to conclude that

$$\lim_{\epsilon \to 0} \| s - \kappa_\epsilon \star s \|_2 = 0. \qquad \qquad \square$$

The conclusion of this analysis is that our spherical CWT, with the modified dilation operator D^a, leads to the same approximation scheme as that developed by Freeden [Fre97,Fre99]. The present approach, however, has the additional advantage of giving a clear geometric meaning to the approximation parameter a. By the same token, it intuitively explains the validity of the Euclidean limit established in [29]. Indeed, taking $a \to 0$ means going to the pointwise limit where curvature becomes unimportant, that is, going to the tangent plane and recovering the flat CWT.

10 Spatio-temporal wavelets and motion estimation

10.1 Introduction

We live in a world where objects (cars, animals, men, birds, aeroplanes, the Sun, etc.) that surround us are constantly in relative motion. One would like to extract the motion information from the observation of the scene and use it for various purposes, such as *detection, tracking* and *identification*. In particular, tracking of multiple objects is of great importance in many real world scenarios. The examples include traffic monitoring, autonomous vehicle navigation, and tracking of ballistic missile warheads. Tracking is a complex problem, often requiring to estimate motion parameters – such as position, velocity – under very challenging situations. Algorithms of this type typically have difficulty in the presence of noise, when the object is obscured, in situations including crossing trajectories, and when highly maneuvering objects are present.

Most motion estimation (ME) techniques such as the ones based on block matching, optical flow, and phase difference [Jah97,280,281] assume that the object is constant from frame to frame. That is, the signature of the object does not change with time. Consequently, these techniques tend to have difficulty handling complex motion, particularly when noise is present.

The time-dependent continuous wavelet transform (CWT) is attractive as a tool for analysis, in that important motion parameters can be compactly and clearly represented. The CWT maps a given image sequence into a six-dimensional representation in which position, time, scale, and velocity (speed and orientation of the velocity) are explicit parameters. In effect, this transform provides a multiscale description of motion – the conventional continuous wavelet transform in two dimensions – with additional operators to provide control over the speed and orientation (i.e., velocity) in space–time [158]. It is a multidimensional filtering in all six variables, position, time, scale, speed, and orientation.

Several authors have already shown how band-pass filtering can be used to evaluate the speed in an image. A large class of models of human motion sensing use this approach [3,218,371]. The early mechanisms involved in human perception of motion appeared to be sensitive to spatial and temporal frequencies. Some neurons of the

343

visual cortex were found to respond best when they were stimulated with temporally modulated stimuli, the temporal frequency being included in a given range. One of these neurons may respond to the presentation of a slowly moving bar, and yet not respond to the same bar moving faster in its receptive field.

The organization of these *filters* in spatial and temporal frequencies show some special characteristics related with our spatio-temporal sensitivity. In fact, our visual system seems to make a trade-off between its spatial and temporal resolution. For instance, if one stares at a train leaving the station, at low speed it is still possible to read the destination of the cars, but at a higher speed it becomes impossible to read those details, only global shapes are still available for our vision. Actually, the interpretation of this example is complicated by the fact that our eyes (and head) may be following the train.

The organization of this chapter is parallel to that of Chapter 2. We start by describing the spatio-temporal signals and motions, emphasizing that motion is in fact orientation in space–time. Next we describe the elementary operations applied to spatio-temporal filters for motion extraction. Five of these operations are the same as the ones introduced in the 1-D and the 2-D CWT (but in a spatio-temporal setting). The sixth, called *speed tuning operator*, is different and acts on space and time in a way that allows us to cope with the presence of motion in a spatio-temporal signal.

This is then cast into the group-theoretical language, in particular, we define the appropriate unitary irreducible representation. Since the latter is square integrable, wavelets in the usual sense may be constructed. Thus we define the (2+1)-D CWT and give its properties. Finally, we describe the CWT tracking algorithm, referred to here as the Mujica–Murenzi–Leduc–Smith (MMLS) tracking algorithm, and apply it to two test sequences that reflect difficulties associated with noise, accelerated motion, temporary occlusion, and time-varying signatures. The first scene corresponds to four moving objects with linear and nonlinear motion. The experiments are performed both for noiseless and for noisy cases. The results are compared with those of block matching algorithms (BMA). The second sequence corresponds to a circularly moving object under the conditions of increasing velocity and acceleration.

10.2   Spatio-temporal signals and their transformations

We will consider (2+1)-dimensional signals s (image sequences) of finite energy, represented by square integrable complex valued functions on $\mathbb{R}^2 \times \mathbb{R}$, i.e., $s \in L^2(\mathbb{R}^2 \times \mathbb{R}, d^2\vec{x}\, dt)$:

$$\|s\|^2 = \iint_{\mathbb{R}^2 \times \mathbb{R}} d^2\vec{x}\, dt\, |s(\vec{x}, t)|^2 < \infty, \tag{10.1}$$

where $\vec{x}$ is a vector of $\mathbb{R}^2$ and t is the time.

In practice, a black and white sequence of images will be represented by a bounded non-negative function: $0 \leqslant s < M < \infty$. As in the spatial case, we will also consider as admissible signals some generalized functions (distributions), such as a delta function $\delta(\vec{x} - \vec{v}_o t)$ or a plane wave $\exp i(\vec{k}_o \cdot \vec{x} + \omega_o t)$.

The Fourier transform of s is defined, as usual, by

$$\hat{s}(\vec{k}, \omega) = (2\pi)^{-3/2} \iint_{\mathbb{R}^2 \times \mathbb{R}} d^2\vec{x} \, dt \, e^{-i(\vec{k}\cdot\vec{x}-\omega t)} s(\vec{x}, t), \tag{10.2}$$

where $\vec{k}$ is the spatial frequency, ω is the temporal frequency, and $\vec{k} \cdot \vec{x}$ is the Euclidean scalar product.

If s moves with a constant speed $\vec{v}_o$, the resulting signal $\tilde{s}$ may be written as

$$\tilde{s}(\vec{x}, t) = s(\vec{x} - \vec{v}_o t, t), \tag{10.3}$$

and its Fourier transform becomes

$$\hat{\tilde{s}}(\vec{k}, \omega) = \hat{s}(\vec{k}, \omega - \vec{k} \cdot \vec{v}_o). \tag{10.4}$$

This constant speed motion does not affect the spatial frequency of the signal, but the components of its Fourier transform corresponding to the spatial frequency $\vec{k}$ are translated by $-\vec{k} \cdot \vec{v}_o$ in the direction of the ω axis. This operation transforms the plane defined by $\omega = 0$ into a new plane defined by $\omega = \vec{k} \cdot \vec{v}_o$.

Consider now a static signal which, in Fourier space, is roughly concentrated around the plane defined by $\omega = 0$. When the same signal moves with constant speed $\vec{v}_o$, its Fourier transform is concentrated around the plane defined by

$$\omega - \vec{k} \cdot \vec{v}_o = 0$$

$$\begin{pmatrix} \vec{v}_o^T & 1 \end{pmatrix} \begin{pmatrix} \vec{k} \\ \omega \end{pmatrix} = 0 \tag{10.5}$$

where $\vec{v}_o^T$ denotes the transposed vector. Thus the velocity plane defined by (10.5) is perpendicular to the velocity vector $\vec{v}_o$.

Therefore a filter which is concentrated around this plane will be sensitive to the moving components of the image which have the same speed $\vec{v}_o$. This idea is the cornerstone of the band-pass filtering approach to visual motion perception. It is directly used in spatio-temporal energy models, such as that of Adelson and Bergen [3]. Their filters can see a pattern moving with a fixed speed if the ratio of their temporal and spatial mean frequency fits with the value of the speed and if the direction of the speed and the spatial frequency are close enough.

As in Chapter 2, we begin by introducing the elementary operations that we want to apply to spatio-temporal signals for motion extraction. From our past experience, we know, however, this is equivalent to transforming the filters, in particular, the wavelets, and to analyzing a given object with help of the transformed filter (this is the so-called passive point of view described in Section 7.1). Thus, from now on, we consider trans-formations to be applied to a given spatio-temporal filter (in particular, a spatio-temporal wavelet), in order to have a basis on which to decompose a given spatio-temporal signal.

As we will see in next section, appropriate filters, which are wavelets in our case, must vanish at zero temporal frequency and zero spatial frequency, (10.35)–(10.36). A *fortiori*, a good wavelet for motion analysis will have its support in the Fourier domain concentrated in a convex cone in the half space $\mathbb{R}^2 \times \mathbb{R}_*^+$, with apex at the origin. It is also supposed to be concentrated around the plane corresponding to a fixed speed vector $\vec{v}_o$.

Let us consider a spatio-temporal filter ψ (of finite energy, as usual). We consider the following elementary operations.

- *Spatio-temporal translations*

  The wavelet is shifted to a given point of space and time ($\mathbb{R}^2 \times \mathbb{R}$). This transformation is denoted by T in direct space and $\widehat{T}$ in the spatio-temporal Fourier domain:

$$(T_{\vec{b},\tau}\psi)(\vec{x}, t) = \psi(\vec{x} - \vec{b}, t - \tau), \quad (\widehat{T_{\vec{b},\tau}}\widehat{\psi})(\vec{k}, \omega) = e^{-i(\vec{k}\cdot\vec{x}+\omega\tau)}\widehat{\psi}(\vec{k}, \omega). \quad (10.6)$$

  This transformation, with parameter $q = (\vec{b}, \tau)$, is used to detect the location of ob-jects. In the Fourier domain, the wavenumber–frequency spectrum, $\widehat{\psi}(\vec{k}, \omega)$, remains concentrated around the $\vec{v}_o$ velocity plane, only a linear phase term is introduced.

- *Rotation*

  This transformation, denoted by R_θ, rotates the wavelet in spatial coordinates around the temporal (or frequency) axis. In this way, filters can be tuned to a particular orientation associated with the velocity $\vec{v}_o$. It is defined by

$$(R_\theta\psi)(\vec{x}, t) = \psi(r_{-\theta}(\vec{x}), t), \quad (\widehat{R_\theta}\widehat{\psi})(\vec{k}, \omega) = \widehat{\psi}(r_{-\theta}(\vec{k}), \omega), \quad (10.7)$$

  where r_θ is the usual 2×2 rotation matrix

$$r_\theta = \begin{pmatrix} \cos\theta & -\sin\theta \\ \sin\theta & \cos\theta \end{pmatrix}, \quad 0 \leqslant \theta < 2\pi. \quad (10.8)$$

  The effect of this transformation is to change the $\vec{v}_o$-plane into a new velocity plane associated with velocity $r_\theta(\vec{v}_o)$, that is,

$$\omega = r_{-\theta}(\vec{k}) \cdot \vec{v}_o = \vec{k} \cdot r_\theta(\vec{v}_o). \quad (10.9)$$

  Thus, the parameter θ allows for orientation changes and is used to estimate velocities in the context of motion estimation, together with the speed tuning parameter c, to be defined next.

- *Scaling*

  This transformation, denoted by D in space–time domain and $\widehat{D}$ in wave-number–frequency domain, is the analog of the one we used in Chapter 2 and is defined by

$$(D_a\psi)(\vec{x}, t) = a^{-3/2}\,\psi(a^{-1}\vec{x}, a^{-1}t), \quad (\widehat{D_a\widehat{\psi}})(\vec{k}, \omega) = a^{3/2}\,\psi(a\vec{k}, a\omega), \quad a > 0. \tag{10.10}$$

The dilation preserves the norm of $L^2(\mathbb{R}^2 \times \mathbb{R}, d^2\vec{x}\,dt)$. Since it operates in the same way on both time and space, the speed of the object is not affected by the dilation. In the Fourier domain, this means that the dilation keeps the wavelet on the same constant-speed plane.

• *Speed tuning transformation*

Unlike the dilation and the shift transformations, which do not affect the concentration of the energy of the wavelet around a plane of constant speed in the Fourier space, the speed tuning transformation allows the concentration of the energy of the wavelet to move from one velocity plane with speed $|\vec{v}_o|$ to a velocity plane with a different speed.

The transformation, denoted by Λ in the space–time domain and $\widehat{\Lambda}$ in wavenumber domain, can be seen as an anisotropic scaling on space and time, that is,

$$(\widehat{\Lambda}_c\psi)(\vec{k}, \omega) = \psi(c^{\alpha}\vec{k}, c^{-\beta}\omega), \quad c > 0. \tag{10.11}$$

We require the transformation to be unitary and to map the $\vec{v}_o$-plane into the $c\vec{v}_o$-plane. This results in a system of two linear equations, that fixes α and β. The first constraint gives

$$\|\widehat{\psi}(\vec{k}, \omega)\|^2 = \|(\widehat{\Lambda}_c\widehat{\psi})(\vec{k}, \omega)\|^2 = c^{-2\alpha+\beta}\|\widehat{\psi}\|^2 \tag{10.12}$$

that is, $2\alpha = \beta$. The second constraint gives

$$c^{-\beta}\omega = c^{\alpha}\vec{k} \cdot \vec{v}_o = \vec{k} \cdot c^{\alpha+\beta}\vec{v}_o = \vec{k} \cdot c\vec{v}_o. \tag{10.13}$$

This requires that $\alpha + \beta = 1$, which together with the unitarity constraint requires that $\alpha = \frac{2}{3}$ and $\beta = \frac{1}{3}$. The speed tuning transformation is thus explicitly defined by

$$(\Lambda_c\psi)(\vec{x}, t) = \psi(c^{-1/3}\vec{x}, c^{2/3}t), \tag{10.14}$$

$$(\widehat{\Lambda}_c\widehat{\psi})(\vec{k}, \omega) = \widehat{\psi}(c^{1/3}\vec{k}, c^{-2/3}\omega). \tag{10.15}$$

The speed tuning transforms a wavelet which can see a speed $\vec{v}_o$ into a wavelet sensitive to $c\vec{v}_o$. Therefore, the transformation allows a direct and natural tuning of the value of the speed that the wavelet is able to analyze. In Fourier space, the transformation generates a family of wavelets which are distorted and shifted along hyperbolas defined by the constancy of the product $|\vec{k}|\,\omega$. The parameter c of the speed tuning allows to adapt the speed analysis independently of the scale analysis. When analyzing a moving pattern, the dilation may be used to adapt the scale of the wavelet to the spatial extension of the pattern without affecting the speed that the wavelet analysis will detect. Therefore, the values of the wavelet transform for a fixed point in time and space $(\vec{b}, \tau)$ can be directly interpreted in terms of scale (with parameter a) and speed (c and θ).

Combining all these operators, one obtains the operator $U(\vec{b}, \tau, \theta; a, c) \equiv T_{\vec{b}, \tau} R_\theta D_a \Lambda_c$, namely,

$$\left[U(\vec{b}, \tau, \theta; a, c) \psi \right](\vec{x}, t) = a^{-3/2} \psi(a^{-1} c^{-1/3} r_{-\theta}(\vec{x} - \vec{b}), a^{-1} c^{2/3}(t - \tau)). \quad (10.16)$$

From the wavelet ψ, one obtains a family of wavelets $\{U(\vec{b}, \tau, \theta; a, c)\psi\}$, in terms of which one can decompose any spatio-temporal signal s.

10.3   The transformation group and its representations

Clearly, the transformations presented in the previous section form a group. In this section we will develop the wavelet machinery associated with that group. As in the case of 2-D spatial wavelets studied in Chapter 2 and in Section 7.2, the key to the construction of spatio-temporal wavelets is to have at one's disposal a unitary representation of this transformation group in the natural space of finite energy signals, namely, $L^2(\mathbb{R}^2 \times \mathbb{R}, d^2\vec{x}\, dt)$.

We are looking for a group G acting on space–time $\mathbb{R}^2 \times \mathbb{R}$, whose restriction to space variables coincides with the usual continuous wavelet group in two dimensions, i.e., the similitude group SIM(2), while the restriction to the time variable coincides with the usual continuous wavelet group in one dimension, i.e., the affine group G_{aff}^+.

Let us consider first the group $G_1 = \text{E}(2) \times \mathbb{R}$, the Euclidean group (rotations and translations) acting on the (2+1)-dimensional space–time. The elements of G_1 are triplets $(\vec{b}, \tau, \theta)$, where $\vec{b}$ is the translation vector, τ is the time translation, and θ is the rotation parameter. The element $(\vec{b}, \tau, \theta) \in G_1$ acts on the point $(\vec{x}, t)$ in the space–time $\mathbb{R}^2 \times \mathbb{R}$ in a natural way:

$$(\vec{b}, \tau, \theta) : (\vec{x}, t) \mapsto (r_\theta(\vec{x}) + \vec{b}, t + \tau), \quad \vec{x} \in \mathbb{R}^2, t \in \mathbb{R}. \quad (10.17)$$

Next, we consider the product $G_2 = \mathbb{R}_*^+ \times \mathbb{R}_*^+$ of two dilation groups with the following action on space–time:

$$(a, c) : (\vec{x}, t) \mapsto (ac^{3/2}\vec{x}, act), \quad a \in \mathbb{R}_*^+, c \in \mathbb{R}_*^+. \quad (10.18)$$

This action allows one to define the semidirect product $G_{\text{mv}} \equiv G_1 \rtimes G_2 = \{g \equiv (\vec{b}, \tau, \theta; a, c)\}$, with multiplication:

$$(\vec{b}, \tau, \theta; a, c)(\vec{b}', \tau', \theta'; a', c') = (\vec{b} + ac^{1/3} r_\theta(\vec{b}'), \tau + ac^{-2/3}\tau', \theta + \theta'; aa', cc'). \quad (10.19)$$

The inverse is given by

$$(\vec{b}, \tau, \theta; a, c)^{-1} = (-a^{-1}c^{-1/3} r_{-\theta}(\vec{b}), -a^{-1}c^{2/3}\tau, -\theta; a^{-1}, c^{-1}), \quad (10.20)$$

and the unit element is $e = (\vec{0}, 0, 0; 1, 1)$. In 4×4 matrix notation, we may write (compare (7.37)):

$$(\vec{b}, \tau, \theta; a, c) \equiv \begin{pmatrix} ac^{1/3} r_\theta & 0 & \vec{b} \\ \vec{0} \cdot & ac^{2/3} & \tau \\ \vec{0} & 0 & 1 \end{pmatrix}. \tag{10.21}$$

The group G_{mv} is locally compact with right Haar measure $d\mu_{\mathrm{R}}$ and left Haar measure $d\mu_{\mathrm{L}}$:

$$d\mu_{\mathrm{R}} = d^2\vec{b}\, d\tau\, d\theta\, \frac{da}{a}\frac{dc}{c}, \quad d\mu_{\mathrm{L}} = d^2\vec{b}\, d\tau\, d\theta\, \frac{da}{a^4}\frac{dc}{c}. \tag{10.22}$$

In the sequel, we will use systematically the left Haar measure $d\mu_{\mathrm{L}}$ and, correspondingly, we will write

$$\int_{G_{\mathrm{mv}}} d\mu_{\mathrm{L}}(\vec{b}, \tau, \theta; a, c) \equiv \int_{\mathbb{R}^2} d^2\vec{b} \int_{\mathbb{R}} d\tau \int_0^{2\pi} d\theta \int_0^\infty \frac{da}{a^4} \int_0^\infty \frac{dc}{c}. \tag{10.23}$$

Having identified the appropriate transformation group G_{mv}, we proceed to find a square integrable representation of it, in the Hilbert space of finite energy signals, according to the general formalism sketched in Section 7.1. For every $g \equiv (\vec{b}, \tau, \theta; a, c) \in G_{\mathrm{mv}}$, consider the operator $U(g)$ on $L^2(\mathbb{R}^2 \times \mathbb{R}, d^2\vec{x}\, dt)$ defined in (10.16), namely,

$$[U(g)\psi](\vec{x}, t) = a^{-3/2}\psi(a^{-1}c^{-1/3}r_{-\theta}(\vec{x} - \vec{b}), a^{-1}c^{2/3}(t - \tau)). \tag{10.24}$$

This may be written in the Fourier space as

$$\left[\widehat{U}(g)\widehat{\psi}\right](\vec{k}, \omega) = a^{3/2}e^{-i(\vec{k}\cdot\vec{b} + \omega\tau)}\widehat{\psi}(ac^{-1/3}r_{-\theta}(\vec{k}), ac^{-2/3}\omega). \tag{10.25}$$

Proposition 10.3.1 *The operator U defines a unitary representation of G_{mv} in $L^2(\mathbb{R}^2 \times \mathbb{R}, d^2\vec{x}\, dt)$.*

This is proven by a straightforward verification.

Proposition 10.3.2 *The operator $\widehat{U}$ defines two unitary irreducible representations of G_{mv} in $\mathfrak{H}_+ \equiv \mathfrak{H}_+(\mathbb{R}^2 \times \mathbb{R})$ and $\mathfrak{H}_- \equiv \mathfrak{H}_-(\mathbb{R}^2 \times \mathbb{R})$, respectively, where*

$$\mathfrak{H}_+(\mathbb{R}^2 \times \mathbb{R}) = \{\psi \in L^2(\mathbb{R}^2 \times \mathbb{R}, d^2\vec{x}\, dt) : \widehat{\psi}(\vec{k}, \omega) = 0,\ \omega < 0\} \tag{10.26}$$

$$\mathfrak{H}_-(\mathbb{R}^2 \times \mathbb{R}) = \{\psi \in L^2(\mathbb{R}^2 \times \mathbb{R}, d^2\vec{x}\, dt) : \widehat{\psi}(\vec{k}, \omega) = 0,\ \omega > 0\} \tag{10.27}$$

with

$$L^2(\mathbb{R}^2 \times \mathbb{R}, d^2\vec{x}\, dt) = \mathfrak{H}_+(\mathbb{R}^2 \times \mathbb{R}) \oplus \mathfrak{H}_-(\mathbb{R}^2 \times \mathbb{R}). \tag{10.28}$$

Proof. The proof follows the same line as that of Proposition 2.1.2. Let $\psi \in \mathfrak{H}_+$ be an arbitrary nonzero vector. We are going to show that ψ is cyclic for the representation

U, i.e., the linear span of the orbit of ψ is dense in $\mathfrak{H}_+$. Let f be orthogonal to the span of $\{U(g)\psi\}$, that is, $\langle U(g)\psi | f \rangle = 0$, $\forall g \in G_{\mathrm{mv}}$. We show that $f = 0$. We have indeed, for all $(\vec{b}, \tau, \theta; a, c) \in G_{\mathrm{mv}}$,

$$\langle U(g)\psi | f \rangle = \langle \widehat{U}(g)\widehat{\psi} | \widehat{f} \rangle$$

$$= a^{3/2} \iint_{\mathbb{R}^2 \times \mathbb{R}} d^2\vec{k} \, d\omega \, e^{i(\vec{k}\cdot\vec{b}+\omega\tau)} \, \overline{\widehat{\psi}(ac^{-1/3}r_{-\theta}(\vec{k}), ac^{-2/3}\omega)}$$

$$= 0.$$

This means that the Fourier transform of $\overline{\widehat{\psi}(ac^{-1/3}r_{-\theta}(\vec{k}), ac^{-2/3}\omega)} \, \widehat{f}(\vec{k}, \omega)$ vanishes (almost everywhere) for all θ, a, c. Then the transitivity of the action of $(\theta, a, c) \in \mathrm{SO}(2) \times \mathbb{R}_+^* \times \mathbb{R}_+^*$ on $\mathbb{R}^2 \times \mathbb{R}_+^*$ implies that $\widehat{f}(\vec{k}, \omega) = 0$ (almost everywhere).

The same holds true for $\mathfrak{H}_-$. □

Proposition 10.3.3 *The representation U is square integrable, that is, there exists a function $\psi \in L^2(\mathbb{R} \times \mathbb{R}, d^2\vec{x}dt)$, $\psi \neq 0$, such that the matrix element $\langle U(\vec{b}, \tau, \theta; a, c)\psi | \psi \rangle$ is square integrable with respect to the measure $d\mu_L(\vec{b}, \tau, \theta; a, c)$, that is, in the notation of (10.23),*

$$I = \int_{G_{\mathrm{mv}}} d\mu_L(\vec{b}, \tau, \theta; a, c) |\langle U(\vec{b}, \tau, \theta; a, c)\psi | \psi \rangle|^2 < \infty. \tag{10.29}$$

In addition,

$$I = c_\psi \|\psi\|^2 \tag{10.30}$$

where

$$c_\psi = (2\pi)^3 \iint_{\mathbb{R}^2 \times \mathbb{R}} \frac{d^2\vec{k} \, d\omega}{|\vec{k}|^2 |\omega|} |\widehat{\psi}(\vec{k}, \omega)|^2. \tag{10.31}$$

Proof. The result follows from a direct calculation:

$$I = \int_{G_{\mathrm{mv}}} d\mu_L |\langle U(\vec{b}, \tau, \theta; a, c)\psi | \psi \rangle|^2$$

$$= \int_{G_{\mathrm{mv}}} d\mu_L |\langle \widehat{U}(\vec{b}, \tau, \theta; a, c)\widehat{\psi} | \widehat{\psi} \rangle|^2$$

$$= \int_{G_{\mathrm{mv}}} d\mu_L \, a^3 \iint_{\mathbb{R}^2 \times \mathbb{R}} d^2\vec{k} \, d\omega \, \widehat{\psi}(ac^{-1/3}r_{-\theta}(\vec{k}), ac^{-2/3}\omega) \, e^{-i(\vec{k}\cdot\vec{b}+\omega\tau)} \, \overline{\widehat{\psi}(\vec{k}, \omega)}$$

$$\times \iint_{\mathbb{R}^2 \times \mathbb{R}} d^2\vec{k}' \, d\omega' \, \overline{\widehat{\psi}(ac^{-1/3}r_{-\theta}(\vec{k}'), ac^{-2/3}\omega')} \, e^{i(\vec{k}'\cdot\vec{b}+\omega'\tau)} \, \widehat{\psi}(\vec{k}', \omega')$$

$$= \int_0^{2\pi} \int_0^\infty \int_0^\infty d\theta \, \frac{da}{a} \frac{dc}{c} \iint_{\mathbb{R}^2 \times \mathbb{R}} d^2\vec{k} \, d\omega \iint_{\mathbb{R}^2 \times \mathbb{R}} d^2\vec{k}' \, d\omega'$$

$$\times \left\{ \iint_{\mathbb{R}^2 \times \mathbb{R}} d^2\vec{b} d\tau e^{-i\{(\vec{k}-\vec{k}')\cdot\vec{b}+(\omega-\omega')\tau\}} \right\} \overline{\widehat{\psi}(\vec{k}, \omega)} \, \widehat{\psi}(\vec{k}', \omega')$$

$$\times \widehat{\psi}(ac^{-1/3}r_{-\theta}(\vec{k}), ac^{-2/3}\omega) \, \overline{\widehat{\psi}(ac^{-1/3}r_{-\theta}(\vec{k}'), ac^{-2/3}\omega')}$$

$$= (2\pi)^3 \int_0^{2\pi} \int_0^\infty \int_0^\infty d\theta \, \frac{da}{a} \frac{dc}{c}$$

$$\times \iint_{\mathbb{R}^2 \times \mathbb{R}} d^2\vec{k} \, d\omega \, |\widehat{\psi}(ac^{-1/3}r_{-\theta}(\vec{k}), ac^{-2/3}\omega)|^2 \, |\widehat{\psi}(\vec{k}, \omega)|^2$$

$$= (2\pi)^3 \iint_{\mathbb{R}^2 \times \mathbb{R}} d^2\vec{k} \, d\omega |\widehat{\psi}(\vec{k}, \omega)|^2$$

$$\times \left[\int_0^{2\pi} \int_0^\infty \int_0^\infty d\theta \, \frac{da}{a} \frac{dc}{c} |\widehat{\psi}(ac^{-1/3}r_{-\theta}(\vec{k}), ac^{-2/3}\omega)|^2 \right].$$

Perform now the following change of variables:

$$\vec{k}' = ac^{-1/3}r_{-\theta}(\vec{k}),$$ $$\omega' = ac^{-2/3}\omega,$$

(10.32)

which is equivalent to

$$\begin{pmatrix} k_1' \\ k_2' \\ \omega' \end{pmatrix} = \begin{pmatrix} k_1 & k_2 & 0 \\ -k_2 & k_1 & 0 \\ 0 & 0 & \omega \end{pmatrix} \begin{pmatrix} ac^{-1/3}\cos\theta \\ ac^{-1/3}\sin\theta \\ ac^{-2/3} \end{pmatrix}.$$

(10.33)

Thus the measure $d\theta \, \frac{da}{a} \frac{dc}{c}$ becomes $\frac{d\vec{k}' d\omega'}{|\vec{k}'|^2|\omega'|}$ and we have

$$I = (2\pi)^3 \iint_{\mathbb{R}^2 \times \mathbb{R}} d\vec{k}' \, d\omega' \frac{|\widehat{\psi}(\vec{k}', \omega')|^2}{|\vec{k}'|^2|\omega'|} \iint_{\mathbb{R}^2 \times \mathbb{R}} d^2\vec{k} \, d\omega \, |\widehat{\psi}(\vec{k}, \omega)|^2$$

$$= (2\pi)^3 \iint_{\mathbb{R}^2 \times \mathbb{R}} d\vec{k} \, d\omega \frac{|\widehat{\psi}(\vec{k}, \omega)|^2}{|\vec{k}|^2|\omega|} \, \|\psi\|^2.$$

It is clear that the integral I converges if and only if ψ satisfies the admissibility condition (10.29). $\qquad \square$

10.4　The spatio-temporal wavelet transform

10.4.1　Spatio-temporal wavelets: definition and examples

By definition, a *spatio-temporal wavelet* [158,Duv91] is a complex-valued function $\psi \in L^2(\mathbb{R}^2 \times \mathbb{R}, d^2\vec{x}\, dt)$ satisfying the condition:

$$c_\psi = (2\pi)^3 \iint_{\mathbb{R}^2 \times \mathbb{R}} \frac{d^2\vec{k}\, d\omega}{|\vec{k}|^2 |\omega|} |\widehat{\psi}(\vec{k}, \omega)|^2 < \infty \tag{10.34}$$

where $\widehat{\psi}$ is the Fourier transform of ψ.

If ψ is regular enough, the admissibility condition (10.34) simply means that the wavelet must be of zero mean with respect to space and time independently:

$$\widehat{\psi}(\vec{0}, \omega) = 0 \iff \int_{\mathbb{R}^2} d^2\vec{x}\, \psi(\vec{x}, t) = 0. \tag{10.35}$$

and

$$\widehat{\psi}(\vec{k}, 0) = 0 \iff \int_{\mathbb{R}} dt\, \psi(\vec{x}, t) = 0. \tag{10.36}$$

Clearly the unitary operators $T_{(\vec{b}, \tau)}$, R_θ, D_a, Λ_c preserve the admissibility condition, and so does therefore $U(\vec{b}, \tau, \theta; a, c)$. Hence any function $\psi_{\vec{b}, \tau, \theta; a, c} = U(\vec{b}, \tau, \theta; a, c)\psi$ obtained from a wavelet ψ by translation, dilation, rotation, or speed tuning is again a wavelet. Thus the given wavelet ψ generates the whole family $\{\psi_{\vec{b}, \tau, \theta; a, c}\}$, indexed by the elements $(\vec{b}, \tau) \in \mathbb{R}^2 \times \mathbb{R}$, $\theta \in [0, 2\pi)$, $a > 0$, $c > 0$, that is, by the elements of G_{mv}.

We will consider only one example of spatio-temporal wavelet, the spatio-temporal Morlet wavelet, characterized by the wavenumber–frequency $(\vec{k}_0, \omega_0)$ and the anisotropy parameter ϵ and defined by

$$\psi_\epsilon(\vec{x}, t) = \left(e^{i\vec{k}_0 \cdot A^{-1}\vec{x}} e^{-\frac{1}{2}|A^{-1}\vec{x}|^2} - e^{-\frac{1}{2}|A^{-1}\vec{x}|^2} e^{-\frac{1}{2}|\vec{k}_0|^2}\right)$$
$$\times \left(e^{i\omega_0 t} e^{-\frac{1}{2}t^2} - e^{-\frac{1}{2}t^2} e^{-\frac{1}{2}\omega_0^2}\right) \tag{10.37}$$

in the spatio-temporal domain and by

$$\widehat{\psi}_\epsilon(\vec{k}, \omega) = \left(e^{-\frac{1}{2}|A\vec{k} - \vec{k}_0|^2} - e^{-\frac{1}{2}(|A\vec{k}|^2 + |\vec{k}_0|^2)}\right)\left(e^{-\frac{1}{2}(\omega - \omega_0)^2} - e^{-\frac{1}{2}(\omega^2 + \omega_0^2)}\right) \tag{10.38}$$

in the wavenumber–frequency domain, where $A = \text{diag}[\epsilon^{-1/2}, 1]$, $\epsilon \geqslant 1$, is the usual anisotropy matrix. As in the pure spatial case, for large $|k_0|$ and $|\omega_0|$, typically $|k_0| \geqslant 6$, $|\omega_0| \geqslant 6$, the counterterms (the second term in each factor of (10.37) and (10.38)) are small enough to be neglected.

This wavelet is a good candidate for motion estimation applications. Its region of support can be appropriately located around a particular plane, $\vec{v}_o$, typically $\vec{v}_o = (1, 0)$, for example in the case where $\vec{k}_0 = (k_0, 0)$, $k_0 = \omega_0$.

Figure 10.1 shows half-energy equisurfaces of the squared amplitude of Morlet wavelets for different values of the parameters, θ, a and c. This illustrates how the

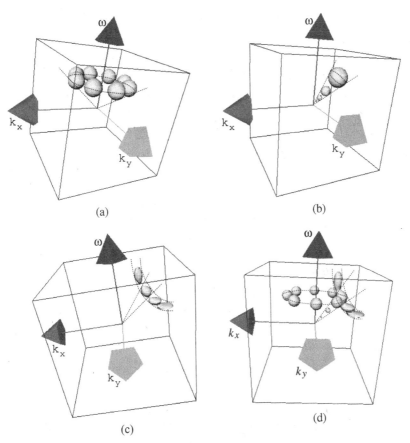

(a)

(b)

(c)

(d)

Fig. 10.1. Wavenumber–frequency domain coverage of the spatio-temporal Morlet wavelet for different parameters of the CWT. (a) Rotation θ; (b) scale a; (c) speed tuning c; (d) rotation, scale, and speed tuning θ, a, c.

wavelet parameters distribute the energy of the resulting filter on the wavenumber–frequency domain. The energy is distributed around a circle by the rotation parameter, along a conic volume by the scale parameter, and along a hyperbolic-like path by the speed tuning parameter. In order to control the variance of the wavelet with respect to the reference velocity plane, an anisotropy parameter ϵ applied to the spatial variables is used. This indeed controls its temporal support. As it can be seen from Figure 10.1, the region of support of the filters is an ellipsoidal cone concentrated around a particular velocity plane. Thus, velocity detection and filtering are possible.

10.4.2 The spatio-temporal wavelet transform

Let now $s \in L^2(\mathbb{R}^2 \times \mathbb{R}, d^2\vec{x}dt)$ be an image sequence. Its $(2+1)$-D *continuous wavelet transform* (with respect to the fixed wavelet ψ), $S \equiv W_\psi s$ is the scalar product of s with the transformed wavelet $\psi_{\vec{b},\tau,a,c,\theta} \equiv U(\vec{b}, \tau, \theta; a, c)\psi$, considered as a function of $(\vec{b}, \tau, a, c, \theta)$:

$$S(\vec{b}, \tau, \theta; a, c) = \frac{1}{\sqrt{c_\psi}} < \psi_{\vec{b},\tau,\theta;a,c}|s > \tag{10.39}$$

$$= \frac{1}{\sqrt{c_\psi}} \iint d^2\vec{x} \, dt \, a^{-3/2} \, \overline{\psi(a^{-1}c^{-1/3} r_\theta(\vec{x} - \vec{b}), \, a^{-1}c^{2/3}(t - \tau))} \, s(\vec{x}, t) \tag{10.40}$$

$$= \frac{1}{\sqrt{c_\psi}} \iint d^2\vec{k} \, d\omega \, a^{3/2} \, e^{-i(\vec{k}\cdot\vec{b}+\omega\tau)} \, \overline{\hat{\psi}(ac^{1/3} r_\theta(\vec{k})ac^{-2/3}\omega)} \, \hat{s}(\vec{k}, \omega). \tag{10.41}$$

The main properties of the (2+1)-D CWT $W_\psi : s \mapsto S$ may be summarized as follows [Com89,Mey91]:

(i) As the purely spatial version, W_ψ is *linear* in the signal s;

(ii) W_ψ is *covariant* under all the operations considered, namely [Com89,Mur90,13].
   Proposition 10.4.1 *The map W_ψ is covariant under translations, rotations and dilations, which means that the correspondence $W_\psi : s(\vec{x}, t) \mapsto S(\vec{b}, \tau, \theta; a, c)$ implies the following ones:*

$$s(\vec{x} - \vec{b}_o, t) \mapsto S(\vec{b} - \vec{b}_o, \tau, \theta; a, c) \tag{10.42}$$

$$s(\vec{x}, t - t_o) \mapsto S(\vec{b}, \tau - t_o\theta; a, c) \tag{10.43}$$

$$s(r_{-\theta_o}(\vec{x}), t) \mapsto S(r_{-\theta_o}(\vec{b}), \tau, \theta - \theta_o; a, c). \tag{10.44}$$

$$a_o^{-1}s(a_o^{-1}\vec{x}, a_o^{-1}t) \mapsto S(a_o^{-1}\vec{b}, a_o^{-1}\tau, \theta; a_o^{-1}a, c) \tag{10.45}$$

$$s(c_o^{-1/3}\vec{x}, c_o^{2/3}) \mapsto S(c_o^{1/3}\vec{b}, c_o^{-2/3}\tau, , \theta; a, c_o^{-1}c) \tag{10.46}$$

It is worth noting that, conversely, the wavelet transform is uniquely determined by the three conditions of linearity, covariance and energy conservation, plus some continuity [Mur90].

(iii) *Energy conservation:*

$$\iint_{\mathbb{R}^2 \times \mathbb{R}} d^2\vec{x} \, dt \, |s(\vec{x}, t)|^2 = \frac{1}{c_\psi} \int_{G_{mv}} d\mu_L(\vec{b}, \tau, \theta; a, c)|\langle U(\vec{b}, \tau, \theta; a, c)\psi \, | \, s\rangle|^2. \tag{10.47}$$

Thus, W_ψ is an isometry from the space of signals into the space of transforms.

(iv) As a consequence, W_ψ is invertible on its range and the inverse transformation is simply the adjoint of W_ψ. Thus one has an exact *reconstruction formula:*

$$s(\vec{x}, t) = \frac{1}{c_\psi} \int_{G_{mv}} d\mu_L(\vec{b}, \tau, \theta; a, c)\psi_{\vec{b},\tau,\theta;a,c}(\vec{x}, t) \, S(\vec{b}, \tau, \theta; a, c). \tag{10.48}$$

In other words, the (2+1)-D wavelet transform, like its 1-D and 2-D counterpart, provides a decomposition of the signal in terms of the analyzing wavelets $\psi_{\vec{b},\tau,\theta;a,c}$, with coefficients $S(\vec{b}, \tau, \theta; a, c)$.

(v) *Redundancy:* Exactly as in the 2-D case discussed in Chapter 2, one has:
   Proposition 10.4.2 *The projection from $L^2(G_{mv}, d\mu_L)$ onto the range $\mathfrak{H}_\psi$ of*

W_ψ, *the space of wavelet transforms, is an integral operator whose kernel* $K(\vec{b}', \tau', \theta'; a', c' \mid \vec{b}, \tau, \theta; a, c)$ *is the autocorrelation function of* ψ, *also called reproducing kernel:*

$$K(\vec{b}', \tau', \theta'; a', c' \mid \vec{b}, \tau, \theta; a, c) = c_\psi^{-1} \langle \psi_{\vec{b}', \tau', \theta'; a', c'} \mid \psi_{\vec{b}, \tau, \theta; a, c} \rangle. \tag{10.49}$$

Therefore, a function $f \in L^2(G, dg)$ *is the wavelet transform of a certain signal iff it satisfies the reproduction property:*

$$f(\vec{b}', \tau', \theta'; a', c')$$
$$= \int_{G_{mv}} d\mu_L(\vec{b}, \tau, \theta; a, c) K(\vec{b}', \tau', \theta'; a', c' \mid \vec{b}, \tau, \theta; a, c) \, f(\vec{b}, \tau, \theta; a, c). \tag{10.50}$$

10.4.3 An alternative: relativistic wavelets

The spatio-temporal wavelets just described, which could be called *kinematical*, may not always be sufficient, depending on the type of signal to be analyzed. One may wish to consider a specific form of movement, i.e., choose a particular relativity group. Three examples may be of interest (we begin again with one space dimension).

(i) *Galilean wavelets*

Here we add to the transformations discussed above the Galilei boosts, thus getting $(x, t) \mapsto (a_1 x + vt + b_1, a_0 t + b_0)$. The resulting group G_1^{aff}, called the affine Galilei group, is quite complicated. It has a natural unitary representation in the space of finite energy signals, which splits into the direct sum of four irreducible ones, and each of these is square integrable, so that wavelets may be constructed in the usual way [25]. In addition, more restricted wavelets may be constructed by taking as parameter space various quotient spaces G_1^{aff}/H, where H is *not* the stability subgroup of the basic wavelet.

(ii) *Schrödinger wavelets*

One obtains an interesting subclass of the previous one by imposing the relation $a_0 = a_1^2$, so that the transformations leave invariant the Schrödinger (or the heat) equation. Then the unitary irreducible representation U_G of G_1^{aff} splits into the direct sum of two square integrable ones of the (Schrödinger) subgroup. Thus again a CWT is at hand, which may prove useful for describing, for instance, the motion of quantum particles on the line.

(iii) *Poincaré wavelets*

In order to get a CWT in the relativistic regime, it suffices to replace Galilei transformations by Poincaré ones, while of course imposing the relation $a_0 = a_1$ to space and time dilations. The result is the affine Poincaré group, that we have discussed at length in Section 7.4. The Poincaré wavelets might be useful, for instance, in the presence of electromagnetic fields.

Of course, this analysis extends in a straightforward way to higher dimensions, just by adding rotations. Details may be found in [Ali00; Section 15.3].

These three types of relativistic wavelets offer additional examples of the general group-based wavelet formalism. They have a definite mathematical interest, but they have not been tested on practical situations, indeed no motion estimation (ME) algorithm based on them has been designed. On the contrary, the kinematical spatio-temporal CWT does lead to an efficient ME algorithm, that we now describe in detail.

10.5 A motion estimation (ME) algorithm

We shall now exploit the general CWT formalism developed in the previous section and describe an algorithm for motion estimation. A complete discussion can be found in [Muj99,281,282].

Velocity filtering approaches for motion estimation allow temporal information to be incorporated in the estimation process. These techniques have performance advantages over two-frame-at-a-time based approaches, like block matching and optical flow [Tek95,353] in nonideal environments. The spatio-temporal CWT facilitates adaptive velocity filtering and offers an elegant framework for motion analysis and estimation. It performs a mapping from the Hilbert space $\mathcal{H}$ to a parameter space meaningful for motion estimation purposes. It can also be seen as a tool for motion-based filtering, where the filter characteristics are appropriately determined by a set of parameters directly associated with motion features. The wavelet basis matches the *motion characteristics* of the object of interest rather than its *spatial features*. General spatial selectivity is taken into account through the scale parameter a.

It is assumed that starting conditions, i.e., position and velocity, of the object of interest are known initially. Our ME algorithm deals with the problem of following time-varying motion parameters on a frame-by-frame basis, which allows us to determine object coordinates at any time [Muj99]. In this sense, our ME algorithm can be viewed as an object tracking algorithm after the initial detection has been performed. This section is organized in two parts. First, we formulate the rôle of the spatio-temporal CWT as a motion parameter estimator and we describe the three *partial energy densities* used for this purpose. In the second part, we describe how the interaction of these energy densities can be used to track a particular object from incoming video. Our CWT-based tracking algorithm is then introduced.

10.5.1 Partial energy densities

The multidimensional nature of the spatio-temporal CWT allows for the definition of a multitude of energy densities either by fixing a subset of the parameter space or, better, by partial integration of the CWT energy,

$$E(g) \equiv E[s](g) = \left| \langle \psi_g | s \rangle \right|^2 = \left| S_\psi(g) \right|^2, \quad g = \{\vec{b}, \tau, \theta; a, c\}, \tag{10.51}$$

on subsets of the parameter space. As discussed at length in the spatial case, in Section 2.3.4, this approach has the nice property that it can result in invariant representations with respect to some parameters. Thus, partial integration on subsets of the parameter space results in different energy representations that can be used to extract relevant features. Three energy densities particularly interesting for motion estimation purposes are studied here.

(1) *Speed-orientation energy density:*

Here integration is performed over spatial translation, $\vec{b} = (b_x, b_y)$, on a region

$$\mathcal{B} : (b_{x_{\min}} < b_x < b_{x_{\max}}) \cap (b_{y_{\min}} < b_y < b_{y_{\max}}), \tag{10.52}$$

while the scale a and the temporal variable τ are fixed. As a result, we obtain the first energy density,

$$E^{\mathrm{I}}_{a_o, \tau_o}(c, \theta) = \int_{\vec{b} \in \mathcal{B}} d^2 \vec{b} \; \left| \langle \psi_{\vec{b}, \tau_o, \theta; a_o, c} | s \rangle \right|^2, \tag{10.53}$$

which can be interpreted as an estimator of *local* velocity. The boundaries of the spatial region $\mathcal{B}$ are updated on a frame-by-frame basis to reflect changes in object location. This allows the algorithm to focus on the object of interest, and reduce interference with other nearby objects.

(2) *Spatial energy density:*

In this representation, the speed tuning parameter c, the orientation θ, the scale a, and the temporal translation τ, are fixed, while the spatial translation, $\vec{b}$ is the variable of interest. The resulting energy density is given by,

$$E^{\mathrm{II}}_{\tau_o, \theta_o; a_o, c_o}(\vec{b}) = \frac{1}{a_o^4} \left| \langle \psi_{\vec{b}, \tau_o, \theta_o; a_o, c_o} | s \rangle \right|^2. \tag{10.54}$$

We can think of this energy density as the output energy of a *velocity selective filter*, where the location of objects moving at a pre-specified velocity, $\vec{v} = c_o \, e^{i\theta_o}$, can be easily determined. In addition, size selectivity (invariability) can be obtained by appropriately choosing (integrating over) the scale parameter a.

(3) *Scale energy density:*

The scale parameter is associated with spatial size. Consequently, there exists an optimum scale a_{opt} that best *matches* the size of the object of interest. A measure of "scale optimality" can be defined by integrating the global energy density, E, over the spatial variables, $\vec{b}$, while fixing the temporal translation τ, and the velocity parameters, c and θ, leading to

$$E^{\mathrm{III}}_{c_o, \theta_o, \tau_o}(a) = \frac{1}{a^4} \int_{\vec{b} \in \mathcal{B}} d^2 \vec{b} \; \left| \langle \psi_{\vec{b}, \tau_o, \theta_o; a, c_o} | s \rangle \right|^2. \tag{10.55}$$

The spatial integration is constrained to the region $\mathcal{B}$ to avoid interference with nearby objects.

These energy densities can be used to derive *local estimates* of the motion parameters. It is noteworthy that these energy densities are different from the *global energy densities* where additional integration is performed over the temporal parameter τ [279]. The global approach can only handle linear motion, while the local approach can deal with accelerated motion and time varying signatures as well.

The energy densities presented here are the computational core of the CWT-based tracking algorithm, which is presented in the next section.

10.5.2 Description of the algorithm

Computer implementation of the CWT requires discretization of the spatial and temporal variables $(\vec{x}, t)$. The energy densities E^I, E^II, and E^III of (10.53), (10.54), and (10.55) must then be discretized by replacing integrations with appropriate summations.

It is assumed the input data is presented as an incoming video signal with one or more objects moving with a constant or accelerated speed. For instance, the l-th object in a given image sequence is denoted $s_l(\vec{x}_{n,m}, t_i)$, where n and m are spatial indices (horizontal and vertical respectively), and i is the temporal index. The resulting model for the input image sequence is then

$$s(\vec{x}_{n,m}, t_i) = \sum_l s_l(\vec{x}_{n,m}, t_i) + w(\vec{x}_{n,m}, t_i), \tag{10.56}$$

where w is assumed to be zero-mean white noise. The sum in (10.56) must be interpreted carefully when occlusions occurs. In this case, if the point $(\vec{x}_{n,m}, t_i)$ is subject to occlusion from two or more objects, the correct expression is

$$s(\vec{x}_{n,m}, t_i) = s_L(\vec{x}_{n,m}, t_i) + w(\vec{x}_{n,m}, t_i), \tag{10.57}$$

where L is the index denoting the object closest to the sensor.

The CWT-based ME algorithm relies on the three energy densities defined above and consists of a *frame-by-frame* optimization of the motion parameters associated with a given object in the image sequence. These motion parameters are gathered in a *state vector* $L(t_i)$ defined as

$$L(t_i) = \left(\vec{v}'_{t_i}, \ \vec{x}'_{t_i}, \ a_{t_i} \right)^T, \tag{10.58}$$

for time $t = t_i$. Note the convention we use for the position variable; $\vec{x}_{n,m}$ represents the spatial *location* indexed by the pair (n, m), while $\vec{x}_{t_i}$ represents the spatial *state* at time $t = t_i$. We assume that the starting position, $\vec{x}_0$, and velocity, $\vec{v}_0$, for the object of interest are known initially.

The CWT energy densities are used here as *optimality criteria* or *cost functions* for updating the state vector L. This update can be done either by searching for the local maximum or by a gradient-based approach like the LMS algorithm, on each of the energy densities. In either case, this process is denoted symbolically by

$$L(t_i) \xrightarrow{\; E^{\mathrm{I}} \; E^{\mathrm{II}} \; E^{\mathrm{III}} \;} L(t_{i+1}). \tag{10.59}$$

The optimization of equation (10.59) is performed *sequentially*. That is, when optimizing one component of the state vector (i.e., $\vec{v}$, $\vec{x}$, or a), the others are kept constant. Indeed, we are searching for an optimal set of motion parameters in the 6-D CWT space. The sequential optimization of the parameters considerably reduces the search space and consequently the computational requirements (with respect to a simultaneous 6-D search). This is possible due to the implicit redundancy of the CWT representation and its ability to isolate motion features. Thus, each of the energy densities corresponds to a 2-D (or 1-D for the scale energy density) "slice" of the CWT parameter space, where one parameter is optimized independently of the others.

The CWT can be seen as a spatio-temporal filtering operation where the filter characteristics are controlled by a set of parameters associated with motion features (i.e., velocity and size). More explicitly, manipulating the inner product of equation (10.40) the CWT can be defined as a convolution sum, i.e.,

$$
\begin{aligned}
S_\psi(g) &= \iint_{\mathbb{R}^2 \times \mathbb{R}} d^2\vec{x}\, dt\, s(\vec{x}, t)\, \overline{\psi_{\theta,a,c}(\vec{x} - \vec{b}, t - \tau)} \\
&= \iint_{\mathbb{R}^2 \times \mathbb{R}} d^2\vec{x}\, dt\, s(\vec{x}, t)\, \psi_{\theta,a,c}^{\#}(\vec{b} - \vec{x}, \tau - t) \\
&= s \otimes \overline{\psi}_{\theta,a,c}(\vec{b}, \tau),
\end{aligned}
$$

where the symbol $\otimes$ represents the (2+1)-D convolution operator,

$$\psi_{\theta,a,c}(\vec{x} - \vec{b}, t - \tau) = \psi_{\vec{b},\tau,\theta;a,c}(\vec{x}, t), \text{ and}$$
$$\psi_{\theta,a,c}^{\#}(\vec{x}, t) = \overline{\psi_{\theta,a,c}(-\vec{x}, -t)}.$$

The indices of the filtered signal correspond to the spatio-temporal translation parameters of the CWT (i.e., $\vec{b}$ and τ). This allows us to take advantage of the Fast Fourier Transform (FFT) to compute the CWT (and its associated energy densities) efficiently in the wavenumber–frequency domain. The separability of the wavelets can be exploited to reduce the required computations to construct the motion and scale selective filters.

A block diagram of the state updating process performed by the CWT-based ME algorithm is depicted in Figure 10.2. The $N \times M \times K$ block of data represents a portion of the incoming video signal with K frames of N columns by M rows. This image sequence is first transformed to the Fourier domain by means of a 3-D DFT. As suggested by Figure 10.2 and equation (10.59) the parameter update is done in a specified order: velocity first, then position, and finally scale. Two arguments support this implementation choice. First, velocity is a motion parameter of higher order than position. For highly maneuvering objects, the relative change in velocity from frame to

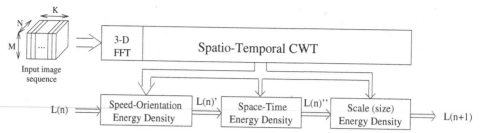

Fig. 10.2. CWT-based tracking algorithm.

frame is smaller than the relative change in position. Second, some of the calculations necessary for determining the position energy density are already done at the end of the speed-orientation energy density stage.

The three update stages embodied in the tracking algorithm are now described in detail. It is important to point out that the diagram shown in Figure 10.2 represents the operations performed in order to update the state vector $L(t_i)$ for each frame of a given data block.

10.5.2.1 Velocity update stage

In this stage the speed-orientation energy density E^I is used as the optimality criterion to update the velocity state at time t_{i+1}. The position and scale states are fixed to their corresponding values at time t_i, which are, $\vec{x}_{t_i}$ and a_{t_i} respectively. The discrete version of the speed-orientation energy density E^I is

$$\mathsf{E}^I_{a_o, \tau_o}(c, \theta) = \sum_{\vec{b}_{n,m} \in \mathcal{B}} \left| \left\langle \psi_{a_o, c, \theta, \vec{b}_{n,m}, \tau_o} | s \right\rangle \right|^2, \tag{10.60}$$

where $a_o = a_{t_i}$, $\tau_o = t_{i+1}$, and the spatial region $\mathcal{B}$, defined in (10.52), corresponds to a local neighborhood around the previous position state, $\vec{x}_{t_i}$. This allows for discrimination of objects moving with similar velocity but spatially separated.

Denoting by $\vec{v}_*$ the value of $\vec{v}$ that maximizes (10.60),

$$\max_{\vec{v} \in \mathcal{V}} \left\{ E \left[\mathsf{E}^I_{a_o, \tau_o}(c(\vec{v}), \theta(\vec{v})) \right] \right\}, \tag{10.61}$$

the updated velocity state is simply $\vec{v}_{t_{i+1}} = \vec{v}_*$. Here $\mathcal{V}$, defined as

$$\mathcal{V}: \ (v_{x_{\min}} < v_x < v_{x_{\max}}) \cap (v_{y_{\min}} < v_y < v_{y_{\max}}), \tag{10.62}$$

is a local neighborhood around the previous velocity state, $\vec{v}_{t_i}$. The limits of $\mathcal{V}$ can be fixed or can depend on the current velocity estimate by defining a maximum acceleration (and de-acceleration). Symbolically, this update operation can be expressed as

$$\vec{v}_{t_i} \xrightarrow{\ \mathsf{E}^I\ } \vec{v}_{t_{i+1}}. \tag{10.63}$$

The maximization of (10.61) requires the evaluation of the speed-orientation energy density of (10.60) at various values of $\vec{v}$ and searching for the local maximum on $\mathcal{V}$.

For evaluating $E^I_{a_o, \tau_o}(c_o, \theta_o)$, we take advantage of the FFT computational efficiency by calculating the CWT convolution sum of (10.60), as follows.

(i) Evaluate $\widehat{\widetilde{\psi}}_{a_o, c_o, \theta_o}(\vec{k}, \omega)$.

(ii) Obtain the product

$$\widehat{z}(\vec{k}, \omega) = \widehat{s}(\vec{k}, \omega)\,\widehat{\widetilde{\psi}}_{a_o, c_o, \theta_o}(\vec{k}, \omega), \tag{10.64}$$

which is a partial result for the (2+1)-D convolution in (10.60).

(iii) Apply a 1-D IFFT to $\widehat{z}$ with respect to the frequency variable and for all wavenumbers $\vec{k}$ to obtain

$$\widetilde{z}(\vec{k}, \tau) = \mathrm{IFFT}_\omega\left\{\widehat{z}(\vec{k}, \omega)\right\}. \tag{10.65}$$

(iv) Calculate the 2-D IFFT of $\widetilde{z}(\vec{k}, \tau)$ in the wavenumber variables at $\tau = \tau_o$,

$$z_{\tau_o}(\vec{b}) = \mathrm{IFFT}_{\vec{k}}\left\{\widetilde{z}(\vec{k}, \tau_o)\right\}. \tag{10.66}$$

(v) Finally, the speed-energy density (10.60) is

$$E^I_{a_o, \tau_o}(c_o, \theta_o) = \sum_{\vec{b} \in \mathcal{B}}\left|z_{\tau_o}(\vec{b})\right|^2. \tag{10.67}$$

Maximizing E^I can be accomplished in many different ways. One possibility is to evaluate E^I on a dense grid and search for a local maximum within the region $\mathcal{V}$. But this is computationally complex. Gradient-based approaches are another possibility but present the problem associated with derivatives when dealing with noisy data. We have chosen to use the Nelder–Mead simplex search algorithm [133] which performs an oriented search and does not require gradients or other derivative information.

10.5.2.2 Position update stage

Here the current velocity state $\vec{v}_{t_{i+1}}$ and the previous scale state a_{t_i} are fixed while the position state, $\vec{x}_{t_{i+1}}$, is updated. The optimality criteria is now the discretized version of the spatial energy density E^{II},

$$E^{II}_{a_o, c_o, \theta_o, \tau_o}(\vec{b}_{n,m}) = \frac{1}{a_o^4}\left|\left\langle \psi_{a_o, c_o, \theta_o, \vec{b}_{n,m}, \tau_o}\,|s\right\rangle\right|^2, \tag{10.68}$$

with $a_o = a_{t_i}$, $c_o = |\vec{v}_{t_{i+1}}|$, $\theta_o = \arg(\vec{v}_{t_{i+1}})$, and $\tau_o = t_{i+1}$.

It is important to note that the spatial energy density E^{II} has already been calculated in the previous stage prior to integration on the spatial translation, $\vec{b}$. Thus,

$$E^{II} = z_{\tau_o}(\vec{b}), \tag{10.69}$$

with z obtained from steps (i)–(iv) of the speed–orientation energy density at the updated velocity $\vec{v}_{t_{i+1}}$. The location of the local maximum in (10.68),

$$\max_{\vec{b} \in \mathcal{B}} \left\{ \mathsf{E}^{II}_{a_o, c_o, \theta_o, \tau_o}(\vec{b}_{n,m}) \right\}, \tag{10.70}$$

is done in two stages. First, direct search on the discrete grid of (10.69) give us an approximate location for the local maximum, $\vec{b}'$. Second, this local maximum is refined by interpolation around the immediate vicinity of $\vec{b}'$. The new local maximum is denoted $\vec{b}_*$.

Once the location of the maximum of (10.68) is identified (i.e., at $\vec{b} = \vec{b}_*$), the position state is updated as $\vec{x}_{t_{i+1}} = \vec{b}_*$. This updating operation is denoted symbolically by,

$$\vec{x}_{t_i} \xrightarrow{\mathsf{E}^{II}} \vec{x}_{t_{i+1}}. \tag{10.71}$$

10.5.2.3 Scale update stage

This stage allows the ME algorithm to handle 3-D object motion, where the 2-D object may experience dramatic changes in its spatial signature. In the simplest case, such variation can be produced by objects moving toward or away from the imaging sensor, i.e., scale changes. In general, 3-D object motion can produce rather complex changes in the object spatial signature, which typically are unknown or difficult to model. The effect of the scale parameter in the Duval–Destin–Murenzi wavelet family [158] is to provide some degree of spatial selectivity. The velocity and position estimation accuracy is not highly sensitive to the particular shape characteristics.

After the velocity and position states are updated, they are used in the scale energy density to obtain the scale that best matches the general spatial characteristic of the object of interest at $\tau_o = t_{i+1}$. The discrete version of the scale energy density takes the form

$$\mathsf{E}^{III}_{c_o, \theta_o, \tau_o}(a) = \frac{1}{a^4} \sum_{\vec{b}_{n,m} \in \mathcal{B}} \left| \left\langle \psi_{a, c_o, \theta_o, \vec{b}_{n,m}, \tau_o} | s \right\rangle \right|^2. \tag{10.72}$$

An optimal value a_* is computed for the scale parameter by maximizing the expression

$$\max_{a \in \mathcal{A}} \left\{ \mathsf{E}^{III}_{c_o, \theta_o, \tau_o}(a) \right\}, \tag{10.73}$$

over $\mathcal{A}$, a range of scales around the previous scale state a_{t_i}. To update the scale state we let $a_{t_{i+1}} = a_*$. This maximization also requires evaluating $\mathsf{E}^{III}_{c_o, \theta_o, \tau_o}(a)$ for different values of a and is performed with the Nelder–Mead simplex search algorithm. This is done using a similar procedure to the one depicted before for the speed-orientation energy density. When the size of the object of interest is known to be constant, the optimization of (10.73) is not necessary.

10.5.3 Application of the algorithm: results

Tracking results of the CWT-based algorithm are presented here for two synthetically generated scenarios. A $64 \times 64 \times 64$ image sequence, of 256-gray levels, representing

Table 10.1. *Parameters of TEST1 sequence* ($t_o = 1$)

Object	$x(t_o)$	$y(t_o)$	$v_x(t_o)$	$v_y(t_o)$	a_x	a_y
1	3	3	0.5	0.5	0.05	0.05
2	3	60	0.5	−0.5	0.05	−0.05
3	56	48	−1.0	−0.5	0.0	−0.075
4	60	50	−1.0	0.0	0.0	0.0

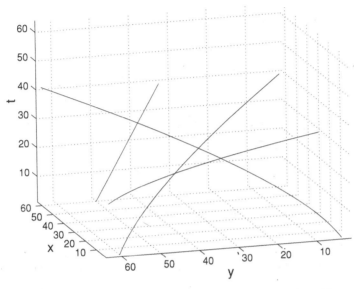

Fig. 10.3. Space–time representation of the trajectories of the four objects on sequence TEST1.

a block of data from incoming video is used in the two examples. The first test scenario (TEST1) includes four Gaussian shaped objects moving with accelerated motion and it is used to assess performance under noisy conditions. The second example (TEST2) contains a Gaussian object following a circular trajectory and is used to estimate the maximum velocity and acceleration that the CWT-based algorithm can handle. It is important to point out that the image sequences are constructed by uniformly sampling the associated (2+1)-D continuous signal, which is obtained from the mathematical expression for the object shape and the continuous motion equation. Thus, the center of the object, which is the location parameter of interest, will in general lie in noninteger locations. It is assumed that the initial conditions are known for every object in the scenes and that the simulation starts ten frames into the sequence. We use the Morlet wavelet (10.38) with $\vec{k}_o = (-6, 0)^T$ and $\omega_o = 6$.

The initial motion parameters of the four objects in the first example are given in Table 10.1. The initial spatio-temporal location and velocity are $(x(t_o), y(t_o), t_o)$ and $(v_x(t_o), v_y(t_o))$ respectively, with $t_o = 1$, while the constant acceleration is (a_x, a_y). Figure 10.3 shows a space–time representation of the trajectories of the four moving

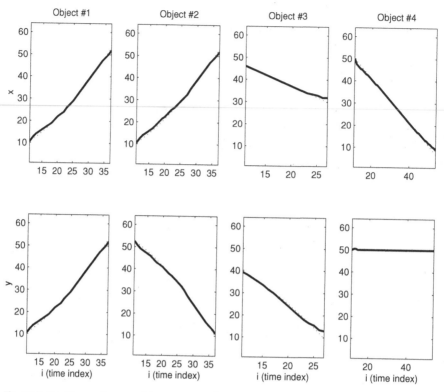

Fig. 10.4. Position estimates with the CWT-based ME algorithm for the noiseless case of the first scenario (TEST1). Dotted lines correspond to the desired trajectories, and the solid lines represent the estimated ones.

objects. Note that only Object 4 moves with linear motion. The other three objects have accelerated motion and the trajectories of Objects 1 and 2 cross each other.

Figures 10.4 and 10.5 show estimates of position and velocity for the four objects (solid lines) along with the desired curves (dotted lines). The algorithm is able to handle accelerated motion well. Even when Object 2 occludes Object 1, around the middle of the sequence, tracking is successful.

We compare the CWT tracking results with a block matching (BM) approach. This particular implementation is based on an exhaustive search BM algorithm and differs from the classical implementation [Jah97,Tek95] in two aspects. First, it includes an object mask to unevenly weight the error in the block in order to improve performance for nonrectangular objects, and second, the starting point for the search is obtained from a temporal filtered version of the previous displacement estimate. To some extent, this takes into consideration the temporal history of the motion. As it can be seen from Figure 10.6, this BM algorithm fails to maintain track at the occlusion point. More specifically, it loses track of Object 2 after the occlusion and tracks Object 1 instead. Also, the tracking error increases for Object 1 in the vicinity of the occlusion point.

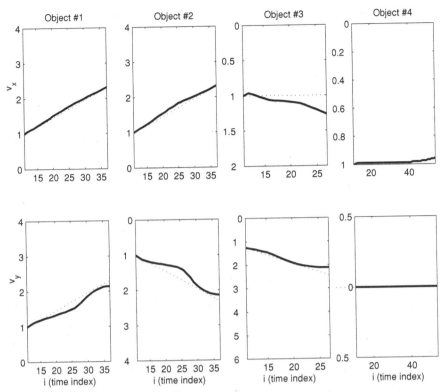

Fig. 10.5. Velocity estimates with the CWT-based ME algorithm for the noiseless case of the first scenario (TEST1). Dotted lines correspond to the desired velocities, and the solid lines represent the estimated ones.

To assess the performance under noisy conditions, we add white Gaussian noise to the first test scenario. We use the peak signal to noise ratio (PSNR) to quantify the noise level, i.e.,

$$PSNR = 10 \log_{10} \frac{255^2}{\sigma^2}, \tag{10.74}$$

where σ^2 is the noise variance. Tracking performance is measured using the Euclidean distance between the real and estimated trajectory, and we define the error,

$$\eta_{e,l}^2 = \sum_{i \in D_l} (x_l(i) - \tilde{x}_l(i))^2 + (y_l(i) - \tilde{y}_l(i))^2, \tag{10.75}$$

$$\sigma_e = \sqrt{\frac{1}{\sum_l N_l} \sum_l \eta_{e,l}^2}, \tag{10.76}$$

where l is the object index, D_l and N_l are respectively the range and number of valid temporal indexes i, (x_l, y_l) is the real trajectory, and $(\tilde{x}_l, \tilde{y}_l)$ is the estimated one.

Figure 10.7 shows plots of tracking performance versus noise level for the first test sequence. Figure 10.7(a) shows a considerably higher error for the BM technique

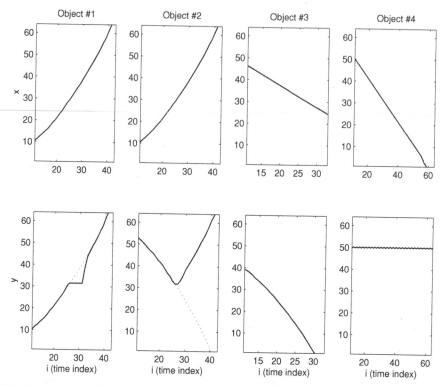

Fig. 10.6. Position estimates with the BM algorithm for the noiseless case of the first scenario (TEST1). Dotted lines correspond to the desired trajectories, and the solid lines represent the estimated ones.

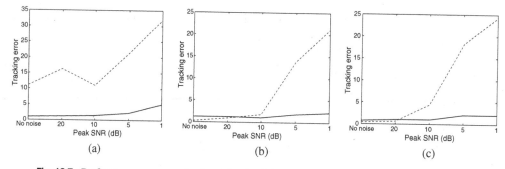

Fig. 10.7. Performance versus noise level for the first scenario (TEST1), for (a) all four objects, (b) object 3, and (c) object 4. Solid lines correspond to the CWT-based ME algorithm and the dashed lines to the BM approach.

(dashed lines) even at high PSNRs. This is mainly caused by the inability of the BM algorithm to handle the occlusion of Objects 1 and 2. In fact, after the occlusion point and even for the noiseless scenario, Object 2 is confused with Object 1. When noise is present, Object 2 is tracked when the algorithm is supposed to track Object 1 and vice versa. At 10, 5, and 1 dB PSNRs, track is lost for all four objects a few frames after

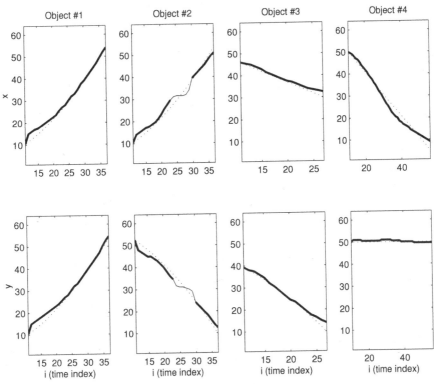

Fig. 10.8. Position estimates with the CWT-based ME algorithm for the first scenario (TEST1) at the 5 dB PSNR noise level. Dotted lines correspond to the desired trajectories, and the solid lines represent the estimated ones.

the initiation point. On the other hand, the CWT-based ME algorithm (solid lines) only mistracked Object 2 for the 1 dB PSNR case, and even then the other three objects were tracked successfully. Figures 10.7(b) and 10.7(c) show tracking results for Objects 3 and 4 respectively, which are not subject to occlusion. We can note that at very high PSNRs, the BM algorithm is able to perform very accurate tracking, provided there are no occlusions. The CWT-approach in this PSNR range is good, but slightly less accurate. This is in part due to the fact the wavelet filters used in the CWT-based approach are tuned to linear motion, while the BM algorithm does not assume any particular motion model. This explains why the curves for the BM and CWT approaches are so close to each other at low PSNRs for Object 4, which is the only one that follows a linear trajectory.

In order to get a qualitative feeling for the performance figures, we present tracking results for the 5 dB noise level. Figures 10.8 and 10.9 correspond respectively to position and velocity estimates using the CWT-based approach. These figures show two levels of reliability for the motion estimates, indicated in the figures by thick and thin lines. The thick curves corresponds to *reliable* estimates and the thin curves represent

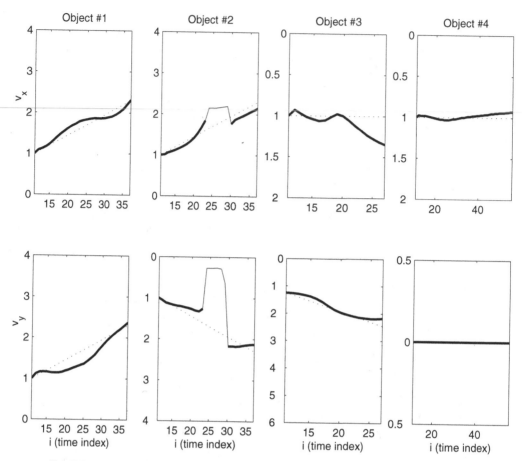

Fig. 10.9. Velocity estimates with the CWT-based ME algorithm for the first scenario (TEST1) at the 5 dB PSNR noise level. Dotted lines correspond to the desired velocities, and the solid lines represent the estimated ones.

partially reliable estimates. These two cases are differentiated by two weighted thresholds applied to the local maximum of the spatial energy density, $E^{II}(\vec{b}_*)$, i.e.,

$$\text{If } E^{II}_t(\vec{b}_*) > th_1 \, \overline{E}^{II}_t(\vec{b}_*) \rightarrow \text{Type 1} \tag{10.77}$$

$$\text{If } E^{II}_t(\vec{b}_*) < th_2 \, \overline{E}^{II}_t(\vec{b}_*) \rightarrow \text{Type 2}, \tag{10.78}$$

where the Type 1 and Type 2 labels correspond to the reliable and partially reliable cases respectively, the thresholds th_1 and th_2 satisfy $th_1 > th_2$, and $\overline{E}^{II}_t(\vec{b}_*)$ is the time-filtered local maximum of the spatial energy density. We use a 4-tap moving average filter and $th_1 = 0.9$ and $th_2 = 0.5$.

The CWT-based algorithm is able to maintain track of all four objects as is evident from Figure 10.8. Estimates for Object 2 become partially reliable around the occlusion point, however tracking is still maintained and the reliable classification is recovered

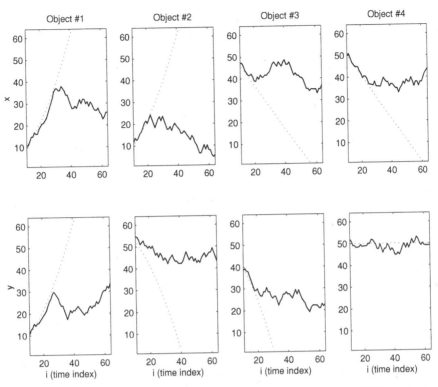

Fig. 10.10. Position estimates with the BM algorithm for the first scenario (TEST1) at the 5 dB PSNR noise level. Dotted lines correspond to the desired trajectories, and the solid lines represent the estimated ones.

a few frames after that. Equivalent results for the BM-based algorithm are shown in Figure 10.10. The BM algorithm on the other hand loses track of all four objects after a few frames. Comparing Figures 10.8 and 10.9 with their counterparts for the noiseless scenario, Figures 10.4 and 10.5, show the robustness of the CWT-based tracking algorithm to noise. Recall this test scenario also includes temporary occlusion and acceleration effects.

Figure 10.11 shows frame # 20 of the TEST1 sequence, at the 5 dB PSNR noise level, along with the spatial energy density results for each of the four objects. Notice that the four objects have been isolated and the background noise considerably reduced. As expected, the shape of the spatial energy density in the vicinity of the object is elongated in the direction of the motion. This is a consequence of the effective convolution of two signals with similar local motion characteristics (see the discussion in Section 10.5.2). It is important to point out that the shape of the energy density is not needed for motion estimation purposes. All that is required is the local maxima, since these indicate the object locations.

The second test scenario (TEST2) attempts to estimate upper bounds for the maximum speed and acceleration that the CWT-based ME algorithm can handle. This is a

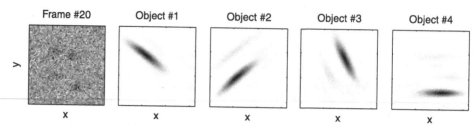

| Frame #20 | Object #1 | Object #2 | Object #3 | Object #4 |

Fig. 10.11. Frame no. 20 of the first test scenario (TEST1) at the 5dB PSNR noise level and filtering results for each of the four objects.

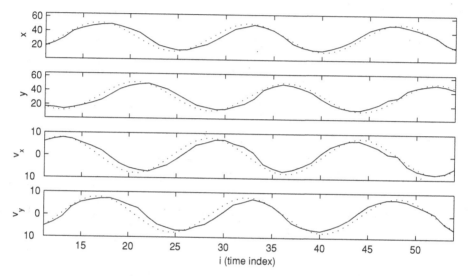

Fig. 10.12. Position (x, y) and velocity (v_x, v_y) estimates with the CWT-based ME algorithm for the second scenario (TEST2) with $|\vec{v}| = 8$. Dotted lines correspond to the desired trajectories, and the solid lines represent the estimated ones.

difficult task, since these bounds depend on a number of factors, including background noise, temporary occlusions, object shape, etc. To obtain an estimate of these bounds, we examine a circularly moving object under conditions of increasing velocity and acceleration (i.e., decreasing frame rate). The object has again a Gaussian shape and the family of trajectories, parameterized by the object speed $|\vec{v}|$, has the form

$$x(t) = 20 \cos\left(|\vec{v}|\, t/20\right) + 32$$
$$y(t) = 20 \sin\left(|\vec{v}|\, t/20\right) + 32. \tag{10.79}$$

Scenarios with object speeds ranging from 1 to 10 pixels-per-frame are used. Equivalently, the associated accelerations range from 1/20 to 5 pixels-per-frame^2. It is found that the CWT ME algorithm is able to successfully handle speeds up to 8 pixels-per-frame and subsequently, accelerations of 3.2 pixels-per-frame^2. This is illustrated in Figure 10.12 for $|\vec{v}| = 8$ where both position and velocity estimates (solid lines) are

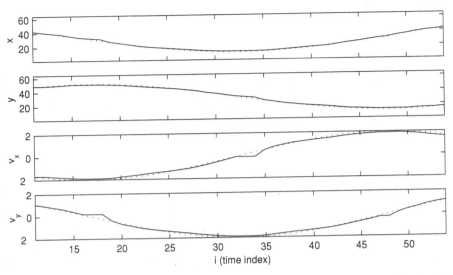

Fig. 10.13. Position (x, y) and velocity (v_x, v_y) estimates with the CWT-based ME algorithm for the second scenario (TEST2) with $|\vec{v}| = 2$. Dotted lines correspond to the desired trajectories, and the solid lines represent the estimated ones.

shown against the desired curves (dashed lines). For comparison purposes, we show in Figure 10.13 the corresponding results for $|\vec{v}| = 2$. Note that tracking performance as defined by (10.76) is 1.1343 for the case $|\vec{v}| = 2$ and 5.1752 for the case $|\vec{v}| = 8$. This last figure might be unacceptable for many applications. A more desirable error figure of less than 3 is maintained for speeds up to 6 pixels-per-frame, which translates to an upper bound for the acceleration of 1.8 pixels-per-frame^2. These upper bounds for the speed and acceleration can be used as a *rule of thumb* to estimate the minimum frame rate necessary for proper operation of the CWT-based ME algorithm.

10.5.4 Conclusions

The CWT-based motion estimation algorithm described in this section dynamically adapts the parameters of a wavelet filter on a frame-by-frame basis and can handle both linear and accelerated motion, even when occlusions and background noise are present. The CWT parameters, which are directly associated with motion features, i.e., velocity, position, and scale, are sequentially updated. This, as compared to a full 6-D search, considerably reduces the computational requirements. Another advantage of these wavelets is their separability. Because the spatial and temporal responses are independent of each other, more efficient implementations can be realized compared to nonseparable approaches like conventional 3-D matched filtering.

Tracking of the object trajectories was shown to be possible even under severe noise conditions, and even when occlusions were present, for both the linear and accelerated motion cases. The presence of noise and occlusions do not impair the accuracy in estimating of object locations. The ability to deal with maneuvering objects is illustrated for a circularly moving object and bounds are giving for the maximum speed and acceleration.

The algorithm represents a general framework for motion analysis. It outlines a strategy for tracking moving objects which leads to the design of efficient discrete filtering approaches suitable for motion estimation purposes.

11 Beyond wavelets

Up to now, we have developed the 2-D CWT and a number of generalizations, relying in each case on the group-theoretical formalism. Given a class of finite energy signals and a group of transformations, including dilations, acting on them, one derives the corresponding continuous WT as soon as one can identify a square integrable representation of that group.

On the other hand, we have also briefly sketched the discrete WT and several transforms intermediate between the two. One conclusion of the study is that the pure DWT is too rigid, whereas redundancy is helpful, in that it increases both flexibility and robustness to noise of the transform. Indeed, the wavelet community has seen in the last few years a growing trend towards more redundancy and the development of tools more efficient than wavelets, such as ridgelets, curvelets, warplets, etc. The key word here is *geometry*: the new transforms and approximation methods take much better into account the geometrical features of the signals. To give a simple example, a smooth curve is in fact a 1-D object and it is a terrible waste (of times or bits) to represent it by a 2-D transform designed for genuine 2-D images.

It is therefore fitting to conclude the book by a chapter that covers these new developments. Actually, we will consider two classes of techniques, that both go beyond our standard group-theoretical paradigm. First we describe two new, geometrically inspired transforms, ridgelets and curvelets, and their application to image compression using anisotropic quadtrees and N-best term approximation. Next we will extend the idea to a general quest of sparse representations of signals using *redundant* dictionaries – a hot topic in approximation theory at this moment. Then, in a second part, we turn back to wavelets, but based on "exotic" sytems of numeration (for instance, with the golden mean τ instead of 2 as basic scaling factor). Here again, there is no obvious group and the driving force is geometry. This is mostly visible in 2-D, where one builds wavelet systems based on the famous Penrose–Robinson tilings of the plane. Hence such developments fit also well in the present chapter.

11.1 New transforms: ridgelets, curvelets, etc.

11.1.1 The ridgelet transform

The main idea behind the continuous ridgelet transform is to efficiently catch and analyze line singularities in images. We have already seen in Chapter 3, Section 3.4.3, that the 2-D CWT is a very efficient tool to this effect, since it is able to localize the singularity and its orientation. On the other hand, if the data to be analyzed is a truly infinitely long ridge, there is no information along its principal direction. There is thus no need to localize the wavelet transform and one could save an unnecessary degree of freedom in the transform. This can be checked on Figure 3.13. Indeed, the modulus of the CWT quickly saturates and reaches a stable constant regime along the regular direction (length of the slab), while precisely detecting the singularity in the orthogonal direction. These considerations led Candès [Can98] to introduce a slightly modified version of the 2-D CWT.

Starting from the original definition (2.18)–(2.20), given in Definition 2.1.3, one requires that a ridgelet behaves like a 1-D wavelet ψ in a given direction, represented by the unit vector $\vec{u}_\theta$ of orientation θ, and be constant along the orthogonal direction. Such an object can then be shifted along its oscillating direction, rotated and scaled in order to yield a family of continuous ridgelets (see Figure 11.1):

$$\psi_{a,\theta,b}(\vec{x}) = a^{-1/2}\psi\left(\frac{\vec{u}_\theta \cdot \vec{x} - b}{a}\right).$$

Equivalently, ridgelets can be seen as usual 2-D wavelets, except they are constant along a preferred direction and will thus never be admissible. The very notion of an admissible ridgelet and the associated ridgelet transform can nevertheless be easily derived as shown by the following result. Note that ψ is a function of one variable only. We manifest this by using a different notation, namely, "a, θ, b" as subscript instead of the traditional "$\vec{b}, a, \theta$" (in that case, the order of the variables was imposed by the structure of the similitude group).

Theorem 11.1.1 (Reconstruction formula) *Any* $f \in L^1(\mathbb{R}^2)$ *such that* $\widehat{f} \in L^1(\mathbb{R}^2)$ *admits the following decomposition:*

$$f = c_\psi^{-1}\int_0^{+\infty}\frac{da}{a^3}\int_0^{2\pi}d\theta\int_{\mathbb{R}}db\,\langle\psi_{a,\theta,b}|f\rangle,$$

with

$$c_\psi = \frac{1}{2}\int_{\mathbb{R}}d\zeta\,\frac{|\widehat{\psi}(\zeta)|^2}{\cdot|\zeta|^2}.$$

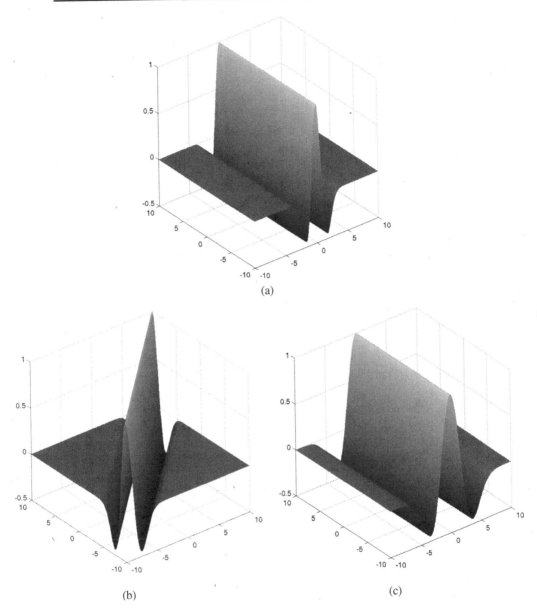

Fig. 11.1. (a) A typical ridgelet; (b and c) a rotated and a scaled version.

Any profile ψ satisfying $c_\psi < +\infty$ is termed admissible ridgelet, or simply ridgelet for short. As in the wavelet case, this condition is essentially equivalent to imposing at least one vanishing moment on ψ.

The definition of the continuous ridgelet transform (CRT) follows that of the CWT and shares most of its properties:

Proposition 11.1.2 (Continuous ridgelet transform)
Let $\Gamma = \{(a,\theta, b),\ a \in \mathbb{R}_*^+,\ \theta \in SO(2),\ b \in \mathbb{R}\}$. *Then the linear map* $R : L^2(\mathbb{R}^2) \to L^2(\Gamma, a^{-3}\, da\, d\theta\, db)$ *defined by*

$$Rf(a, \theta, b) = \langle \psi_{a,\theta,b} | f \rangle$$

is an L^2 *isometry.*

The CRT is a stable representation, since it satisfies the following Plancherel formula.

Theorem 11.1.3 (Plancherel formula) *For any* $f \in L^1 \cap L^2(\mathbb{R}^2)$ *and* ψ *admissible, one has*

$$\|f\|_2^2 = c_\psi^{-1} \int_\Gamma \frac{da}{a^3}\, d\theta\, db\ |\langle \psi_{a,\theta,b} | f \rangle|^2.$$

It is very interesting to wonder about the existence of frames of ridgelets and what would be their properties. Such a construction was proposed by Candès [Can98] and we simply highlight here the main ideas. As usual, the problem is to find a discretization $(a_j, \theta_{j,\ell}, b_{j,k})$ of the parameter set Γ such that one can sample the CRT in a stable manner, that is, satisfying a norm equivalence:

$$A\|f\|^2 \leqslant \sum_{j,k,\ell} |Rf(a_j, \theta_{j,\ell}, b_{j,k})|^2 \leqslant B\|f\|^2. \qquad (11.1)$$

First, the scale and position variables are sampled exactly as in the case of wavelets or Littlewood–Paley analysis, i.e., $a_j = a_0\, 2^{-j}$, $b_{j,k} = k\, b_0\, a_0\, 2^{-j}$. The difference with respect to usual 2-D wavelet frames comes in the particular sampling of the rotation parameter. In order to reach the equivalence (11.1), Candès found out that the angular resolution should increase at finer scales, $\theta_{j,\ell} = 2\pi\, \ell\, 2^{-j}$. The difference between the ridgelet and wavelet discretization of the frequency plane is illustrated in Figure 11.2.

More information about the properties of the CRT can be found in the seminal reference [Can98] and in the papers [93–95]. Applications to statistical estimation and image processing can be found in [97].

11.1.2   Links with the Radon and wavelet transforms

A very handy way of understanding the CRT consists in seeing it as a combination of a Radon transform followed by a classical 1-D wavelet transform. The Radon transform [Ang92] of a bivariate function f is the collection of line integrals:

$$\mathcal{R}f(\theta, t) = \int_{\mathbb{R}^2} dx\, dy\, f(x, y)\, \delta(x \cos\theta + y \sin\theta - t).$$

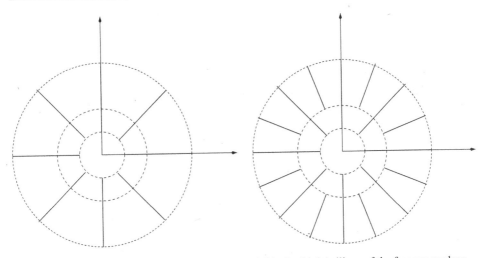

Fig. 11.2. Comparison between the wavelet (left) and ridgelet (right) tilings of the frequency plane.

The CRT coefficients are then simply given by taking the 1-D CWT of $\mathcal{R}f(\theta, t)$ as a function of the line variable t:

$$Rf(a, \theta, b) = \frac{1}{\sqrt{a}} \int_{\mathbb{R}} dt \, \overline{\psi \left(\frac{t - b}{a} \right)} \mathcal{R}f(\theta, t).$$

If the analyzed object is singular along a single line, the Radon coefficients at this angle will be constant. Because of the vanishing moments of the wavelet ψ, the CRT coefficients will then vanish for this direction.

Actually, the 2-D CWT bridges the gap between the Radon and the ridgelet transforms [Hol95]. Let us first consider the following analyzing wavelet:

$$\psi(\vec{x}) = \delta(\vec{e} \cdot \vec{x}). \tag{11.2}$$

This a δ-distribution along the line $\vec{e} \cdot \vec{x} = 0$, i.e., perpendicular to $\vec{e}$. It is a very singular wavelet, but we recall from Section 2.1.2 the generalized cross-admissibility condition (2.30) which allows the use of such objects:

$$c_{\psi\chi} = (2\pi)^2 \int_{\mathbb{R}^2} \frac{d^2\vec{k}}{|\vec{k}|^2} \, \overline{\widehat{\psi(\vec{k})}} \, \widehat{\chi(\vec{k})} < \infty.$$

Applying translations, rotations and dilations to this particular wavelet, we obtain:

$$\begin{aligned} T_{\vec{b}} \, R_\theta \, D_a \psi(\vec{x}) &= T_{\vec{b}} \, R_\theta \, a^{-1} \delta \big(\vec{e} \cdot a^{-1} \vec{x} \big) \\ &= a\delta \big(r_\theta \vec{e} \cdot (\vec{x} - \vec{b}) \big) \\ &= a\delta \big(r_\theta \vec{e} \cdot \vec{x} - r_\theta \vec{e} \cdot \vec{b} \big), \end{aligned}$$

where we have used the homogeneity of the δ function $\langle D_a \delta, s \rangle = a \langle \delta, s \rangle$. Using this wavelet, the 2-D CWT becomes a Radon transform:

$$\langle T_{\vec{b}}\, R_\theta\, D_a \psi | s \rangle = a \int_{\mathbb{R}^2} d^2\vec{x}\, \delta\big(r_\theta \vec{e} \cdot \vec{x} - r_\theta \vec{e} \cdot \vec{b}\big)\, s(\vec{x})$$
$$= a\, \mathcal{R}s\big(\theta, r_\theta \vec{e} \cdot \vec{b}\big).$$

This result tells us that we may use the wavelet inversion formula for inverting the Radon transform, provided we choose a suitable synthesis wavelet χ:

$$s(\vec{x}) = \int_{\mathbb{R}_*^+} \frac{da}{a^3} \int_0^{2\pi} d\theta \int_{\mathbb{R}^2} d^2\vec{b}\, a^{-1} \overline{\chi\big(a^{-1} r_{-\theta}(\vec{x} - \vec{b})\big)}\, a\, \mathcal{R}s\big(\theta, r_\theta \vec{e} \cdot \vec{b}\big). \qquad (11.3)$$

The following theorem, due to Holschneider [224], shows that χ is a ridgelet-like synthesis wavelet.

Theorem 11.1.4 *The relation (11.3) holds strongly in $L^2(\mathbb{R}^2)$ provided the synthesis wavelet χ satisfies the condition*

$$c_\chi = \int_{\mathbb{R}} du\, \frac{\widehat{\chi}(u\vec{e})}{u^2} < \infty. \qquad (11.4)$$

We will just sketch the proof here. The interested reader can check [Hol95,224] for more details. The right-hand side of (11.3) is a convolution between the rotated and scaled synthesis wavelet χ and the Radon transform $\mathcal{R}s$. We can interpret the latter quantity as a 2-D wavelet transform that is constant in the direction orthogonal to $r_\theta \vec{e}$. One degree of freedom can thus be removed. In the coordinate system (α_1, α_2) defined by $\alpha_1 = r_\theta \vec{e}$, the reconstruction formula is again a convolution of $\mathcal{R}s(\theta, \alpha_1)$ with the kernel defined by integrating the synthesis wavelet over α_2:

$$\kappa(\alpha_1) = \int_{\mathbb{R}} d\alpha_2\, \chi(r_{\vec{e}}\, \vec{\alpha})\,, \qquad (11.5)$$

where $r_{\vec{e}}$ is the rotation matrix corresponding to the angle defined by the unit vector $\vec{e}$. Equation (11.5) can also be written:

$$\kappa(\alpha_1) = \int_{\mathbb{R}^2} d^2\vec{u}\, \chi(\vec{u})\, \delta\big(\vec{e} \cdot \vec{u} - \alpha_1\big). \qquad (11.6)$$

Plugging this definition into (11.3) yields:

$$\widetilde{s}(\vec{x}) = \int_{\mathbb{R}_*^+} \frac{da}{a^3} \int_0^{2\pi} d\theta \int_{\mathbb{R}} d\alpha_1\, a^{-1} \overline{\kappa\big(a^{-1}(\vec{e} \cdot r_{-\theta}\vec{x} - \alpha_1)\big)}\, a\, \mathcal{R}s(\theta, \alpha_1).$$

Notice that this is almost the inverse ridgelet transform with ridgelet profile κ. Let us now focus on the 1-D integral over α_1 on the right-hand side and apply the Fourier convolution theorem:

$$\tilde{s}(\vec{x}) = \int_{\mathbb{R}^+_*} \frac{da}{a^3} \int_0^{2\pi} d\theta \, a \int_{\mathbb{R}} d\zeta \, \mathbb{F}\Big[\mathcal{R}s(\theta, \cdot)\Big](\zeta) \, \hat{\chi}(a\zeta\vec{e}) \, e^{i\zeta(\vec{e}\cdot r_{-\theta}\vec{x})}$$

$$= c_\chi \int_0^{2\pi} d\theta \int_{\mathbb{R}} d\zeta \, e^{i\zeta(\vec{e}\cdot r_{-\theta}\vec{x})} \, \zeta \cdot \mathbb{F}\Big[\mathcal{R}s(\theta, \cdot)\Big](\zeta),$$

where $\mathbb{F}\Big[\mathcal{R}s(\theta, \cdot)\Big]$ stands for the 1-D Fourier transform of $\mathcal{R}s$ with respect to its second argument. Noticing that this last equation is exactly the inverse Radon transform completes the proof. The admissibility condition (11.4) can also be rephrased in terms of the 1-D profile κ using (11.6):

$$c_\kappa = \int_{\mathbb{R}} du \, \frac{\hat{\kappa}(u)}{u^2} < \infty. \tag{11.7}$$

This condition is a little bit stronger than the usual ridgelet admissibility condition of Theorem 11.1.1 because the analysis ridgelet used is a very singular object (a δ function).

11.1.3 Ridgelets and 2-D singularities

One of the most interesting properties of ridgelets lies in their ability to sparsify objects with linear singularities in higher dimensions. It is well known that wavelet (bi)-orthogonal bases are ideal for efficiently representing piecewise smooth functions [138]. To illustrate this, consider the example of a univariate function f whose smoothness is measured by its belonging to the Sobolev class H^s, s a non-negative integer. Expressing f in a wavelet basis and keeping only the largest n coefficients would yield an approximation error of order $O(n^{-s})$:

$$\|f - \tilde{f}_n\|_2 \sim O(n^{-s}).$$

This result is not particularly impressive and could also be obtained in the Fourier basis, for instance, but a little miracle happens when one adds singularities to f. In this case indeed, the approximation error of trigonometric series reduces to:

$$\|f - \tilde{f}_n\|_2 \sim O(n^{-1/2}),$$

which means that Fourier techniques don't see the local regularity of the signal. This is the celebrated Gibbs effect. What about wavelet expansions? This is where the miracle occurs: wavelet bases allow for representing these singular objects with a rate of approximation of order $O(n^{-s})$. In a nutshell, the nonlinear approximation power of wavelets is not spoiled by the presence of singularities [Mal99]. Moreover, this result is optimal, in the sense that no other orthonormal basis would yield a better rate of approximation. Thus wavelets are the most economical system for representing piecewise smooth objects and this further explains their success in statistical estimation and compression [148].

Unfortunately, when moving to higher dimensions, the nice properties of wavelet orthonormal bases degrade suddenly. Consider the simple example of a 2-D function on the unit square $[0, 1] \times [0, 1]$, that is singular along a line, but regular everywhere else. Expanding in an orthonormal basis of compactly supported wavelets, it is easy to check that there will be $O(n)$ coefficients greater than $1/n$. To see this, it suffices to notice that the number of wavelets overlapping the singularity at scale 2^j is of order $O(2^j)$. Such an amount of significant coefficients drives the rate of nonlinear approximation to $O(n^{-1/2})$, independently of the regularity of the signal far from the singular ridge! In other words, wavelets face a curse of dimensionality: they are unable to catch the regularity of the object. This should not be a surprise, since discrete wavelets were mainly developed in 1-D and extended to 2-D by tensor product. They use an isotropic refinement mechanism which only allows for catching point-like singularities. But in dimensions larger than one, there is a whole geometry of possible singular behaviors.

What is the advantage of ridgelets? The main difference between wavelets and ridgelets lies in the fact that the latter uses an *anisotropic* refinement mechanism. Consider a linear singularity again. Once the orientation of the ridgelet has been tuned to that of the line, one can arbitrarily refine in the orthogonal (singular) direction and obtain a nice estimation of the singularity, essentially as good as the corresponding wavelet estimation for this 1-D profile. Using 2-D wavelets would require us to pave the whole line, while a few ridgelets (ultimately one if using the correct angle) will do the job. This translates into a rate of nonlinear approximation of order $O(n^{-s/2})$ for objects which are s-Sobolev smooth away from a linear singularity: ridgelets are nearly optimal at representing such objects [95].

11.1.4 The curvelet transform

When turning to efficiently representing natural images with wavelets, one faces the problem that typical images contain curved singularities, namely, contours. Wavelet compression algorithms suffer from the curse of dimensionality explained above and this is the main cause of annoying ringing artifacts in low bit rate image compression, for example. Ridgelets alone are not sufficient to deal efficiently with curved singularities, but intuitively a correct localization of ridgelet elements might help. This heuristic led Candès and Donoho [96,97] to introduce a new multiscale system particularly well suited for curved singularities: *curvelets*.

The first step is to localize the ridgelet transform. This can be done by building a smooth partition of the image into dyadic squares of size $2^s \times 2^s$. These squares are then translated in order to cover the whole image. Following [96,346] we denote by $\mathcal{Q}_s = \mathcal{Q}_s(k_1, k_2) = [k_1/2^s, (k_1 + 1)/2^s) \times [k_2/2^s, (k_2 + 1)/2^s)$ such a square at scale s. In practice we use smooth partitioning windows w_s essentially supported on $\mathcal{Q}_s$ in order to build a partition of identity:

$$\sum_{k_1, k_2} w_s^2 = 1.$$

We also denote by T_{Q_s} the transport operator:

$$\left(T_{Q_s} f\right)(x, y) = 2^s f\left(2^s x - k_1, 2^s y - k_2\right).$$

It was then shown in [97] that an image f can be reconstructed as

$$f = \sum_{k_1, k_2} \int_\Gamma \frac{da}{a^3}\, d\theta\, db\; \langle w_s T_{Q_s} \psi_{a,\theta,b} | f \rangle\; w_s T_{Q_s} \psi_{a,\theta,b}. \qquad (11.8)$$

Equation (11.8) shows that any signal can be expressed as a superposition of elements of the form $w_s T_{Q_s} \psi_{a,\theta,b}$, that is, ridgelets localized in the square Q_s. Now letting the scale 2^s of the partition vary would yield a tremendously overcomplete dictionary of localized ridgelets, and no reconstruction formula would exist. But the inner refinement scale $2^s a$ and the localization scale 2^s should not be independent. Indeed consider a C^2 curve $y(x)$ and represent it by its Taylor series:

$$y(x) \sim y(0) + x y'(0) + \frac{x^2}{2} y''(0) + O(x^3).$$

Using the translation and rotation degrees of freedom of ridgelets and neglecting higher order terms, we can always assume the reduced form:

$$y(x) \sim \frac{\kappa}{2} x^2,$$

where $\kappa = y''(0)$ is the curvature at the analysis point. Such a curve has a natural parabolic scaling: dilating along the x axis by a factor a amounts to dilating along the y axis by a factor a^2. Furthermore, if we model a localized basis function by a rectangular box aligned along these axes and use it to approximate the curve, it should have an aspect ratio given by

$$\text{width} \sim \frac{\kappa}{2} \text{length}^2.$$

We will see later on that trying to find an economical representation of simple image models somehow forces us to use building blocks with the same aspect.

Restricting our attention to signals smooth away from C^2 edges, the ideal representation would thus consist in a system looking like localized ridgelets, but obeying a parabolic scaling law and looking like anisotropic needles of aspect width $\sim$ length^2. At the same time, these constraints allow to reduce the redundancy of the original multiscale ridgelet dictionary by imposing some interscale orthogonality. Candès and Donoho proposed a very simple way to achieve this using band-pass filtering. Pick up a set of filters Φ and Ψ, such that $\widehat{\Phi}$ is a low-pass filter concentrated at frequencies $|\vec{k}| \leqslant 1$ and $\widehat{\Psi}(2^{2s}\vec{k})$, $s = 0, 1, 2\dots$ is a set of isotropic band-pass filters concentrated in the radial coronas $|\vec{k}| \in [2^{2s}, 2^{2s+2}]$. Furthermore, we impose the following partition of frequencies:

$$|\widehat{\Phi}(\vec{k})|^2 + \sum_{s=0}^{+\infty} |\widehat{\Psi}(2^{2s}\vec{k})|^2 = 1, \; \forall \vec{k}.$$

These requirements can be easily fulfilled by designing a dyadic wavelet transform, as described in Section 2.4.4. The full algorithm for computing the curvelet transform reads then:

- Subband decomposition using Φ and the set of $\Psi(2^{-2s}\vec{x})$.
- Smooth partitioning of each subband into squares of scale 2^s.
- Localized ridgelet analysis, as described above.

Putting it all together, we end up with an analyzing dictionary consisting of highly anisotropic needle-like probes. Each probe has an aspect ratio satisfying width $\sim$ length^2 and due to the combination of band-pass filtering and smooth partitioning, the whole system exhibits a parabolic scaling law well suited for representing curved singularities. Fast algorithms, based on fast digital Radon transforms or fast slant stack, exist. We refer the interested reader to [96,97,346] for more details. Curvelets provide a near optimal rate of nonlinear approximation for images that are smooth away from $\mathcal{C}^2$ edges.

11.1.5 Other wavelet-like decompositions

Besides the ridgelet and the curvelet transforms, several new types of wavelet-like representations have emerged recently. Here again the leitmotiv is to adapt the decomposition to the local geometry of images, in order to circumvent the limitations of the DWT. We may quote, for instance:

- *Contourlets*

  The aim of this transform is again to construct an efficient linear expansion of 2-D signals, which are smooth away from discontinuities across smooth curves. Unlike curvelets, however, this is a purely discrete construction, due to Do and Vetterli [142], that combines a standard multiscale decomposition (Laplacian pyramid) and a directional filter bank. The result is an image expression using basic elements like contour segments – hence the name – and it turns out to be quite efficient.

- *Complex wavelets*

  In the standard orthogonal DWT, and in the biorthogonal scheme as well, filter coefficients are almost always taken as real. This results in a number of inconvenients, for instance, lack of translation covariance ("shift invariance"), nonsymmetric wavelets, etc. These difficulties can be circumvented, however, by admitting *complex* coefficients. The idea was pioneered by Lina and Mayrand [256], for designing a basis of orthogonal, symmetric wavelets on the line or a finite interval. More recently, Kingsbury [243,244] came back to the same idea, in a different context, namely, trying to achieve approximate shift invariance. His so-called Dual Tree DWT consists in omitting the down-sampling by 2 at each level (or at least, at the first level) of the DWT pyramid, thus obtaining a redundant transform. The structure is essentially equivalent to having two parallel fully decimated trees (hence the name "dual tree"). Since the two sets of coefficients may be taken as real and imaginary parts of a single coefficient, we do have a complex WT, and it has good shift invariance properties. More interesting, the construction extends to 2-D and yields both approximately rotation-invariant wavelets, and wavelets with good directional properties (somewhat related to Simoncelli's steerable filters [341,342] described in Section 2.7). Once again, we verify that the advantages of redundancy outweigh its drawbacks! Section 11.3 below will go further in the same direction.

Other new methods have been proposed for overcoming the difficulties of the DWT. For instance, *wavelet footprints* is a concept designed by Dragotti and Vetterli [153] for allowing a perfect reconstruction of non-band-limited functions. The basic tool here is the WT of a single Dirac peak, the footprint being the set of wavelet coefficients lying in the support of the wavelets ψ_{jk} (there are finitely many of these at each scale, since ψ is supposed to have compact support). The technique is quite efficient but, since it is purely 1-D, we will not go any further in its description.

In order to get a feeling of the vitality of these new approaches generalizing wavelets, we refer the reader to the proceedings of the last SPIE Wavelet conference (San Diego, 2003) [Un03]. Each one of the approaches mentioned above is illustrated by several papers, even a dedicated parallel session!

11.2   Rate-distortion analysis of anisotropic approximations

Let us now turn to the analysis of these new wavelet-like decompositions for image compression. Our aim in this section is to perform a rather simple theoretical study of the asymptotic performances of anisotropic decompositions for a class of "toy" images called the *horizon model*. This model is composed of images defined on the unit square $[0, 1]^2$, and such that

$$I(x_1, x_2) = 1_{x_2 \geqslant y(x_1)} \qquad 0 \leqslant x_1, x_2 \leqslant 1, \tag{11.9}$$

where $y(x_1) \in C^p$ is p-times continuously differentiable and has finite length. One way to represent this image is through a quadtree decomposition, which is in fact a toy model for wavelets. Do, Dragotti, Shukla and Vetterli [143] already demonstrated that the rate-distortion behavior of this model is characterized by a power law decay of the form $D(R) \simeq R^{-1}$, where the rate R is the number of bits used to represent the image and the distortion D is measured by the mean square error with respect to the original object. We will now show that using some anisotropic subdivision clearly improves this result [168].

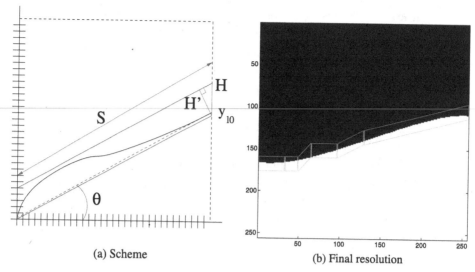

(a) Scheme (b) Final resolution

Fig. 11.3. Anisotropic quadtree with rotation.

11.2.1 An anisotropic quadtree

Let us highlight now how we can build a recursive partition of an object belonging to the horizon model. Let us take a closer look at the edge curve defined in $[0, 1]^2$ and join the two extreme points with a line that represents its average slope. This line can be then moved up and down in such a way that it does not cross the edge anymore. This defines a rectangular box that models a coarse basis function. This procedure is repeated iteratively, continuing to split the x axis inside the previous box in a dyadic way (see Figure 11.3(b)). As can readily be noticed, the x-partition is a fixed dyadic grid, while the y-partition basically depends on the edge. Let us now compute the distortion, that is, the error we commit when replacing the fine edge by this crude approximation. At each iteration j $(0 \leqslant j \leqslant J)$ and for each box k, the distortion is limited by the area of the box that encloses the edge (see Figure 11.3(a)):

$$D_j^k \leqslant S_j^k H_j'^k = S_j^k H_j^k \cos \theta = 2^{-j} H_j^k. \tag{11.10}$$

Inside every box, the edge function will be approximated by its second-order Taylor expansion at the central point of the partition, taking as initial partition the unit interval. Defining x_- as the lowest point on the x axis which is inside the interval to be analyzed and x_+ as the highest one, the coordinates of the two extreme points of the curve, quantized on a dyadic grid, will be $(x_-, Q[y(x_-)])$ and $(x_+, Q[y(x_+)])$. The line which joins these two points is:

$$y_{LQ}(x) = Q[y(x_-)] + \frac{Q[y(x_+)] - Q[y(x_-)]}{x_+ - x_-}(x - x_-), \tag{11.11}$$

where $Q[\cdot]$ stands for uniform quantization. The parallelogram cannot cross the edge, therefore its superior and inferior distances to the line are:

$$d_+ = \max\{0, \sup(y - y_{LQ})\} \geq 0$$
$$d_- = \max\{0, \sup(y_{LQ} - y)\} \geq 0. \tag{11.12}$$

Then the height of the parallelogram confining the edge will be:

$$H = Q[d_+ + d_-]. \tag{11.13}$$

Three cases have to be considered: d_+ and d_- are both larger than zero, one of them is equal to zero, and finally $d_+ = d_- = 0$. The distortion equals at most the area of the parallelogram that contains the edge, as already shown in (11.10). So, when the evolution of H with the number of bits is found, the evolution of the distortion as a function of the iteration number will be known as well.

11.2.1.1 Case $d_+ > 0$ and $d_- > 0$

Let us first compute the distances d_+ and d_- in order to determine H. If x_{d_+} is the point on the x axis where $y_{LQ} - y$ is maximum, we get:

$$d_+ = y(x_{d_+}) - y_{LQ}(x_{d_+}). \tag{11.14}$$

The above expression, when approximating the curve and $y(x_-)$ of (11.11) by its second-order Taylor expansion at the central point of the interval being analyzed, turns into

$$d_+ = \left[y'\left(\frac{x_+ + x_-}{2}\right) - \frac{y(x_+) - y(x_-)}{x_+ - x_-} \right] (x_{d_+} - x_-)$$
$$\pm 2^{-J}\left(\frac{x_{d+} - x_-}{x_+ - x_-}\right) \pm \frac{2^{-J}}{2}$$
$$+ \frac{1}{2}y''\left(\frac{x_+ + x_-}{2}\right)\left[\left(x_{d_+} - \frac{x_+ + x_-}{2}\right)^2 - \left(\frac{x_+ - x_-}{2}\right)^2\right] \tag{11.15}$$
$$+ O\left(\left(x_{d_+} - \frac{x_+ + x_-}{2}\right)^3\right) + O\left(\left(\frac{x_- - x_+}{2}\right)^3\right).$$

It is possible to show that the first term in (11.15) is $O\left(2^{-3j}\right)$ [168] and that the fourth term is bounded by $\frac{1}{2}\left|y''\left(\frac{x_+ + x_-}{2}\right)\right| 2^{-2(j+1)}$. This expression is related to the second derivative computed in the middle point of each interval k at each iteration j. From now on, to simplify the notation, it will be referred to as K_j^k:

$$K_j^k = \left| y''\left(\left(k + \frac{1}{2}\right) 2^{-j}\right) \right|, \tag{11.16}$$

with $0 \leqslant k \leqslant 2^j - 1$. Since the curvature of a function $y(x)$ is $(1 + y'^2)^{-3/2} y''(x)$, K can be considered as its approximation. As the edge is a C^2 curve, the set of K_j^k is bounded:

$$\beta = \max_{0 \leqslant k \leqslant 2^J - 1} K_j^k < \infty. \tag{11.17}$$

From (11.15) it follows that the asymptotic behavior of d_+ is given by

$$d_+ \sim \frac{1}{2} K_j^k \, 2^{-J - \log_2 \beta} + \frac{3}{2} \cdot 2^{-J}. \tag{11.18}$$

In fact, the other terms are $O(2^{-3j})$ (one order of magnitude smaller), so they can be rejected when computing the asymptotic behavior. Finally d_- can be found with exactly the same method and has identical behavior. The iterative algorithm is going to stop when the requested resolution is reached, i.e., when $H_j = 2^{-J}$. Substituting in (11.13), the number of iterations needed to reach this parallelogram height can be obtained as a function of the resolution and of the curvature of the edge:

$$j_{\text{stop}} = \max \left\{ 0, \left\lceil \frac{1}{2} \left(J + \lceil \log_2 \beta \rceil + 1 \right) \right\rceil \right\}. \tag{11.19}$$

Notice that now the parallelogram has the a/a^2 anisotropy present in curvelets [96]:

$$\frac{\text{width}}{\text{height}} = \frac{2^{-j_{\text{stop}}}}{2^{-J}} = 2^{-\log_2 \beta} \frac{2^{-\frac{J}{2}}}{2^{-J}} \sim \frac{\text{width}}{\text{width}^2}. \tag{11.20}$$

The final distortion ($j = j_{\text{stop}}$) is

$$D = \sum_{k=0}^{2^{j_{\text{stop}}} - 1} (d_+ + d_-) \, 2^{-j_{\text{stop}}}$$

$$= 2^{-2 j_{\text{stop}}} \left(\sum_{k=0}^{2^{j_{\text{stop}}} - 1} K_{j_{\text{stop}}}^k 2^{-j_{\text{stop}}} + \sum_{k=0}^{2^{j_{\text{stop}}} - 1} 3 \cdot 2^{-j_{\text{stop}}} \right).$$

On the right-hand side of this equation, the second sum gives a constant, while the first, when $j_{\text{stop}} \to \infty$, converges to the Riemann integral of the second derivative of the curve, which can be seen as an approximation of the total variation TV of the edge, with the only difference that we have a sum of K_j^k instead of the Riemann integral of the curvature. Calling it $\widetilde{TV}$, the final expression of the distortion turns into

$$D \sim \left(\widetilde{TV} + 3 \right) \cdot 2^{-2 j_{\text{stop}}} \sim \left(\widetilde{TV} + 3 \right) \cdot 2^{-J - \log_2 \beta}. \tag{11.21}$$

Each rotated box is coded by means of H_j and a left and a right vertex. At iteration j, as at least two of the vertices of the following parallelogram will be inside the previous one, the number of bits needed to code one vertex of the box k will evolve as follows:

$$N_{\text{bits\_v}}^k = J - 2(j - 2) + \lceil \log_2 \left(K_{j-1}^k \right) \rceil. \tag{11.22}$$

Therefore, the total rate will be:

$$R = \sum_{j=0}^{j_{\text{stop}}} (2N_{\text{bits\_V}} + N_{\text{bits\_H}}) \cdot 2^j, \tag{11.23}$$

where $N_{\text{bits\_H}} = N_{\text{bits\_V}}$ is the number of bits needed to code the height of each box. Simplifying,

$$R \leqslant 3J + 2 + \sum_{j=1}^{j_{\text{stop}}} \left((J - 2(j-1) + \lceil \log_2 \beta \rceil) \cdot 3 + 2 \right) \cdot 2^j.$$

This is an arithmetico-geometrical progression, whose sum, for J large enough, can be approximated by [Gra94]:

$$R \sim 2^{\frac{1}{2}(J + \lceil \log_2 \beta \rceil)}. \tag{11.24}$$

Combining this equation with (11.21) we obtain the asymptotic R-D behavior:

$$D(R) \sim \left(\widetilde{TV} + 3 \right) \cdot 2^{-2\log_2 R} \sim \left(\widetilde{TV} + 3 \right) \cdot R^{-2}. \tag{11.25}$$

11.2.1.2 Case $d_+ > 0, d_- = 0$ or vice versa

This case turns to have the same R-D as the previous one, because the evolution of the rectangle height is led by the positive distance.

11.2.1.3 Case $d_- = d_+ = 0$

This case is very favorable to our coding scheme, because it means that the minimum distortion requirement is reached with just one iteration. The parallelogram height will be $H = 2^{-J}$, and the rate will consist of the bits needed to code the two vertices and the box height, $R = 3J$. This makes a R-D behavior coherent with the results obtained in [143]:

$$D(R) = 2^{-R/3}. \tag{11.26}$$

The anisotropic quadtree with rotation has a good R-D decay, but it has the drawback that the reconstructed edge may lose its original continuity. The introduction of refinement solves this problem. In this case, the procedure is exactly the same as that explained in the previous section but, when the minimum resolution has been achieved, a refinement is performed inside the last resolution rectangle by splitting the x axis into intervals of size 2^{-J}. The effect of adding refinement in the anisotropic quadtree with rotations does not change the slope of the R-D decay, but it allows a better PSNR given a certain rate, shifting the R-D line to the left (see Figure 11.4). Following the procedure that has been adopted previously, the distortion found for the case $d_+ > 0$ and/or $d_- > 0$ is

$$D \sim \widetilde{TV} \cdot 2^{-2J} + 3 \cdot 2^{-J-M}. \tag{11.27}$$

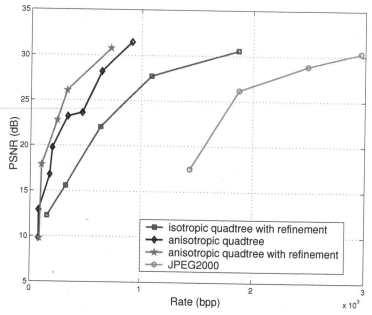

Fig. 11.4. Comparison among JPEG2000, isotropic quadtree with refinement and anisotropic quadtree for an image of 1024×1024 pixels.

The rate now has to take into account the number of refinements performed inside each parallelogram, the number of parallelograms to refine and the number of bits to perform the refinement. Including the refinement bits in (11.24) and taking $M = J$, we find

$$R = 2^{\frac{1}{2}(J + \log_2 \beta)} + M \cdot 2^J. \tag{11.28}$$

From (11.27) and (11.28), it is easy to deduce the final R-D expression:

$$D(R) \sim \widetilde{TV} \cdot \frac{\log_2 R}{R^2}. \tag{11.29}$$

The R-D found in the case where both distances vanish (i.e., the edge is a straight line) is very similar to the one obtained in the case without refinement:

$$D(R) = 2^{-R/2}. \tag{11.30}$$

We display in Figure 11.4 a comparison between this scheme, the classical isotropic quadtree that models wavelet expansions for the horizon model, and the isotropic quadtree with refinement studied by Do *et al.* [143]. For the sake of comparison, we also include the results obtained by using a JPEG2000 encoder [237] on the image, although it is evident that such an encoder is not well suited for bi-level images. It should be seen as a reference.

11.3 Further into redundancy: sparse approximations in redundant dictionaries

11.3.1 Redundant approximations

Browsing this book in search of guiding lines, the reader may identify several candidates. An obvious one is given by wavelets, in all different incarnations presented before, but there is an alternative candidate that opens new perspectives: flexible redundant representations. It is somewhat remarkable that wavelet representations stemmed from continuous, i.e., fully redundant expansions, but gained success through the widely used orthogonal bases of wavelets. At the very core of the wavelet foundation lies the idea of obtaining a sparse approximation of the signal. This is really the key idea that leads to efficient techniques in such diverse fields as compression, statistical estimation or even scientific computing: concentrating the interesting characteristics of the signal in few coefficients. We have already shown that, when moving to higher dimensions, classical wavelet bases fail to give very sparse representations because they cannot cope with the geometry that is inherent to the nature of images for example. This is the reason why curvelets were invented and their most promising property lies in their ability to provide sparse representations of 2-D piecewise smooth images with edges. On the other hand, we have seen that including redundancy in the representation brings robustness with respect to noise and more flexibility in the design of wavelets. A natural question arises then at this point: is it possible to obtain highly sparse representations of signals using redundant families of wavelet-like functions? This question is presently a very hot topic in approximation theory and, in the following, we intend to give only a flavor of that boiling field for the interested reader, pointing to precise technical references when necessary.

Let us first pose the general problem more clearly. We want to obtain sparse approximations of signals defined as elements of a given Hilbert space $\mathcal{H}$. In what follows, we will mostly restrict ourselves to the finite dimensional case, i.e., $\mathcal{H} = \mathbb{R}^N$ or $\mathcal{H} = \mathbb{C}^N$. The infinite dimensional case leads to some technical difficulties, at least for a part of the algorithms described below as we will see later.

Definition 11.3.1 *A dictionary $\mathcal{D}$ in $\mathcal{H}$ is a collection of unit norm vectors, called atoms, $\{g_\gamma, \ \gamma \in \Gamma\}$, that spans $\mathcal{H}$, where Γ is a given index set.*

The index γ could be a mere number used to identify an atom but in most cases it carries a particular significance. The atom g_γ could for example be well localized around the position $\gamma = (t, \omega)$ in the time–frequency plane. Another possiblity would be to have atoms that are well concentrated around a particular time instant such that

$$|\langle g_x, g_y \rangle| \leqslant C \, (1 + |x - y|)^{-s}, \ s \in \mathbb{R}_*^+, \ s > 1.$$

A dictionary composed of coherent states, for instance, is localized by the reproducing kernel of the CS system:

$$\langle g_\gamma, g_{\gamma'} \rangle = \mathcal{K}_g(\gamma, \gamma').$$

We seek to find a very sparse representation of any signal $s \in \mathcal{H}$ in terms of atoms in $\mathcal{D}$, in other words, we would like to find a set of indexes and associated coefficients such that the following equality holds in the strong sense:

$$s = \sum_{m=0}^{M-1} c_m g_{\gamma_m}. \tag{11.31}$$

In the following we will call this representation a M-sparse approximation. It may happen that this definition is too restrictive and we will also consider the slightly relaxed definition:

$$s_M = \sum_{m=0}^{M-1} c_m g_{\gamma_m}, \quad \|s_M - s\| \leqslant \epsilon, \tag{11.32}$$

where s_M is called (ϵ, M)-sparse. Our ultimate goal would be to find the best, that is the sparsest, possible representation of the signal. In other words, we would like to solve the following problem:

$$\text{minimize } \|c\|_0 \text{ subject to } s = \sum_{m=0}^{M-1} c_m g_{\gamma_m},$$

where $\|c\|_0$ is just the number of nonzero entries in the sequence $\{c_n\}$. If the dictionary is well adapted to the signal, there are high hopes that this kind of representation exists and would actually be sparser than a nonlinear wavelet-based approximation.

In [128], Davis *et al.* seemed to give a fatal blow to such a hope by showing that finding the best (ϵ, M)-sparse approximation in an arbitrary dictionary leads to a NP-hard problem! Nevertheless, this result holds for an *unconstrained* dictionary and it is not hopeless to look for such a solution given a particular dictionary, as we will see now.

11.3.2 Algorithms

11.3.2.1 *N*-term approximations

When dealing with orthonormal bases, we have seen that best M-term approximations are obtained by means of a fairly simple algorithm: the best approximant is found by first looking for the M strongest Fourier coefficients $|\langle \phi_n, s \rangle|$ and then constructing:

$$\tilde{s}_M = \sum_{m \in \Lambda} \langle \phi_m, s \rangle \phi_m,$$

where Λ is an index set of cardinality M that corresponds to the best coefficients selected above. Now that we have the freedom of considering arbitrary dictionaries, such a simple procedure does not hold anymore and other types of algorithms have to be considered. In the following we will mainly consider two classes of such algorithms: greedy algorithms [137] and basis pursuit [102].

11.3.2.2 Greedy algorithms

Greedy algorithms iteratively construct an approximant by selecting the element of the dictionary that best matches the signal at each iteration. The pure greedy algorithm is known as *Matching Pursuit* (MP). Assuming as before that all atoms in $\mathcal{D}$ have norm one, we initialize the algorithm by setting $R_0 = s$ and we first decompose the signal as

$$R_0 = \langle g_{\gamma_0}, R_0 \rangle g_{\gamma_0} + R_1.$$

Clearly g_{γ_0} is orthogonal to R_1 and we thus have

$$\|R_0\|^2 = |\langle g_{\gamma_0}, R_0 \rangle|^2 + \|R_1\|^2.$$

If we want to minimize the energy of the residual R_1 we must maximize the projection $|\langle g_{\gamma_0}, R_0 \rangle|$. At the next step, we simply apply the same procedure to R_1, which yields

$$R_1 = \langle g_{\gamma_1}, R_1 \rangle g_{\gamma_1} + R_2,$$

where g_{γ_1} maximizes $|\langle g_{\gamma_1}, R_1 \rangle|$. Iterating this procedure, we thus obtain an approximant after M steps:

$$s = \sum_{m=0}^{M-1} \langle g_{\gamma_m}, R_m \rangle g_{\gamma_m} + R_M,$$

where the norm of the residual (approximation error) satisifies

$$\|R_M\|^2 = \|s\|^2 - \sum_{m=0}^{M-1} |\langle g_{\gamma_m}, R_m \rangle|^2.$$

Some variations around this algorithm are possible. An example is given by the weak greedy algorithm, which consists in modifying the atom selection rule by allowing to choose a slightly suboptimal candidate:

$$|\langle R_m, g_{\gamma_m} \rangle| \geqslant t_m \sup_{g \in \mathcal{D}} |\langle R_m, g \rangle|, \quad t_m \leqslant 1.$$

One can easily show that Matching Pursuit converges exponentially in the strong topology in finite dimension, see [263] for a proof. Unfortunately this is not true in general in infinite dimension, even though this property holds for particular dictionaries [365]. However, DeVore and Temlyakov [137] constructed a dictionary for which even a good signal, i.e., a sum of two dictionary elements, has a very bad rate of approximation: $\|s - s_M\| \geqslant CM^{-1/2}$. In this case a very sparse representation of the signal exists, but the algorithm dramatically fails to recover it!

A clear drawback of the pure greedy algorithm is that the expansion of s on the linear span of the selected atoms is not the best possible one, since it is not an orthogonal projection. Orthogonal Matching Pursuit [304,127] solves this problem by recursively orthogonalizing the set of selected atoms using a Gram–Schmidt procedure. The best M-term approximation on the set of selected atoms is thus computed and the algorithm can be shown to converge in a finite number of steps, but at the expense of a much bigger computational complexity.

11.3.2.3 Basis pursuit

As previously stated, the problem of finding a sparse expansion of a signal in a generic dictionary leads to a daunting NP hard combinatorial optimization problem. In order to overcome this, Chen *et al.* [102] proposed to solve the following slightly different problem:

$$\text{minimize } \|c\|_1 \text{ subject to } s = \sum_{m=0}^{M-1} c_m g_{\gamma_m}.$$

Minimizing the ℓ_1 norm helps finding a sparse approximation because it prevents diffusing the energy of the signal over a lot of coefficients. While keeping the essential property of the original problem, this subtle modification leads to a tremendous change in the very nature of the optimization challenge. Indeed, this ℓ_1 problem, called *Basis Pursuit* or BP, is a much simpler linear programming problem, that can be efficiently solved using, for example, interior point methods. Another very important difference with respect to greedy algorithms is that BP seeks for a *global* solution to the problem (while MP constructs a step by step approximation). What are the performances of BP? In particular can BP recover a sparse approximation of a signal if the latter exists? As we will see now, it turns out that the answer to this question is positive in certain circumstances.

In order to obtain constructive results concerning highly nonlinear approximation in dictionaries, we must first restrict our attention to *particular* dictionaries. In the following we will thus focus on incoherent dictionaries, i.e., dictionaries such that

$$M = \sup_{i,j} |\langle g_i, g_j \rangle|, \tag{11.33}$$

where M is the coherence of $\mathcal{D}$. Coherence is another possible measure of the redundancy of the dictionary and (11.33) shows that $\mathcal{D}$ is not too far from an orthogonal basis (although it may be highly overcomplete). Let us first concentrate on a dictionary $\mathcal{D}$ that is given by the union of two orthogonal bases in $\mathbb{R}^N$, i.e., $\mathcal{D} = \{\psi_i\} \cup \{\phi_j\}$, $1 \leqslant i, j \leqslant N$. Building on early results of Donoho and Huo [149], Elad and Bruckstein have shown the following two striking theorems [160].

Theorem 11.3.1 *Let $\mathcal{D}$ be the concatenated dictionary described above with coherence M. If a signal $s \in \mathbb{R}^N$ has a representation*

$$s = \sum_i c_i g_i$$

such that $\|c\|_0 < 1/M$, this representation is the unique sparsest expansion of s in $\mathcal{D}$.

This first result shows that, although redundancy rules out uniqueness of expansions, a sufficiently sparse solution is unique provided the dictionary is incoherent! We still do not know how to find this solution though, and remember that the generic problem is NP hard. The next theorem shows that incoherence helps us in an astonishing way.

Theorem 11.3.2 *If the signal s has a sparse representation in the concatenated dictionary such that*

$$\|c\|_0 < \frac{\sqrt{2} - 0.5}{M}, \tag{11.34}$$

then the ℓ_1 minimization problem has a unique solution and it coincides with the unique solution of the original ℓ_0 problem.

This result means that you can replace the original combinatorial optimization problem of finding the sparsest representation of s by the much simpler ℓ_1 problem solved by BP. In other words the incoherence of $\mathcal{D}$ helps you solve a problem that is NP hard in general! These results have been extended to arbitrary dictionaries by Gribonval and Nielsen [202] who showed the following

Theorem 11.3.3 *Let $\mathcal{D}$ be an arbitrary dictionary of coherence M in a finite dimensional Hilbert space. If the signal s has a sparse expansion satisfying:*

$$\|c\|_0 < \frac{1}{2}\left(1 + \frac{1}{M}\right),$$

then this is the unique solution of both the ℓ_0 and ℓ_1 problems.

So far the results obtained are not constructive: they essentially tell us that if a sufficiently sparse solution exists in a sufficiently incoherent dictionary, it can be found by solving the ℓ_1 optimization problem. Practically, given a signal, one does not know whether such a solution can be found and the only possibility at hand would be to run BP and check *a posteriori* that the algorithm finds a sufficiently sparse solution.

Greedy algorithms, discussed in the previous section, offer constructive procedures for computing highly nonlinear N-term approximations. Although the mathematical analysis of their approximation properties is complicated by their nonlinear nature, interesting results are emerging (see for example [190]). Let us illustrate this by an example dealing with Orthogonal Matching Pursuit (OMP):

Theorem 11.3.4 *Let s_{opt} be the optimal (sparsest) approximation of s in a dictionary $\mathcal{D}$ of coherence μ. Assume that $M \leq \frac{1}{3}\mu^{-1}$. Then OMP generates M-term approximants which satisfy*

$$\|s - s_M\|_2 \leqslant \sqrt{1 + 6M} \|s - s_{\text{opt}}\|.$$

Other modified greedy algorithms exist and show better performances, but their description is beyond the scope of this book.

11.3.3 Applications

Highly nonlinear approximation techniques provide us with the basic ingredient for many applications, namely, a sparse data representation. Stemming from this point, numerous applications are flourishing and, since these techniques are just emerging, we expect more successes in the near future. Some typical examples where redundant approximations have been applied include source separation [203] and, of course, data compression [296, 139, 320]. We will focus on the latter in order to give a flavor of the advantages of using redundant approximations.

11.3.3.1 Image compression using redundant dictionaries

As already explained previously, 2-D orthogonal bases of wavelets suffer from a *curse of dimensionality*: the nonlinear approximation properties of wavelets are severely impaired by their lack of geometrical adaptivity. We already discussed ridgelets and curvelets that try to solve this problem, but an obvious alternative solution would be to decompose the image over a redundant dictionary built with some structure.

A good dictionary can be built by acting on a generating function of unit L^2 norm by means of a family of unitary operators U_γ:

$$\mathcal{D} = \left\{ U_\gamma, \ \gamma \in \Gamma \right\}, \tag{11.35}$$

for a given set of indexes Γ. Basically this set must contain three types of operations.
- Translations $\vec{b}$, to move the atom all over the image.
- Rotations θ, to orient the atom locally along contours.
- Anisotropic scaling (a_1, a_2), to adapt to contour smoothness.

A possible action of U_γ on the generating atom g is thus given by:

$$U_\gamma g = \mathcal{U}(\vec{b}, \theta) D(a_1, a_2) g \tag{11.36}$$

where $\mathcal{U}$ is a representation of the Euclidean group,

$$\mathcal{U}(\vec{b}, \theta) g(\vec{x}) = g\left(r_{-\theta}(\vec{x} - \vec{b}) \right), \tag{11.37}$$

r_θ is the usual rotation matrix, and D acts as an anisotropic dilation operator:

$$D(a_1, a_2) g(x, y) = \frac{1}{\sqrt{a_1 a_2}} g\left(\frac{x}{a_1}, \frac{y}{a_2} \right). \tag{11.38}$$

It is easy to prove that such a dictionary is overcomplete using the fact that, when $a_1 = a_2$ one gets the 2-D similitude group of $\mathbb{R}^2$ and thus 2-D continuous wavelets. It is also worth stressing that, avoiding rotations, the parameter space is a group studied by Bernier and Taylor [69]. The advantage of such a parametrization is that the full dictionary is invariant under translations and rotations. Moreover, it is also invariant under isotropic scaling, that is, $a_1 = a_2$.

The choice of the generating atom g is driven by the idea of efficiently approximating contour-like singularities in 2-D. To achieve this, the atom must be a smooth low resolution function in the direction of the contour and must behave like a wavelet in the orthogonal (singular) direction. In our experiments, we chose a combination of a Gaussian and its second derivative, that is,

$$g(x, y) = (4 x^2 - 2) \exp(-(x^2 + y^2)). \tag{11.39}$$

This choice is motivated by the optimal joint spatial and frequency localization of the Gaussian kernel. We also noticed that degradations caused by truncating the MP expansion are visually less disturbing with this choice.

Based on this dictionary, a candidate compression algorithm could be the one depicted in Figure 11.5, see also [177]. The image is first recursively decomposed, by MP, in a series of coefficients and atoms chosen from a redundant dictionary. The coefficients are then quantized by means of an exponentially bounded uniform quantization method adapted to MP characteristics [178]. Coefficients and atom indexes are finally entropy coded with a context adaptive arithmetic coder, and the reverse operations are performed at decoder. Matching Pursuit provides meaningful image representations even with very short descriptions of the input signal. The reconstructed image is then successively refined by each additional bit of information.

Due to the structured nature of our dictionary, the MP stream provides spatial scalability features. Indeed, when the two scaling parameters are equal, the group law of the similitude group of $\mathbb{R}^2$ applies and allows for covariance with respect to *isotropic* scaling, rotation and translation. When the image to be decomposed is submitted to any combination of these transforms (denoted here by the group element $\gamma \equiv (\vec{b}, a, \theta)$), the indexes of the MP stream are simply transformed according to the group law:

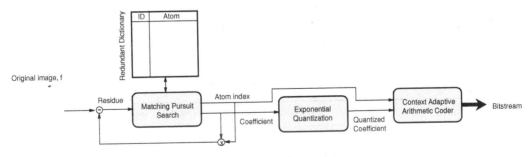

Fig. 11.5. Block diagram of the Matching Pursuit image coder.

$$U(\gamma)f = \sum_{n=0}^{+\infty} \langle g_{\gamma_n} | R_n f \rangle \, g_{\gamma \circ \gamma_n},$$ (11.40)

where $U(\cdot)$ is the familiar representation (7.40) of the similitude group SIM(2) and R_n is the nth order residual of the MP algorithm.

The decoder can thus apply those transformations to the reconstructed image by simply modifying the parameter strings of the atoms. For example, the bitstream of an image of size $X \times Y$ can be decoded at any resolution $\alpha X \times \alpha Y$ by multiplying positions and scales by the corresponding factor α and re-scaling all coefficients. The scaled image f^α is given by

$$f^\alpha = \alpha \sum_{i=0}^{N} c_{\gamma_i} g_{\gamma_i}^\alpha.$$ (11.41)

The transcoded atoms $g_{\gamma_i}^\alpha$ are given by simply scaling the original indexes γ_i in the decomposition. Atoms that are too small after transcoding are discarded, which allows to reduce further the bit rate. Note that the scaling factor α can take any real value, as long as the scaling is isotropic. Finally, image editing manipulations, such as rotating the image, or zooming on a region of interest, can be performed following the same principle. Hierarchical coding can also be efficiently implemented using this flexible image representation.

Results of spatial transcoding are summarized in Table 11.1, where a comparison with the JPEG-2000 standard clearly shows the gain of using redundant dictionaries for image coding. PSNR values are computed with reference to the original 512×512 *lena* image downsampled to 128×128 pixels. This is but one possibility for computing such a reference and other more complex techniques involving, for example, filtering and interpolation, could be adopted. However we don't think this could significantly change the comparison. A simple example of this procedure is given in Figure 11.6.

Due to the very good match between the dictionary and image structures, MP quickly recovers visually appealing image features and, specially at very low bit rates, the quality of the compressed image is much better than with wavelet techniques as can be seen on Figures 11.7 and 11.8. Moreover, MP generates naturally *progressive* streams, which means that we can obtain a lower bit-rate image by simply truncating the bit stream (i.e., considering the first B bits is more or less equivalent to discarding the lowest energy atoms). This is also illustrated on Figures 11.7 and 11.8.

Table 11.1. *Comparison of spatial scalability of the Matching Pursuit (MP) encoder and JPEG-2000; the numbers given are in (dB/bpp)*

Encoder	32×32	64×64	128×128	256×256	512×512
MP	16.7/4	18.97/2.25	30.37/1.7	26.05/0.41	25.94/0.1
JPEG-2000	16.98/6.5	19.18/4	33.8/1.7	—	—

Fig. 11.6. *lena* image of size 128×128 encoded with MP at 1.6 bpp (center), and decoded with a scaling factor of $1/\sqrt{2}$ (left) and with a scaling factor of $\sqrt{2}$ (right).

Fig. 11.7. *lena* image of size 128×128 encoded with MP at 1.7 bpp (top left), and truncated at 0.1 bpp (top right), 0.4 bpp (bottom left), and 0.8 bpp (bottom right).

Fig. 11.8. *lena* image of size 128 × 128 encoded with JPEG-2000 at 1.7 bpp (top left), and truncated at 0.1 bpp (top right), 0.4 bpp (bottom left), and 0.8 bpp (bottom right).

11.4 Algebraic wavelets

We conclude the chapter with a topic of a different nature. The term "algebraic wavelets" refers a class of wavelets obtained by replacing the usual natural numbers, and the dyadic numeration underlying the multiresolution approach, by another system of numeration. Although this is actually a generalization of the discrete WT, it provides another example of wavelets adapted to a specific geometry; hence it is not out of place in this volume.

In order to give the reader a feeling for this type of construction, we will begin with the 1-D case, and first with the simplest example, based on the *golden mean* $\tau = \frac{1}{2}(1 + \sqrt{5})$, and we shall describe it in some detail in Section 11.4.1.1, following [184] and [Ali00; Section 13.4]. Two interesting aspects emerge. One is a construction that is a genuine generalization of the standard multiresolution, as described in Section 1.5, and yields a corresponding orthonormal wavelet basis, the so-called τ-Haar basis. The other is the occurrence of a quasiperiodic structure at each multiresolution level, instead of the usual lattice structure. Then, more general cases will be indicated in Section 11.4.1.2, namely wavelet systems based on particular Pisot numbers, first studied in [185], or adapted spline wavelets, following [Ber98, 70, And02, 11].

In a second step, we will turn to the 2-D case and consider wavelet bases corresponding to particular tilings of the plane, namely 2-D τ-Haar or spline wavelets based on successive subdivisions of the so-called Penrose–Robinson triangles.

11.4.1 Algebraic wavelets on the line

11.4.1.1 τ-wavelets of Haar

In this section, we will construct the Haar basis obtained by replacing as scaling factor the usual factor of 2 by the golden mean $\tau = \frac{1}{2}(1 + \sqrt{5})$ [184]. The algebraic nature of τ, based on the equation $\tau^2 = \tau + 1$, implies the τ-adic property

$$\tau^j = \tau^{j-1} + \tau^{j-2}, \quad \forall j \in \mathbb{Z}, \tag{11.42}$$

which is similar to a certain extent to the binary splitting $2^j = 2^{j-1} + 2^{j-1}$. This fact has an immediate consequence on the form of the τ-wavelets. Whereas the discrete dyadic wavelets constructed in Section 1.5 are functions $\psi(x) \in L^2(\mathbb{R})$ such that the family

$$\{\psi_{j,k} = 2^{j/2}\psi(2^j x - k), \ j, k \in \mathbb{Z}\}$$

is an orthonormal basis in the Hilbert space $L^2(\mathbb{R})$, the τ-wavelet(s) will be function(s) $\psi_\mu \in L^2(\mathbb{R})$, μ in some finite labeling set, such that the family

$$\{\tau^{j/2}\psi_\mu(\tau^j x - b), \ j \in \mathbb{Z}\}$$

is an orthonormal basis (or at least a Riesz basis) in $L^2(\mathbb{R})$, provided the numbers b belong to an appropriate discrete number set, namely, the so-called τ-integers.

To begin with, (11.42) provides a subdivision of the unit interval into two parts

$$A = [0, 1] = \left[0, \frac{1}{\tau}\right] \cup \left[\frac{1}{\tau}, 1\right], \tag{11.43}$$

of lengths $1/\tau$ and $1/\tau^2$, respectively. Then the equation (11.43) is the starting point of an iterative sequence of subdivisions of A into intervals of the type

$$A_{j,b} = \left[\frac{b}{\tau^j}, \frac{b+1}{\tau^j}\right], \quad j \in \mathbb{N}, \, A_{0,0} = A, \tag{11.44}$$

$$A_{j,b} = A_{j+1,\tau b} \cup A_{j+2,\tau^2 b + \tau}, \tag{11.45}$$

where b is an appropriate τ-*integer* [184]. This concept derives from the system of numeration based on the irrational τ [71,325] and can be described as follows. Each nonnegative real number x can be expanded in powers of τ with coefficients ξ_l equal to 0 or 1:

$$x = \xi_j \tau^j + \xi_{j-1}\tau^{j-1} + \ldots + \xi_l \tau^l + \ldots, \quad \xi_l = 0, 1. \tag{11.46}$$

The coefficients ξ_l satisfy the relation $\xi_l \, \xi_{l-1} = 0$, which means that no partial sum like $\tau^l + \tau^{l-1}$ occurs in (11.46), since it equals τ^{l+1}. The coefficients ξ_l are computed with the aid of the so-called Rényi algorithm. First, the exponent j in (11.46) is the highest integer such that

$$\tau^j \leqslant x < \tau^{j+1}, \tag{11.47}$$

so that $\xi_j = \lfloor x/\tau^j \rfloor =$ integer part of $x/\tau^j = 1$. Then we put $r_j = \{x/\tau^j\} =$ fractional part of x/τ^j, and the other digits ξ_l, $l < j$, are determined recursively from (ξ_j, r_j):

$$\xi_l = \lfloor \tau r_{l+1} \rfloor, \quad r_l = \{\tau r_{l+1}\}. \tag{11.48}$$

In this system of numeration based on τ, the set of non-negative τ-integers is the (lexicographically ordered) strictly increasing sequence of real numbers having only non-negative powers of τ in their τ-expansion:

$$\mathbb{Z}_\tau^+ = \{0, 1, \tau, \tau^2, \tau^2 + 1, \tau^3, \tau^3 + 1, \tau^3 + \tau, \tau^4, \ldots\}. \tag{11.49}$$

The set $\mathbb{Z}_\tau$ of τ-integers is then defined by

$$\mathbb{Z}_\tau = \mathbb{Z}_\tau^+ \cup (-\mathbb{Z}_\tau^+). \tag{11.50}$$

The numbers in $\mathbb{Z}_\tau^+$ are the nodes of a quasiperiodic chain, called the Fibonacci tiling of the positive real line with two types of tiles, the long one with length 1 and the short one with length $1/\tau$. A similar quasiperiodic sequence will occur at each multiresolution level j, $\mathbb{Z}_\tau^+$ itself corresponding to $j = 0$.

We now come to the explicit construction of the τ-Haar basis of $L^2(\mathbb{R})$. For that purpose, we need a "mother" wavelet, called here a τ-Haar wavelet, namely the function on the real line $\mathbb{R}$ defined as:

$$h^\tau(x) = \begin{cases} \tau^{-1/2}, & \text{for } 0 \leqslant x \leqslant 1/\tau, \\ -\tau^{1/2}, & \text{for } 1/\tau < x \leqslant 1, \\ 0, & \text{otherwise.} \end{cases} \tag{11.51}$$

As in the usual dyadic case, the basis wavelet h^τ may be obtained from a scaling function by the action of particular elements of the affine group, under the representation $U(b, a)$ given in (6.8) (see Section 6.1.1). Here the scaling function is the characteristic function χ_A of the unit interval $A \equiv [0, 1]$ and one has (almost everywhere)

$$h^\tau(x) = \tau^{-1/2} \chi_A(\tau x) - \tau^{1/2} \chi_A(\tau^2 x - \tau)$$
$$= \left(\tau^{-1} U(0, \tau^{-1}) - \tau^{-1/2} U(\tau^{-1}, \tau^{-2})\right) \chi_A(x). \tag{11.52}$$

Using the basis wavelet h^τ, one constructs first an orthogonal basis in $L^2[0, 1]$, then combining the latter with translations by τ-integers, one obtains finally.

Theorem 11.4.1 *The system*

$$\{\tau^{j/2} h^\tau(\tau^j x - b), \ \tau^{j/2} h^\tau(\tau^j x + b + 1); \ j \in \mathbb{Z}, b \in \tau \mathbb{Z}_\tau^+\} \tag{11.53}$$

is an orthonormal basis of $L^2(\mathbb{R})$, called the τ-Haar basis.

One may note that the construction above is not based on the standard concept of multiresolution. However, the latter may be generalized to the τ-adic environment, as follows.

Definition 11.4.2 *A τ-multiresolution analysis is an increasing sequence $(V_j)_{j\in\mathbb{Z}}$ of closed subspaces of $L^2(\mathbb{R})$*

$$\ldots \subset V_{-2} \subset V_{-1} \subset V_0 \subset V_1 \subset V_2 \subset \ldots,$$

with $\bigcap_{j\in\mathbb{Z}} V_j = \{0\}$ and $\bigcup_{j\in\mathbb{Z}} V_j$ dense in $L^2(\mathbb{R})$, and such that
(1) $f(x) \in V_j$ if and only if $f(\tau^{-j}x) \in V_0$;
(2) There exists a function $\Phi \in V_0$, called a scaling function, such that the system $\{\Phi(x - b), \ \Phi(x + b + 1); \ b \in \tau\mathbb{Z}_\tau^+\} \cup \{\Phi(\tau x - b), \ \Phi(\tau x + b + 1); \ b \in \tau\mathbb{Z}_\tau^{+\,odd}\}$, where $\mathbb{Z}_\tau^{odd} = \mathbb{Z}_\tau \setminus \tau\mathbb{Z}_\tau$, is an orthonormal basis in V_0 (or at least a Riesz basis).

The main difference with respect to the standard dyadic multiresolution of Definition 1.5.1 is the condition (2), the counterpart of the invariance of V_0 under translations by integers. Indeed, one cannot assert that $f(x) \in V_0$ if and only if $f(x - b)$ and $f(x + b + 1) \in V_0$ for all $b \in \tau\mathbb{Z}_\tau^+$, since the latter is not closed under addition. Instead, $\mathbb{Z}_\tau$ has only the so-called Meyer property [Mey72,271], since

$$\mathbb{Z}_\tau + \mathbb{Z}_\tau \subset \mathbb{Z}_\tau + \left\{0, \pm\frac{1}{\tau}, \pm\frac{1}{\tau^2}\right\}. \tag{11.54}$$

Motivated by this discussion, the following definition of τ-wavelets has been proposed in [Ali00], where further details on the whole procedure may be found.

Definition 11.4.3 *A τ-wavelet is a function $\psi(x) \in L^2(\mathbb{R})$ such that the family of functions*

$$\{\psi_{j,b}(x) \equiv \tau^{j/2} \psi(\tau^j x - b), \ \psi_{j,-b-1}(x) \equiv \tau^{j/2} \psi(\tau^j x + b + 1)\},$$

where $j \in \mathbb{Z}$ and $b \in \tau\mathbb{Z}_\tau^+$, is an orthonormal basis (or at least a Riesz basis) in the Hilbert space $L^2(\mathbb{R})$.

An example of such a τ-wavelet is, of course, the τ-Haar wavelet. A less trivial example is that of τ-spline wavelets constructed by Bernuau [Ber98,70] (a full account may also be found in [And02]).

11.4.1.2   Pisot wavelets, β-spline wavelets, etc.

Of course, the procedure that was followed in the previous section could tentatively be extended to *any* real number $\beta > 1$. Indeed, for such a β, there exists a numeration system based on the so-called Rényi β-expansion of real numbers, analogous to (11.46)

$$\mathbb{R}_+ \ni x = \sum_{l=-\infty}^{j} \xi_l \beta^l \equiv \xi_j \xi_{j-1} \ldots \xi_0 \xi_{-1} \ldots, \tag{11.55}$$

where j is the largest integer such that $\beta^j \leqslant x < \beta^{j+1}$. The positive integers ξ_l take their values in the alphabet (for β noninteger) $\{0, 1, 2, \ldots, \lfloor \beta \rfloor\}$ and are computed with the so-called greedy algorithm. One defines recursively

$$\xi_j = \lfloor x/\beta^j \rfloor, \quad r_j = \{x/\beta^j\}, \text{ and for } l < j, \; \xi_l = \lfloor \beta r_{l+1} \rfloor, \quad r_l = \{\beta r_{l+1}\}. \tag{11.56}$$

The set of real numbers which have a zero fractional part in their β-expansion is called the set of β-integers and is denoted by

$$\mathbb{Z}_\beta = \{\pm(\xi_j \beta^j + \xi_{j-1}\beta^{j-1} + \ldots + \xi_1 \beta + \xi_0)\} = \mathbb{Z}_\beta^+ \cup (-\mathbb{Z}_\beta^+). \tag{11.57}$$

As in the τ-case, some configurations $\xi_j \xi_{j-1} \ldots \xi_0 \xi_{-1} \ldots$ in this definition are not possible, but the situation is clearly more complicated here.

The countable set $\mathbb{Z}_\beta$ is naturally self-similar and symmetric with respect to the origin:

$$\beta \mathbb{Z}_\beta \subset \mathbb{Z}_\beta, \quad \mathbb{Z}_\beta = -\mathbb{Z}_\beta. \tag{11.58}$$

It tiles the line with intervals separating two nearest neighbors $x_i < x_{i+1}$. The length of the intervals, $l_i = x_{i+1} - x_i \leqslant 1$, may then take its value in a finite or countably infinite set, depending on the nature of the number β. The simplest case is that where β is a Pisot–Vijayaraghavan (for short, PV or Pisot) number, i.e., an algebraic integer which is a solution of the equation

$$X^m = a_{m-1} X^{m-1} + \ldots + a_1 X + a_0, \quad a_i \in \mathbb{Z}, \tag{11.59}$$

such that all other solutions $\beta^{(i)}$ of (11.59) (the Galois conjugates of β) have a modulus strictly smaller than 1,

$$\beta^{(0)} = \beta > 1, \quad |\beta^{(i)}| < 1, \; i = 1, \ldots, m - 1. \tag{11.60}$$

Examples of Pisot numbers are $1 + \sqrt{2}$, $\tau = \frac{1}{2}(1 + \sqrt{5})$, $1 + \sqrt{3}$, $\frac{1}{2}(3 + \sqrt{5})$, and $2 + \sqrt{3}$; on the other hand, neither $\frac{3}{2}$, nor $\sqrt{2}$ is Pisot.

The crucial result is that, when β is a Pisot number, then $\mathbb{Z}_\beta$ is a self-similar tiling of the line with a *finite* set of different tiles (in general, one may need infinitely many of them). This result makes it possible to extend to Pisot numbers the algebraic approach described in the previous section and construct "Pisot wavelets". For instance, the following definition is the exact counterpart of Definition 11.4.2.

Definition 11.4.4 *A β-multiresolution analysis is an increasing sequence $(V_j)_{j \in \mathbb{Z}}$ of closed subspaces of $L^2(\mathbb{R})$*

$$\ldots \subset V_{-2} \subset V_{-1} \subset V_0 \subset V_1 \subset V_2 \subset \ldots ,$$

with $\bigcap_{j \in \mathbb{Z}} V_j = \{0\}$ and $\bigcup_{j \in \mathbb{Z}} V_j$ dense in $L^2(\mathbb{R})$, and such that
(1) $f(x) \in V_j$ if and only if $f(\beta^{-j}x) \in V_0$;
(2) There exists a function $\Phi \in V_0$, called a scaling function, such that the collection of all its suitably translated–dilated copies on the set of β-integers is an orthonormal basis (or at least a Riesz basis) in V_0.

Similarly, a definition of β-wavelets can be given that parallels Definition 11.4.3. Examples of β-wavelets are again the β-Haar wavelets and the β-spline wavelets constructed by Andrle *et al.* [And02,11]).

11.4.2   Two-dimensional algebraic wavelets

11.4.2.1   Introduction

As we mentioned already in Section 4.5.2, 2-D wavelet analysis may have interesting applications in the field of quasicrystals. The latter have characteristic dilation factors $\tau = \frac{1}{2}(1 + \sqrt{5})$ (symmetry of order 5 or 10), $1 + \sqrt{2}$ (symmetry 8), and $2 + \sqrt{3}$ (symmetry 12), and these are all Pisot numbers, as we have seen above.

As soon as quasicrystals were discovered, in 1984, one recognized the importance of geometric models like the one-dimensional Fibonacci chain (as toy model) or two- and three- dimensional Penrose tilings in the theoretical developments. The former was described in the previous section, here we shall concentrate on the 2-D case, more precisely on the τ-invariant case.

A moment's reflection on the construction of Section 11.4.1 shows that the possibility of having a multiresolution analysis rests upon the presence of a hierarchy of tilings of the line. At each resolution j, the line is covered by infinitely many nonoverlapping segments, which are all translated copies of a finite set of basic ones. In addition, going to the next finer resolution is achieved by splitting each basic segment into several smaller ones, which are contracted copies of the previous ones by a fixed scaling factor. For the usual dyadic case, there is only one basic tile, e.g. the interval $[0, 1]$, and the scaling factor is, of course, 2; in the τ-case, there are two of them, of lengths $1/\tau$ and $1/\tau^2$, respectively, and the scaling factor is τ; in the β-case, two or more, etc. This gives a hint as how to proceed in the 2-D case, we simply need a similar hierarchy of tilings (the 1-D τ-case is described in detail in [Ali00; Section 13.4]).

For the sake of precision, we first recall that a tiling of $\mathbb{R}^d$ is a covering with nonoverlapping pieces (tiles) which are all congruent (in sense of Euclidean group action, i.e., translation and rotation) to tiles belonging to a predetermined finite set of *prototiles* or *tile alphabet*:

$$\mathbb{R}^d = \cup_{i \in \mathbb{N}} P_i, \quad P_i \cong T_l, \quad T_l \in \{T_1, \cdots, T_N\}, \quad P_i \cap P_j \subset \mathbb{R}^{d'}, \quad i \neq j, \ d' < d. \quad (11.61)$$

In addition, one needs matching rules prescribing which neighbors are allowed for a given prototile. In the $d = 2$ case that we consider here, tiles will be just nonoverlapping polygons, usually triangles.

In the Penrose tiling picture [309,310], tiles represent distinct clusters of atoms, and matching rules represent atomic interactions. As a quasiperiodic alternative to crystalline lattices, the set Λ of vertices (or nodes) of a Penrose tiling is the ideal illustration of a point set with the following properties [276].

- It is a Delone (or Delaunay) set, which means it is *uniformly discrete* (the distances between any pair of points in Λ are greater than a fixed $r > 0$) <u>and</u> *relatively dense* (there exists $R > 0$ such that $\mathbb{R}^2$ is covered by balls of radius R centered at points of Λ).
- It has locally finite complexity.
- It displays repetitivity.
- It displays a rotational symmetry ($n = 5$) which is forbidden in crystallography.

Additional examples are the quasiperiodic point sets described in Section 4.5.2.2.

11.4.2.2   Penrose–Robinson tiling of the plane

The construction of 2-D τ-wavelets described here is based on a unique iterative decomposition process of the so-called Penrose–Robinson triangles, which leads to a triangular covering of the plane. At any given step of the iteration, the resulting patch of triangular tiles is a finite subset of a Penrose–Robinson tiling. The principle of this construction has been given in [186,187,189], but without consideration of uniqueness of the process. The latter was achieved in [188], that we follow.

We denote by $\mathcal{A}$ the triangle with vertices $0, 1, \xi \equiv e^{i\pi/5}$ and by $\mathcal{B}$ the triangle with vertices $0, 1, \xi^3 = e^{i3\pi/5}$ (see Figure 11.9). These two triangles, called the *Penrose–Robinson* triangles, are at the heart of five-fold planar tilings first considered by these authors [309,310]. They may be used as prototiles in the construction of several classes of nonperiodic tilings with pentagonal symmetry. An example of such a tiling given through a deflation–inflation iteration procedure (see below), is depicted in Figure 11.10. This tiling can be transformed by means of substitution rules into the famous Penrose *kites and darts* tiling given in Figure 11.11 (for details see [Gru74]).

The segmentation of the Penrose–Robinson triangles given in [188] may be formulated in terms of an action of the affine group SIM(2), following the standard pattern

$$(b, a)X = b + aX, \quad a, b \in \mathbb{C}, \ a \neq 0,$$

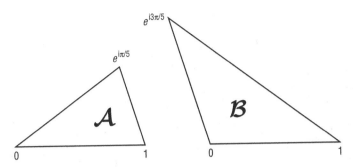

Fig. 11.9. The Penrose–Robinson triangles.

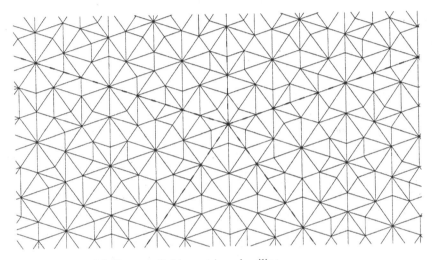

Fig. 11.10. Part of the Penrose–Robinson triangular tiling.

where X is an arbitrary set in $\mathbb{C}$, so that $(b, a)(b', a')X = b + ab' + aa'X$. (Note that here we use the complex viewpoint, identifying $\mathbb{R}^2$ with the complex plane $\mathbb{C}$, e.g. $\vec{b} \equiv b \in \mathbb{C}$, $(\lambda, r_\theta) \equiv a \in \mathbb{C}$.)

It turns out that there are two distinct ways of decomposing the triangles $\mathcal{A}$ and $\mathcal{B}$ with the self-similarity factor τ, as shown in Figure 11.12. The choice between two types of division of each triangle corresponds to a mirror symmetry with respect to the main bisector of the latter. Hence, we shall distinguish between the triangles $\mathcal{A}$ and $\mathcal{B}$, and their reflected partners, denoted $\bar{\mathcal{A}}$ and $\bar{\mathcal{B}}$, respectively.

The key to the uniqueness of the procedure is to formulate these segmentations in the language of affine transformations, using a tensor-like notation: μ^ν means that a triangle of type μ is obtained through affine action on a triangle of type ν. The indices μ, ν take the values α, β, $\widehat{\alpha}$, $\widehat{\beta}$, corresponding to the triangles $\mathcal{A}, \mathcal{B}, \bar{\mathcal{A}}, \bar{\mathcal{B}}$, respectively (tensor rule). The four possible decompositions are the following:

Fig. 11.11. Substitution rules which leads to the Kites and Darts Penrose tiling.

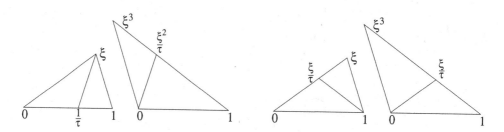

Fig. 11.12. The two decompositions of the triangles $\mathcal{A}$ and $\mathcal{B}$.

$$\mathcal{A} = \alpha^\alpha \mathcal{A} \cup \alpha^\beta \mathcal{B}, \text{ with } \alpha^\alpha = \left(\xi, \frac{-\xi^2}{\tau}\right), \ \alpha^\beta = \left(\frac{1}{\tau}, \frac{\xi^2}{\tau}\right) \tag{11.62}$$

$$\mathcal{B} = \beta^\beta \mathcal{B} \cup \beta^{\bar\alpha} \bar{\mathcal{A}}, \text{ with } \beta^\beta = \left(\frac{\xi^2}{\tau}, \frac{\xi^4}{\tau}\right), \ \beta^{\bar\alpha} = \left(1, \xi^4\right) \tag{11.63}$$

$$\bar{\mathcal{A}} = \bar\alpha^{\bar\alpha} \bar{\mathcal{A}} \cup \bar\alpha^{\bar\beta} \bar{\mathcal{B}}, \text{ with } \bar\alpha^{\bar\alpha} = \left(1, \frac{\xi^3}{\tau}\right), \ \bar\alpha^{\bar\beta} = \left(\frac{\xi}{\tau}, \frac{-\xi}{\tau}\right) \tag{11.64}$$

$$\bar{\mathcal{B}} = \bar\beta^\alpha \mathcal{A} \cup \bar\beta^{\bar\beta} \bar{\mathcal{B}}, \text{ with } \bar\beta^\alpha = (\xi^3, -\xi^3), \ \bar\beta^{\bar\beta} = \left(\frac{\xi}{\tau}, \frac{-\xi}{\tau}\right). \tag{11.65}$$

In this notation, the mirror symmetries mentioned above read as

$$\bar{\mathcal{A}} = \bar{\alpha}^{\alpha}\mathcal{A}, \quad \mathcal{A} = \alpha^{\bar{\alpha}}\bar{\mathcal{A}}, \quad \bar{\mathcal{B}} = \bar{\beta}^{\beta}\mathcal{B}, \quad \mathcal{B} = \beta^{\bar{\beta}}\bar{\mathcal{B}}.$$

If the four segmentations of $\mathcal{A}$ and $\mathcal{B}$ are performed in a random way, clearly they cannot lead to a deterministic decomposition process, by which we mean a step-by-step segmentation of all triangles starting from a finite patch of triangles. In particular, since the affine transformations $\beta^{\bar{\alpha}}$, $\bar{\beta}^{\alpha}$ are not deflations (that is, contraction by a factor τ), we obtain after a rather small number of iterations a set of triangles with a large number of different sizes. The following rule prevents this defect: at any given stage j of a decomposition process, we allow only two types of triangles, either congruent to $\tau^{-j}\mathcal{A}$, or congruent to $\tau^{-j-1}\mathcal{B}$.

For example, at $j = 1$, starting from the triangle $\mathcal{A}$, we decompose it into two triangles congruent to $\mathcal{A}/\tau$, and one triangle congruent to $\mathcal{B}/\tau^2$, following a composition chain respecting the markings of the tensorial symbols in (11.62)–(11.65):

$$\begin{aligned}
\mathcal{A} &= \alpha^{\alpha}\mathcal{A} \cup \alpha^{\beta}\beta^{\beta}\mathcal{B} \cup \alpha^{\beta}\beta^{\bar{\alpha}}\bar{\mathcal{A}}, \\
&= (\xi, -\tau^{-1}\xi^2)\mathcal{A} \cup (\tau^{-2}\xi, -\tau^{-2}\xi)\mathcal{B} \cup (\xi, -\tau^{-1}\xi)\bar{\mathcal{A}}.
\end{aligned} \tag{11.66}$$

On the other hand, if we start from the triangle $\mathcal{B}/\tau$, such a first decomposition process is the same as (11.63), up to the similarity $1/\tau$. This procedure can advantageously be reformulated in a matrix form. Using the vector notation $\mathcal{V}^T = (\mathcal{A}, \mathcal{B}/\tau, \bar{\mathcal{A}}, \bar{\mathcal{B}}/\tau)$, we may write (11.66) as

$$\mathcal{A} = \sum_{j=1}^{4} \mathcal{S}_{1j}\mathcal{V}_j,$$

where addition is understood as set union, and similarly for the other decompositions (11.63)–(11.65). Globally, this yields the eigenvalue equation

$$\mathcal{V} = \mathcal{S}\mathcal{V}, \tag{11.67}$$

where the first row of the matrix $\mathcal{S}$ is

$$(\mathcal{S}_{1j}) = \left((\xi, -\tau^{-1}\xi^2), (\tau^{-2}\xi, -\tau^{-2}\xi), (\xi, -\tau^{-1}\xi), 0\right).$$

There are other ways of formalizing the gradual decomposition process just described, using, for instance, the notion of 5-coloring of the ring $\mathbb{Z}[\xi]$ of cyclotomic integers (see [188] for details). But in any case, the crucial result is that the decomposition process of the triangles $\mathcal{A}$, $\mathcal{B}$ generated by the segmentations (11.62)–(11.65) is *unique*, provided one respects the tensorial rule at each stage. Of course, other decomposition procedures exist, like that one leading to the so-called triangle tiling [189].

We now come to the construction of the whole Penrose–Robinson tiling. To that effect we need an *inflation* procedure, which allows to inflate the patch obtained after the action of the segmentation, in order to restore the original scale. The general notion is the following [59,119]. A tiling $\mathcal{T}$ of $\mathbb{R}^2$ is said to have the *stone-inflation symmetry* if everyone of its tiles, when rotated and scaled by a given $R\lambda$, $R \in \mathrm{SO}(2)$, $\lambda > 1$, and translated by a given $b \in \mathbb{R}^2$, can be packed face-to-face with copies of the original ones (a similar definition can be given in $\mathbb{R}^d$, $d > 2$). More precisely, suppose we are given a tiling $\mathcal{T}$ of $\mathbb{R}^2$ built from a finite set of prototiles $\{T_1, \cdots, T_n\}$,

$$\mathcal{T} = \bigcup_{l=1}^{n} \bigcup_{\gamma_l \in \Gamma_l} \gamma_l \cdot T_l, \tag{11.68}$$

where $\gamma_l = (\vec{b}_l, r_{\theta_l})$ is a translation–rotation and $\Gamma_l \subset \mathbb{R}^2 \rtimes SO(2)$ consists of all those transformations $\gamma_l = (\vec{b}_l, r_{\theta_l})$ which bring the prototile T_l to one of its congruent companions appearing in the tiling. The stone-inflation symmetry based on the affine-linear inflation $\sigma = (\vec{b}, a, \theta)$ then means that for each l and each $\gamma_l \in \Gamma_l$, the following finite patch

$$\sigma \gamma_l \cdot T_l = \bigcup_{m=1}^{n} \bigcup_{\sigma_{lm} \in \Gamma_m} \sigma_{lm} \cdot T_m, \tag{11.69}$$

is present in the tiling.

In the present case, choosing as inverse stone-inflation the affine transformation α^α from (11.66), $\sigma^{-1} = \alpha^\alpha = (\xi, -\tau^{-1}\xi^2)$, we obtain a Penrose–Robinson tiling $\mathcal{T}$ as the following inductive limit, in the notation of (11.67):

$$\mathcal{T} = \lim_{j \to \infty} \sigma^j \sum_{l=1}^{4} (\mathcal{S}^j)_{1l} \mathcal{V}_l. \tag{11.70}$$

The first steps of this specific stone-inflation are shown in Figure 11.13.

11.4.2.3 Multiresolution scheme for stone-invariant tilings

With the geometric tools just described, we are now able construct an adapted wavelet analysis, following the usual procedure. Indeed, in accordance with the Penrose tiling discretization (11.68) of $\mathbb{R}^2$, there exists a multiresolution analysis of the Hilbert space $L^2(\mathbb{R}^2)$. The appropriate definition is the exact counterpart of Definitions 11.4.2 and 11.4.4 (and, of course, subsumes them both), namely,

Definition 11.4.5. *Let $\mathcal{T}$ be a tiling of $\mathbb{R}^2$, generated by the prototiles $\{T_1, \ldots, T_n\}$ and stone-inflation invariant with respect to the affine transformation σ. Then,*

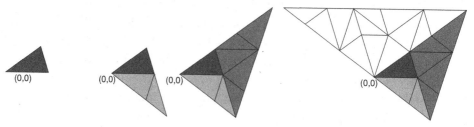

Fig. 11.13. First three steps of stone-inflation.

a σ-multiresolution analysis is an increasing sequence $(V_j)_{j \in \mathbb{Z}}$ of closed subspaces of $L^2(\mathbb{R}^2)$

$$\cdots \subset V_{-2} \subset V_{-1} \subset V_0 \subset V_1 \subset V_2 \subset \cdots,$$

with $\bigcap_{j \in \mathbb{Z}} V_j = \{0\}$ and $\bigcup_{j \in \mathbb{Z}} V_j$ dense in $L^2(\mathbb{R}^2)$, and such that
(1) $f(\vec{x}) \in V_0 \Leftrightarrow f(\sigma^j \vec{x}) \in V_j$;
(2) There exist n scaling functions $(\phi_1(\vec{x}), \cdots, \phi_n(\vec{x})) \equiv \Phi(\vec{x})^T$, such that all their linear affine transforms $\{\phi_l(\gamma_l^{-1}\vec{x}), 1 \leqslant l \leqslant n, \gamma_l \in \Gamma_l\}$ form an orthonormal basis (or at least a Riesz basis) in V_0.

The rest is standard. Let W_0 denote the orthogonal complement of V_0 in V_1, $V_0 \subset V_1 = V_0 \oplus W_0$. Then a wavelet set is given by $(\psi_1(\vec{x}), \ldots, \psi_n(\vec{x})) \equiv \Psi(\vec{x})^T$, such that all linear-affine transforms $\{\psi_l(\gamma_l^{-1} \cdot \vec{x}), 1 \leqslant l \leqslant n, \gamma_l \in \Gamma_l\}$ form an orthonormal basis of W_0. Then the set

$$\{\psi_l(\gamma_l^{-1}\sigma^j \vec{x}), j \in \mathbb{Z}, 1 \leqslant l \leqslant n, \gamma_l \in \Gamma_l\}$$

is an orthonormal basis of $L^2(\mathbb{R}^2)$.

Now, the way to obtain such a wavelet basis from the scaling function set $\{\phi_l(\vec{x})\}$ rests on the scaling equation resulting from the inclusion $V_0 \subset V_1$:

$$
\begin{aligned}
\phi_l(\gamma_{l_0}^{-1} \cdot \vec{x}) &= a \sum_{m=1}^{n} \sum_{\gamma_m \in \Gamma_m^l} c_{l\gamma_m} \phi_m(\gamma_m^{-1}\sigma \cdot \vec{x}) \\
&= \sum_{m=1}^{n} \sum_{\gamma_m \in \Gamma_m^l} c_{l\gamma_m} [U(\sigma^{-1}\gamma_m)\phi_m](\vec{x}),
\end{aligned}
\tag{11.71}
$$

where U is the usual unitary representation (7.39) of the affine group SIM(2) in $L^2(\mathbb{R}^2)$, written in complex notation, that is,

$$(U(b, \Lambda)\phi)(\vec{x}) = |\det \Lambda|^{-1/2} \phi(\Lambda^{-1} \cdot (\vec{x} - b)), \ b \in \mathbb{C} \equiv \vec{b} \in \mathbb{R}^2.$$

The "scaling admissible sets" Γ_m^l are defined according to consistency requirements, namely that (11.71) be also valid for all $\phi_l(\gamma_l^{-1} \cdot \vec{x}), \gamma_l \in \Gamma_l$:

$$\Gamma_m^l = \{\gamma_m \in \Gamma_m \mid \gamma_m^{-1}\sigma\gamma_{l_0}\gamma_l^{-1} \in \Gamma_m \text{ for all } \gamma_l \in \Gamma_l\},\tag{11.72}$$

and these sets should not be trivial! This is a restriction on the original tiling $\mathcal{T}$, which is satisfied by the present Penrose tiling.

Introducing as above a vector notation, (11.71) can be rewritten as follows:

$$\Phi(\vec{x}) = \mathcal{M}(\sigma, \gamma_1, \ldots, \gamma_n)\Phi(\vec{x}).\tag{11.73}$$

This is an eigenvalue equation for the refinement operator $\mathcal{M}$. The latter is an $n \times n$ scaling matrix with entries belonging to the unitary representation of the group algebra of SIM(2) associated to U. More precisely, we have from (11.71):

$$\mathcal{M}_{lm} = \sum_{\gamma_m \in \Gamma_m^l} c_{l\gamma_m} U(\sigma^{-1}\gamma_m). \tag{11.74}$$

Taking Fourier transforms, we obtain

$$\widehat{\Phi}(\vec{k}) = \widehat{\mathcal{M}}(\sigma, \gamma_1, \dots, \gamma_n)\widehat{\Phi}(\vec{k}), \tag{11.75}$$

where $\widehat{\mathcal{M}}_{lm} = \sum_{\gamma_m \in \Gamma_m^l} c_{l\gamma_m} \widehat{U}(\sigma^{-1}\gamma_m)$ and $\widehat{U}$ is the Fourier-transformed representation (7.40). Iterating (11.73) or (11.75) provides a way of finding the set of scaling functions. Of course, convergence conditions have to be stated in a rigorous way. Similarly, the corresponding wavelet set in W_0 can be found also via the scaling equation resulting from $W_0 \subset V_1$. The counterpart of (11.73) has the form

$$\Psi(\vec{x}) = \mathcal{N}(\sigma, \gamma_1, \dots, \gamma_n)\Phi(\vec{x}), \tag{11.76}$$

where $\mathcal{N}$ is also an $n \times n$ scaling matrix with entries belonging to the group algebra of the unitary representation group U. Specific conditions involving both $\mathcal{M}$ and $\mathcal{N}$ stem from the orthogonality $V_0 \perp W_0$. They allow in principle to solve the problem of finding the orthonormal wavelet family.

11.4.2.4   Wavelet bases for a stone-inflation invariant tiling

The simplest example is provided by the Haar wavelets, for which the scaling functions are the normalized characteristic functions of the prototiles:

$$\Phi^T \equiv (\phi_1, \phi_2, \phi_3, \phi_4) = \frac{1}{\sqrt{\Omega_{\mathcal{A}}}}(\chi_{\mathcal{A}}, \sqrt{\tau}\chi_{\mathcal{B}/\tau}, \chi_{\bar{\mathcal{A}}}, \sqrt{\tau}\chi_{\bar{\mathcal{B}}/\tau}),$$

where $\Omega_{\mathcal{A}} \equiv |\text{area } \mathcal{A}|$. It is easy to check that these functions indeed generate a multiresolution in the sense of Definition 11.4.5 (see [And02] for a proof). The scaling equations (11.71) or (11.73) just reproduce the geometric scaling equation (11.67) $\mathcal{V} = \mathcal{S}\mathcal{V}$ at the group representation level. For instance, the first row of the matrix $\mathcal{M}$ in (11.68) is given in the present case (in the complex notation) by

$$(\mathcal{M}_{1j}) = \frac{1}{\tau\sqrt{\Omega_{\mathcal{A}}}}\left(U(\xi, -\tau^{-1}\xi^2), \; \frac{1}{\sqrt{\tau}}U(\tau^{-2}\xi, -\tau^{-1}\xi), \; U(\xi, -\tau^{-1}\xi), \; 0\right).$$

This means that

$$\phi_1 = \frac{1}{\tau\sqrt{\Omega_{\mathcal{A}}}}\left(U(\xi, -\tau^{-1}\xi^2)\chi_{\mathcal{A}} + U(\tau^{-2}\xi, -\tau^{-1}\xi)\chi_{\mathcal{B}/\tau} + U(\xi, -\tau^{-1}\xi)\chi_{\bar{\mathcal{A}}}\right).$$

One recognizes easily the decomposition of a tile $\mathcal{A}$ into three tiles of type $\mathcal{A}$, $\mathcal{B}$ and $\bar{\mathcal{A}}$, respectively, as shown in Figure 11.13. One possible Haar wavelet system is $\Psi^T = (\psi_1, \psi_2, \psi_3, \psi_4)$ is given by the corresponding matrix $\mathcal{N}$ [188]. This yields, for instance,

$$\psi_1 = \frac{1}{\tau\sqrt{\Omega_{\mathcal{A}}}}\left(\frac{1}{\sqrt{\tau}}U(\xi, -\tau^{-1}\xi^2)\chi_{\mathcal{A}} + \frac{1}{\tau}U(\tau^{-2}\xi, -\tau^{-1}\xi)\chi_{\mathcal{B}/\tau}\right.$$

$$\left. - \sqrt{\tau}U(\xi, -\tau^{-1}\xi)\chi_{\bar{\mathcal{A}}}\right).$$

One verifies indeed that the functions ψ_1 and ϕ_1 are normalized and orthogonal to each other. The other three basic wavelets (protowavelets) ψ_2, ψ_3, ψ_4 are obtained in the same way. Note that all four of them vanish a.e. outside of the initial tile $\mathcal{A}$. The subdivision and the linear decomposition can then be repeated on every tile at the next scale. All the new functions will be orthogonal to their predecessors. In the limit one obtains an orthogonal wavelet basis for the space $L^2(\mathcal{A})$. Using the inflation procedure, one can then generate a tiling of the full plane $\mathbb{R}^2$ and a corresponding orthogonal wavelet basis of $L^2(\mathbb{R}^2)$.

Further details on the construction of Penrose wavelet bases may be found in [And02], including another explicit construction for the Haar wavelets, based on a Gram orthogonalization process, and the principle of construction of a spline basis adapted to the tiling. Furthermore, the method can be extended easily to higher dimensions, for instance, the 3-D Danzer tiling based on four tetrahedral tiles with a global icosahedral symmetry.

Comparing now the results derived in this section with the discussion of quasiperiodic point sets given in Section 4.5.2.2, one is led to the following strategy. When faced with a set, such as the diffraction pattern of a quasicrystal, one may first determine the geometric type of the pattern, using the CWT-based technique of Section 4.5.2. Then one may build an orthonormal wavelet basis in the manner outlined here. Such a wavelet basis should then provide a better representation for the corresponding class of patterns. Preliminary results in that direction have been obtained in [And02].

Epilogue

We may now conclude our overview of the "world according to (2-D) wavelets" [Bur98]. We have thoroughly analyzed the 2-D continuous wavelet transform, given some ideas about the discrete or discretized versions, discussed a large number of applications and generalizations (3-D, sphere, space–time). Where do we go now?

Why wavelets in the first place? When should one use them instead of other methods? Suppose we are facing a new signal or image. The very first question to ask is, what do we want to know or to measure from it? Depending on the answer, wavelets will or will not be useful. If we think they might be, we must next (i) choose a wavelet technique, discrete or continuous; (ii) then select a wavelet well adapted to the signal/image at hand, and (iii) determine the relevant parameter ranges. We emphasize that this approach is totally different from the standard one, based on Fourier methods. There is indeed no parameter to adjust here, the Fourier transform is universal. Wavelets on the other hand are extremely flexible, and the tool must be adapted each time to the situation at hand.

As for the first choice, discrete versus continuous WT, it is a fact that the vast majority of authors use the former, in particular if some data compression is required. However, we hope to have demonstrated throughout the book that the CWT often offers a viable alternative, and in a number of cases (e.g. feature detection) it is the only efficient method. This is already true in 1-D, but the situation gets worse in 2-D, because the discrete approach has intrinsic limitations, characterized by the catchword "curse of dimensions." The reason is that the DWT does not properly take into account the geometry of the image to be analyzed. A plane curve is basically a one-dimensional object and the conventional 2-D DWT simply ignores this fact. This was in fact the motivation for more specific techniques such as ridgelets, curvelets or other contourlets, and indeed there is in recent times a growing trend in the wavelet community toward more redundancy and hybrid methods. Once again the SPIE Proceedings volume [Uns03] is a vivid testimony of this situation.

Today the wavelet world continues to grow at a rapid pace. As we said in the beginning, wavelet techniques are spreading into virtually all corners of physics and signal processing, and they also occupy a sizable niche in applied mathematics. The recent relaunch of the popular electronic news bulletin, the *Wavelet Digest*, with a readership of more than 25 000 people, attests of the good health of the subject. As for tomorrow, one can safely predict that no slowdown is to be expected. The trend toward more diversity of techniques will surely continue. In particular, nonlinear extensions and geometrically adapted transforms probably represent the incoming wave. Only the future will tell.

Appendix   Some elements of group theory

This appendix has been put in place mainly for the benefit of readers who may not be entirely familiar with group theory, or with square integrable group representations. To this end, we have collected here some essential abstract notions and results which underlie the concrete examples discussed in the book. However, this is *by no means* intended to be a "crash course" on group theory. There is a vast, indeed bewildering, amount of literature on groups, their representations and applications, pertinent to the subject matter of this book. The interested reader may wish to browse some of it, of which the following is just a small sampling: [Bar77,Cor84,Cor97,Gaa73] or [Gil74].

A.1   Groups

We begin with some basic notions, at a purely algebraic level. Examples appear in the next subsection.

A.1.1   Definitions

(1) A *group* is a set G on which there is defined a binary operation, usually called the *group multiplication* or *group product* mapping $G \times G$ to G, $(g, g') \mapsto gg'$, and obeying the following three axioms.

(G1) *Associativity*: for any $g_1, g_2, g_3 \in G$, one has

$$(g_1 g_2)g_3 = g_1(g_2 g_3).$$

(G2) *Neutral element*: there exists a (necessarily unique) element $e \in G$ such that

$$ge = eg = g, \ \forall g \in G.$$

(G3) *Inverse*: every element $g \in G$ possesses a unique inverse g^{-1}, such that

$$gg^{-1} = g^{-1}g = e.$$

If in addition,

$$g_1 g_2 = g_2 g_1, \ \forall g_1, g_2 \in G,$$

the group G is said to be *abelian* or *commutative*.

(2) Given a group G, a subset $H \subset G$ is called a *subgroup* if it is stable under all group operations:

$$h_1, h_2 \in H \implies h_1 h_2 \in H$$
$$h \in H \implies h^{-1} \in H$$

(this implies of course, that $e \in H$). Thus a subgroup is itself a group under the same law of multiplication.

(3) Given $k \in G$, the *conjugation* by k is the map $g \mapsto kgk^{-1}$. A subgroup H is called *invariant* or *normal* if it is invariant under conjugation:

$$h \in H \implies ghg^{-1} \in H, \forall g \in G.$$

This may also be expressed as

$$gHg^{-1} = H.$$

(4) Given a subgroup H of G, the *left cosets* of G mod H are the subsets $gH = \{gh \mid h \in H\}$. It is easy to see that two different left cosets are always disjoint, and thus the subgroup H induces a partition of G into left cosets:

$$G = H \cup g_1 H \cup g_2 H \cup \ldots$$

(where, of course, $H \equiv eH$ is the only left coset containing the unit element e). Similarly, one defines *right cosets* Hg and a corresponding partition of G. Note that these two partitions are in general distinct.

(5) Given a subgroup H of G, the set of left cosets is called a *quotient* of G by H and is denoted G/H. Similarly the set of all right cosets is the quotient $H \backslash G$.

(6) The two partitions of G into left and right cosets coincide if given $g \in G$, there is an element $g' \in G$ such that

$$gH = Hg',$$

that is, if and only if H is an invariant subgroup. In that case, the quotient G/H (and also $H \backslash G$) is itself a group, with respect to the following operations:

$$(g_1 H)(g_2 H) = g_1 g_2 H, \ \forall g_1, g_2 \in G,$$
$$(gH)^{-1} = g^{-1} H, \ \forall g \in G,$$

and the neutral element is $eH \equiv H$.

(7) Given two groups G, G', a *homomorphism* $\sigma : G \to G'$ is a map that preserves the group properties:

$$\sigma(g_1)\sigma(g_2) = \sigma(g_1 g_2), \ \forall g_1, g_2 \in G,$$
$$\sigma(g)^{-1} = \sigma(g^{-1}), \ \forall g \in G,$$

which implies $\sigma(e_G) = e_{G'}$. The *kernel* of the homomorphism σ is the subset

$$\text{Ker}\,\sigma = \{g \in G \mid \sigma(g) = e_{G'}\}.$$

The homomorphism σ is an *isomorphism* if it is bijective and its inverse σ^{-1} : $G' \to G$ is also a homomorphism. If $G' = G$, the isomorphism σ is called an *automorphism*.

One has the following fundamental result:

- if H is an invariant subgroup of G, the map $\sigma : g \mapsto gH$ is a homomorphism (called canonical) of G onto G/H and $\text{Ker}\,\sigma = H$;
- conversely, if $\sigma : G \to G'$ is a homomorphism, the kernel $\text{Ker}\,\sigma$ is an invariant subgroup of G and $G/\text{Ker}\,\sigma$ is isomorphic to G'.

(8) Given two groups G, G', their *direct product* $G \times G'$ is the set of pairs (g, g') with the group law

$$(g_1, g_1')(g_2, g_2') = (g_1 g_2, g_1' g_2').$$

Then both G, identified with $\{(g, e'),\ g \in G\}$, and G' (similarly identified) are invariant subgroups of $G \times G'$. Let K be an abelian group (with composition denoted by the addition symbol) and G an arbitrary group, and let σ be a homomorphism of G into the automorphisms of K (thus, for each $g \in G$, $\sigma(g) : K \to K$ is an automorphism). Then the *semidirect product* $K \rtimes G$ of K by G (with respect to σ) is the set of pairs (k, g) with the product law

$$(k, g)(k', g') = (k + \sigma(g)k', gg').$$

Thus K is an invariant subgroup of $K \rtimes G$, and G is a subgroup which is invariant iff $\sigma(g)$ is the identity for all $g \in G$, and then the semidirect product is the direct product.

A.1.2 Examples

- $\mathbb{R}^n (n \geqslant 1)$, with vector addition $(a, b) \mapsto a + b$, is an abelian group.
- SO(2), the group of plane rotations around the origin of $\mathbb{R}^2$, is an abelian group. It may also be realized by the 2×2 rotation matrices,

$$r_\theta = \begin{pmatrix} \cos\theta & -\sin\theta \\ \sin\theta & \cos\theta \end{pmatrix}, \quad 0 \leqslant \theta < 2\pi.$$

One has

$$r_{\theta_1}\, r_{\theta_2} = r_{\theta_2}\, r_{\theta_1} = r_{\theta_1 + \theta_2}, \quad r_\theta^{-1} = r_{-\theta}, \quad r_0 = I.$$

- SO(3), the group of rotations of $\mathbb{R}^3$, is a nonabelian group. It may be realized as the set of all 3×3 real orthogonal matrices of determinant 1:

$$R^T R = R R^T \ (\text{thus, } R^{-1} = R^T), \quad \det R = 1.$$

- $SO_o(1, 3)$, the connected component of the Lorentz group is a nonabelian group, that may be realized as the set of all real 4×4 pseudo-orthogonal matrices Λ, i.e., matrices with the property:

$$\Lambda^T \eta \Lambda = \eta, \quad \det \Lambda = 1, \quad \Lambda_{00} \geqslant 1, \quad \eta = \begin{pmatrix} -1 & 0 & 0 & 0 \\ 0 & 1 & 0 & 0 \\ 0 & 0 & 1 & 0 \\ 0 & 0 & 0 & 1 \end{pmatrix}.$$

- Subgroups
  - $\mathbb{R}^k$ is a subgroup of $\mathbb{R}^m$, $\forall k < m$.
  - $SO(2)$ is a subgroup of $SO(3)$, but a noninvariant one. Indeed the quotient space $SO(3)/SO(2)$ may be identified with the 2-sphere S^2, which is not a group.
  - $SO(3)$ is a noninvariant subgroup of $SO_o(1, 3)$.
- The affine subgroup G_{aff} of the line is the set of transformations

$$(b, a) : x \mapsto ax + b, \quad b \in \mathbb{R}, \ a \neq 0.$$

The group G_{aff} may be realized as the set of 2×2 matrices of the form

$$(b, a) \equiv \begin{pmatrix} a & b \\ 0 & 1 \end{pmatrix}.$$

Thus the group law is

$$(b, a)(b', a') = (b + ab', aa')$$
$$(b, a)^{-1} = (-a^{-1}b, a^{-1}).$$

The subgroup of translations $(b, 1), b \in \mathbb{R}$, is an invariant subgroup and G_{aff} is the semidirect product

$$G_{\text{aff}} = \mathbb{R} \rtimes \mathbb{R}_*.$$

- The $ax + b$ group, also noted G_{aff}^+, is the subgroup of G_{aff} corresponding to $a > 0$. The group law is the same and

$$G_{\text{aff}}^+ = \mathbb{R} \rtimes \mathbb{R}_*^+.$$

- The similitude group $SIM(2)$ of the plane is the set of transformations consisting of translations, dilations and rotations

$$(\vec{b}, a, \theta) : \vec{x} \mapsto a r_\theta(\vec{x}) + \vec{b}, \quad a > 0, \ r_\theta \in SO(2), \ \vec{b} \in \mathbb{R}^2,$$

or, in 3×3 matrix form,

$$(\vec{b}, a, \theta) = \begin{pmatrix} a r_\theta & \vec{b} \\ \vec{0}^T & 1 \end{pmatrix}.$$

The group law is thus

$$(\vec{b}, a, \theta)(\vec{b}', a', \theta') = (\vec{b} + a\, r_\theta(\vec{b}'), aa', \theta + \theta'),$$

corresponding to the semidirect product structure:

$$\text{SIM}(2) = \mathbb{R}^2 \rtimes (\mathbb{R}_*^+ \times \text{SO}(2)).$$

Indeed dilations and rotations commute, and the subgroup of translations is invariant.
• One defines in the same way the similitude group for $\mathbb{R}^3$:

$$\text{SIM}(3) = \mathbb{R}^3 \rtimes (\mathbb{R}_*^+ \times \text{SO}(3)).$$

• *Groups of transformations*
Groups may be defined abstractly or by a matrix representation, as above. They can also be defined geometrically, as groups of transformations of certain manifolds. Let X be a (smooth) manifold, such as the space $\mathbb{R}^n$, the 2-sphere S^2, space–time $\mathbb{R} \times \mathbb{R}$ or $\mathbb{R}^2 \times \mathbb{R}$, etc. Then a group G is said to *act* on X if each element $g \in G$ defines a (smooth) map $x \mapsto g[x]$ of X into itself, in such a way that $g[g'[x]] = gg'[x]$, and $e[x] = x$, $\forall g, g' \in G$, $x \in X$. The action is called *transitive* if, for any pair $x, x' \in X$, there is at least one $g \in G$ such that $g[x] = x'$. One usually writes the action simply as $x \mapsto gx$.

Trivial examples of a transitive group action are the action of the group G on itself by left translation, $L_{g_0} : g \mapsto g_0^{-1}g$ or by right translation, $R_{g_0} : g \mapsto gg_0$. Another one is the left action of G on the quotient G/H by a (closed) subgroup H, namely, $L_{g_0} : gH \mapsto g_0^{-1}gH$. Concrete examples are, for instance, the action of G_{aff}^+ on $\mathbb{R}$; that of SIM(2) on $\mathbb{R}^2$; that of SO(3) on the 2-sphere S^2; or, less intuitively, that of the connected Lorentz group $\text{SO}_o(1, 3)$ on S^2, which underlies the construction of the continuous wavelet transform on S^2, described in Section 9.2.

If the action of G on X is transitive, and if H denotes the subgroup of G that leaves invariant an arbitrary point of X (it is easily seen that all invariance subgroups are conjugate, hence isomorphic, to each other), then the manifold X may be identified with (is homeomorphic to) the quotient G/H. The manifold X is then called a *homogeneous space* for G, or a (transitive) G-space. The classical example is $S^2 \simeq \text{SO}(3)/\text{SO}(2)$.

Let X be a transformation space for G and $x \in G$. The *orbit* of x under G is the set

$$Gx = \{y = gx | g \in G\} \subset X. \tag{A.1}$$

If X is a homogeneous space, it corresponds to a single orbit under G.

(Remark: Mathematicians usually define group actions in terms of the map $G \times Y \to Y$ given by $(g, y) \mapsto g[y]$, under suitable smoothness assumptions. Group actions may also be considered on arbitrary sets.)

A.1.3   Integration on groups

All the groups encountered in this book are locally compact – in fact, they are Lie groups (further information on Lie groups and Lie algebras is given in Section A.3). Now, every locally compact group G carries a left and a right invariant Haar measure, both unique up to equivalence. We shall denote the left Haar measure by μ and use it systematically. The right Haar measure, when used, will be denoted by μ_r. Here "invariance" means invariance under left or right translation by G:

$$\mu(g^{-1}E) = \mu(E) \qquad \text{and} \qquad \mu_r(Eg) = \mu_r(E), \tag{A.2}$$

for any Borel subset $E \subset G$ and all $g \in G$. Equivalently,

$$d\mu(g_o^{-1}g) = d\mu(g) \qquad \text{and} \qquad d\mu_r(gg_o) = d\mu_r(g), \ \forall \, g, g_o \in G. \tag{A.3}$$

If $\mu = \mu_r$, the group is called *unimodular*. Examples of unimodular groups are all abelian groups, and all compact groups, such as SO(2) or SO(3).

In general, μ and μ_r are different, but equivalent measures; i.e., they have the same null sets. Thus, there exists a measurable function $\Delta : G \to \mathbb{R}^+$, such that

$$d\mu(g) = \Delta(g)d\mu_r(g). \tag{A.4}$$

This function, called the *modular function* of the group, is an $\mathbb{R}^+$-valued character, which means that, for μ-almost all $g, g_1, g_2 \in G$,

$$\begin{aligned} &\Delta(g) > 0, \\ &\Delta(e) = 1, \quad e = \text{identity element of } G, \\ &\Delta(g_1 g_2) = \Delta(g_1)\Delta(g_2). \end{aligned} \tag{A.5}$$

Furthermore, for μ-almost all $g, g' \in G$, the following relations hold:

$$\begin{aligned} d\mu_r(g) &= \Delta(g^{-1}) \, d\mu(g) \ = \ d\mu(g^{-1}), \\ d\mu(gg') &= \Delta(g') \, d\mu(g). \end{aligned} \tag{A.6}$$

While the group itself always carries a left (and a right) invariant Haar measure, the homogeneous space X need not carry any measure invariant under the action $x \mapsto gx$. *Quasi-invariant* measures, however, always exist on X. The measure ν on X is said to be quasi-invariant if ν and ν_g are equivalent measures, for all $g \in G$, where ν_g is defined to be the measure obtained by the natural action of g on ν:

$$\nu_g(E) = \nu(gE), \quad \text{for any Borel subset } E \subset X. \tag{A.7}$$

The Radon–Nikodym derivative of ν_g with respect to ν,

$$\lambda(g, x) = \frac{d\nu_g(x)}{d\nu(x)}, \tag{A.8}$$

is then a *cocycle*, $\lambda : G \times X \to \mathbb{R}^+$, with the properties

$$\lambda(g_1 g_2, x) = \lambda(g_1, x)\lambda(g_2, g_1^{-1}x),$$
$$\lambda(e, x) = 1,$$

(A.9)

(these equations may always be assumed to hold for all $g_1, g_2 \in G$, and all $x \in X$). Note that all the measures ν_g, $g \in G$, have the same measure-zero sets.

A.1.4   Convolution on groups

The convolution of two functions on a locally compact group is a well-defined operation that shares many properties with its well-known Euclidean counterpart. We have the following definition:

Definition A.1.1 (Group convolution) *Let G be a locally compact group with left Haar measure dy, and let $f, g : G \to \mathbb{C}$ be two measurable functions. The convolution product of f and g is defined (almost everywhere) by the integral:*

$$(f \star g)(x) \equiv \int_G dy \; f(xy)g(y^{-1}) = \int_G dy \; f(y)g(y^{-1}x).$$

(A.10)

When G is a commutative group, one has $f \star g = g \star f$. In general, however, convolution is a noncommutative operation and we have the following relations:

$$(f \star g)(x) = \int_G dy \; f(xy^{-1})g(y)\Delta(y^{-1}),$$

where $\Delta(x)$ is the modular function on G.

One of the most interesting properties of the convolution integral is its regularizing effect on L^p-elements. This is embodied in a number of inequalities, which all stem from the following general statement.

Proposition A.1.1 (Young's inequality) *Let G be a locally compact group with left Haar measure dx. Let $p, q, r \geqslant 1$ and $1/p + 1/q + 1/r = 2$. Let $f \in L^p(G, dx)$, $g \in L^q(G, dx)$, and $h \in L^r(G, dx)$. Then*

$$\left| \int_G dx \; (f \star g)(x)\, h(x) \right| = \left| \int_G dy \int_G dx \; f(y)\, g(y^{-1}x)\, h(x) \right|$$
$$\leqslant \|f\|_p \|g\|_q \|h\|_r.$$

(A.11)

Equivalently,

$$\|f \star g\|_r \leqslant \|f\|_p \|g\|_q, \; with \; 1/p + 1/q = 1 + 1/r.$$

(A.12)

A proof may be found in [35], following that of the corresponding theorem on $\mathbb{R}^n$ given in [Lie97; Theorem 4.2], which itself generalizes [Gaa73; Proposition V.4.6]. The above result extends to homogeneous spaces, as mentioned in [Gaa73; Section V.4] for the particular case $p = 1, r = p'$. First we have to define the proper notion on quotient spaces.

Definition A.1.2 (Group convolution on a homogeneous space) *Let G be a locally compact group, H a closed subgroup such that the quotient space G/H has the left invariant measure $d\zeta$. Let g, h be two measurable functions defined on G/H. Then the spherical convolution product of g and h, with respect to G, is the function on G defined (almost everywhere) by the integral*

$$(g \widetilde{\star} h)(y) = \int_{G/H} g(y^{-1}\zeta) h(\zeta) d\zeta. \tag{A.13}$$

Similarly, if f and h are two measurable functions defined on G and G/H, respectively, their convolution is the function on G/H defined (almost everywhere) by the integral

$$(f \star g)(\zeta) = \int_G f(y) g(y^{-1}\zeta) dy. \tag{A.14}$$

Following the pattern of Proposition A.1.1, we may now state the following generalization.

Proposition A.1.2 (Young's inequality on homogeneous spaces) *Let G be a locally compact group with left Haar measure dy, H a closed subgroup such that the quotient space G/H has the left invariant measure $d\zeta$. Let $p, q, r \geqslant 1$ and $1/p + 1/q + 1/r = 2$. Let $f \in L^p(G, dx)$, $g \in L^q(G/H, d\zeta)$, and $h \in L^r(G/H, d\zeta)$. Then*

$$\left| \int_{G/H} d\zeta \, (f \star g)(\zeta) h(\zeta) \right| = \left| \int_{G/H} d\zeta \int_G dy \, f(y) g(y^{-1}\zeta) h(\zeta) \right|$$
$$\leqslant \|f\|_p \|g\|_q \|h\|_r. \tag{A.15}$$

Equivalently, $f \in L^p(G, dx)$, $g \in L^q(G/H, d\zeta)$ implies $f \star g \in L^r(G/H, d\zeta)$ with $1/p + 1/q = 1 + 1/r$ and

$$\|f \star g\|_r \leqslant \|f\|_p \|g\|_q. \tag{A.16}$$

Similarly, $g \in L^q(G/H, d\zeta)$, $h \in L^r(G/H, d\zeta)$ implies $g \widetilde{\star} h \in L^p(G, dx)$, with $1/q + 1/r = 1 + 1/p$, and

$$\|g \widetilde{\star} h\|_p \leqslant \|g\|_q \|h\|_r. \tag{A.17}$$

The proof is an easy adaptation of that of Proposition A.1.1, and may also be found in [35].

Note that sharper constants, smaller than 1, may be put in the upper bounds on the right-hand sides of all the inequalities, as shown in detail for $\mathbb{R}^n$ in [Lie97]. These inequalities are used in Chapter 9, Section 9.3, for $G = SO(3)$, $G/H = SO(3)/SO(2) = S^2$, under the following continuous inclusions:

$$L^2(SO(3), d\varrho) \star L^1(S^2, d\mu) \hookrightarrow L^2(S^2, d\mu) \tag{A.18}$$

$$L^2(S^2, d\mu) \widetilde{\star} L^1(S^2, d\mu) \hookrightarrow L^2(SO(3), d\varrho). \tag{A.19}$$

A.2 Group representations

A.2.1 Definitions

(1) Given a group G and a vector space V, a *representation* of G in V is a homomorphism $T : G \to GL(V)$ of G into the group $GL(V)$ of invertible linear operators on V; thus:

$$
\begin{aligned}
T(g_1 g_2) &= T(g_1)T(g_2), \quad \forall g_1, g_2 \in G, \\
T(g^{-1}) &= T(g)^{-1}, \quad \forall g \in G, \\
T(e) &= I.
\end{aligned}
\tag{A.20}
$$

The dimension of the representation T is the dimension of V.

(2) Most interesting is the case where the representation space is a Hilbert space $\mathfrak{H}$, the operators $T(g)$ are bounded with bounded inverse, and the operator $T(g)$ depends continuously on g, e.g.,

$$
\|(T(g) - I)\phi\| \to 0 \text{ for } g \to e, \ \forall \phi \in \mathfrak{H}.
$$

In particular, the representation U of G into $\mathfrak{H}$ is *unitary* if every operator $U(g)$ is unitary, that is

$$
U(g)U(g)^* = U(g)^*U(g) = I, \quad \forall g \in G,
$$

which implies

$$
U(g)^{-1} = U(g^{-1}) = U(g)^*.
$$

(3) Two unitary representations U_1 and U_2 of G in Hilbert spaces $\mathfrak{H}_1$ and $\mathfrak{H}_2$, respectively, are *unitarily equivalent* if there is a unitary operator $S : \mathfrak{H}_1 \to \mathfrak{H}_2$ such that

$$
U_2(g) = SU_1(g)S^{-1}, \quad \forall g \in G.
$$

In that case, U_1 and U_2 may in general be identified.

(4) A subspace $\mathfrak{K} \subset \mathfrak{H}$ is *invariant* under U if $h \in \mathfrak{K}$ implies $U(g)h \in \mathfrak{K}, \ \forall g \in G$. The restriction of U to an invariant subspace is called a *subrepresentation*. We consider only Hilbert (i.e., closed) subspaces.

The representation U is said to be *irreducible* if there are no nontrivial subspaces invariant under U (that is, subspaces different from $\{0\}$ and $\mathfrak{H}$ itself). Otherwise U is said to be *reducible*.

(5) The representation U is said to be *completely reducible* if the orthogonal complement $\mathfrak{K}^\perp$ of any invariant subspace $\mathfrak{K}$ is also invariant. In that case, $\mathfrak{H}$ decomposes into a direct sum

$$
\mathfrak{H} = \mathfrak{K} \oplus \mathfrak{K}^\perp,
$$

of subspaces, each carrying a subrepresentation. A fundamental result is that every reducible *unitary* representation is completely reducible.

A.2.2   Examples

(1) If G is compact, all its unitary irreducible representations (UIR) are finite dimensional, and any unitary representation of G decomposes as a direct sum of irreducible subrepresentations.

• For SO(2), all UIRs are one-dimensional and of the form

$$U_k(\theta)\phi = e^{ik\theta}\phi, \ k \in \mathbb{Z}, \ \theta \in \text{SO}(2).$$

• For SO(3), the UIRs are of dimension $(2l + 1)$, and denoted D_l, for $l = 0, 1, 2 \ldots$; in appropriate (spherical) coordinates, a convenient basis of D_l is the family of spherical harmonics $Y_l^m(\theta, \varphi)$, $m = -l, \ldots, l$.

(2) The *left regular* representation U_L of G is defined by left translation in the Hilbert space $L^2(G, d\mu)$:

$$(U_L(g_o)\phi)(g) = \phi(g_o^{-1}g), \ \ \phi \in L^2(G, d\mu), \ g \in G.$$

The *right regular* representation U_R is defined in an analogous way in $L^2(G, d\mu_r)$:

$$(U_R(g_o)\psi)(g) = \phi(gg_o), \ \ \phi \in L^2(G, d\mu_r), \ g \in G.$$

If H is a maximal compact subgroup of G, and $X = G/H$ carries a left invariant measure ν, then the (left) *quasi-regular* representation U_{qL} of G is defined by left translation in $L^2(X, d\nu)$:

$$(U_{qL}(g)\phi)(x) = \phi(g^{-1}x), \ \ \phi \in L^2(X, d\nu), \ g \in G.$$

Two standard examples are given by SO(2) and SO(3):

(i) The regular representation of SO(2) acting in $L^2(S^1, d\theta)$ decomposes into the infinite direct sum of all the one-dimensional representations U_k:

$$U_L(\theta) = \bigoplus_{k \in \mathbb{Z}} U_k(\theta).$$

Actually this is nothing but the theory of Fourier series expressed in group-theoretical terms!

(ii) The quasi-regular representation U_{qL} of SO(3) acts in $L^2(S^2, d\zeta)$ and decomposes into the direct sum of all UIRs D_l, each appearing once:

$$U_{qL}(g) = \bigoplus_{l=0}^{\infty} D_l(g), \ g \in G$$

Considering polar spherical coordinates $\zeta = (\theta, \varphi)$ on the 2-sphere S^2, one obtains in this way the orthonormal basis of $L^2(S^2, d\zeta)$ consisting of all spherical harmonics $Y_l^m(\theta, \varphi)$, $l = 0, 1, 2, \ldots, m = -l, \ldots, l$.

(3) The affine group of the line G_{aff} has, up to unitary equivalence, only one UIR, acting in $L^2(\mathbb{R}, dx)$, namely,

$$[U(b, a)\phi](x) = |a|^{-1/2}\phi(a^{-1}(x - b)).\tag{A.21}$$

Upon restriction to the connected subgroup G_{aff}^+, corresponding to $a > 0$, the representation U decomposes into two inequivalent representations $U_\pm$, acting in the so-called Hardy spaces $\mathfrak{H}_\pm$:

$$\mathfrak{H}_+(\mathbb{R}) = \{f \in L^2(\mathbb{R}, dx) \mid \widehat{f}(\xi) = 0, \ \forall \xi \leqslant 0\},\tag{A.22}$$
$$\mathfrak{H}_-(\mathbb{R}) = \{f \in L^2(\mathbb{R}, dx) \mid \widehat{f}(\xi) = 0, \ \forall \xi \geqslant 0\},\tag{A.23}$$

and, of course $L^2(\mathbb{R}, dx) = \mathfrak{H}_+(\mathbb{R}) \oplus \mathfrak{H}_-(\mathbb{R})$ (here $\widehat{f}(\xi)$ denotes the Fourier transform of $f(x)$). These representations underlie the one-dimensional CWT.

(4) The similitude group SIM(2) has, up to unitary equivalence, only one UIR, which acts in $L^2(\mathbb{R}^2, d\vec{x})$, namely:

$$(U(\vec{b}, a, \theta)f)(\vec{x}) = a^{-1}f(a^{-1}r_\theta(\vec{x} - \vec{b}))\tag{A.24}$$

This representation generates the 2-D CWT.

(5) For the Lorentz group $SO_o(1, 3)$, which is noncompact (and simple), all the nontrivial UIRs are infinite dimensional. An interesting class are the representations acting in $L^2(S^2, d\zeta)$ (see Section 9.2), which underlie the construction of the CWT on S^2.

A.2.3   Square integrable representations

Let U be a UIR of the locally compact group G in the Hilbert space $\mathfrak{H}$. A vector $\eta \in \mathfrak{H}$ is said to be *admissible* if

$$I(\eta) = \int_G |\langle U(g)\eta|\eta\rangle|^2 \, d\mu(g) < \infty\tag{A.25}$$

(one can replace the left invariant Haar measure μ by the right one μ_r in this definition). Equivalently, η is admissible iff

$$\int_G |\langle U(g)\eta|\phi\rangle|^2 \, d\mu(g) < \infty, \ \forall \phi \in \mathfrak{H}.\tag{A.26}$$

Let $\mathcal{A}$ denote the set of all admissible vectors. Then, it follows from the irreducibility of U and (A.26) that either $\mathcal{A} = \{0\}$, or $\mathcal{A}$ is *dense* in $\mathfrak{H}$. The representation U is said to be *square integrable* if $\mathcal{A} \neq \{0\}$.

Theorem A.2.1. *Let U be a square integrable unitary representation of the locally compact group G in the Hilbert space $\mathfrak{H}$. Then, for any $\eta \in \mathcal{A}$, the map $W_\eta : \mathfrak{H} \to L^2(G, d\mu)$ defined by*

$$(W_\eta\phi)(g) = [c(\eta)]^{-\frac{1}{2}}\langle \eta_g|\phi\rangle, \quad \phi \in \mathfrak{H}, \ g \in G, \ \text{where } c(\eta) \equiv I(\eta)/\|\eta\|^2,\tag{A.27}$$

is a linear isometry onto a (closed) subspace $\mathfrak{H}_\eta$ of $L^2(G, d\mu)$. In other words, the resolution of the identity

$$\frac{1}{c(\eta)} \int_G |\eta_g\rangle\langle\eta_g|\, d\mu(g) = I \qquad (A.28)$$

holds on $\mathfrak{H}$. The subspace $\mathfrak{H}_\eta = W_\eta \mathfrak{H} \subset L^2(G, d\mu)$ is a reproducing kernel Hilbert space, so that the corresponding projection operator

$$\mathbb{P}_\eta = W_\eta W_\eta^*, \qquad \mathbb{P}_\eta L^2(G, d\mu) = \mathfrak{H}_\eta, \qquad (A.29)$$

has the reproducing kernel K_η,

$$(\mathbb{P}_\eta \widetilde{\Phi})(g) = \int_G K_\eta(g, g')\widetilde{\Phi}(g')\, d\mu(g'), \qquad \widetilde{\Phi} \in L^2(G, d\mu),$$

$$K_\eta(g, g') = \frac{1}{c(\eta)}\langle\eta_g|\eta_{g'}\rangle, \qquad (A.30)$$

as its integral kernel. Furthermore, W_η intertwines U and the left regular representation U_L,

$$W_\eta U(g) = U_L(g) W_\eta, \quad g \in G, \qquad (A.31)$$

and therefore, U is unitarily equivalent to a subrepresentation of the left and the right regular representation of G.

A characteristic property of square integrable representations is the existence of the following *orthogonality relations*.

Theorem A.2.2. *Let G be a locally compact group and U a square integrable representation of G on the Hilbert space $\mathfrak{H}$. Then there exists a unique positive, self-adjoint, invertible operator C in $\mathfrak{H}$, the domain $\mathcal{D}(C)$ of which is dense in $\mathfrak{H}$ and is equal to $\mathcal{A}$, the set of all admissible vectors; if η and η' are any two admissible vectors and ϕ, ϕ' are arbitrary vectors in $\mathfrak{H}$, then*

$$\int_G \overline{\langle\eta_g'\,|\,\phi'\rangle}\langle\eta_g\,|\,\phi\rangle\, d\mu(g) = \langle C\eta\,|\,C\eta'\rangle\,\langle\phi'\,|\,\phi\rangle. \qquad (A.32)$$

Furthermore, $C = \lambda I$, $\lambda > 0$, if and only if G is unimodular.

Therefore, one may write $c(\eta) = \|C\eta\|^2$. The operator C, which is in most cases a multiplication operator, is called the Duflo–Moore operator.

Examples

• If G is compact, every UIR is square integrable. In fact, square integrable representations generalize to noncompact groups the UIRs of compact groups.

- The representation (A.21) of G_{aff} in $L^2(\mathbb{R}, dx)$ is square integrable, and so are its restrictions $U_{\pm}$ to G_{aff}^+, acting in $\mathfrak{H}_{\pm}$. In the case of U_+, for instance, a vector $\eta \in \mathfrak{H}_+(\mathbb{R})$ is admissible if its Fourier transform satisfies

$$c_\eta = 2\pi \int_0^\infty |\widehat{\eta}(\xi)|^2 \, \frac{d\xi}{\xi} < \infty. \tag{A.33}$$

The proofs of these statements may be found in Chapter 6, Section 6.1.1, in the context of the 1-D CWT.

- The same situation prevails for SIM(2). The natural UIR (A.24) in $L^2(\mathbb{R}^2, d\vec{x})$ is square integrable, and a vector ψ is admissible iff

$$c_\psi \equiv (2\pi)^2 \int_{\mathbb{R}^2} \frac{d^2\vec{k}}{|\vec{k}|^2} \, |\widehat{\psi}(\vec{k})|^2 < \infty, \tag{A.34}$$

as discussed in Sections 2.1.2 and 7.2 ($\widehat{\psi}$ is the Fourier transform of ψ).

- The representation of $SO_o(1, 3)$ in $L^2(S^2, d\zeta)$ that underlies the spherical CWT is square integrable (and this is precisely what allows the construction to proceed, as in the previous cases), see Section 9.2.2.

A.3   Lie groups and Lie algebras

All the groups we have encountered in this book are in fact Lie groups, that is, groups whose elements depend in a smooth way on finitely many parameters. More precisely, a *Lie group* is a group G that is at the same time a $\mathcal{C}^2$ manifold in such a way that the two group operations, $(g, g') \mapsto gg' : G \times G \to G$ (multiplication) and $g \mapsto g^{-1} : G \to G$ (inversion) are $\mathcal{C}^2$ mappings. Note that the $\mathcal{C}^2$ conditions, together with the group structure, imply $\mathcal{C}^\infty$ conditions: G is a smooth manifold and group operations are smooth.

The simplest example is the group $\text{GL}(n, \mathbb{R})$ of all $n \times n$ nonsingular real matrices. Other examples we have met are the translation group $\mathbb{R}^n$, the rotation groups SO(2) and SO(3), the proper Lorentz group $SO_o(1,3)$, the $ax + b$ group, the similitude group SIM(2), and all these groups are matrix groups (that is, subgroups of $\text{GL}(n, \mathbb{R})$ for some n), the group multiplication being simply matrix multiplication. The theory of Lie groups is a monument of mathematics, for which we refer the reader to standard textbooks such as [Bar77], [Dui00], [Gil74] or [Hel78].

The characteristic feature of a Lie group G is that is possesses a unique *Lie algebra* $\mathfrak{g}$, defined as the tangent plane at the identity element e of the group (this is well-defined, since G is a smooth manifold). In more intuitive terms, this can be visualized as follows. Let d be the dimension of G (as a manifold), that is, the number of independent parameters necessary to describe G. Next, consider d linearly independent one-parameter subgroups $g_j(s)$, $s \in \mathbb{R}$, $j = 1, 2, \ldots, d$, with $g_j(s)g_j(t) = g_j(s + t)$, $s, t \in \mathbb{R}$, i.e., curves on G that intersect at e only. For each such subgroup g_j, take the tangent vector, X_j at e, which generates the subgroup g_i by exponentiation, $g_j(s) = \exp s X_j$ (while this operation can be given a general abstract meaning, it is best understood in a matrix realization). Then the Lie algebra $\mathfrak{g}$ is simply the real vector space generated by $X_1, \ldots, X_d$, equipped with the so-called Lie bracket $(X, Y) \mapsto [X, Y]$. This is a bilinear map from $\mathfrak{g} \times \mathfrak{g}$ to $\mathfrak{g}$, which is antisymmetric, $[X, Y] = -[Y, X]$ and satisfies the Jacobi identity

$$[X, [Y, Z]] + [Y, [Z, X]] + [Z, [X, Y]] = 0, \quad \forall X, Y, Z \in \mathfrak{g}.$$

Once again, things are easy in a matrix realization, the Lie bracket is the commutator, $[X, Y] = XY - YX$. A Lie algebra $\mathfrak{g}$ is abelian, and stems from an abelian Lie group, iff its Lie bracket vanishes identically, $[X, Y] = 0$, $\forall X, Y \in \mathfrak{g}$. Thus the Lie bracket measures the noncommutativity of the Lie algebra (and thus also of the corresponding Lie group).

Instead of pursuing the general theory, let us give a few concrete examples, using throughout matrix realizations. We may remark that the conventions are different in the physical and in the mathematical literature. Whereas the latter uses the exponentiation law $g_j(s) = \exp s X_j$, the former considers the law $g_j(s) = \exp -is\widetilde{X}_j$, so that the generators are represented by self-adjoint operators in a unitary representation of the group. In Chapter 6, we have followed the physicists' convention.

(1) *The rotation group SO(2)*

In the familiar 2×2 matrix realization

$$r_\theta = \begin{pmatrix} \cos\theta & -\sin\theta \\ \sin\theta & \cos\theta \end{pmatrix}, \quad 0 \leqslant \theta < 2\pi,$$

the infinitesimal generator is clearly the matrix

$$J = \begin{pmatrix} 0 & -1 \\ 1 & 0 \end{pmatrix}$$

(there is only one, since SO(2) is a one-parameter group). One has indeed $r_\theta = \exp \theta J$.

(2) *The rotation group SO(3)*

From the standard realization as 3×3 rotation matrices, parametrized in terms of the three Euler angles, one sees easily that the infinitesimal generators are the three matrices J_1, J_2, J_3, with the cyclic commutation relations $[J_1, J_2] = J_3$, etc. (the corresponding hermitian generators $\widetilde{J}_j$ represent the three components of angular momentum in quantum mechanics).

(3) *The ax + b group*

Starting from the 2 × 2 matrix realization

$$(b, a) \equiv \begin{pmatrix} a & b \\ 0 & 1 \end{pmatrix}, \quad a > 0, b \in \mathbb{R},$$

we readily obtain the two infinitesimal generators:

$$D = \begin{pmatrix} 1 & 0 \\ 0 & 0 \end{pmatrix} \text{ (dilation)}, \quad P = \begin{pmatrix} 0 & 1 \\ 0 & 0 \end{pmatrix}, \text{ (translation)},$$

with commutation relation $[D, P] = P$.

(4) *The similitude group SIM(2)*

From the 3 × 3 matrix form,

$$(\vec{b}, a, \theta) = \begin{pmatrix} a r_\theta & \vec{b} \\ \vec{0}^T & 1 \end{pmatrix},$$

we obtain the infinitesimal generators

$$D = \begin{pmatrix} 1 & 0 & 0 \\ 0 & 1 & 0 \\ 0 & 0 & 0 \end{pmatrix} \text{ (dilation)}, \quad J = \begin{pmatrix} 0 & -1 & 0 \\ 1 & 0 & 0 \\ 0 & 0 & 0 \end{pmatrix} \text{ (rotation)},$$

$$P_1 = \begin{pmatrix} 0 & 0 & 1 \\ 0 & 0 & 0 \\ 0 & 0 & 0 \end{pmatrix}, \quad P_2 = \begin{pmatrix} 0 & 0 & 0 \\ 0 & 0 & 1 \\ 0 & 0 & 0 \end{pmatrix} \text{ (translations)},$$

with commutation relations

$$[D, J] = 0, \quad [D, P_j] = P_j, \quad [P_1, P_2] = 0, \quad [J, P_1] = P_2, \quad [J, P_2] = -P_1.$$

A.4   Some useful formulas from harmonic analysis on the sphere

In Chapter 9 we used formulas involving representations of the rotation group $SO(3)$, for the analysis of wavelets on a sphere. We collect some of these expressions here; for details, the reader may wish to consult [Bar77, Tal68]. The group $SO(3)$ consists of 3×3 real, orthogonal matrices, of determinant one. Its unitary irreducible representations are in one-to-one correspondence with the integers, $l = 0, 1, 2, \ldots$, and for each l the corresponding representation is realized by unitary matrices $\mathcal{D}^l(\varrho)$, on a $2l + 1$ dimensional Hilbert space $\mathfrak{R}^l$. Here ϱ denotes an $SO(3)$ variable, which we will represent by the set of three Euler angles, χ, θ, φ.

Let Y_l^m denote a spherical harmonic, $m = -l, -l+1, \ldots l+1$. These form an orthonormal basis in $\mathfrak{R}^l$, satisfying the orthogonality relations,

$$\int_0^\pi d\theta \int_0^{2\pi} d\varphi \ Y_l^m(\theta, \varphi) Y_{l'}^{m'}(\theta, \varphi) \ \sin\theta = \delta_{ll'} \delta_{mm'},$$ (A.35)

and the addition theorem,

$$\sum_{m=-l}^l \overline{Y_l^m}(\zeta) Y_l^m(\zeta') = \frac{2l+1}{4\pi} P_l\left(\widehat{\zeta} \cdot \widehat{\zeta'}\right)$$ (A.36)

where ζ, ζ' denote unit vectors with polar angles θ, φ and θ', φ', respectively. Denoting by $\mathcal{D}^l(\varrho)_{mn}$ the matrix elements of $\mathcal{D}^l(\varrho)$ in the Y_l^m-basis (called *Wigner functions* in the physics literature), it can be shown that

$$\mathcal{D}^l_{m0}(\varrho) = \sqrt{\frac{4\pi}{2l+1}} Y_l^m(\theta, \varphi), \quad \text{with } \varrho \equiv (\chi, \theta, \varphi).$$ (A.37)

Furthermore, the matrix elements satisfy the orthogonality relation

$$\int_{SO(3)} d\varrho \ \overline{\mathcal{D}^l_{mn}(\varrho)} \mathcal{D}^{l'}_{m'n'}(\varrho) = \frac{8\pi^2}{2l+1} \delta_{ll'} \delta_{mm'} \delta_{nn'},$$ (A.38)

where $d\varrho$ is the invariant measure on $SO(3)$, normalized as

$$\int_{SO(3)} d\varrho = 8\pi^2.$$

Let $L^2(S^2)$ denote the Hilbert space of all square integrable functions on the sphere (with respect to the usual surface measure). As a consequence of (A.38), the action of a rotation ϱ on a function $f \in L^2(S^2)$ can be analyzed into "Fourier components" in the manner

$$f(\varrho^{-1}\zeta) = \sum_{l=0}^\infty \sum_{m=-l}^l \sum_{n=-l}^l \mathcal{D}^l_{mn}(\varrho) \widehat{f}(l, n) Y_l^m(\zeta), \quad \forall f \in L^2(S^2),$$ (A.39)

where $\widehat{f}(l, n)$ is a "Fourier coefficient":

$$\widehat{f}(l, n) = \langle Y_l^n | f \rangle.$$

References

A. Books and Theses

[Abr97] P. Abry. *Ondelettes et turbulences — Multirésolutions, algorithmes de décomposition, invariance d'échelle et signaux de pression.* (Paris: Diderot. 1997).

[Add02] P. S. Addison. *The Illustrated Wavelet Transform Handbook — Introductory Theory and Applications in Science, Engineering, Medicine and Finance.* (Bristol and Philadelphia: Institute of Physics Publishing. 2002).

[Ald96] A. Aldroubi & M. Unser (eds.). *Wavelets in Medicine and Biology.* (Boca Raton, FL: CRC Press. 1996).

[Ali00] S. T. Ali, J.-P. Antoine & J.-P. Gazeau. *Coherent States, Wavelets and Their Generalizations.* (New York: Springer. 2000).

[And02] M. Andrle. Model sets and adapted wavelet transform. Ph.D. Thesis, Czech Technical University, Prague. (2002).

[Ang92] B. Anger & C. Portenier. *Radon Integrals.* (Boston, MA: Birkhäuser. 1992).

[Arn95] A. Arnéodo, F. Argoul, E. Bacry, J. Elezgaray & J. F. Muzy. *Ondelettes, multifractales et turbulences – De l'ADN aux croissances cristallines.* (Paris: Diderot. 1995).

[Bar94] D. Barache. Propriétés algébriques et spectrales des structures apériodiques. Ph.D. Thesis, Université Paris 7, Paris. (1994).

[Bar77] A. O. Barut & R. Rączka. *Theory of Group Representations and Applications.* (Warszawa: PWN. 1977).

[Ber99] J. C. van den Berg (ed.). *Wavelets in Physics.* (Cambridge: Cambridge University Press. 1999).

[Ber98] G. Bernuau. Propriétés spectrales et géométriques des quasicristaux. Ondelettes adaptées aux quasicristaux. Ph.D. Thesis, Ceremade, Université Paris IX Dauphine. (1998).

[Bha99] S. K. Bhattacharjee. A computational approach to image retrieval. Ph.D. Thesis, EPFL, Lausanne. (1999). http://ltswww.epfl.ch/pub_files/sushil

[Bor72] A. Borel. *Représentations des groupes localement compacts,* Lecture Notes in Mathematics, Vol. 276 (Berlin: Springer. 1972).

[Bou97] E. Bournay Bouchereau. Analyse d'images par transformées en ondelettes. Application aux images sismiques. Ph.D. Thesis, Université Joseph Fourier – Grenoble I. (1997). http://cepax6.lis.inpg.fr/these/Th_Bournay.html

[Bou93] K. Bouyoucef. Sur des aspects multirésolution en reconstruction d'image. Application au télescope spatial de Hubble. Ph.D. Thesis, Université Paul Sabatier – Toulouse III. (1993). http://www.cerfacs.fr/dsp/papers/phd2.html

[Bra86] R. N. Bracewell. *The Fourier Theory and its Applications.* (New York: McGraw-Hill. 1986).

[Bur98]   B. Burke Hubbard. *The World According to Wavelets*, 2nd edn. (Wellesley, MA: A.K. Peters. 1998).

[Can98]   E. J. Candès. Ridgelets: theory and applications. Ph.D. Thesis, Department of Statistics, Stanford University. (1998).

[Ces97]   R. M. Cesar. Jr. Análise multi-escala de formas bidimensionais (Multiscale analysis of bidimensional shapes). Ph.D. Thesis, IFSC–Universidade de São Paulo, São Carlos. (1997).

[Chu92]   C. K. Chui. *An Introduction to Wavelets*. (San Diego, CA: Academic Press. 1992).

[Cle99]   M. Clerc. Analyse par ondelettes de processus localement dilatés, et application au gradient de texture. Ph.D. Thesis, Ecole Polytechnique, Palaiseau. (1999). http://cermics.enpc.fr/~maureen/main.ps.gz

[Coh89]   C. Cohen-Tannoudji, B. Diu & F. Laloë. *Mécanique Quantique, Tome I*. (Paris: Hermann. 1977).

[Com89]   J.-M. Combes, A. Grossmann & Ph. Tchamitchian (eds.). *Wavelets, Time-Frequency Methods and Phase Space (Proc. Marseille 1987)*. (Berlin: Springer. 1989; 2nd edn. 1990).

[Cor84]   J. F. Cornwell. *Group Theory in Physics. I. II* (Orlando, New York and London: Academic Press. 1984).

[Cor97]   J. F. Cornwell. *Group Theory in Physics. An Introduction*. (San Diego, CA: Academic Press. 1984).

[Cos01]   L. da F. Costa & R. M. Cesar, Jr. *Shape Analysis and Classification — Theory and Practice*. (Boca Raton, FL: CRC Press. 2001).

[Dau92]   I. Daubechies. *Ten Lectures on Wavelets*. (Philadelphia, PA: SIAM. 1992).

[Dec00]   N. Decoster. Analyse multifractale d'images de surfaces rugueuses à l'aide de la transformation en ondelettes. Ph.D. Thesis, Université de Bordeaux I. (1999).

[Do01]   M. N. Do. Directional multiresolution image representations. Ph.D. Thesis, EPFL, Lausanne. (2001). http://www.ifp.uiuc.edu/~minhdo/publications/thesis.pdf

[DeV88]   R. De Valois & K. De Valois. *Spatial Vision*. (New York: Oxford University Press. 1988) (see in particular Chapter 4, pp. 137–43).

[Dui00]   J. J. Duistermaat & J. A. C. Kok. *Lie Groups*. (Berlin, Heidelberg, New York: Springer. 2000).

[Duv91]   M. Duval-Destin. Analyse spatiale et spatio-temporelle de la stimulation visuelle à l'aide de la transformée en ondelettes. Ph.D. Thesis, Université d'Aix-Marseille II. (1991).

[Ead71]   W. T. Eadie, D. Drijard, F. E. James, M. Ross & B. Sadoulet. *Statistical Methods in Experimental Physics*. (Amsterdam, London: North Holland. 1971).

[Fea90]   J.-C. Feauveau. Analyse multirésolution par ondelettes non orthogonales et bancs de filtres numériques. Ph.D. Thesis, Université Paris-Sud. (1990).

[Fei98]   H. G. Feichtinger & T. Strohmer (eds.). *Gabor Analysis and Algorithms – Theory and Applications*. (Boston, MA: Birkhäuser. 1998).

[Fis87]   N. I. Fisher, T. Lewis & B. J. J. Embleton. *Statistical Analysis of Spherical Data*. (Cambridge: Cambridge University Press. 1987, 1993).

[Fou94]   E. Foufoula-Georgiou & P. Kumar. *Wavelets in Geophysics*. (San Diego, CA: Academic Press. 1994).

[Fla93]   P. Flandrin. *Temps-Fréquence*. (Paris: Hermès. 1993); English translation: *Time-Frequency/Time-Scale Analysis*. (San Diego, CA: Academic Press. 1998).

[Fra91]   M. Frazier, B. Jawerth & G. Weiss. *Littlewood-Paley Theory and the Study of Function Spaces*, CBMS-Conference Lecture Notes **79**. (Providence, RI: American Mathematical Society. 1991).

[Fre97] W. Freeden, M. Schreiner & T. Gervens. *Constructive Approximation on the Sphere, with Applications to Geomathematics.* (Oxford: Clarendon Press. 1997).

[Fre99] W. Freeden. *Multiscale Modelling of Spaceborne Geodata.* (Stuttgart: Teubner. 1999).

[Gai00] Ph. Gaillot. Ondelettes continues en Sciences de la Terre – Méthodes et applications. Ph.D. Thesis, Université Paul Sabatier – Toulouse III. (2000). http:// renass.u-strasbg.fr/~philippe/pubpgfr.html

[Gaa73] S. A. Gaal. *Linear Analysis and Representation Theory.* (Berlin: Springer. 1973).

[Gil74] R. Gilmore. *Lie Groups, Lie Algebras, and Some of Their Applications.* (New York and London: Wiley. 1974).

[Got66] K. Gottfried. *Quantum Mechanics. Vol. I: Fundamentals.* (New York and Amsterdam: Benjamin. 1966).

[Gra94] I. S. Gradshteyn & I. M. Ryzhik. *Table of Integrals, Series, and Products*, 5th edn. (New York: Academic Press. 19794).

[Gro01] K. Gröchenig. *Foundations of Time-Frequency Analysis.* (Boston, Basel, Berlin: Birkhäuser. 2001).

[Gru74] B. Grunbaum & G. C. Shephard. *Tilings and Patterns.* (New York: Freeman. 1987).

[Hel78] S. Helgason. *Differential Geometry, Lie Groups, and Symmetric Spaces.* (New York: Academic Press. 1978).

[Hol95] M. Holschneider. *Wavelets, an Analysis Tool.* (Oxford: Oxford University Press. 1995).

[Jac04] L. Jacques. Ondelettes, repères et couronne solaire. Ph.D. Thesis, Université Catholiquele Louvain, Louvain-la-Neuve. (2004).

[Jah97] B. Jahne. *Digital Image Processing.* (Berlin: Springer. 1997).

[Jan00] M. Jansen. Wavelet thresholding and noise reduction, Ph.D. Thesis, Katholicke Universiteit. Leuven (2000).

[Jan01] M. Jansen. *Noise reduction by wavelet thresholding*, Lecture Notes in Statistics, Vol. 161. (Berlin: Springer. 2001).

[Jen01] A. Jensen & A. la Cour-Harbo. *Ripples in Mathematics — The Discrete Wavelet Transform.* (Berlin, Heidelberg, New York: Springer. 2001).

[Kir76] A. A. Kirillov. *Elements of the Theory of Representations.* (Berlin: Springer. 1976).

[Kla85] J. R. Klauder & B. S. Skagerstam. *Coherent States – Applications in Physics and Mathematical Physics.* (Singapore: World Scientific. 1985).

[Kna96] A. W. Knapp. *Lie Groups Beyond an Introduction.* (Basel: Birkhäuser. 1996).

[Kol97] E. Kolaczyk. Wavelet methods for the inversion of certain homogeneous linear operators. Ph.D. Thesis, Stanford University. (1997).

[Kou00] Y. B. Kouagou. Transformations en ondelettes discrètes : une approche utilisant des pseudodilatations. Ph.D. Thesis, IMSP, Univ. Nat. du Bénin, Porto Novo. (2000).

[Kut99] M. Kutter. Digital image watermarking: hiding information in images. Ph.D. Thesis, EPFL, Lausanne. (1999).

[Lie97] E. H. Lieb & M. Loss. *Analysis.* (Providence, RI: American Mathematical Society. 1997).

[Lyn82] P. A. Lynn. *An Introduction to the Analysis and Processing of Signals*, 2nd edn. (London: MacMillan. 1982).

[Mal99] S. G. Mallat. *A Wavelet Tour of Signal Processing*, 2nd edn. (San Diego, CA: Academic Press. 1999).

[Mar82] D. Marr. *Vision.* (San Francisco, CA: Freeman. 1982).

[Mey72] Y. Meyer. *Algebraic Numbers and Harmonic Analysis.* (Amsterdam: North Holland. 1972).

[Mey91] Y. Meyer. (ed.). *Wavelets and Applications.* (*Proc. Marseille 1989*) (Berlin: Springer and Paris: Masson. 1991).

[Mey94]   Y. Meyer. *Les Ondelettes, Algorithmes et Applications*, 2nd edn. (Paris: Armand Colin. 1994); English translation of the 1st edn. Y. Meyer and R. D. Ryan, *Wavelets, Algorithms and Applications*. (Philadelphia, PA: SIAM. 1993).

[Mey93]   Y. Meyer & S. Roques (eds.). *Progress in Wavelet Analysis and Applications (Proc. Toulouse 1992)*. (Gif-sur-Yvette: Ed. Frontières. 1993).

[Mic27]   A. A. Michelson. *Studies in Optics*. (Chicago, IL: University of Chicago Press. 1927).

[Moh97]   M. J. Mohlenkamp. A fast transform for spherical harmonics. Ph.D. Thesis, Yale University, New Haven, CT. (1997).

[Mor02]   M. Morvidone. Etude et comparaison d'algorithmes de détection optimale pour les signaux modulés en amplitude et en fréquence; application aux ondes gravitationnelles. Ph.D. Thesis, Université de Provence (Aix–Marseille I), Marseille. (2002).

[Muj99]   F. A. Mujica. Spatio-temporal continuous wavelet transform for motion estimation. Ph.D. Thesis, Georgia Institute of Technology, Atlanta, GA. (1999).

[Mur90]   R. Murenzi. Ondelettes multidimensionnelles et applications à l'analyse d'images. Ph.D. Thesis, Université Catholique de Louvain, Louvain-la-Neuve. (1990).

[Oon00]   P. J. Ooninckx. Mathematical signal analysis: wavelets, Wigner distribution and a seismic application. Ph.D. Thesis, University of Amsterdam. (2000).

[Oui95]   G. Ouillon. Application de l'analyse multifractale et de la transformée en ondelettes anisotropes à la caractérisation géométrique multi-échelle des réseaux de failles et de fractures. Ph.D. Thesis, Université de Nice. (1995).

[Pap77]   A. Papoulis. *Signal Analysis*. (New York: McGraw Hill. 1977).

[Pau85]   Th. Paul. Ondelettes et mécanique quantique. Ph.D. Thesis, Université d'Aix-Marseille II. (1985).

[Pav77]   T. Pavlidis. *Structural Pattern Recognition*. (New York: Springer. 1977).

[Per86]   A. Perelomov. *Generalized Coherent States and Their Applications*. (Berlin: Springer. 1986).

[Rus92]   M. B. Ruskai, G. Beylkin, R. Coifman, I. Daubechies, S. Mallat, Y. Meyer & L. Raphael (eds.). *Wavelets and Their Applications*. (Boston, MA: Jones and Bartlett. 1992).

[Scu97]   M. O. Scully & M. S. Zubairy. *Quantum Optics*. (Cambridge: Cambridge University Press. 1997).

[Sem97]   D. Semwogerere. The use of the two-dimensional continuous wavelet transform for classification of targets in FLIR imagery. Master's Thesis, Clark Atlanta University. (1997).

[Sta98]   J.-L. Starck, F. Murtagh & A. Bijaoui. *Image Processing and Data Analysis. The Multiscale Approach*. (Cambridge: Cambridge University Press. 1998).

[Ste71]   E. M. Stein & G. Weiss. *Introduction to Fourier Analysis on Euclidean Spaces*. (Princeton, NJ: Princeton University Press. 1971).

[Str64]   R. F. Streater & A. S. Wightman. *PCT, Spin and Statistics, and All That*. (New York: Benjamin. 1964).

[Tal68]   J. D. Talman. *Special Functions — A Group Theoretic Approach*. (New York: W. A. Benjamin. 1968).

[Tek95]   A. M. Tekalp. *Digital Video Processing*. (Englewood Cliffs, NJ: Prentice-Hall. 1995).

[Tho98]   G. Thonet. New aspects of time-frequency analysis for biomedical signal processing. Ph.D. Thesis, EPFL, Lausanne. (1998).

[Tor95]   B. Torrésani. *Analyse continue par ondelettes*. (Paris: InterÉditions/CNRS Éditions, 1995), English translation, Philadelphia, PA: SIAM. (to appear).

[Uns03]   M. A. Unser, A. Aldroubi & A. F. Laine (eds). *Proc. SPIE, vol. 5207: Wavelets: Applications in Signal and Image Processing X*. (Bellingham, WA: SPIE. 2003).

[Vdg98]   P. Vandergheynst. Ondelettes directionnelles et ondelettes sur la sphère. Ph.D. Thesis, Université Catholique de Louvain, Louvain-la-Neuve. (1998).

[Vet95]   M. Vetterli & J. Kovačević. *Wavelets and Subband Coding*. (Englewood Cliffs, NJ: Prentice Hall. 1995).

[Wic94]   M. V. Wickerhauser. *Adapted Wavelet Analysis from Theory to Software*. (Wellesley, MA: A.K. Peters. 1994).

[Win95]   U. Windheuser. Sphärishe Wavelets: Theorie und Anwendung in der Physikalischen Geodäsie. Ph.D. Thesis, Universität Kaiserslautern. (1995).

[Wis93]   W. Wisnoe. Utilisation de la méthode de transformée en ondelettes 2D pour l'analyse de visualisation d'écoulements. Ph.D. Thesis, ENSAE, Toulouse. (1993).

[Woj97]   P. Wojtaszczyk. *A Mathematical Introduction to Wavelets*. (Cambridge: Cambridge University Press. 1997).

[Yar67]   A. L. Yarbus. *Eye Movements and Vision*. (New York: Plenum Press. 1967).

B. Articles

[1]   P. Abry, R. Baraniuk, P. Flandrin, R. Riedi & D. Veitch. Multiscale nature of network traffic. *IEEE Signal Process. Magazine*, 28–46, (2002).

[2]   M. D. Adams & F. Kossentini. Reversible integer-to-integer wavelet transforms for image compression: Performance evaluation and analysis. *IEEE Trans. Image Process.*, **9**:1010–24, (2000).

[3]   E. H. Adelson & R. Bergen. Spatio-temporal energy models for the perception of motion. *J. Opt. Soc. Amer. A*, **2**:285–99, (1985).

[4]   M. Alexandrescu, D. Gibert, G. Hulot, J.-L. Le Mouel & G. Saracco. Worldwide wavelet analysis of geomagnetic jerks. *J. Geophys. Res. B*, **101**:21975–94, (1996).

[5]   S. T. Ali. Stochastic localization, quantum mechanics on phase space and quantum space-time. *Rivista Nuovo Cim.*, **8**:1–128, (1985).

[6]   S. T. Ali, J.-P. Antoine, J.-P. Gazeau & U. A. Mueller. Coherent states and their generalizations: a mathematical overview. *Reviews Math. Phys.*, **7**:1013–104, (1995).

[7]   S. T. Ali, N. M. Atakishiyev, S. M. Chumakov & K. B. Wolf. The Wigner function for general Lie groups and the wavelet transform. *Ann. H. Poincaré*, **1**:685–714, (2000).

[8]   S. T. Ali, A. E. Krasowska & R. Murenzi. Wigner functions for the two-dimensional wavelet group. *J. Opt. Soc. Amer. A*, **17**:2277–87, (2000).

[9]   S. T. Ali, A. E. Krasowska & H. Führ. Plancherel inversion as unified approach to wavelet transforms and Wigner functions. *Ann. H. Poincaré*, to appear.

[10]   W. L. Anderson & H. Y. Diao. Two-dimensional wavelet transform and application to holographic particle velocimetry. *Applied Optics*, **34**:249–255, (1995).

[11]   M. Andrle, Č. Burdík & J.-P. Gazeau. Bernuau spline wavelets and Sturmian sequences. *J. Fourier Anal. Appl.*, **10**: (2004) (to appear).

[12]   J.-P. Antoine, M. Duval-Destin, R. Murenzi & B. Piette. Image analysis with 2D wavelet transform: detection of position, orientation and visual contrast of simple objects. In [Mey91], pp. 144–59.

[13]   J.-P. Antoine, P. Carrette, R. Murenzi & B. Piette. Image analysis with two-dimensional continuous wavelet transform. *Signal Process.*, **31**:241–72, (1993).

[14]   J.-P. Antoine & F. Bagarello. Wavelet-like orthonormal bases for the lowest Landau level. *J. Phys. A: Math. Gen.*, **27**:2471–81, (1994).

[15] J.-P. Antoine & R. Murenzi. The continuous wavelet transform, from 1 to 3 dimensions. In *Subband and Wavelet Transforms: Design and Applications*, pp. 149–87, eds. A. N. Akansu & M. J. T. Smith (Dordrecht: Kluwer. 1995).

[16] J.-P. Antoine, P. Vandergheynst, K. Bouyoucef & R. Murenzi. Target detection and recognition using two-dimensional continuous isotropic and anisotropic wavelets. *Automatic Object Recognition V, Proc. SPIE*, **2485**:20–31, (1995).

[17] J.-P. Antoine, P. Vandergheynst, K. Bouyoucef & R. Murenzi. Alternative representations of an image via the 2D wavelet transform. Application to character recognition. *Visual Information Processing IV, Proc. SPIE*, **2488**:486–97, (1995).

[18] J.-P. Antoine & R. Murenzi. Two-dimensional directional wavelets and the scale-angle representation. *Signal Process.*, **52**:259–81, (1995).

[19] J.-P. Antoine, R. Murenzi & P. Vandergheynst. Two-dimensional directional wavelets in image processing. *Int. J. of Imaging Systems and Technology*, **7**:152–65, (1996).

[20] J.-P. Antoine & P. Vandergheynst. Contrast enhancement in images using the two-dimensional wavelet transform. In *Proc. IWISP '96 (3rd Int. Workshop Image & Signal Processing.* pp. 65–68, eds. B. G. Mertzios and P. Liatsis (Amsterdam: Elsevier. 1996).

[21] J.-P. Antoine, D. Barache, R. M. Cesar, Jr & L. da F. Costa. Shape characterization with the wavelet transform. *Signal Process.*, **62**:265–90, (1997).

[22] J.-P. Antoine & R. Murenzi. Two-dimensional continuous wavelet transform as linear phase space representation of two-dimensional signals. In *Wavelet Applications in Signal and Image Processing IV* eds. A. Aldroubi, A. Laine & M. Unser, *Proc. SPIE*, **3078**:206–17, (1997).

[23] J.-P. Antoine & P. Vandergheynst. Wavelets on the n-sphere and related manifolds. *J. Math. Phys.*, **39**:3987–4008, (1998).

[24] J.-P. Antoine, R. Murenzi & P. Vandergheynst. Directional wavelets revisited: Cauchy wavelets and symmetry detection in patterns. *Appl. Comput. Harmon. Anal.*, **6**:314–45, (1999).

[25] J.-P. Antoine & I. Mahara. Galilean wavelets: coherent states for the affine Galilei group. *J. Math. Phys.* **40**:5956–71, (1999).

[26] J.-P. Antoine, L. Jacques & P. Vandergheynst. Penrose tilings, quasicrystals, and wavelets. In *Wavelet Applications in Signal and Image Processing VII*, eds. A. Aldroubi, A. Laine and M. Unser, *Proc. SPIE*, **3813**:28–39, (1999).

[27] J.-P. Antoine, L. Jacques & R. Twarock. Wavelet analysis of a quasiperiodic tiling with fivefold symmetry. *Phys. Lett. A*, **261**:265–74, (1999).

[28] J.-P. Antoine & P. Vandergheynst. Wavelets on the 2-sphere and related manifolds. *Reports Math. Phys.*, **43**:13–24, (1999).

[29] J.-P. Antoine & P. Vandergheynst. Wavelets on the 2-sphere: A group-theoretical approach. *Appl. Comput. Harmon. Anal.*, **7**:262–91, (1999).

[30] J.-P. Antoine, Ph. Antoine & B. Piraux. Wavelets in atomic physics. In *Spline Functions and the Theory of Wavelets.* pp. 261–76, eds. S. Dubuc & G. Deslauriers, CRM Proceedings and Lecture Notes, **18**, (Providence, RI: AMS, 1999).

[31] J.-P. Antoine, Ph. Antoine & B. Piraux. Wavelets in atomic physics and in solid state physics. In [Ber99]. Chap. 8.

[32] J.-P. Antoine, D. Lambert, Y. B. Kouagou & B. Torrésani. An algebraic approach to discrete dilations. Application to discrete wavelet transforms. *J. Fourier Anal. Appl.*, **6**:113–41, (2000).

[33] J.-P. Antoine, A. Coron & J-M. Dereppe. Water peak suppression: Time-frequency vs. time-scale approach. *J. Magn. Reson.*, **144**:189–94, (2000).

[34] J.-P. Antoine & A. Coron. Time-frequency and time-scale approach to magnetic resonance spectroscopy. *J. Comput. Meth. in Sci. & Engrg. (JCMSE)*, **1**:327–52, (2001).

[35] J.-P. Antoine, L. Demanet, L. Jacques & P. Vandergheynst. Wavelets on the sphere: implementation and approximations. *Appl. Comput. Harmon. Anal.*, **13**:177–200, (2002).

[36] J.-P. Antoine & F. Bagarello. Localization properties and wavelet-like orthonormal bases for the lowest Landau level. In *Advances in Gabor Analysis*, pp. 221–57, eds. H. G. Feichtinger and T. Strohmer (Boston: Birkhäuser. 2002).

[37] J.-P. Antoine, L. Demanet, J.-F. Hochedez, L. Jacques, R. Terrier & E. Verwichte. Application of the 2-D wavelet transform to astrophysical images. *Physicalia Mag.*, **24**:93–116, (2002).

[38] J.-P. Antoine & L. Jacques. Angular multiselectivity analysis of images. In [Uns03], pp. 196–207.

[39] Ph. Antoine, B. Piraux, D. B. Milošević & M. Gajda. Generation of ultrashort pulses of harmonics. *Phys. Rev. A*, **54**:R1761–4, (1996).

[40] Ph. Antoine, B. Piraux, D. B. Milošević & M. Gajda. Temporal profile and time control of harmonic generation. *Laser Phys.*, **7**:594–601, (1997).

[41] M. Antonini, M. Barlaud, P. Mathieu & I. Daubechies. Image coding using wavelet transform. *IEEE Trans. Image Process.*, **1**:205–20, (1992).

[42] F. T. Arecchi, E. Courtens, R. Gilmore & H. Thomas. Atomic coherent states in quantum optics. *Phys. Rev. A*, **6**:2211–37, (1972).

[43] F. Argoul, A. Arnéodo, J. Elezgaray, G. Grasseau & R. Murenzi. Wavelet analysis of the self-similarity of diffusion-limited aggregates and electrodeposition clusters. *Phys. Rev. A*, **41**:5537–60, (1990).

[44] A. Arnéodo, F. Argoul, E. Bacry, J. Elezgaray, E. Freysz, G. Grasseau, J. F. Muzy & B. Pouligny. Wavelet transform of fractals. In [Mey91], pp. 286–352.

[45] A. Arnéodo, G. Grasseau & M. Holschneider. Wavelet transform of multifractals. *Phys. Rev. Lett.*, **61**:2281–4, (1988).

[46] A. Arnéodo, F. Argoul, J. F. Muzy, B. Pouligny & E. Freysz. The Optical Wavelet Transform. In [Rus92], pp. 241–73.

[47] A. Arnéodo, E. Bacry & J. F. Muzy. The thermodynamics of fractals revisited with wavelets. *Physica A*, **213**:232–75, (1995); and in [Ber99], pp. 339–90.

[48] A. Arnéodo, E. Bacry & J. F. Muzy. Oscillating singularities in locally self-similar functions. *Phys. Rev. Lett.*, **74**:4823–6, (1995).

[49] A. Arnéodo, E. Bacry, P. V. Graves & J. F. Muzy. Characterizing long-range correlations in DNA sequences from wavelet analysis. *Phys. Rev. Lett.*, **74**:3293–6, (1996).

[50] A. Arnéodo, E. Bacry, S. Jaffard & J. F. Muzy. Oscillating singularities on Cantor sets. A grand canonical multifractal formalism. *J. Stat. Phys.*, **87**:179–209, (1997).

[51] A. Arnéodo, E. Bacry, S. Jaffard & J. F. Muzy. Singularity spectrum of multifractal functions involving oscillating singularities. *J. Fourier Anal. Appl.*, **4**:159–74, (1998).

[52] A. Arnéodo, N. Decoster & S. G. Roux. Intermittency, log-normal statistics, and multifractal cascade process in high-resolution satellite images of cloud structure. *Phys. Rev. Lett.*, **83**:1255–8, (1999).

[53] A. Arnéodo, N. Decoster & S. G. Roux. A wavelet-based method for multifractal image analysis. I. Methodology and test applications on isotropic and and anisotropic random rough surfaces. *Eur. Phys. J. B*, **15**:567–600, (2000).

[54] J. Arrault, A. Arnéodo, A. Davis & A. Marshak. Wavelet based multifractal analysis of rough surfaces: application to cloud models and satellite data. *Phys. Rev. Lett.*, **879**:75–8, (1997).

[55] E. W. Aslaksen & J. R. Klauder. Unitary representations of the affine group. *J. Math. Phys.*, **9**:206–11, (1968); Continuous representation theory using the affine group. *ibid.*, **10**:2267–75, (1969).

[56] D. Astruc, L. Plantié, R. Murenzi, Y. Lebret & D. Vandromme. On the use of the 3-D wavelet transform for the analysis of computational fluid dynamics results. In [Mey93], pp. 463–70.

[57] F. Bagarello. Multi-resolution analysis and fractional quantum Hall effect: an equivalence result. *J. Math. Phys.*, **42**:5116–29, (2001).

[58] F. Bagarello. Multi-resolution analysis and fractional quantum Hall effect: more results. *J. Phys. A: Math. Gen.*, **36**:123–38, (2003).

[59] C. Bandt. In *The Mathematics of Long-Range Aperiodic Order*, NATO ASI Series, Vol. 382, pp. 45–83, ed. R. V. Moody (Dordrecht: Kluwer. 1997).

[60] D. Barache, J.-P. Antoine & J.-M. Dereppe. The continuous wavelet transform, a tool for NMR spectroscopy. *J. Magn. Reson.*, **128**:1–11, (1997).

[61] D. Barache, B. Champagne & J.-P. Gazeau. Pisot-cyclotomic quasilattices and their symmetry semigroups. In *Quasicrystals and Discrete Geometry*, Fields Institute Monograph Series, Vol. 10, ed. J. Patera (Providence, RI: American Mathematical Society. 1998).

[62] R. B. Barreiro, M. P. Hobson, A. N. Lasenby, A. J. Banday, K. M. Gorski & G. Hinshaw. Testing the Gaussianity of the COBE DMR data with spherical wavelets. *Mon. Not. R. Astron. Soc.*, **318**:475–81, (2000).

[63] M. J. Bastiaans. Wigner distribution functions and its application to first-order optics. *J. Opt. Soc. Am.*, **69**:1710–16, (1979).

[64] M. J. Bastiaans. The Wigner distribution function applied to optical signals and systems. *Opt. Comm.*, **25**:26–30, (1978).

[65] G. Battle. Wavelets: A renormalization group point of view. In [Rus92], pp. 323–49.

[66] G. Battle. Klein–Gordon propagation of Daubechies wavelets. *J. Math. Phys.*, **34**:1095–109, (1993).

[67] J. J. Benedetto & H.-C. Wu. Non-uniform sampling and spiral MRI reconstruction. In *Wavelet Applications in Signal and Image Processing VIII*, eds. A. Aldroubi, A. Laine & M. Unser, *Proc. SPIE*, **4119**:130–41, (2000).

[68] C. A. Berenstein & D. F. Walnut. Wavelets and local tomography. In [Ald96], pp. 231–61.

[69] D. Bernier & K. F. Taylor. Wavelets from square-integrable representations. *SIAM J. Math. Anal.*, **27**:594–608, (1996).

[70] G. Bernuau. Wavelet bases adapted to a self-similar quasicrystal. *J. Math. Phys.* **39**:4213–25, (1998).

[71] A. Bertrand. Développements en base de Pisot et répartition modulo 1. *C. R. Acad. Sci. Paris* **285**:419–421, (1977).

[72] J. Bertrand & P. Bertrand. Microwave imaging of time-varying radar targets. *Inverse Problems*, **13**:621–45, (1997).

[73] J. Bertrand & P. Bertrand. A class of Wigner functions with extended covariance properties. *J. Math. Phys.*, **33**:2515–27, (1992).

[74] J. Bertrand & P. Bertrand. Représentations temps-fréquence des signaux. *C.R. Acad. Sc. Paris* **299**, Série I:635–8, (1984).

[75] N. Bethoux, G. Ouillon & M. Nicolas. The instrumental seismicity of the western Alps: spatio-temporal patterns analysed with the wavelet transform. *Geophys. J. Intern.*, **135**:177–94, (1998).

[76] S. K. Bhattacharjee & P. Vandergheynst. End-stopped wavelets for detecting low-level structures. In *Wavelet Applications in Signal and Image Processing VIII*, eds. A. Aldroubi, A. Laine & M. Unser, *Proc. SPIE*, **4119**:732–41, (2000).

[77]   S. K. Bhattacharjee. Detection of feature points using an end-stopped wavelet. EPFL internal report, unpublished (1999).

[78]   A. Bijaoui & F. Rué. A multiscale vision model adapted to the astronomical images. *Signal Process.*, **46**:345–62, (1995).

[79]   A. Bijaoui, E. Slezak, F. Rué & E. Lega. Wavelets and the study of the distant Universe. *Proc. IEEE*, **84**:670–9, (1996).

[80]   A. Bijaoui. Wavelets and astrophysical applications. In [Ber99], pp. 77–115.

[81]   A. Bilgin, G. Zweig & M. W. Marcellin. Three-dimensional image compression with integer wavelet transforms. *Applied Optics*, **39**:1799–814, (2000).

[82]   P. Bosch, private commun. and Thèse annexe, Université Catholique de Louvain, (1997).

[83]   E. Boureranne, M. Paindavoine & F. Truchetet. An improvement of Canny-Deriche filter for ramp edge detection (in French). *Trait. du Signal*, **10**:297–310, (1993).

[84]   M. Bourgeois, F. T. A. W. Wajer, D. van Ormondt & D. Graveron-Demilly. Reconstruction of MRI images from non-uniform sampling and its application to intrascan motion correction in Functional MRI. In *Modern Sampling Theory – Mathematics and Applications*, pp. 343–63; eds. J. J. Benedetto & P. J. S. G. Ferreira (Boston: Birkhäuser. 2001).

[85]   K. Bouyoucef, D. Fraix-Burnaix & S. Roques. Interactive Deconvolution with Error Analysis (IDEA) in astronomical imaging: Application to aberrated HST images on SN 1987A, M 87 and 3C 66B. *Astron. Astroph. Suppl.*, **121**:575–85, (1997).

[86]   C. Bowman & A. C. Newell. Natural patterns and wavelets. *Rev. Mod. Phys.*, **70**:289–301, (1998).

[87]   C. M. Brislawn. Fingerprints go digital. *Notices Amer. Math. Soc.*, **42**:1278–83, (1995).

[88]   A. Bruce, D. Donoho & H. Y. Gao. Wavelet analysis. In: *IEEE Spectrum*, October 1996, pp. 26–35.

[89]   T. Bülow & K. Daniilidis. Surface representations using spherical harmonics and Gabor wavelets on the sphere. Techn. Report MS-CIS-01-37, GRASP Lab., U. Pennsylvania, Philadelphia, PA (2001).

[90]   T. Bülow. Spherical diffusion. Techn. Report MS-CIS-01-38, GRASP Lab., U. Pennsylvania, Philadelphia, PA (2001).

[91]   P. Burt & E. Adelson. The Laplacian pyramid as a compact image code. *IEEE Trans. Comm.*, **31**:482–540, (1983).

[92]   A. R. Calderbank, I. Daubechies, W. Sweldens & B. L. Yeo. Wavelets that map integers to integers. *Appl. Comput. Harmon. Anal.*, **5**:332–69, (1998).

[93]   E. J. Candès. Harmonic analysis of neural networks. *Appl. Comput. Harmon. Anal.*, **6**:197–218, (1999).

[94]   E. J. Candès & D. L. Donoho. Ridgelets: A key to higher-dimensional intermittency? *Phil. Trans. R. Soc. Lond. A.*, **357**:2495–509, (1999).

[95]   E. J. Candès. Ridgelets and the representation of mutilated Sobolev functions. *SIAM J. Math. Anal.*, **33**:347–68, (2001).

[96]   E. J. Candès & D. L. Donoho. Curvelets – A surprisingly effective nonadaptive representation for objects with edges. In *Curves and Surfaces*, eds. L. L. Schumaker, *et al.* (Nashville, TN: Vanderbilt University Press. 1999).

[97]   E. J. Candès & D. L. Donoho. Curvelets, multiresolution representation, and scaling laws. In *Wavelet Applications in Signal and Image Processing VIII*, eds. A. Aldroubi, A. Laine & M. Unser, *Proc. SPIE*, **4119**:1–12, (2000).

[98]   J. Canny. A computational approach to edge detection. *IEEE Trans. Pattern Anal. Machine Intell.*, **8**:679–98, (1986).

[99] R. Carmona, W.-L. Hwang & B. Torrésani. Characterization of signals by the ridges of their wavelet transform. *IEEE Trans. Signal Process.*, **45**:2586–90, (1997).

[100] L. Cayòn, J. L. Sanz, R. B. Barreiro, E. Martinez-Gonzalez, P. Vielva, L. Toffolati, J. Silk, J. M. Diego & F. Argüeso. Isotropic wavelets: A powerful tool to extract point sources from cosmic microwave background maps. *Mon. Not. R. Astron. Soc.*, **315**:757–61, (2000).

[101] L. Cayòn, J. L. Sanz, E. Martinez-Gonzalez, A. J. Banday, F. Argüeso, J. E. Gallegos, K. M. Gorski & G. Hinshaw. Spherical Mexican Hat wavelet: an application to detect non-Gaussianity in the COBE-DMR maps. *Mon. Not. R. Astron. Soc.*, **326**:1243–9, (2001).

[102] S. Chen, D. L Donoho & M. A. Saunders. Atomic decomposition by basis pursuit. *SIAM J. Scientific Comp.*, **20**:33–61 (1999).

[103] C. K. Chui & X. Shi. Inequalities of Littlewood–Paley type for frames and wavelets. *SIAM J. Math. Anal.*, **24**:263–77, (1993).

[104] C. Cishahayo. Private commun. and Thèse annexe, Université Catholique de Louvain (1995).

[105] E. Clarkson. Angular channels in a multidimensional wavelet transform. *SIAM J. Math. Anal.*, **32**:80–102, (2000).

[106] M. Clerc & S. Mallat. Shape from texture and shading with wavelets. In *Dynamical Systems, Control, Coding, Computer Vision, Progress in Systems and Control Theory*, **25**:393–417, (1999).

[107] M. Clerc & S. Mallat. The Texture Gradient Equation for recovering shape from texture. *IEEE Trans. Pattern Anal. Machine Intell.*, **4**:536–49, (2002).

[108] A. Cohen, I. Daubechies & J.-C. Feauveau. Biorthogonal bases of compactly supported wavelets. *Commun. Pure Appl. Math.*, **45**:485–560, (1992).

[109] L. Cohen. General phase-space distribution functions. *J. Math. Phys.*, **7**:781–6, (1966).

[110] R. R. Coifman, Y. Meyer & M. V. Wickerhauser. Wavelet analysis and signal processing. In [Rus92], pp. 153–78.

[111] R. R. Coifman, Y. Meyer, S. Quake & M. V. Wickerhauser. Signal processing and compression with wavelet packets. In [Mey93], pp. 77–93.

[112] R. R. Coifman & D. Donoho. Translation-invariant de-noising. In *Wavelets and Statistics, Lecture Notes in Statistics*, **103**, pp. 125–50, eds. A. Antoniadis & G. Oppenheim (Heidelberg: Springer. 1995).

[113] J. Cui & W. Freeden. Equidistribution on the sphere. *SIAM J. Sci. and Stat. Comp.*, **18**:595–609, (1997).

[114] S. Dahlke, W. Dahmen, E. Schmitt & I. Weinreich. Multiresolution analysis and wavelets on S^2 and S^3. *Num. Funct. Anal. and Optimiz.*, **16**:19–41, (1995).

[115] F. Damiani, A. Maggio, G. Micela & S. Sciortino. A method based on wavelet transforms for source detection in photon-counting detector images. I. Theory and general properties. *Astroph. J.* **483**:350–69, (1997); II. Application to *ROSAT* PSPC images. *Astroph. J.*, **483**:370–89, (1997).

[116] S. Dahlke & P. Maass. The affine uncertainty principle in one and two dimensions. *Computers Math. Applic.*, **30**:293–305, (1995).

[117] T. Dallard & G. R. Spedding. 2-D wavelet transforms: generalisation of the Hardy space and application to experimental studies. *Eur. J. Mech., B/Fluids*, **12**:107–34, (1993).

[118] S. Daly. The visible differences predictor: an algorithm for the assessment of image fidelity. In *Digital Images and Human Vision*, pp. 179–206, ed. A. B. Watson (Cambridge, MA: MIT Press. 1993).

[119] L. Danzer. Quasiperiodicity: local and global aspects. *Lecture Notes in Physics*, **382**, pp. 561–72, eds. V. V. Dodonov & V. I. Manko (Berlin: Springer. 1990).

[120]   J. Darrozes, P. Gaillot, M. de Saint Blanquat & J.-L. Bouchez. Software for multi-scale image analysis: the normalized optimized anisotropic wavelet coefficient method. *Computers & Geosc.*, **23**:889–95, (1997).

[121]   I. Daubechies, A. Grossmann & Y. Meyer. Painless nonorthogonal expansions. *J. Math. Phys.*, **27**:1271–83, (1986).

[122]   I. Daubechies. The wavelet transform, time-frequency localization and signal analysis. *IEEE Trans. Inform. Theory*, **36**:961–1005, (1990).

[123]   I. Daubechies & S. Maes. A nonlinear squeezing of the continuous wavelet transform based on auditory nerve models. In [Ald96], pp. 527–46.

[124]   J. G. Daugman. Two-dimensional spectral analysis of cortical receptive field profiles. *Vision Res.*, **20**:847–56, (1980).

[125]   J. G. Daugman. Six formal properties of two-dimensional anisotropic visual filters: structural principles and frequency/orientation selectivity. *IEEE Trans. Systems, Man Cyb.*, **13**:882–6, (1983).

[126]   J. G. Daugman. Complete discrete 2-D Gabor transforms by neural networks for image analysis and compression. *IEEE Trans. Acoust., Speech, Signal Process.*, **36**:1169–79, (1988).

[127]   G. M. Davis, S. Mallat & Z. Zhang. Adaptive time-frequency decompositions. *SPIE J. of Opt. Eng.*, **33**:2183–91, (1994).

[128]   G. Davis, S. Mallat & M. Avellaneda. Greedy adaptive approximation. *J. Constr. Approx.* **13**:57–98, (1997).

[129]   N. Decoster, S. G. Roux & A. Arnéodo. A wavelet-based method for multifractal image analysis. II. Applications to synthetic multifractal rough surfaces. *Eur. Phys. J. B*, **15**:739–64, (2000).

[130]   J.-P. Delaboudinière, *et al.* EIT: Extreme-ultraviolet imaging telescope for the SoHO mission. *Solar Physics*, **175**:291–312, (1995).

[131]   N. Delprat, B. Escudié, Ph. Guillemain, R. Kronland-Martinet, Ph. Tchamitchian & B. Torrésani. Asymptotic wavelet and Gabor analysis: Extraction of instantaneous frequencies. *IEEE Trans. Inform. Theory*, **38**:644–64, (1992).

[132]   B. Delyon & A. Juditsky. On the computation of wavelet coefficients. *J. Approx. Theory*, **88**:47–79, (1997).

[133]   J. E. Dennis, Jr & D. J. Woods. Optimization on microcomputers: The Nelder-Mead simplex algorithm. In *New Computing Environments: Microcomputers in Large-Scale Computing*, pp. 116–22; ed. A. Wouk (Philadelphia, PA: SIAM. 1987).

[134]   S. D. Der & R. Chellappa. Probe-based automatic target recognition in infrared imagery. *IEEE Trans. Image Process.*, **6**:92–102, (1997).

[135]   R. Deriche. Détection optimale de contours avec une mise en oeuvre récursive. In *Onzième Colloque GRETSI, Nice*, pp. 483–6 (1987).

[136]   R. Deriche. Using Canny's criteria to derive a recursively implemented optimal edge detector. *Intern. J. Computer Vision*, **1**:167–87, (1987).

[137]   R. A. DeVore & V. N. Temlyakov. Some remarks on greedy algorithms. *Adv. Comp. Math.*, **5**:173–87, (1996).

[138]   R. A. DeVore. Nonlinear approximation. *Acta Numerica*, **7**:51–150, (1998).

[139]   O. Divorra Escoda & P. Vandergheynst. Video coding using a deformation compensation algorithm based on adaptive matching pursuit image decompositions. In *Proc. IEEE Intl. Conf. Image Process.*, Barcelona, Spain (2003) (to appear).

[140]   M. N. Do, A. C. Lozano & M. Vetterli. Rotation invariant texture retrieval using steerable wavelet-domain hidden Markov models. In *Wavelet Applications in Signal and*

Image Processing VIII, eds. A. Aldroubi, A. Laine & M. Unser, *Proc. SPIE*, **4119**:274–85, (2000).

[141] M. N. Do & M. Vetterli. Rotation invariant texture characterization and retrieval using steerable wavelet-domain hidden Markov models. *IEEE Trans. Multimedia*, **4**:517–27, (2002).

[142] M. N. Do & M. Vetterli. Contourlets. In *Beyond Wavelets,* pp. 1–27; eds. J. Stoeckler & G. V. Welland (San Diego, CA: Academic Press. 2001).

[143] M. N. Do, P. L. Dragotti, R. Shukla & M. Vetterli. On the compression of two dimensional piecewise smooth functions. In *Proc. IEEE International Conference on Image Processing (ICIP01*, Thessaloniki), 2001.

[144] D. L. Donoho. Nonlinear wavelet methods for recovery of signals, densities, and spectra from indirect and noisy data. In: *Different Perspectives on Wavelets*, pp. 173–205, Proc. Symp. Appl. Math., **38**, ed. I. Daubechies (Providence, RI: American Mathematical Society. 1993).

[145] D. L. Donoho. Unconditional bases are optimal bases for data compression and statistical estimation. *Appl. Comput. Harmon. Anal.*, **1**:100–5, (1993).

[146] D. L. Donoho. De-Noising by soft-Threshholding. *IEEE Trans. Inform. Theory*, **41**:613–27, (1995).

[147] D. L. Donoho & I. M. Johnstone. Ideal spacial adaptation by wavelet shrinkage. *Biometrika*, **81**:425–55, (1994).

[148] D. L. Donoho, M. Vetterli, R. DeVore & I. Daubechies. Data compression and harmonic analysis. *IEEE Trans. Inform. Theory*, **44**:391–432, (1998).

[149] D. L Donoho & X. Huo. Uncertainty principles and ideal atomic decompositions. *IEEE Trans. Inform. Theory*, **47**:2845–62, (2001).

[150] A. H. Dooley. Contractions of Lie groups and applications to analysis. In *Topics in Modern Harmonic Analysis*, Vol. I, pp. 483–515 (Rome: Istituto Nazionale di Alta Matematica Francesco Severi. 1983).

[151] A. H. Dooley & J. W. Rice. Contractions of rotation groups and their representations. *Math. Proc. Camb. Phil. Soc.*, **94**:509–17, (1983).

[152] A. H. Dooley & J. W. Rice. On contractions of semisimple Lie groups. *Trans. Amer. Math. Soc.*, **289**:185–202, (1985).

[153] P. L. Dragotti & M. Vetterli. Wavelet footprints: theory, algorithms and applications. *IEEE Trans. Signal Process,* **51**:1306–23, (2003).

[154] J. R. Driscoll & D. M. Healy. Computing Fourier transforms and convolutions on the 2-sphere. *Adv. Appl. Math.*, **15**:202–50, (1985).

[155] D. E. Dudgeon & R. T. Lacoss. An overview of automatic target recognition. *Linc. Lab. J.*, **6**:3–10, (1993).

[156] R. J. Duffin & A. C. Schaefer. A class of nonharmonic Fourier series. *Trans. Amer. Math. Soc.*, **72**:341–66, (1952).

[157] P. Dutilleux. An implementation of the "algorithm à trous" to compute the wavelet transform. In [Com89], pp. 298–304.

[158] M. Duval-Destin & R. Murenzi. Spatio-temporal wavelets: application to the analysis of moving patterns. In [Mey93], pp. 399–408.

[159] M. Duval-Destin, M.-A. Muschietti & B. Torrésani. Continuous wavelet decompositions, multiresolution, and contrast analysis. *SIAM J. Math. Anal.*, **24**:739–55, (1993).

[160] M. Elad & A. M. Bruckstein. A generalized uncertainty principle and sparse representation in pairs of bases. *IEEE Trans. Inform. Theory*, **48**:2558–67, (2002).

[161] E. Escalera & A. Mazure. Wavelet analysis of subclustering – an illustration, Abell-754. *Astroph. J.*, **388**:23–32, (1992).

[162] E. Escalera & H. T. MacGillivray. Topics in galaxy distributions: method for a multi-scale analysis. A use of the wavelet transform. *Astron. Astrophys.*, **298**:1–21, (1995).

[163] M. Farge & G. Rabreau. Transformée en ondelettes pour détecter et analyser les structures cohérentes dans les écoulements turbulents bidimensionnels. *C.R. Acad. Sci. Paris*, **307**, Série II: 1479–86, (1988).

[164] M. Farge. Wavelet transforms and their applications to turbulence, *Annu. Rev. Fluid Mech.*, **24**:395–457, (1992).

[165] M. Farge, N. K.-R. Kevlahan, V. Perrier & K. Schneider. Turbulence analysis, modelling and computing using wavelets. In [Ber99], pp. 117–200.

[166] N. Fatemi-Ghomi, P. L. Palmer & M. Petrou. Performance evaluation of texture segmentation algorithms based on wavelets. In *Proc. Workshop "Performance Characterization of Vision Algorithms – ECCV-96"*, Cambridge, UK (1996).

[167] H. G. Feichtinger, K. Gröchenig & T. Strohmer. Efficient numerical methods in nonuniform sampling theory. *Num. Math.*, **69**:423–40, (1995).

[168] R. M. Figueras i Ventura, L. Granai & P. Vandergheynst. R-D analysis of adaptive edge representations. In *Proc. IEEE Intl. Workshop on Multimedia Signal Processing (MMSP02)*, St. Thomas, US Virgin Islands, December 2002.

[169] W. Freeden & U. Windheuser. Combined spherical harmonic and wavelet expansion – a future concept in Earth's gravitational determination. *Applied Comput. Harm. Anal.*, **4**:1–37, (1997), and papers quoted therein.

[170] W. T. Freeman & E. H. Adelson. The design and use of steerable filters, *IEEE Trans. Pattern Anal. Machine Intell.*, **13**:891–906, (1991).

[171] E. Freysz, B. Pouligny, F. Argoul & A. Arnéodo. Optical wavelet transform of fractal aggregates. *Phys. Rev. Lett.*, **64**:745–53, (1990).

[172] P. Frick, A. Grossmann & Ph. Tchamitchian. Wavelet analysis of signals with gaps. *J. Math. Phys.*, **39**:4091–107, (1998).

[173] P. Frick, R. Beck, A. Shukurov, D. Sokoloff, M. Ehle & J. Kamphuis. Magnetic and optical spiral arms in the galaxy NGC 6946. *Mon. Not. R. Astron. Soc.*, **318**:925–37, (2000).

[174] P. Frick, R. Stepanov, A. Shukurov & D. Sokoloff. Structures in the rotation measure sky. *Mon. Not. R. Astron. Soc.*, **325**:649–64, (2001).

[175] P. Frick, R. Beck, E. M. Berkhuijsen & I. Patrickeyev. Scaling and correlation analysis of galactic images. *Mon. Not. R. Astron. Soc.*, **327**:1145–57, (2001).

[176] J. Froment & S. Mallat. Arbitrary low bit rate image compression using wavelets. In [Mey93], pp. 413–18, and references therein.

[177] P. Frossard, P. Vandergheynst & R. M. Figueras i Ventura. High flexibility scalable image coding. In *Proc. SPIE VCIP*, Lugano, Switzerland (2003) (to appear).

[178] P. Frossard, P. Vandergheynst, R. M. Figueras i Ventura & M. Kunt. A posteriori quantization of progressive matching pursuit streams. *IEEE Trans. Sig. Process.*, (2003) (to appear).

[179] H. Führ, W.-L. Hwang & B. Torrésani. Shape from texture using continuous wavelet transform. In *Wavelet Applications in Signal and Image Processing VIII*, eds. A. Aldroubi, A. Laine & M. Unser, *Proc. SPIE*, **4119**:93–107, (2000).

[180] D. Gabor. Theory of communication. *J. Inst. Electr. Engrg. (London)*, **93**:429–57, (1946).

[181] P. Gaillot, J. Darrozes, M. de Saint Blanquat & G. Ouillon. The normalised optimised anisotropic wavelet coefficient (NOAWC) method: an image processing tool for multi-scale analysis of rock fabric. *Geophys. Res. Lett.*, **24**:1819–22, (1997).

[182] P. Gaillot, J. Darrozes & J.-L. Bouchez. Wavelet transform: a future of rock fabric analysis? *J. Struct. Geol.*, **21**:1615–21, (1999).

[183] P. Gaillot, J. Darrozes, P. Courjault-Rade & D. Amorese. Structural analysis of hypocentral distribution of an earthquake sequence using anisotropic wavelets: method and application. *J. Geophys. Res. – Solid Earth*, **107 (B10)**: art. no. 2218, (2002).

[184] J.-P. Gazeau & J. Patera. Tau-wavelets of Haar. *J. Phys. A: Math. Gen.*, **29**:4549–59, (1996).

[185] J.-P. Gazeau & V. Spiridonov. Toward discrete wavelets with irrational scaling factor. *J. Math. Phys.* **37**:3001–13, (1996).

[186] J.-P. Gazeau, J. Patera & E. Pelantova. Tau-wavelets in the plane. *J. Math. Phys.*, **39**:4201–12, (1998).

[187] J.-P. Gazeau & R. Krejcar. Tau-wavelets for Penrose-Robinson tilings in the complex plane. In *Symmetries & Structural Properties of Condensed Matter, Zajaczkowo 1998*, ed. T. Lulek, B. Lulek & A. Wal, pp. 418–34 (Singapore: World Scientific. 1999).

[188] J.-P. Gazeau & R. Krejcar. Penrose tiling wavelets and quasicrystals. In *Complex Geometry 98, Paris, July 1998*, eds. F. Norguet & S. Ofman (Hermann, *"Actualités scientifiques et industrielles"*, 2004).

[189] J.-P. Gazeau & P. Kramer. From quasiperiodic tilings with τ-inflation to τ-wavelets. In *Proceedings of the VIIth Int. Conference on Quasicrystals, Stuttgart, 1999, Material Science and Engineering A*, **A294-6**:421–4, (2000).

[190] A. C. Gilbert, S. Muthukrishnan & M. J. Strauss. In: *14th Annual ACM-SIAM Symposium on Discrete Algorithms* (2003).

[191] R. Gilmore. Geometry of symmetrized states. *Ann. Phys. (NY)*, **74**:391–463, (1972).

[192] R. Gilmore. On properties of coherent states. *Rev. Mex. Fis.*, **23**:143–87, (1974).

[193] M. Girardi, E. Escalera, D. Fadda, G. Giuricin, F. Mardirossian & M. Mezzetti. Optical substructures in 48 galaxy clusters: new insights from a multiscale analysis. *Astroph. J.*, **482**:41–62, (1997).

[194] J. P. Gollub & J. S. Langer. Pattern formation in nonequilibrium physics. *Rev. Mod. Phys.*, **71**:S396–403, (1999).

[195] C. Gonnet. Caractérisation de textures à l'aide de la transformée en ondelettes, Rapport de DEA INPG; Internal report CPT-92/P.2730, Marseille (1992).

[196] C. Gonnet & B. Torrésani. Local frequency analysis with two-dimensional wavelet transform. *Signal Process.*, **37**:389–404, (1994).

[197] J. Göttelmann. Locally supported wavelets on manifolds, with applications to the 2D sphere. *Appl. Comput. Harmon. Anal.*, **7**:1–33, (1999).

[198] J. Goutsias & H. J. A. M. Heijmans. Nonlinear multiresolution signal decomposition schemes - Part I: Morphological pyramids. *IEEE Trans. Image Process.*, **9**:1862–76, (2000).

[199] P. Goupillaud, A. Grossmann & J. Morlet. Cycle-octave and related transforms in seismic signal analysis. *Geoexploration*, **23**:85–102, (1984).

[200] V. K. Goyal & M. Vetterli. Quantized overcomplete expansions in $\mathbb{R}^N$: Analysis, synthesis, and algorithms. *IEEE Trans. Inform. Theory*, **44**:16–31, (1998).

[201] S. A. Grebenev, W. Forman, C. Jones & S. Murray. Wavelet transform analysis of the small-scale X-ray structure of the cluster Abell 1367. *Astroph. J.*, **445**:607–23, (1995).

[202] R. Gribonval & M. Nielsen. Sparse representation in unions of bases. *Preprint IRISA*, **1499** (2002).

[203] R. Gribonval. Sparse decomposition of stereo signals with Matching Pursuit and application to blind separation of more than two sources from a stereo mixture. In *Proc. Int. Conf. Acoust. Speech Signal Process. (ICASSP'02)*, Orlando, FL (2002).

[204] A. Grossmann. Wavelet transform and edge detection. In *Stochastic Processes in Physics and Engineering*, pp. 149–57, eds. S. Albeverio, Ph. Blanchard, L. Streit & M. Hazewinkel (Dordrecht: Reidel. 1988).

[205] A. Grossmann & J. Morlet. Decomposition of Hardy functions into square integrable wavelets of constant shape. *SIAM J. Math. Anal.*, **15**:723–36, (1984).

[206] A. Grossmann & J. Morlet. Decomposition of functions into wavelets of constant shape, and related transforms. In: *Mathematics + Physics, Lectures on recent results. I.*, pp. 135–66, ed. L. Streit (Singapore: World Scientific. 1985).

[207] A. Grossmann, J. Morlet & T. Paul. Integral transforms associated to square integrable representations. I. General results. *J. Math. Phys.*, **26**:2473–9, (1985).

[208] A. Grossmann, J. Morlet & T. Paul. Integral transforms associated to square integrable representations. II. Examples. *Ann. Inst. H. Poincaré*, **45**:293–309, (1986).

[209] A. Grossmann, R. Kronland-Martinet & J. Morlet. Reading and understanding continuous wavelet transforms. In [Com89], pp. 2–20.

[210] Ph. Guillemain, R. Kronland-Martinet & B. Martens. Estimation of spectral lines with the help of the wavelet transform. Applications in N.M.R. spectroscopy. In [Mey91], pp. 38–60.

[211] C. R. Hagelberg & N. K. K. Gamage. Applications of structure preserving wavelet decompositions to intermittent turbulence: a case study. In [Fou94], pp. 45–80.

[212] C. Hagelberg & J. Helland. Thin-line detection in meteorological radar images using wavelet transforms. *J. Atmos. Ocean Tech.*, **12**:633–42, (1995).

[213] R. C. Hartman, *et al.*, The third EGRET catalog of high-energy gamma-ray sources, *Astroph. J. Suppl.*, **123**:79–202, (1999).

[214] F. Hartung & M. Kutter. Multimedia watermarking techniques. In *Proc. IEEE: Special Issue on Identification and Protection of Multimedia Information*, **87**:1079–107, (1999).

[215] H. J. Haubold. Wavelet analysis of the new solar neutrino capture rate data for the Homestake experiment. *Astrophys. Space. Sci.*, **258**:201–18, (1998).

[216] D. M. Healy, Jr, D. Rockmore & S. S. B. Moore. FFTs for the 2-sphere – Improvements and variations. Technical Report PCS-TR96-292, Dartmouth College (Hanover, NH, 1996).

[217] D. M. Healy, Jr & J. B. Weaver. Adapted wavelet techniques for encoding Magnetic Resonance Images. In [Ald96], pp. 297–352.

[218] D. J. Heeger. Model for the extraction of image flow. *J. Opt. Soc. Am. A*, **4**:1455–70, (1987).

[219] H. J. A. M. Heijmans & J. Goutsias. Nonlinear multiresolution signal decomposition schemes - Part II: Morphological wavelets. *IEEE Trans. Image Process.*, **9**:1897–913, (2000).

[220] C. Heil & D. Walnut. Continuous and discrete wavelet transforms. *SIAM Review*, **31**:628–66, (1989).

[221] M. Holschneider. On the wavelet transformation of fractal objects, *J. Stat. Phys.*, **50**:963–93, (1988).

[222] M. Holschneider, R. Kronland-Martinet, J. Morlet & Ph. Tchamitchian. A real-time algorithm for signal analysis with the help of the wavelet transform. In [Com89], pp. 286–97.

[223] M. Holschneider & Ph. Tchamitchian. Pointwise analysis of Riemann's 'nondifferentiable' function. *Invent. Math.*, **105**:157–75, (1991).

[224] M. Holschneider. Inverse Radon transforms through inverse wavelet transforms. *Inverse Problems*, **7**:853–61, (1991).

[225] M. Holschneider. Continuous wavelet transforms on the sphere. *J. Math. Phys.*, **37**:4156–65, (1996).

[226] D. H. Hubel & T. Wiesel. Receptive fields, binocular interaction and function architecture in the cat's visual cortex. *J. Physiol.*, **160**:106–54, (1962).

[227] L. Hudgins, C. A. Friehe & M. E. Mayer. Wavelet transforms and atmospheric turbulence. *Phys. Rev. Lett.*, **71**:3279–82, (1993).

[228] W.-L. Hwang & S. Mallat. Characterization of self-similar multifractals with wavelet maxima. *Appl. Comput. Harmon. Anal.*, **1**:316–28, (1994).

[229] W.-L. Hwang, C.-S. Lu & P.-C. Chung. Shape from texture: estimation of planar surface orientation through the ridge surfaces of continuous wavelet transform. *IEEE Trans. Image Process.*, **7**:773–80, (1998).

[230] W.-L. Hwang & F. Chang. Character extraction from documents using wavelet maxima. *Image Vis. Comput.*, **16**:307–15, (1998).

[231] W.-L. Hwang, C.-S. Lu & P.-C. Chung. Segmentation of perspective textured planes through the ridges of continuous wavelet transform. *J. Visual Commun. Image Repres.*, **12**:201–16, (2001).

[232] E. Inönü & E. P. Wigner. On the contraction of groups and their representations. *Proc. Nat. Acad. Sci. U. S.*, **39**:510–24, (1953).

[233] F. Jaillet & B. Torrésani. Sculpture temps-fréquence des sons par modification de leur transformée en ondelettes. *Actes du Congrès Français d'Acoustique* (2002).

[234] A. K. Jain & F. Farrokhnia. Unsupervised texture discrimination using Gabor filters. *J. Patt. Recogn.*, **24**:1167–86, (1991).

[235] B. Josso, D. R. Burton & M. J. Lalor. Wavelet strategy for surface roughness analysis and characterisation. *Comput. Methods Appl. Mech. Engrg.*, **191**:829–42, (2001).

[236] S. Journaux, P. Gouton & G. Thauvin. Evaluating creep in metals by grain boundary extraction using directional wavelets and mathematical morphology. *J. Mat. Proc. Techn.*, **117**:132–45, (2001).

[237] ISO/IEC 15444-1:2000, Information technology - JPEG2000 image coding system, December 2000.

[238] K. Kadooka, K. Kunoo, N. Uda, K. Ono & T. Nagayasu. Strain analysis for Moiré interferometry using the two-dimensional continuous wavelet transform. *Exp. Mech.*, **43**:45–51, (2003).

[239] G. Kaiser & R. F. Streater. Windowed Radon transform, analytic signals, and the wave equation. In *Wavelets: A Tutorial in Theory and Applications*, pp. 399–441, ed. C. K. Chui (Boston: Academic Press. 1992).

[240] L. Kaplan & R. Murenzi. SAR target feature extraction using the 2-D continuous wavelet transform. In *Proc. SPIE Conference*, Orlando, Florida, April 21–25 (1997).

[241] L. Kaplan & R. Murenzi. Pose estimation of SAR imagery using the two-dimensional continuous wavelet transform. *Pattern Recogn. Lett.*, **24**:2269–80, (2003).

[242] A. Khodakovsky, P. Schröder & W. Sweldens. Progressive geometry compression. In: *Computer Graphics Proc. (SIGGRAPH 2000)*, ACM Siggraph 2000, pp. 271–8.

[243] N. Kingsbury. Image processing with complex wavelets. *Philos. Trans. Roy. Soc. London Ser. A*, **357**:2543–60, (1999).

[244] N. Kingsbury. Complex wavelets for shift invariant analysis and filtering of signals. *Appl. Comput. Harmon. Anal.*, **10**:234–53, (2001).

[245] J. R. Klauder. Continuous-representation theory. I. Postulates of continuous-representation theory. *J. Math. Phys.*, **4**:1055–8, (1963); II. Generalized relation between quantum and classical dynamics, *ibid.*, 1058–73.

[246] J. R. Klauder. Path integrals for affine variables. In *Functional Integration, Theory and Applications*, pp. 101–19, eds. J.-P. Antoine & E. Tirapegui (New York and London: Plenum Press. 1980).

[247] P. J. Kolodzy. Multidimensional automatic target recognition system evaluation. *Linc. Lab. J.*, **6**:117–46, (1993).

[248] A. E. Krasowska & S. T. Ali. Wigner functions for a class of semi-direct product groups. Preprint, arXiv:math-ph/0201007 v1 Jan 3 2002.

[249] P. Kumar. A wavelet based methodology for scale-space anisotropic analysis. *Geophys. Research Lett.*, **22**:2777–80, (1995).

[250] P. Kumar & E. Foufoula-Georgiou. Wavelet analysis for geophysical applications. *Rev. Geophys.*, **35**:385–412, (1997).

[251] M. Kutter & S. Winkler. Spread–spectrum watermarking using the human visual system. *IEEE Trans. Image Process.*, submitted (2000).

[252] J. Lardies & S. Gouttebroze. Identification of modal parameters using the wavelet transform, *Int. J. Mech. Science*, **44**:2263–83, (2002).

[253] T. S. Lee. Image representation using 2-D Gabor wavelets. *IEEE Trans. Patt. Analysis Mach. Intelligence*, **18**:959–71, (1996).

[254] E. Lega, A. Bijaoui, J. M. Alimi & H. Scholl. A morphological indicator for comparing simulated cosmological scenarios with observations. *Astron. Astrophys.*, **309**:23–9, (1996).

[255] Y. Li & Y. Y. Zeevi. Applications of limited-extent waves: an introduction. *Applied Optics*, **33**:5239–40, (1994), and subsequent papers in the same issue.

[256] J.-M. Lina & M. Mayrand. Complex Daubechies wavelets. *Appl. Comput. Harmon. Anal.*, **2**:219–29, (1995).

[257] J. Lubin. A visual discrimination model for imaging system design and evaluation. In *Vision Models for Target Detection and Recognition*, pp. 245–83, ed. E. Peli (Singapore: World Scientific. 1995).

[258] P. Magain. private communication.

[259] S. G. Mallat. Multifrequency channel decompositions of images and wavelet models. *IEEE Trans. Acoust., Speech, Signal Process.*, **37**:2091–110, (1989).

[260] S. G. Mallat. A theory for multiresolution signal decomposition: the wavelet representation. *IEEE Trans. Pattern Anal. Machine Intell.*, **11**:674–93, (1989).

[261] S. G. Mallat. Zero-crossing of a wavelet transform. *IEEE Trans. Inform. Theory*, **37**:1019–33, (1991).

[262] S. Mallat & W.-L. Hwang. Singularity detection and processing with wavelets. *IEEE Trans. Inform. Theory*, **38**:617–43, (1992).

[263] S. Mallat & Z. Zhang. Matching pursuit with time-frequency dictionaries. *IEEE Trans. Signal Process.*, **41**:3397–415 (1993).

[264] S. Mallat & S. Zhong. Wavelet maxima representation. In [Mey91], pp. 207–84.

[265] S. G. Mallat & S. Zhong. Characterization of signals from multiscale edges. *IEEE Trans. Pattern Anal. and Mach. Intell.*, **14**:710–32, (1992).

[266] D. Marr & E. Hildreth. Theory of edge detection. *Proc. R. Soc. Lond. B*, **207**:187–217, (1980).

[267] E. Martinez-Gonzalez, J. E. Gallegos, F. Argüeso, L. Cayòn & J. L. Sanz. The performance of spherical wavelets to detect non-Gaussianity in the cosmic microwave background sky. *Mon. Not. R. Astron. Soc.*, **336**:22–32, (2002).

[268] J. R. Mattox, *et al.* The likelihood analysis of EGRET data. *Astroph. J.*, **461**:396–407, (1996).

[269] D. Mendlovic, I. Ouzieli, I. Kiryuschev & E. Marom. Two-dimensional wavelet transform achieved by computer-generated multireference matched-filter and Dammann grating. *Applied Optics*, **34**:8213–19, (1995).

[270] G. Menegaz, A. Rivoldini & J.-Ph. Thiran. Dyadic frames of directional wavelets as texture descriptors. In *Wavelet Applications in Signal and Image Processing VIII*, eds. A. Aldroubi, A. Laine & M. Unser *Proc. SPIE*, **4119**:263–73, (2000).

[271] Y. Meyer. Quasicrystals, Diophantine approximation and algebraic numbers. In *Beyond Quasicrystals, Les Houches 1994*, pp. 3–16, eds. F. Axel and D. Gratias (Paris: Les Editions de Physique and Berlin: Springer. 1995).

[272] J. Michelsson & J. Niederle. Contraction of representations of de Sitter groups. *Commun. Math. Phys.*, **27**:167–80, (1972).

[273] B. Ph. van Milligen. Wavelets, non-linearity and turbulence in fusion plasmas. In [Ber99], pp. 227–62.

[274] M. J. Mohlenkamp. A fast transform for spherical harmonics. *J. Fourier Anal. Appl.*, **5**:159–84, (1999).

[275] F. Mokhtarian & A. K. Mackworth. A theory of multiscale, curvature-based shape representation for planar curves. *IEEE Trans. Pattern Anal. and Mach. Intell.*, **14**:789–805, (1992).

[276] R. V. Moody. Model sets: a survey. In *From Quasicrystals to More Complex Systems, Les Houches 1998*, eds. F. Axel, F. Denoyer, and J.-P. Gazeau, pp. 145–66 (Paris: EDP Sciences. and Berlin: Springer. 2000).

[277] M. Morvidone & B. Torrésani. Time scale approach for chirp detection. *Int. J. Wavelets, Multires. and Inform. Process.*, **1**: (2003) (to appear).

[278] D. Moses, *et al.* EIT observations of the Extreme Ultraviolet Sun. *Solar Physics*, **175**:571–99, (1997).

[279] F. Mujica, J.-P. Leduc, R. Murenzi & M. J. T. Smith. Spatio-temporal continuous wavelets applied to missile warhead detection and tracking. *Visual Communications and Image Processing, Proc. SPIE*, **3024**:787–98, (1997).

[280] F. Mujica, J.-P. Leduc, R. Murenzi & M. J. T. Smith. A new motion estimation algorithm based on the continuous wavelet transform. *IEEE Trans. Signal Process.*, **9**:873–88, (2000).

[281] F. Mujica, J.-P. Leduc, R. Murenzi & M. J. T. Smith. Robust object tracking in compressed image sequences. *J. Electron. Imaging*, **7**:746–54, (1998).

[282] F. Mujica, J.-P. Leduc, R. Murenzi & M. J. T. Smith. Target tracking in presence of flare decoys using spatio-temporal wavelet transform. In *Proc. ARL (Army Research Laboratory) Symposium on Sensors and Electron Devices* (College Park, Maryland, February 1998).

[283] R. Murenzi. Wavelet transforms associated to the n-dimensional Euclidean group with dilations: signals in more than one dimension. In [Com89], pp. 239–46.

[284] R. Murenzi & J.-P. Antoine. Two-dimensional continuous wavelet transform as linear phase space representation of two-dimensional signals. In *Wavelet Applications IV*, eds. M. Unser, A. Aldroubi & A. Laine, *Proc. SPIE*, **3078**:206–17, (1997).

[285] R. Murenzi, L. Kaplan, F. Mujica & S. Der. Feature extraction in FLIR imagery using the continuous wavelet transform (CWT). In *Proc. ARL (Army Research Laboratory) Symposium on Sensors and Electron Devices* (College Park, Maryland, February 1997).

[286] R. Murenzi, L. Kaplan, J.-P. Antoine & F. Mujica. Computational complexity of the continuous wavelet transform in two dimensions. In *Proc. ARL (Army Research Laboratory) Symposium on Sensors and Electron Devices* (College Park, Maryland, February 1997).

[287] R. Murenzi, D. Semwogere, D. Johnson, L. Kaplan & K. Namuduri. Detection and classification of targets in FLIR imagery using two-dimensional directional wavelets. In *Proc. Second Annual Fedlab Symposium* (College Park, Maryland, February 1998).

[288] R. Murenzi, K. Namuduri, W. Zhai & L. Kaplan. Scale-angle invariant features and target classification. In *Proc. Third Annual ARL (Army Research Laboratory) Federated Laboratory Symposium on Sensors and Electron Devices* (College Park, Maryland, February 1999).

[289] R. Murenzi, W. Zhai, K. Namuduri & L. Kaplan. Scale-angle CWT features: application in object recognition. In *Wavelet Applications in Signal and Image Processing VII*, eds. M. Unser, A. Aldroubi and A. Laine, *Proc. SPIE*, **3813**:16–27, (1999).

[290] R. Murenzi, K. R. Namuduri & N. V. Rao. Image feature extraction and analysis using continuous wavelet transforms. In *Proc. Conf. "Wavelets and Applications", Anna Univ., Chennai, India*, pp. 217–30; eds. M. Krishna, B. Radha & S. Thangavelu (India: Allied Publishers. 2002).

[291] M.-A. Muschietti & B. Torrésani. Pyramidal algorithms for Littlewood-Paley decompositions. *SIAM J. Math. Anal.*, **26**:925–43, (1995).

[292] J. F. Muzy, E. Bacry & A. Arnéodo. Multifractal formalism for fractal signals: The structure-function approach versus the wavelet-transform modulus-maxima method. *Phys. Rev. E*, **47**:875–84, (1993).

[293] J. F. Muzy, E. Bacry & A. Arnéodo. The multifractal formalism revisited with wavelets. *Int. J. Bifurcation Chaos*, **4**:245–302, (1994).

[294] F. J. Narcowich & J. D. Ward. Nonstationary wavelets on the m-sphere for scattered data. *Applied Comput. Harm. Anal.*, **3**:1324–36, (1996).

[295] G. Nason & B. W. Silverman. The discrete wavelet transform in S. *J. Comput. Graph. Stat.*, **3**:163–91, (1994).

[296] R. Neff & A. Zakhor. Modulus quantization for matching pursuit video coding. *IEEE Trans. Circuits and Systems for Video Technology*, **10**:895–912, (2000).

[297] P. J. Ooninckx. A wavelet method for detecting S-waves in seismic data. *Comput. Geosc.*, **3**:111–34, (1999).

[298] G. Ouillon, D. Sornette & C. Castaing. Organization of joints and faults from 1-cm to 100-km scales revealed by optimized anisotropic wavelet coefficient method and multifractal analysis. *Nonlin. Process. in Geophys.*, **2**:158–77, (1995).

[299] G. Ouillon, C. Castaing & D. Sornette. Hierarchical geometry of faulting. *J. Geophys. Res. – Solid Earth*, **101**:5477–87, (1996).

[300] J. Pando & L.-Z. Fang. Discrete wavelet transform power spectrum estimator. *Phys. Rev. E*, **57**:3593–601, (1998).

[301] Sang-Il Park, M. J. T. Smith & R. Murenzi. Multidimensional wavelets for target detection and recognition. In *SPIE International Symposium on Aerospace/Defense Sensing and Controls*, Orlando, Florida, April 8–12, (1996).

[302] W. Parry. On the β-expansions of real numbers. *Acta Math. Acad. Sci. Hungary*, **11**:401–16, (1960).

[303] J. Patera & R. Twarock. Affine extension of noncrystallographic Coxeter groups and quasicrystals. *J. Phys. A: Math. Gen.*, **35**:1551–74, (2002).

[304] Y. C. Pati, R. Rezaifar & P. S. Krishnaprasad. Adaptive time-frequency decompositions. In *27th Asilomar Conf. on Signals, Systems and Comput.* (1993).

[305] Th. Paul & K. Seip. Wavelets in quantum mechanics. In [Rus92], pp. 303–22.

[306] S. C. Pei & S.-B. Jaw. Two-dimensional general fan-type FIR digital filter design. *Signal Process.*, **37**:265–74, (1994).

[307] E. Peli. In search of a contrast metric: matching the perceived contrast of Gabor patches at different phases and bandwidths. *Vision Res.*, **37**: 3217–24, (1997).

[308] E. Peli. Contrast in complex images. *J. Opt. Soc. Am. A*, **7**:2032–40, (1990).

[309] R. Penrose. The role of aesthetics in pure and applied mathematical research. *Bull. Inst. Math. & its Applications*, **10**:266–71, (1974).

[310] R. Penrose. Pentaplexity, a class of non-periodic tilings of the plane. *Mathematical Intelligencer*, **2**:32–7, (1979) (reproduced from *Eureka*, No. 39).

[311] P. Perona. Steerable-scalable kernels for edge detection and junction analysis. *Image and Vision Computing*, **10**:663–72, (1992).

[312] A. M. Perelomov. Coherent states for arbitrary Lie group. *Commun. Math. Phys.*, **26**:222–36, (1972).

[313] J. Perrin, B. Torrésani & P. Fuchs. Une fonction de corrélation localisée pour la mise en correspondance des images stéréoscopiques. *Traitement du Signal*, **16**:3–14, (1999).

[314] J. Perrin, B. Torrésani & P. Fuchs. Stereoscopic images matching using a localized correlation function. *IEEE Trans. Image Process.*, submitted (1999) `<http://citeseer.nj.nec.com/378749.html>` `<http://www.cmi.univ-mrs.fr/~torresan/publi.html#IP>`

[315] F. Peyrin & M. Zaim. Wavelet transform and tomography: continuous and discrete approaches. In [Ald96], pp. 209–30.

[316] F. Portier-Fozzani, B. Vandame, A. Bijaoui & A. J. Maucherat. A Multiscale Vision Model applied to analyze EIT images of the solar corona. *Solar Physics*, **201**:271–88, (2001).

[317] J. Portilla & E. P. Simoncelli. A parametric texture model based on joint statistics of complex wavelet coefficients. *Int. J. Computer Vision*, **40**:49–71, (2000).

[318] D. Potts & M. Tasche. Interpolatory wavelets on the sphere. In *Approximation Theory VIII*, pp. 335–42, eds. C. K. Chui and L. L. Schumaker, (Singapore: World Scientific. 1995)

[319] D. Potts, G. Steidl & M. Tasche. Kernels of spherical harmonics and spherical frames. In *Advanced Topics in Multivariate Approximation*, pp. 1–154, eds. F. Fontanella, K. Jetter & P.-J. Laurent (Singapore: World Scientific. 1996).

[320] H. R. Rabiee, R. L. Kashyap & S. R. Safavian. Adaptive multiresolution image coding with matching and bases pursuit. In *Proc. IEEE Intl. Conf. Image Process.*, **1**:273–6, Lausanne, Switzerland (1996).

[321] J. M. Radcliffe. Some properties of spin coherent states. *J. Phys. A: Math. Gen.*, **4**:313–23, (1971).

[322] C. Radin. Symmetry and tilings. *Notices Amer. Math. Soc.*, **42**:26–31, (1995).

[323] A. R. Rao & B. G. Schunck. Computing oriented texture fields. *CVGIP: Graph. Models Image Process.*, **53**:157–85, (1991).

[324] J. Reichel, G. Menegaz, M. J. Nadenau & M. Kunt. Integer wavelet transform for embedded lossy to lossless image compression. *IEEE Trans. Image Process.*, **10**:383–92, (2001).

[325] A. Rényi. Representations for real numbers and their ergodic properties. *Acta Math. Acad. Sci. Hungary*, **8**:477–93, (1957).

[326] O. Rioul & M. Vetterli. Wavelets and signal processing. *IEEE Signal Process. Magazine*, October 1991, 14–38.

[327] O. Rioul & P. Duhamel. Fast algorithms for discrete and continuous wavelet transforms. *IEEE Trans. Inform. Theory*, **2**:569–86, (1992).

[328] S. Roques, F. Bourzeix & K. Bouyoucef. Soft-thresholding technique and restoration of 3C273 jet. *Astrophys. and Space Science*, **239**:297–304, (1996).

[329] S. G. Roux, A. Arnéodo & N. Decoster. A wavelet-based method for multifractal image analysis. III. Applications to high-resolution satellite images of cloud structure. *Eur. Phys. J. B*, **15**:765–86, (2000).

[330] E. B. Saff & A. B. J. Kuijlaars. Distributing many points on a sphere. *Math. Intell.*, **19**:5–11, (1997).

[331] J. Saletan. Contraction of Lie groups. *J. Math. Phys.*, **2**:1–21, (1961).

[332] J. L. Sanz, R. B. Barreiro, L. Cayòn, E. Martinez-Gonzalez, G. A. Ruiz, F. J. Diaz, F. Argüeso, J. Silk & L. Toffolati. Analysis of CMB maps with 2D wavelets, *Astron. Astroph., Suppl. Series*, **140**:99–105, (1999).

[333] J. L. Sanz, F. Argüeso, L. Cayòn, E. Martinez-Gonzalez, R. B. Barreiro & L. Toffolati. Wavelets applied to cosmic microwave background maps: A multiresolution analysis for denoising. *Mon. Not. R. Astron. Soc.*, **309**:672–80, (1999).

[334] G. Saracco, A. Grossmann & Ph. Tchamitchian. Use of wavelet transforms in the study of propagation of transient acoustic signals across a plane interface between two homogeneous media. In [Com89], pp. 139–46.

[335] K. Schneider & M. Farge. Computing and analyzing turbulent flows using wavelets. In *Wavelet Transforms and Time-Frequency Signal Analysis*, pp. 181–216; ed. L. Debnath (Boston: Birkhäuser. 2001).

[336] P. Schröder & W. Sweldens. Spherical wavelets: efficiently representing functions on the sphere. In: *Computer Graphics Proc. (SIGGRAPH95)*, ACM Siggraph 1995, pp. 161–75.

[337] P. Schröder & W. Sweldens. Spherical wavelets: texture processing. In *Rendering Techniques '95*, pp. 252–63, eds. P. Hanrahan & W. Purgathofer (Wien, New York: Springer. 1995).

[338] D. Shechtman, I. A. Blech, D. Gratias & J. W. Cahn. Metallic phase with long-range orientational order and no translational symmetry. *Phys. Rev. Lett.*, **53**:1951–3, (1984).

[339] M. J. Shensa. The discrete wavelet transform: wedding the à trous and Mallat algorithms, *IEEE Trans. Signal Process.*, **40**:2464–82, (1992).

[340] J.-Y. Sheu, R.-S. Chang, C.-H. Lin & P.-L. Fan. Analysis of wave-aberration by use of the wavelet transform, *Chinese J. Phys.*, **40**:20–30, (2002).

[341] E. P. Simoncelli, W. T. Freeman, E. H. Adelson & D. J. Heeger. Shiftable multiscale transforms. *IEEE Trans. Inform. Theory*, **38**:587–607, (1992).

[342] E. P. Simoncelli & H. Farid. Steerable wedge filters for local orientation analysis. *IEEE Trans. Image Process.*, **5**:1377–82, (1996).

[343] E. Slezak, A. Bijaoui & G. Mars. Identification of structures from galaxy counts. Use of the wavelet transform. *Astron. Astrophys.*, **227**:301–16, (1990) and in [Mey91], pp. 175–80.

[344] E. Slezak, V. de Lapparent & A. Bijaoui. Objective detection of voids and high-density structures in the 1st CfA redshift survey slice. *Astroph. J.*, **409**:517–29, (1993).

[345] J.-L. Starck & M. Pierre. Structure detection in low-intensity X-ray images. *Astron. Astroph. Suppl. Ser.*, **128**:397–407, (1998).

[346] J.-L. Starck, E. J. Candès & D. L. Donoho. The curvelet transform for image denoising. *IEEE Trans. Image Process.*, **11**:670–84, (2002).

[347] C. F. Stromeyer, III & S. Klein. Evidence against narrow-band spatial frequency channels in human vision: the detectability of frequency modulated gratings. *Vision Res.*, **15**:899–910, (1975).

[348] B. J. Super & A. C. Bovik. Planar surface orientation from texture spatial frequencies. *Pattern Recognit.*, **28**:729–43, (1995).

[349] W. Sweldens. The lifting scheme: a custom-design construction of biorthogonal wavelets. *Appl. Comput. Harmon. Anal.*, **3**:1186–200, (1996).

[350] W. Sweldens. The lifting scheme: a construction of second generation wavelets. *SIAM J. MATH. Anal.*, **29**:511–46, (1998).

[351] R. Takahashi. Sur les représentations unitaires des groupes de Lorentz généralisés. *Bull. Soc. Math. France*, **91**:289–433, (1963).

[352] V. N. Temlyakov. Weak greedy algorithms. *Adv. Comp. Math.*, **12**: 213–27, (2000).

[353] P. Thevenaz. Motion analysis. In *Pattern Recognition in Image Processing in Physics (Proc. 37th Scott. Univ. Summer School in Physics)*, pp. 129–66, ed. R. A. Vaughan (1990).

[354] G. Thonet, O. Blanc, P. Vandergheynst, E. Pruvot, J.-M. Vesin & J.-P. Antoine. Wavelet-based detection of ventricular ectopic beats in heart rate signals. *Appl. Signal Process.*, **5**:170–81, (1998).

[355] B. Torrésani. Phase space decompositions: local Fourier analysis on spheres. Preprint CPT-93/P.2878, Marseille (1993).

[356] B. Torrésani. Position-frequency analysis for signals defined on spheres. *Signal Process.*, **43**:341–6, (1995).

[357] M. Unser. Fast Gabor-like windowed Fourier and continuous wavelet transforms. *IEEE Signal Process. Lett.*, **42**:3519–23, (1994).

[358] P. Vandergheynst, E. Van Vyve, A. Goldberg, J.-P. Antoine & I. Doghri. Modeling and simulation of an impact test using wavelets, analytical solutions and finite elements. *Int. J. Solids Structures*, **38**:5481–508, (2001).

[359] P. Vandergheynst, M. Kutter & S. Winkler. Wavelet-based contrast computation and application to digital image watermarking. In *Wavelet Applications in Signal and Image Processing VIII*, eds. M. Unser, A. Aldroubi and A. Laine, *Proc. SPIE*, **4119**:82–92, (2000).

[360] P. Vandergheynst & J.-F. Gobbers. Directional dyadic wavelet transforms: design and algorithms. *IEEE Trans. Image Process.*, **11**:363–72, (2002).

[361] P. Vielva, E. Martinez-Gonzalez, L. Cayòn, J. M. Diego, J. L. Sanz & L. Toffolati. Predicted Planck extragalactic point-source catalogue. *Mon. Not. R. Astron. Soc.*, **326**:181–91, (2001).

[362] P. Vielva, R. B. Barreiro, M. P. Hobson, E. Martinez-Gonzalez, A. N. Lasenby, J. L. Sanz & L. Toffolati. Combining maximum-entropy and the Mexican hat wavelet to reconstruct the microwave sky. *Mon. Not. R. Astron. Soc.*, **328**:1–16, (2001).

[363] P. Vielva, E. Martinez-Gonzalez, J. E. Gallegos, L. Toffolati & J. L. Sanz. Point source detection using the Spherical Mexican Hat Wavelet on simulated all-sky Planck maps. *Mon. Not. R. Astron. Soc.*, **344**:89–104, (2003).

[364] J. Ville. Théorie et applications de la notion de signal analytique. *Câbles et Transm.*, **2ème A**: 61–74, (1948).

[365] L. F. Villemoes. Nonlinear approximation with Walsh atoms. In *Surface Fitting and Multiresolution Methods*, eds. A. Méhauté, C. Rabut, L. L. Schumaker, pp. 329–36, (Nashville, TN: Vanderbilt University Press, 1997).

[366] D. Vu Giang & F. Móricz. Hardy spaces on the plane and double Fourier transforms. *J. Fourier Anal. Appl.*, **2**:487–505, (1996).

[367] B. D. Wandelt & K. M. Górski. Fast convolution on the sphere. *Phys. Rev. D*, **63**:123002-1-6, (2001).

[368] Y.-P. Wang. Image representations using multiscale differential operators. *IEEE Trans. Image Process.*, **8**:1757–71, (1999).

[369] A. B. Watson. The Cortex Transform: rapid computation of simulated neural images. *Comput. Vision Graph. Image Process.*, **39**:311–27, (1987).

[370] A. B. Watson. Efficiency of a model human image code. *J. Opt. Soc. Am. A*, **4**:2401–17, (1987).

[371] A. Watson & A. Ahumada, Jr. Model of human visual-motion sensing. *J. Opt. Soc. Am. A*, **2**:322–42, (1985).

[372] M. A. Webster & R. L. De Valois. Relationship between spatial frequency and orientation tuning of striate cortex cells. *J. Opt. Soc. Am. A*, **2**:1324–36, (1985).

[373] E. P. Wigner. On the quantum correction for thermodynamic equilibrium. *Phys. Rev.*, **40**:749–59, (1932).

[374] S. Winkler. Issues in vision modeling for perceptual video quality assessment. *Signal Process.*, **78**:231–52, (1999).

[375] W. Wisnoe, P. Gajan, A. Strzelecki, C. Lempereur & J.-M. Mathé. The use of the two-dimensional wavelet transform in flow visualization processing. In [Mey93], pp. 455–8.

[376] K. B. Wolf. Wigner distribution function for paraxial polychromatic optics. *Opt. Comm.*, **132**:343–52, (1996).

Index